UNIFORM SUPERSONIC FLOWS IN CHEMICAL PHYSICS

Chemistry Close to Absolute Zero Studied Using the CRESU Method

UNIFORM SUPERSONIC FLOWS IN CHEMICAL PHYSICS

Chemistry Close to Absolute Zero Studied Using the CRESU Method

Edited by

Bertrand R. Rowe
Rowe Consulting, France

André Canosa
CNRS, University of Rennes 1, France

Dwayne E. Heard
University of Leeds, UK

 World Scientific

NEW JERSEY · LONDON · SINGAPORE · BEIJING · SHANGHAI · HONG KONG · TAIPEI · CHENNAI · TOKYO

Published by

World Scientific Publishing Europe Ltd.

57 Shelton Street, Covent Garden, London WC2H 9HE

Head office: 5 Toh Tuck Link, Singapore 596224

USA office: 27 Warren Street, Suite 401-402, Hackensack, NJ 07601

Library of Congress Cataloging-in-Publication Data

Names: Rowe, Bertrand R., editor. | Canosa, André, editor. | Heard, Dwayne E., editor.
Title: Uniform supersonic flows in chemical physics : chemistry close to absolute zero
    studied using the CRESU method / Bertrand R. Rowe, Rowe Consulting, France, André Canosa,
    CNRS, University of Rennes 1, France, Dwayne E. Heard, University of Leeds, UK, editors.
Description: New Jersey : World Scientific, [2022] | Includes bibliographical references and index.
Identifiers: LCCN 2021038195 | ISBN 9781800610989 (hardcover) |
    ISBN 9781800610996 (ebook) | ISBN 9781800611009 (ebook other)
Subjects: LCSH: Chemistry, Physical and theoretical. | Aerodynamics, Supersonic.
Classification: LCC QD450 .U55 2022 | DDC 541--dc23/eng/20211004
LC record available at https://lccn.loc.gov/2021038195

British Library Cataloguing-in-Publication Data
A catalogue record for this book is available from the British Library.

Cover design by Lali Abril.
Background image: T. A. Rector/University of Alaska Anchorage, WIYN and NOIRLab/NSF/AURA.

For any available supplementary material, please visit
https://www.worldscientific.com/worldscibooks/10.1142/Q0324#t=suppl

Desk Editors: Christina Ramalingam/Michael Beale/Shi Ying Koe

Typeset by Stallion Press
Email: enquiries@stallionpress.com

Dedication

The editors want to dedicate this book to the memory of Professor Ian WM Smith, who passed away in November 2016. Ian played a key role in extending the CRESU technique beyond the field of ion–molecule reactions. In the mid-eighties, Bertrand Rowe was invited to give a seminar for Ian's group at the University of Birmingham concerning the reaction of the atomic nitrogen ion with molecular hydrogen studied using the CRESU apparatus. Immediately after the seminar, Ian proposed to Bertrand to start a collaborative program in order to extend the technique far beyond its use at that time. The impetus given by Ian, his scientific insight, and his enthusiasm made this program a considerable success, and it was awarded the first European Descartes prize in 2000.

Ian was not only a great scientific collaborator but also a great and close friend who contributed much to transform a successful scientific program into a great human adventure, which will stay in our memory forever! We would also like to remember Ian's wife Sue.

Foreword

The quantitative analysis of complex reaction systems requires information on the nature of the contributing "elementary" reactions and their rate coefficients. A variety of static and flow reactors have provided this information for temperatures above room temperature. In contrast to this, conditions below room temperature are much less well covered. For example, conventional flow system experiments are generally limited to temperatures above about 100 Kelvin. In addition, in such studies, there may be uncertainties about the precise, sometimes non-thermal, conditions and the preparation of reactants may be difficult. Although being very advanced today, theoretical studies do not always provide a solution to this dilemma, as they also may encounter difficulties. For example, the open electronic shell character of many reactants results in electronically nonadiabatic reaction effects whose relevance generally is underestimated; the transition between long-range electrostatic and short-range valence properties of the interaction potential between reactants remains difficult to handle. The comparison of experimental and theoretical rate constants then still may be inconclusive. In this situation, methods for studying gas phase reactions under well-defined low-temperature conditions appear highly desirable.

A variety of strategies have been developed to prepare cold or even ultracold molecules and to study their properties. Providing well-defined thermal low-temperature conditions for studying reactive processes provides an additional challenge. Although cooling of a gas by adiabatic expansion is a long-known principle for attaining low temperatures, it needs more than that. It was an ingenious idea to note that cooling of supersonic flows by expansion through Laval nozzles produces a place where local thermodynamic equilibrium is generated. This place turned out to provide an ideal environment for studying low-temperature reactions. Wind tunnels, such as those used in many aerodynamics laboratories, then found a new application. The corresponding technique was termed CRESU (Cinétique de Réaction en Ecoulement Supersonique Uniforme). Since its first realization in the 1980s, instruments dedicated especially to reaction studies have been designed and used to provide a wealth of data, not only for rate coefficients but also for other features of cold molecules, for which precise knowledge of the temperature of the system is required. It appears, therefore, most timely to devote a complete monograph to the achievements of the technique. It is written by the inventors of the method and by a series of successful practitioners. It demonstrates the large range of possible applications and opportunities which modern CRESU facilities offer.

Kineticists have established critically evaluated banks of rate coefficient data to be used in atmospheric and combustion chemistry. Extrapolation of the recommended results to lower temperatures, such as of astrochemical relevance, unfortunately is often only of limited value, because the simple Arrhenius law breaks down and simple power laws are also inadequate. In this situation, the CRESU method plays a particularly important role. It closes the gap between the sub-Kelvin range of "cold and ultracold molecules" and the range of conventional rate studies above about 100 Kelvin. Naturally, it is of central importance for applications in astrochemistry, when temperatures between about 10 and 100 Kelvin are of interest and reaction conditions need to be well defined. But, it not only "closes the gap" but it also opens up access to a range of particular low-temperature effects which are hidden at higher temperatures. Quantum effects of some coordinates of the reactants and the influence of small energy barriers on reaction rates become detectable. Under the considered conditions, chemical reactions sample the transition between long-range and short-range contributions of the interaction potential. Low-temperature rate measurements then may be considered as a kind of "spectroscopy of the potential energy surface" under otherwise difficult accessible conditions. Precise rate coefficients for well-defined low-temperature conditions, therefore, are of immense value for the test of theoretical treatments, and the CRESU method here plays a central role. Nevertheless, this method also has its limitations. The attachment of electrons to neutral molecules may serve as an example. On the one hand, CRESU measurements have extended the accessible temperature range to conditions which have not been accessible by other methods. In this way, the existence of small energy barriers of the crossing of neutral and anionic potential energy surfaces could be proven. On the other hand, the method apparently did not provide enough time for electrons to be completely thermalized such that differences between electron and gas temperatures had to be accounted for. Nevertheless, the possibility to sample small energy barriers of the process stands out and documents the unique value of the technique.

The concluding chapter of the book by Bertrand Rowe, André Canosa, and Dwayne Heard beautifully summarizes the past achievements of the CRESU technique and provides an outlook for future research. One can hardly add anything else to this authoritative chapter. Reaction kineticists like myself are looking forward to new results which will help to improve collections of rate data for use in various fields of kinetics. Further stimulating exchange of ideas between experiment and theory can be expected.

Jürgen Troe
Göttingen
March 2021

About the Editors

Bertrand R. Rowe, born in 1947, is a former Director of Research at the CNRS (Centre National de la Recherche Scientifique). He started his scientific career at the laboratory of "Aérothermique" of the CNRS in Meudon (France), on subjects devoted to problems with spacecraft re-entry. He obtained his "Doctorat ès Sciences" in 1975 and then started to develop techniques allowing the study of chemical reactions in rarefied flow jets. He spent a year in the Aeronomy Laboratory, directed then by Eldon Ferguson at Boulder (Colorado, USA) in 1979–1980. He is the pioneer of CRESU, which he first developed in a large wind tunnel in Meudon, for applications to ion-molecule reactions. After moving to Rennes where he became director of the "Département de Physique Atomique et Moléculaire", a joint laboratory of CNRS and University of Rennes 1, he developed together with Professor Ian W.M. Smith and their respective teams the extension of the CRESU method toward the study of radical-molecule neutral-neutral reactions and other topics. He was awarded the Descartes Prize in 2000 for this collaborative work: "Chemistry close to the absolute zero", with Ian W.M. Smith as the coordinator.

He is the inventor of a pulsing technique known as the "Aerodynamic Chopper" allowing the generation of high quality uniform supersonic flows which is used at the University of Castilla-La Mancha (Ciudad Real, Spain) in the framework of a collaborative program with Rennes University led by André Canosa for the French side.

Besides academic research he has been a scientific consultant for major industrial companies for more than twenty years. He leads Rowe Consulting, a small French company promoting applications and research linked to his former activities.

 **André Canosa** was born in Lorient, Bretagne, France. He studied Physics at the University of Rennes 1 (France) and got his Master's degree from the University of Paris VII (Jussieu) in "Astrophysics and Space Techniques" prepared at the Observatoire de Meudon (France) in 1988. He obtained his PhD in 1991 at the Atomic and Molecular Physics Department of the University of Rennes 1 under the supervision of Dr. Bertrand Rowe. During that time he did pioneering studies regarding the quantification of branching ratios of gas phase dissociative recombination reactions of polyatomic cations with electrons in collaboration with Professors David Smith and Nigel Adams from the University of Birmingham (UK). In 1992, he was hired to a permanent position at the French CNRS (Centre National de la Recherche Scientifique) in the group led by Bertrand Rowe in Rennes. Since then, his interest has mainly focused on experimental gas phase reaction kinetics of astrophysical and atmospheric interest with particular attention given to radical-neutral reactions taking place at very low temperatures and studied with the CRESU technique. He became an expert in uniform supersonic flows and contributed to the more widespread use of the technique by improving a pulsed version initially developed at the University of Rennes 1 which he then helped to setup at the University of Castilla-La Mancha (Ciudad Real, Spain) in Elena Jiménez's group within the framework of a very fruitful collaboration started in 2010 and still ongoing. He got an invited research fellowship for 5 months in 2015 and 6 months in 2017 in this group.

He was granted the first European Descartes Prize in 2000 in collaboration with Dr. Bertrand Rowe and Professor Ian W.M. Smith and Dr. Ian R. Sims, both from the University of Birmingham (UK) for the development of the CRESU technique. He has published 78 articles in peer-reviewed journals in various fields including physical chemistry, astrochemistry and environmental sciences. He was also a co-editor of the book *Dissociative Recombination: Theory, Experiments and Applications* with Bertrand Rowe and J. Brian A. Mitchell; Nato Asi Series B, volume 313, Plenum Press New York (1993).

He was the team leader of the "Experimental Astrochemistry" group at the Institut de Physique de Rennes from 2005 to 2011 and Assistant Director of the French national program PCMI (Physics and Chemistry of the Interstellar Medium) from 2005 to 2013. He is presently Director of Research at the CNRS in the University of Rennes 1.

Dwayne E. Heard was born in Barnstaple, North Devon, UK, and grew up on a dairy farm near Hartland. He attended Shebbear College, Devon, before studying at Magdalen College, Oxford, where he received a B.A. in Chemistry in 1986. Remaining in Oxford and under the supervision of Professor Gus Hancock, he completed his D.Phil. in 1990 in the Physical Chemistry Laboratory, where he built a time-resolved step-scan FTIR emission spectrometer for the study of chemical kinetics and dynamics. He then moved to SRI International in Menlo Park, California, where he undertook a postdoc with David Crosley, working both in laser-based combustion diagnostics and atmospheric spectroscopy.

In 1992 he moved to the School of Chemistry at Macquarie University, Sydney, Australia, as a lecturer, before coming to the School of Chemistry at the University of Leeds in March 1994, where he held a Royal Society University Research Fellowship until 2002. He was a Visiting Fellow at JILA, University of Colorado in 2000, where he first performed experiments measuring low temperature rate coefficients using a CRESU apparatus in collaboration with Professor Steve Leone, and also with Professor Ian W.M. Smith who was also visiting JILA at the time. He became Professor of Atmospheric Chemistry in 2004, and was Head of the School of Chemistry at the University of Leeds from 2009–2013.

His research interests include quantitative field measurements of hydroxyl (OH) and other short-lived intermediates in the atmosphere from surface- and aircraft-based platforms using laser-induced fluorescence spectroscopy, which are compared with calculations from constrained atmospheric models using the *Master Chemical Mechanism*. His group has participated in over 30 field campaigns worldwide since 1996. His other research interests are laboratory and atmospheric chamber studies of the reaction kinetics and photochemistry of gas phase and aerosol processes pertinent to Earth's atmosphere, and very low temperature studies of gas phase reactions relevant to interstellar space and planetary atmospheres, studied using a pulsed Laval nozzle within a CRESU apparatus. His astrochemical interests have focused on the low temperature kinetics of neutral-neutral reactions, particularly those of small radical species with complex organic molecules (COMs), studied experimentally using the CRESU method and also using theoretical calculations. He has published over 200 scientific articles in peer-reviewed journals in the areas of physical chemistry, chemical physics, atmospheric chemistry and astrochemistry. He was awarded the Environment Prize from the Royal Society of Chemistry in 2017.

List of Contributors

Abdessamad Benidar
University of Rennes 1, France
abdessamad.benidar@univ-rennes1.fr

André Canosa
University of Rennes 1, France
andre.canosa@univ-rennes1.fr

Barbara Wyslouzil
Ohio State University, USA
wyslouzil.1@osu.edu

Bertrand R. Rowe
Rowe Consulting, France
bertrand.rowe@gmail.com

Dwayne E. Heard
University of Leeds, UK
d.e.heard@leeds.ac.uk

Elena Jiménez
University of Castilla-La Mancha, Spain
elena.jimenez@uclm.es

Eric Herbst
University of Virginia, USA
eh2ef@virginia.edu

Eszter Dudás
University of Rennes 1, France
eszter.dudas@univ-rennes1.fr

Fabien Goulay
West Virginia University, USA
fabien.goulay@mail.wvu.edu

François Lique
University of Rennes 1, France
francois.lique@univ-rennes1.fr

Ian R. Sims
University of Rennes 1, France
ian.sims@univ-rennes1.fr

Ilsa R. Cooke
University of British Columbia, Canada
icooke@chem.ubc.ca

James Brian Mitchell
Merl-Consulting, France
merlrennes@gmail.com

Kevin M. Hickson
University of Bordeaux, France
kevin.hickson@u-bordeaux.fr

Lucile Rutkowski
University of Rennes 1, France
lucile.rutkowski@univ-rennes1.fr

Ludovic Biennier
University of Rennes 1, France
ludovic.biennier@univ-rennes1.fr

Nicolas Suas-David
University of Leiden, Netherlands
suas@strw.leidenuniv.nl

Philippe Halvick
University of Bordeaux, France
philippe.halvick@u-bordeaux.fr

Robert Georges
University of Rennes 1, France
robert.georges@univ-rennes1.fr

Ruth Signorell
ETH Zurich, Switzerland
rsignorell@ethz.ch

Sébastien D. Le Picard
University of Rennes 1, France
sebastien.le-picard@univ-rennes1.fr

Sophie Carles
University of Rennes 1, France
sophie.carles@univ-rennes1.fr

Thierry Stoecklin
University of Bordeaux, France
thierry.stoecklin@u-bordeaux.fr

Contents

3.   Ion Chemistry in Uniform Supersonic Flows   173

Ludovic Biennier, Sophie Carles, François Lique
and James Brian Mitchell

4. Study of Electron Attachment Reactions Using the CRESU Technique $\qquad$ **243**

Fabien Goulay and Bertrand R. Rowe

Barbara Wyslouzil and Ruth Signorell

Ilsa R. Cooke and Ian R. Sims

11. Theoretical Rate Constants   583

Philippe Halvick and Thierry Stoecklin

12. Conclusions: Future Challenges and Perspectives 639

Bertrand R. Rowe, André Canosa and Dwayne E. Heard

Chapter 1

Gas Phase Reaction Kinetics at Low and Very Low Temperatures: Fundamentals and Contexts

Bertrand R. Rowe[*,‡] and André Canosa[†,§]

Rowe Consulting, 22 chemin des moines, 22750 Saint Jacut de la Mer, France
†*CNRS, IPR (Institut de Physique de Rennes)-UMR 6251, Université de Rennes, F-35000 Rennes, France*
‡*bertrand.rowe@gmail.com*
§*andre.canosa@univ-rennes1.fr*

Abstract

The present book deals with more than thirty years of research performed in Chemical Physics with uniform supersonic flows and CRESU-like apparatuses. As such, it is linked to processes occurring at temperature close to absolute zero, studied for their theoretical interest or for their key role in ultracold media. The goal of this first chapter is to introduce a few concepts and equations that are ubiquitous within this context. The leading thread is the question of Local Thermodynamic Equilibrium (LTE) in dilute gases, complete or partial, and how it is implied in kinetic processes (Section 1.2), cold astrophysical media (Section 1.3), laboratory methods (Section 1.4), and the derivation of hydrodynamic equations (Section 1.5). It has been chosen to use the Boltzmann equations to derive the link between microscopic quantities (as state-to-state cross sections) and macroscopic quantities (as rate coefficients). How they are used in the derivation of hydrodynamic equations is also highlighted. The key role of distribution functions for translational or internal energies either in astrophysical media (Section 1.3) or laboratory methods (Section 1.4) is discussed. Note that the concept of temperature is itself subtle and is discussed at length in Section 1.2.

Keywords: Reaction kinetics; Temperature; Ultracold media; Local thermodynamic equilibrium; Boltzmann equation; Cross sections; Rate coefficients; Uniform supersonic flows; CRESU.

1.1   Introduction

Matter can exist in a variety of states, the most common being solid, liquid, gases, and plasmas. Besides their common character of being able to fill a container and to conform to its shape, gases and plasma are most often so diluted that the intermolecular distance is much larger than the molecular/atomic size. Intermolecular distance is defined as the mean distance between neighboring molecules/atoms and is grossly proportional to density$^{-1/3}$. The molecular size can be viewed as a typical distance at which molecules/atoms can interact. Although this latter concept remains sometimes vague, a good order of magnitude is the van der Waals radius [1] Hereafter, in this chapter, we most often use the term "particle" as generic for molecule, atom, or electron.

The macroscopic behavior of diluted gases and plasmas is directly linked to microscopic properties of particles and intermolecular forces (either long or short range) involved in the encounter of two particles, most often referred to as binary collisions. At the microscopic level, the collision cross sections in their various forms are key data not only for the fundamental understanding of the encounter process but also because they allow the calculation of some macroscopic properties, such as rate or transport coefficients, by integration over distribution functions.

In the case of plasmas, the long-range coulombic interactions between charged particles can induce some collective effects and correlations involving more than two partners (ions and/or electrons). However, it can be shown that many plasmas (including interstellar clouds and planetary ionosphere) can be mostly described by the sole consideration of binary encounters [2].

Binary collisions can be classified as elastic (when the nature and internal states of colliders are unchanged after collisions), inelastic (change occurs for internal states), or reactive (change in nature).

A large part of chemical physics is interested by the study of what happen in such reactive and inelastic collisions.

Equations describing the behavior of macroscopic properties such as mass, momentum, and energy in evolving multicomponent media can be derived directly by expressing fundamental laws of mechanics and thermodynamics in the form of conservation equations for a small volume of the Cartesian space. Besides the total time derivative of macroscopic properties, they can include two kinds of terms: either local production terms (for example, mass production in a reactive medium) or flux terms linked to transport phenomena (for example, diffusion terms in a gaseous mixture). These terms can be defined phenomenologically and studied as such. Equations describing the flows of gases are presented in Section 1.5 and in Chapter 2.

Macroscopic quantities can be derived from microscopic quantities Ψ_{Ai} relative to particles A in internal state i by their integration and summation over velocity and internal states distribution functions, respectively. For a dilute gas or plasmas, there is no need to consider another velocity (or similarly momentum) distribution function than the one particle distribution function $f(\vec{v}, A, i, \vec{r}, t)$ in the six dimensions phase space $(\vec{r}, \vec{v})$, which is defined such that $f(\vec{v}, A, i, \vec{r}, t)d\vec{r}d\vec{v}$ is the number of particles of species A in internal state i (as defined below) at time t in a volume element $d\vec{r}d\vec{v}$ around position $(\vec{r}, \vec{v})$ in the one particle phase space. In fact, the internal state i refers to a variety of quantum numbers corresponding to various modes of energy, i.e.

$$i = (i_1, i_2, \ldots, i_n) = \{i_j\} \tag{1.1}$$

and, depending on the situation, a summation can be made on one mode of energy only, for example, molecular rotation. For the sake of simplicity, when not specified, summation over subscript i indicates summation over all modes of energy. Also, f_{Ai} will be used for $f(\vec{v}, A, i, \vec{r}, t)$ hereafter.

Integration of f_{Ai} over $\vec{v}$ and summation over i define the number density of species A:

$$n_A(\vec{r}, t) = \sum_i \int f_{Ai} d\vec{v} \equiv n_A \tag{1.2}$$

From where the mass density of the medium can be defined,

$$\rho(\vec{r}, t) = \sum_A n_A(\vec{r}, t) m_A = \sum_A \rho_A \equiv \rho \tag{1.3}$$

It is also possible to define the mean velocity for species A_i as

$$\vec{v}_{0Ai} = \frac{1}{n_{Ai}} \int f_{Ai} \vec{v} d\vec{v} \tag{1.4}$$

If the velocity distribution itself is independent of i, then it also represents the mean velocity $\vec{v}_{0A}$ of species A and can be considered as a barycentric velocity as well since mass is independent of internal state. For a multicomponent mixture, a barycentric velocity $\vec{v}_b$ can then be defined by

$$\rho \vec{v}_b = \sum_A \rho_A \vec{v}_{0A} \tag{1.5}$$

By using a Cartesian frame moving with the mean velocity $\vec{v}_{0Ai}$, we can define a microscopic quantity with a new relative velocity $\vec{v}^*$ [3, 4]:

$$\vec{v}^* = \vec{v} - \vec{v}_{0Ai} \tag{1.6}$$

Then, the mean value $\widehat{\Psi_{Ai}^*}(\vec{r}, t)$ of a microscopic quantity $\Psi(v^*, A, i)$ can be written as

$$\widehat{\Psi_{Ai}^*}(\vec{r}, t) = \frac{\int f_{Ai} \Psi(v^*, A, i) d\vec{v}}{\int f_{Ai} d\vec{v}} \equiv \widehat{\Psi_{Ai}^*} \tag{1.7}$$

and their mean values over i and A can be expressed as

$$\widehat{\Psi_A^*} = \frac{1}{n_A} \sum_i n_{Ai} \widehat{\Psi_{Ai}^*} \text{ and } \widehat{\Psi^*} = \frac{1}{n} \sum_A n_A \widehat{\Psi_A^*} \text{ with } n = \sum_A n_A \tag{1.8}$$

Note that it is possible and usual [3] to employ a frame moving with the total barycentric velocity $\vec{v}_b$ to define a relative velocity $\vec{V} = \vec{v} - \vec{v}_b$ and the corresponding mean values of microscopic quantities.

Correspondingly, conservation equations can be derived from the kinetic equations which govern the evolution of the distribution functions, most often in the form of the Boltzmann equations, in which collision cross sections play again a key role [3–6]. It is then shown that local production terms can be obtained straightforwardly by integration and summation after multiplying Boltzmann equations by the convenient respective microscopic quantities. The velocity distribution solution of the Boltzmann equation for equilibrium is usually known as the Maxwell-Boltzmann (hereafter MB) distribution.

Hence, at equilibrium, integration is performed assuming an MB distribution and summation over the internal states distribution. On the contrary, transport phenomena need to consider a small departure of the velocity distribution function from MB and can be obtained from the so-called Chapman–Enskog solution of Boltzmann equations [5]. In this case, transport equations are of the Navier-Stokes (NS) kind (at least in the terminology of fluid mechanics and aerothermochemistry). For strictly MB distributions, transport flux terms are zero and conservation equations are of the Euler type. Following this procedure (and as detailed below in this chapter), it can also be shown that integration over relative velocity and summation over internal states of reactive and inelastic cross sections over distribution functions lead to quantities known as rate coefficients, which are central in the studies of gas phase reaction kinetics. It is beyond the scope of the present chapter to present the complete derivation of conservation equations following this procedure. The interested reader could refer to references [3, 4].

Binary collisions are essential in the kinetic equations of the distribution function and besides their characterization as elastic, inelastic, or reactive, they can also be classified by the range of their center of mass collision energy (hereafter E_{cm} and defined below), and in the last decades a strong interest arose for the understanding of what happens when this energy tends toward zero. Such studies have fundamental interest for highlighting specific quantum effects, for example. Another concern is to derive by integration rate coefficients at low and very low temperatures. These rate coefficients can also be measured directly as a function of temperature with convenient apparatuses. Such data are essential for the description of numerous "cold" and "ultracold" reactive media such as planetary atmospheres and interstellar clouds. We define cold conditions in the usual way, i.e. by a medium temperature below our normal room temperature.

In undertaking very low temperature or collision energy experiments, it is necessary to cool down molecules both from the translational and internal point of view and this leads to the following problem: As temperature is going down (from moderately low to ultralow T) toward absolute zero, most gas phase species undergo phase transition toward liquid and then solid state. For example, at the typical temperature of dense interstellar clouds (around $10\,\mathrm{K}$), only helium and hydrogen keep an appreciable vapor pressure allowing their use in a static cell at such a temperature.

However, matter can sometimes ignore the state in which it should be, i.e. it remains gaseous when it should be liquid or solid, or liquid instead of solid. The state of matter is then said to be metastable and for a gas this corresponds to supersaturation, i.e. to a gas vapor pressure higher than the saturation one. This is due to the fact that in the homogeneous gas phase, there is a barrier which hinders nucleation, as shown by the classical nucleation theory (CNT) or by the consideration at the microscopic level of the network of physical reactions leading to nucleation (see Chapter 6). If there is a seed to nucleation, either walls, aerosol grains, or ions, the nucleation is much faster, which prevents the possibility of direct cryogenic cooling of most gaseous species down to the lowest temperatures. Therefore, fast cooling of the bulk of a gas by expansion is the easiest way to obtain a supersaturated vapor and this is easily done using supersonic expansions where the thermal energy of a gas is converted into a bulk kinetic energy leading to strong temperature drop.

Cooling a gas in this way is useful for a variety of purposes: by skimming the flow, it can be the source of a molecular beam (i.e. a beam without molecular collisions) with the best energy resolution, in spectroscopy it can lead to strong simplification of spectra by internal states cooling (see Chapter 9), and it can be used as a flow reactor for the study of chemical reactivity and relaxation processes at the lowest temperatures (see Chapters 3–8 and 11).

At this stage, it is probably useful to provide some insight on the variety of supersonic/hypersonic flows that have and can be used in chemical physics and that are summarized in Fig. 1.1. In practice, a gas is introduced from a high-pressure reservoir P_0 toward a low-pressure chamber P_{ch}. If the pumping is sufficient to establish a critical ratio of pressure at a sufficient mass flow rate, the gas expands at a velocity higher than the local sound velocity. As discussed in Section 1.4.3.1, depending on the static pressure at the nozzle exit P_e versus the pressure in the chamber P_{ch}, the resulting jet is considered as underexpanded $P_e > P_{ch}$, perfectly expanded $P_e = P_{ch}$, and overexpanded $P_e < P_{ch}$, respectively.

The most commonly used hypersonic flow in chemical physics has been in fact the underexpanded free jet which basically is the flow obtained by expanding a gas from a reservoir to a chamber with a high-pressure ratio through a hole or a thin slit with or without a

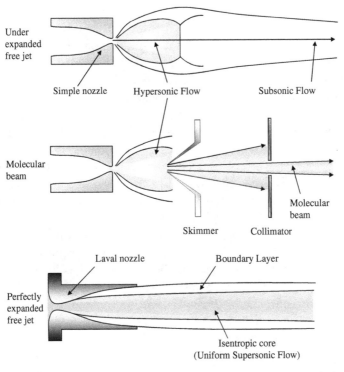

Fig. 1.1.   The most used supersonic/hypersonic tools in chemical physics.

simple converging nozzle. Very high Mach numbers (see its definition in Section 1.5.2) can be obtained in this way leading to very strong cooling. By skimming such a flow toward a very-low-pressure chamber, a molecular beam of good velocity resolution can be obtained.

Although the underexpanded free jet allows very strong cooling, it suffers of some strong drawbacks from the point of view of local thermodynamic equilibrium (see next sections) and of flow gradients which prevent its generalized use as a flow reactor. It was first realized by Rowe *et al.* in the early eighties [7] that the uniform supersonic flow issued from a Laval nozzle (see Chapter 2 for a historical perspective of Laval nozzles) was much more suitable for this purpose and the technique was named CRESU standing for *Cinétique de Réactions en Ecoulement Supersonique Uniforme*, a French acronym meaning "kinetic of reactions in a uniform supersonic flow". Besides its use for the measurement of rate coefficients (Chapters 3–5), the uniform supersonic flow has also been used in spectroscopy (Chapter 9) and nucleation (Chapter 6) studies. In fact, this kind of flow is a perfectly expanded free jet with a nozzle well designed to obtain spatial uniformity downstream of the nozzle exit. More information on the terminology, laws, and calculation of supersonic expansions can be found in Sections 1.4 and 1.5 and in Chapter 2.

It must be noticed that in a race to cool down atoms and molecules at ultimate values of energy, very close to absolute zero, physicists have developed very specific techniques such as laser cooling [8] and Stark deceleration [9]. This was done with great success leading to the observation at the end of the 20th century of the Bose-Einstein condensations, a state of matter predicted theoretically as early as 1924/25 by A. Einstein [10] following the work of S. Bose. Although studies of chemical reactivity in the sub-milli-Kelvin range in the last two decades have clearly become a target for such techniques, it has in fact been restricted to very specific cases (see Section 1.4.5). These techniques are not in the scope of the present book, the aim of which is to review a large part of the work that has been performed for more than thirty years using the CRESU technique and uniform supersonic flows in chemical physics. It is intended also to open new perspectives in the field.

The following sections present some basic concepts that are mostly assumed to be known in the following chapters. The guiding thread is to introduce the link between microscopic and macroscopic quantities, highlighting the question of thermodynamic equilibrium. Of course, the reader familiar with these questions is invited to skip the lecture further down.

1.2   Some Basic Concepts in Gas Phase Chemistry and Kinetic Processes

This section is not an extensive review of the concepts used in gas phase chemical physics (see for instance [11, 12]). Rather, it is mostly devoted to the problem of energy (kinetic or internal) distributions in a population of gaseous atoms and molecules and its links with cross sections and rate coefficients. This is a fundamental problem at low and ultralow temperatures, because, by essence and as stated in the introduction, matter is not gaseous at thermodynamic equilibrium close to absolute zero, even at extremely low pressures. This implies some disequilibrium either in natural media or in experiments in these temperature conditions for the gas phase. Theory of course allows (and prefers) state-to-state cross-section calculations for binary collisions at a given center of mass kinetic energy that can then be used to derive kinetic data for any distributions of real life and close to absolute zero.

1.2.1   *The concept of temperature*

In low- and very-low-temperature studies, a first step is to define the notion of temperature itself and its link to thermodynamic equilibrium.

Historically, the temperature is related to "cold" and "hot", which is a common experience to human kind, although our body is sensitive to heat flux rather than temperature. This is the reason why we find metal to be colder (or hotter) than wood at a temperature below (or above) our hand's temperature. In fact, thermometry came long before the development of thermodynamics and statistical mechanics science. Grossly, to quantify cold and hot, the temperature is then defined from the variation of a physical quantity (dilatation of mercury, for example) with heating or cooling (as we perceive it). To define the temperature scale, physical events such as freezing and boiling of water at sea level are chosen. Following the work of Galileo Galilei and his contemporaries on the thermoscope which used air dilatation, most of the first thermometers were developed at the end of the 17th and early 18th century and based on the dilatation of liquids. They remain one of the most used kinds of thermometers nowadays in ordinary life.

Even if they are related, heat and temperature are very different quantities being, respectively, extensive and intensive (i.e. depending, respectively, on the quantity of matter or not). The temperature itself is an intensive property of a system closely linked to thermodynamic equilibrium and is meaningless far from equilibrium (which is not the case for energy). However, it is sometimes used by extension of statistical laws in two-level systems (in laser level inversion, for example), leading to "negative" temperature in non-equilibrium conditions. Such claim of negative absolute "temperature" is also made in some experiments in optical traps where atoms in a lattice are switched very fast to an upper state of energy. Therefore, such states of matter do not correspond to "colder" but in fact to "hotter" media [13].

Macroscopic thermodynamics, which was mainly developed throughout the 19th century, deals with systems at complete equilibrium and defines temperature through a variety of concepts and principles, using state functions as energy, entropy, and state laws. For example, entropy S and heat Q are linked through the well-known relationship which can be used as a definition of absolute temperature:

$$dS = \frac{\delta Q}{T} \tag{1.9}$$

which applies to a closed system exchanging an amount of heat δQ with an external source. For a perfect gas, the state law is given as

$$PV = NRT \text{ or equivalently } \frac{P}{\rho} = \frac{R}{M}T = r_m T \tag{1.10}$$

which relies pressure P and volume V to absolute temperature T and the number of moles N (R being the universal gas constant). Introducing $r_m = R/M$ as the specific constant of a perfect gas of molar mass M, the state law can also be written in terms of mass density ρ since $\rho = NM/V$.

Macroscopic thermodynamics also leads to the definition of absolute zero [14] as the point at which the internal energy of a system is minimal and where no more heat can be removed from the system. Historically, the idea of absolute zero was derived from the behavior of a perfect gas as defined by Eq. (1.10).

Statistical mechanics and thermodynamics, which historically came mostly after the first development of macroscopic thermodynamics, are developed considering that a macroscopic system consists of a large quantity of microscopic systems in given states of energy, as well kinetic than internal. In a gas or dilute plasma, these microscopic systems are their constituting atoms and molecules (and ions and electrons if any). The states of energy can be continuous or discrete and their distributions at equilibrium are directly given by the value of the absolute temperature. In fact, temperature can be defined in this way and a variety of spectroscopic measurements of temperature rely on this property.

1.2.2 *Energy distributions, chemical composition, and radiation in a gas at thermodynamic equilibrium*

In macroscopic thermodynamics, equilibrium relates to an isolated system, i.e. a system defined by a surface boundary which does not exchange mass (closed system) and energy (insulated system) with its surroundings and therefore will not evolve further after reaching this equilibrium at a given time. Of course, in the evolution between two equilibrium states, exchanges can occur but macroscopic thermodynamics deals only with initial and final states and global value of exchanged quantities (heat, work etc.). The system is assumed to be fully described at any time by macroscopic extensive and intensive properties (volume, pressure, temperature, elemental composition, etc.), and no gradients of these properties or kinetics of change are considered. For such a system at equilibrium, statistical thermodynamics yields distribution functions of its constituting particles, given the values of macroscopic properties with at first rank the temperature.

In the classical kinetic theory of gases, atoms and molecules are considered as geometrical points of mass m and velocity $\vec{v}$. The velocity distribution function has been defined in the introduction and at equilibrium has the form of Maxwell–Boltzmann, which is isotropic and, for a gas component A in state i having a macroscopic mean barycentric velocity $\vec{v}_{0Ai}$, can be written as

$$f_{Ai}^0\left(\vec{v^*}, \vec{r}, t\right) = n_{Ai}(\vec{r}, t)\left(\frac{m_A}{2\pi k_B T_{Ai}(\vec{r}, t)}\right)^{3/2} exp\left(-\frac{m_A \vec{v^*}^2(\vec{r}, t)}{2k_B T_{Ai}(\vec{r}, t)}\right)$$

$$(1.11)$$

where $\vec{v^*}$ has been defined by Eq. (1.5) and k_B is the Boltzmann constant. Of course, at equilibrium, velocity is independent of the internal state of particle A and most often the kinetic temperature T_{Ai} (also named translational) defined by Eq. (1.10) is independent of the state of particle A. Further, as discussed in Section 1.2.3, only species of very different masses can have different translational temperatures, and therefore most often there exists a unique translational temperature (hence $T_A = T \,\forall\, A$). Recognizing the isotropicity of the distribution, integration over angle in spherical coordinate, yields in terms of the modulus of velocity $v = |\vec{v^*}|$:

$$f_A(v) = 4\pi n_A \left(\frac{m_A}{2\pi k_B T}\right)^{3/2} v^2 exp\left(-\frac{m_A v^2}{2k_B T}\right) \qquad (1.12)$$

from where the mean thermal velocity $\widehat{v_{Ath}}$ can be defined:

$$\widehat{v_{Ath}} = \frac{\int f_A(v)\, v dv}{\int f_A(v)\, dv} = \sqrt{\frac{8k_B T}{\pi m_A}} \qquad (1.13)$$

Then, it follows the well-known fact that the mean thermal velocity is much higher for light than for heavy species.

Besides their kinetic energy, molecules and atoms have internal energy levels corresponding to their internal states. Molecules and atoms share an internal state known as electronic (linked to the motion of electrons around the nucleus) and spin-orbit (due to the magnetic interaction between the orbital angular momentum and the spin of the electron) [12, p. 448]. The energy difference between spin-orbit levels is much less than between electronic levels. In spectroscopy, fine structures of the spectra are linked to spin-orbit states. Molecules also have states linked to their internal vibration and rotation. Below, letter i in minuscule refers to any form of internal state.

Following, respectively, quantum mechanics and statistical thermodynamics, internal states i are quantified and distributed at equilibrium following Boltzmann distributions (the subscript related to the chemical species A, B, etc., is omitted for the sake of simplicity):

$$\frac{n_i}{n} = \frac{exp - \frac{E_i}{k_B T}}{Z} \qquad (1.14)$$

where n_i is the density of state i, n the total density of the considered species, and Z the partition function defined by

$$Z = \sum_i g_i exp - \frac{E_i}{k_B T} \qquad (1.15)$$

where E_i is the energy level with statistical weight g_i and $E_{i=0}$ is therefore the ground state energy. Due to the form of Eqs. (1.11) and (1.12), the choice of its value has no influence on the calculated level population. $E_{i=0}$ can be taken as zero, the residual zero-point energy, or include an energy of formation when considering a reactive mixture.

In a gas with a variety of species, chemically reactive absolute temperature also governs the chemical composition through the law of chemical equilibrium. It can be found by writing that the Gibbs free energy of the mixture is at a minimum. If components can be considered as perfect gases, it can be written as

$$\prod_A \left(\frac{P_A}{P_\emptyset} \right)^{\nu_A} = exp - \frac{\Delta G^\phi}{k_B T} \tag{1.16}$$

where ν_A are the stoichiometric coefficients of the chemical reaction written as

$$\sum_A \nu_A A = 0 \tag{1.17}$$

and are taken negative for the reactants and positive for the products. P_A is the partial pressure of species A. The subscript $\emptyset$ refers to standard state. More details on the calculation of equilibrium are beyond the scope of this section and can be found in classical textbooks such as [12] or in JANAF tables [15].

Photons are also a constitutive part of matter and radiation may also be at thermodynamic equilibrium. In an insulated enclosure, i.e. opaque in the case of photons, the local radiation field is completely specified by the local temperature value through the usual radiation laws (Planck law). Such an enclosure pierced by a small hole constitutes a black body source and the radiation escaping from the hole is called the black body radiation and is completely determined by the enclosure temperature.

It can be retained from the present section that **at equilibrium**, knowing the specific microscopic properties of the constituent species of a gas mixture and at a given pressure, the value of the absolute temperature fixes everything: chemical composition and distribution functions.

1.2.3   *Local thermodynamic equilibrium, relaxation times, and Knudsen number*

As stated above, equilibrium of a closed and insulated macroscopic system does not consider spatial and temporal evolution. Describing such evolution of a gas (or dilute plasma) mixture could be done by following very small volumes of matter inside the medium flowing with a barycentric velocity $\vec{v}_b (\vec{r}, t)$ and exchanging matter and

energy with their surroundings while experiencing internal evolution
(for example, chemical change). Such description of small volumes'
evolution while following their movements is known as Lagrangian.
However, it is much simpler to adopt the so-called Eulerian description where everything is described by fields of local properties and
primarily the barycentric velocity field. Fields of local properties are
governed by partial differential equations established in the way discussed in the first section.

In a small volume of the medium, equilibrium must be established at least partially in order to have the possibility or not of
using an absolute temperature as one of the fundamental local properties. From the point of view of distribution functions, this means
that they must be very close to their equilibrium forms, although a
small deviation can be admitted and is indeed necessary to describe
correctly some phenomena such as transport properties.

Assume that for some reason (for example, flow through a shock
wave) a swarm of particles experiences a strong deviation from equilibrium distribution functions; it will take some time to recover a new
equilibrium, generally at a different temperature than the initial one.
For a quasi-exponential recovery, if $f(t)$ is the distribution function
and f_0 its equilibrium form, then

$$f(t) = g(t) exp\left(-\frac{t}{\tau}\right) + f_0 \qquad (1.18)$$

where $g(t)$ changes are much less than exponential. τ can be considered as the relaxation time toward equilibrium. For an evolution time
t, the distribution function is said to be frozen if $t \ll \tau$ and comes at
equilibrium for $t \gg \tau$.

In a flowing medium of typical velocity $v_b = |\vec{v_b}|$ and typical
length L, the evolution time t_h defined by

$$t = t_h = \frac{L}{v_b} \qquad (1.19)$$

is known as the *hydrodynamic time* and must be used for comparison
with various relaxation times—which can be defined for any kind of
energy—in order to know if equilibria have been achieved. Of course,
for a transient flow, the transient time of evolution must be considered too. Hereafter, t_{evol} is used as the characteristic time of any
evolution of a gaseous medium.

In collisions, particles exchange energy in various forms, either for a given species or in an intra-species mode, and this leads locally to establish various equilibrium distributions with a different relaxation time τ_r where the subscript r refers to a determined kind of energy exchange of the particles. If it can be considered that in any point of space

$$\tau_r \ll t_{evol} \qquad (1.20)$$

then the medium can be considered as being in the condition of *local thermodynamic equilibrium* (hereafter LTE), i.e. a local absolute temperature $T(\vec{r}, t)$ can be defined at a given location and time. This temperature fully determines translational velocity and internal energies' distributions of particles by the equilibrium laws defined in Section 1.2.2.

Chemical relaxation times can also be defined if reaction kinetics is known. If conditions such as Eq. (1.17) hold for them, then the medium can be in *local chemical equilibrium* and Eq. (1.16) holds everywhere.

Concerning photons, if the radiation field is also determined in any point at any time by the local temperature, then conditions of **Complete** *local thermodynamic equilibrium* (hereafter CLTE) hold. This is normally not the case for a tenuous cold medium, but can be found, for example, in stellar interiors at very high temperatures. Moreover, at the laboratory scale, for many applications in gases at medium pressure, the behavior of the system does not need to take radiation into account and is completely dominated by collisional processes from the thermal point of view. This is especially the case in supersonic flows which generally use buffer gases which are not optically active and for which photon absorption is negligible for optically active minor species mixed in the flow in small concentration (i.e. the optical depth is zero). Of course, in planetary atmospheres or the interstellar medium where typical lengths are very large versus photon mean free path at some wavelengths (see below), radiation can play an important role on the thermal behavior of the system and on the energy budget. It follows that a complete discussion of the radiation problem has not been considered in the scope of the present book devoted mainly to chemical physics in laboratory supersonic flows.

Note that it is possible to establish, at a given station of a medium, different equilibrium distributions for various modes of energy and species with different temperatures (for example, an electron temperature T_e and a vibrational temperature T_v distinct from the main translational/rotational temperature T). These kinds of situations are most often referred to as situation of *Partial local thermodynamic equilibrium* (hereafter PLTE). They prevail mostly in the rarefied uniform supersonic flow issued from Laval nozzles as well as in astrophysical media at low pressure and temperature. Given below are some insights about the order of magnitude and value of various relaxation times.

Translational energy for particles of the same mass is the most readily exchanged quantity in collisions leading to MB distributions for velocity and kinetic energy. This translational relaxation time τ_t has a value (depending on the choice of the initial distribution) that is of the order of the inverse of the collision frequency ν_c which can be defined by

$$\nu_c = \frac{\widehat{\upsilon_{Ath}}}{\lambda} \tag{1.21}$$

where λ is the mean free path and $\widehat{\upsilon_{Ath}}$ the thermal mean velocity of molecules A defined by Eq. (1.12), with

$$\lambda = \frac{1}{\sqrt{2}n\sigma} \tag{1.22}$$

where σ is the elastic cross section as defined in the next section.

For two species A and B with particle B in small admixture with respect to A which is at translational equilibrium, relaxation time for B can be written as [16]

$$\tau_{BA} = \frac{3\sqrt{2}}{8} \frac{(m_A + m_B)^2}{m_A m_B} \frac{1}{\upsilon_{AB}} \tag{1.23}$$

where ν_{AB} is the inter-species collision frequency defined by

$$\upsilon_{AB} = \frac{\widehat{\upsilon_{Ath}}}{\lambda_{BA}} \tag{1.24}$$

When $m_A = m_B$, this yields a value very close to τ_t defined above. Therefore, only for particles of very different masses, like electrons versus neutrals and ions in a plasma, MB distribution functions with distinct temperatures can be locally established [2] (see also Chapter 4).

If an MB distribution is established locally, i.e. if $\tau_t \ll t_{evol}$, a local temperature (said to be translational) can be considered as defined. For a lot of molecules, with the notable exception of very light molecules, especially H_2, a Boltzmann distribution for the rotation is established with a rotational relaxation time only a few times the translational one. Then, the local temperature is a translational/rotational one. Note also that at very low T, the ortho-para conversion being an extremely slow process, two manifolds of rotational levels can be established at equilibrium for the ortho and para states [17], respectively. For other kinds of energy, equilibrium is often not achieved. For instance, Chapter 4 will show how the non-equilibrium of nitrogen vibrational energy hinders electron relaxation in nitrogen flows.

Besides the temporal aspect of the problem of LTE or PLTE establishment, there is a spatial aspect to consider as well. The walls at the border of a flow reflect molecules in a variety of ways (specular, diffuse or a mix). This can strongly disturb the MB distribution function of velocity for the distance to the wall of the same order of magnitude as compared to the mean free path. There, it is no longer possible to handle problems by using a local temperature. For a flow of given external or internal geometric borders, it is possible to define a characteristic macroscopic length L and it is usual to define the Knudsen number by

$$K_n = \frac{\lambda}{L} \tag{1.25}$$

If $K_n \ll 1$, the flow is said to be in continuous regime. Although consecrated by use, this appellation can be, in our opinion, sometimes misleading since macroscopic quantities such as density or barycentric velocity can be spatially defined as continuous fields even if the Knudsen number is much larger than unity. Only for a description at a typical distance close to or lower than the interparticle distance (which is much less than the mean free path), a gaseous medium cannot be considered as continuous. In fact, $K_n \ll 1$ implies mainly that LTE and local temperature can be used to describe the considered medium [18].

Kinetic regimes are characterized by $K_n = O(1)$; there are not enough collisions per unit of time for a local thermodynamic equilibrium to be reached. The medium is adequately described by

$f = f(\vec{v}, \vec{r}, t)$, the single particle velocity function, solution of the Boltzmann equation. Macroscopic quantities are computed as in the continuous regime by averaging the corresponding microscopic quantity for a single particle over the distribution function [18].

When $K_n \gg 1$, the flow is in the molecular free regime and the velocity distribution has a discontinuous behavior in the velocity space, although as discussed previously many macroscopic quantities are spatially continuous [19].

From the prior discussion and equations, it is seen that densities and cross sections play a key role in the relaxation toward equilibrium due to collision processes. When the collisional relaxation leads to a colder state, it is generally called **collisional cooling**. A special case of collisional cooling is when a given species is cooled down by another, which has itself been cooled by any means evoked in Section 1.1: this kind of cooling is quite often called **sympathetic cooling**. In a way, the various molecules seeded in small amount in a buffer gas which experiences supersonic expansion can be considered to have been cooled by sympathetic cooling.

1.2.4 *Notions of collision cross section*

As discussed, collisions can be described at a time and space where two particles A and B interact resulting in a large number of events: changes in their movements (direction and velocity), internal states or nature (chemical reactions), etc. These events can be described either in classical mechanics or from the quantum point of view. A complete introduction to the theory of molecular collisions is far beyond the scope of the present section. Only a few concepts and in particular the one on cross section, useful for further developments, will be presented below.

In order to define a cross section in a quite general way, consider a beam of mono kinetic (unidirectional velocity of modulus g) particles of density n_A impinging on a target of particles at rest and of density n_B; the flux of particle A is

$$\Phi_A = n_A g \qquad (1.26)$$

where Φ_A is a number of particles A per unit time and unit area.

Collisions can produce a given event at a rate R_{coll} per unit volume and unit time which is proportional to the density n_B and to the

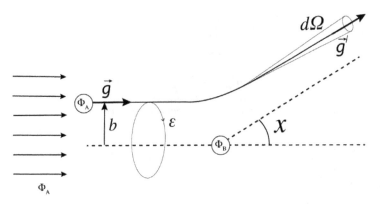

Fig. 1.2.   Elastic collision between two particles A and B.

impinging flux Φ_A.

$$R_{coll} = \sigma n_B \Phi_A \qquad (1.27)$$

The constant of proportionality σ has the dimension of an area and is usually named cross section.

It must be noticed that, since a variety of events can be defined, there is a diversity of cross sections for an encounter between a particle A and a particle B.

One kind of event is an elastic collision with scattering in a defined differential solid angle $d\Omega$ in a direction defined by the angles ε and χ as shown in Fig. 1.2.

In many collisions, the angle ε does not play any role. From this figure, we can also define the impact parameter b. The angle χ depends on the velocity g and also the impact parameter b shown in Fig. 1.2.

$$\chi = \chi(b, g) \qquad (1.28)$$

From this, it is important to emphasize that b can be related with an inverse function to χ and g.

Following the above discussion, the number of particles A scattered per unit volume and unit time in the defined angles ε and χ in the differential solid angle $d\Omega$ is

$$R_{diff} = \sigma_{diff}(\chi) n_A n_B g \qquad (1.29)$$

which defines the differential cross section σ_{diff}.

Note that for a given value of g the number of particles scattered in a ring between χ and $\chi + d\chi$ (solid angle $2\pi\chi$) is the same as the one entering in the ring $2\pi b db$, hence

$$\sigma_{diff}(g, \chi) \sin \chi d\chi = b db \qquad (1.30)$$

from where the total elastic cross section can be defined as

$$\sigma_{AB}^{Tot}(g) = \int_0^{b_{max}} 2\pi b db = \pi b_{max}^2 = 2\pi \int_0^\pi \sigma_{diff}(g, \chi) \sin \chi d\chi \quad (1.31)$$

where b_{max} corresponds to the value of the impact parameter for which $\chi = 0$.

Note that b_{max} depends on the intermolecular potential and then the total cross section can sometimes diverge, for example, in the coulombic interaction. In the case of the hard sphere model, the cross section assimilates to the surface of a disk of the same radius as the hard sphere.

The hard sphere model corresponds to an intermolecular potential $\varphi(r)$ with a "wall" form as a function of distance r between point particles [5]:

$$\varphi(r) = \infty \text{ for } r < b_{max} \text{ and } \varphi(r) = 0 \text{ for } r > b_{max} \qquad (1.32)$$

But, clearly, a variety of more realistic intermolecular potentials can be used in place of the hard sphere. For a complete presentation of elastic collisions, see [5, pp. 30–36].

Note that a momentum transfer cross section can also be defined. This is mostly useful for transport phenomena, but not for rate coefficients.

Extension of the notion of cross section to inelastic and reactive processes has been discussed by a variety of authors [3, 20, 21]. As mentioned earlier, if elastic collisions are defined mostly through the scattering angle χ, inelastic or reactive collisions include the probability of a given event.

1.2.5   *Laboratory and center of mass frame*

A collision between two particles A and B can be described in different frames. The center of mass frame (hereafter CMF) is a Galilean frame which has for origin the center of mass of the two colliding particles. Due to the conservation of momentum, the center of mass moves at a constant velocity V in any other Galilean frame and is unchanged after collision [5]. One of the most used frames besides CMF is the laboratory frame for practical reasons, especially in beam experiments, for example, when a beam of particles collides with an immobile target.

The description of collisions is much easier in the CMF. The total kinetic energy in any frame is conserved for elastic collisions, but for inelastic or reactive collisions only the kinetic energy in the CMF can be used for the change of internal states or chemical nature of the colliders. It can be shown that this kinetic energy, most often referred to as KE_{cm}, is

$$KE_{cm} = \frac{1}{2}\mu v^2 \tag{1.33}$$

where v is the modulus of relative velocity

$$v = \left|\vec{v_A} - \vec{v_B}\right| \tag{1.34}$$

and μ the reduced mass defined by

$$\mu = \frac{m_A m_B}{(m_A + m_B)} \tag{1.35}$$

1.2.6   *Rate coefficient and its links to cross section viewed through the Boltzmann equation*

A chemical reaction, or a change of internal state, can involve a given number of partners: atoms, molecules, ions, electrons, or photons. Most usually, in the gas phase, reactions involve one, two, or three particles and are termed as unimolecular, bimolecular, or termolecular, respectively. In this section, we shall consider only bimolecular reactions, keeping in mind that unimolecular or termolecular reactions can be viewed as involving in fact bimolecular collisions.

A reaction between species A and B leading to two products, respectively, C and D can be written as

$$A + B \underset{k_f}{\overset{k_r}{\rightleftarrows}} C + D \tag{1.36}$$

where k_f and k_r are, respectively, the forward and reverse rate coefficients which govern the kinetics of the forward and reverse reaction, respectively, following the equation

$$\frac{dn_A}{dt} = -k_f n_A n_B + k_r n_C n_D \tag{1.37}$$

with of course

$$\frac{dn_A}{dt} = \frac{dn_B}{dt} = -\frac{dn_C}{dt} = -\frac{dn_D}{dt} = K_R \tag{1.38}$$

These equations do not consider variations of n_A (respectively, B, C, D) due to compressible and diffusion effects discussed in Section 1.5. In LTE conditions, the internal states of products and reactants in Eq. (1.36) are at equilibrium at the local temperature T and the rate coefficients defined above are a function of this temperature alone. In fact, pressure plays a role in the case of unimolecular and termolecular reactions.

There are a large number of theories devoted to calculation of rate coefficients ranging from transition state theory (hereafter TST also named activated complex) first developed by Eyring and Polanyi nearly ninety years ago [22, 23] to *ab initio* calculations of the reactive collision. In general, they involve first the knowledge of potential energy surfaces. TST and its modern versions [24] remain the most used way to calculate rate coefficients nowadays. Nevertheless, they are only likely to be accurate for systems for which the energy distribution can be considered as statistical. It is far beyond the scope of the present section to overview this topic (see Chapter 11). Note that it has been the standard in chemical kinetics to represent the rate coefficient evolution with temperature by an Arrhenius law:

$$k(T) = A \, exp - \frac{E_a}{k_B T} \tag{1.39}$$

where E_a is the activation energy. The Arrhenius law is clearly inappropriate at very low temperature since it predicts that reactions with a positive activation energy do not proceed ($k(T) \to 0$ when $T \to 0K$).

We will restrict ourselves to the link between cross sections and rate coefficients which can be developed using the Boltzmann equation. This point has been discussed extensively in [3, 20, 25]. Following these authors, momentum $\vec{p}$ instead of velocity $\vec{v}$ distribution functions is used below with $\vec{p} = m\vec{v}$. Also, distribution functions normalized to unity are used. The Boltzmann equation for the momentum distribution function of a species A in quantum state i can be written as

$$\frac{d\{n_{Ai} f_{Ai}(\overrightarrow{p_{Ai}}, r, t)\}}{dt} = J_{Ai} \tag{1.40}$$

Note that, as discussed above, the subscript i in minuscule refers to the specification of the quantum state of A for several modes of energy (mostly rotation, spin-orbit, vibration, and electronic).

The left-hand side of the Boltzmann equation is the total time derivative in the phase space and the right-hand side a production term due to collisions that can be broken down into three terms, corresponding to elastic, inelastic, and reactive collisions:

$$J_{Ai} = J_{Ai}^{el} + J_{Ai}^{inel} + J_{Ai}^{reac} \tag{1.41}$$

As stated above in this chapter, multiplying Boltzmann equations by microscopic quantities ψ_{Ai} relative to particles and integration over velocity and summation on quantum states leads to macroscopic equations of conservation [3–5]. If this is directly done for the Boltzmann equation (i.e. $\Psi_{Ai} = 1$), equations governing the evolution of chemical species densities are obtained. If the velocity distribution function is exactly MB, these latter equations are of Euler type, i.e. the diffusion terms (see Section 1.5) reduce to zero.

Since elastic and inelastic collisions conserve the total number of particles A, it results that

$$\sum_i J_{Ai}^{el} = \sum_i J_{Ai}^{inel} = 0 \tag{1.42}$$

and the equation for density of A then reads as

$$\frac{dn_A}{dt} = \sum_i \int_{\overrightarrow{p_{Ai}}} J_{Ai}^{reac} d\overrightarrow{p_{Ai}} \tag{1.43}$$

For a mixture of four species and a reaction $A + B \rightarrow C + D$ which can be decomposed into elementary steps (i.e. state to state reactions), J_{Ai}^{reac} can be written as

$$J_{Ai}^{reac} = \sum_{jlm} \int \frac{p_{AB}}{\mu_{AB}} \sigma_{AB}^{reac}$$
$$\times \left\{ n_{Cl} n_{Dm} f_{Cl}' f_{Dm}' - n_{Ai} n_{Bj} f_{Ai} f_{Bj} \right\} d\overrightarrow{p_{Bj}} d\Omega \tag{1.44}$$

where $p_{AB} = |\vec{p_A} - \vec{p_B}|$ and superscript $'$ refer to collisions leading back to the state of the direct collision (see the above references and [5, pp. 444–449]).

Define the relative population of quantum states as

$$x_{Wi} = \frac{n_{Wi}}{n_W} \quad \text{with} \quad W = A, B, C \text{ or } D \tag{1.45}$$

Considering that the gas mixture behaves as incompressible and without diffusion (see Section 1.5), the macroscopic conservation equation for A then reads as

$$\frac{dn_A}{dt} = -n_A n_B \left\{ \sum_{ijlm} x_{Ai} x_{Bj} \int f_{Ai} f_{Bj} \frac{p_{AB}}{\mu_{AB}} \sigma_{AB}^{reac} d\vec{p_{Ai}} d\vec{p_{Bj}} d\Omega \right\}$$

$$+ n_C n_D \left\{ \sum_{ijlm} x_{Cl} x_{Dm} \int f'_{Cl} f'_{Dm} \frac{p_{AB}}{\mu_{AB}} \sigma_{AB}^{reac} d\vec{p_{Ai}} d\vec{p_{Bj}} d\Omega \right\} \tag{1.46}$$

It results by comparison of Eqs. (1.46) and (1.37) that the forward reaction rate coefficient can be written as

$$k_f(T) = \sum_{ijlm} x_{Ai} x_{Bj} \int f_{Ai} f_{Bj} \frac{p_{AB}}{\mu_{AB}} \sigma_{AB}^{reac}(ij \rightarrow lm) d\vec{p_{Ai}} d\vec{p_{Bj}} d\Omega \tag{1.47}$$

It is then possible to define a total reaction cross section by

$$\sigma_R(p) = \sum_{ijlm} x_{Ai} x_{Bj} \int \sigma^{reac}(ij \rightarrow lm) d\Omega \tag{1.48}$$

which is clearly dependent on the population distribution of internal levels. Note that relative velocity can be used instead of momentum in this equation.

If the population states are at equilibrium (Eq. (1.14)), then the total reaction cross section is a function of the relative velocity between A and B and depends also on temperature T through the population of energy levels. Then a straightforward calculation [3, p. 296] shows that Eq. (1.47) can be expressed as

$$k_f(T) = \sqrt{\frac{8k_BT}{\pi\mu_{AB}}} \int_0^\infty \frac{E}{k_BT}\sigma_R(E)\,exp^{-\frac{E}{k_BT}}\frac{dE}{k_BT} \qquad (1.49)$$

where, as stated above, the total reactive cross section is also a function either of temperature (at equilibrium) or of the distribution of internal states.

In the same way, it is also clearly possible to define reactive cross sections for specified internal states (i,j) of the reactants and then the corresponding rate coefficient. It is also possible to define inelastic cross sections and rate coefficients for inelastic processes in the same way.

1.2.7 *Microscopic reversibility, detailed balance, and mass action law*

Microscopic reversibility is linked to the symmetry of the Schrödinger equation or of classical mechanics equations when time is inversed, implying that microscopic detailed dynamics of particles and fields are time reversible. It can be written as [3]

$$p_{AB}^2 g_{Ai}g_{Bj}\sigma_f(ij \rightarrow lm, \vec{v}, \Omega) = p_{CD}'^2 g_{Cl}g_{Dm}\sigma_r\left(lm \rightarrow ij, \vec{v'}, \Omega\right) \qquad (1.50)$$

where the g_W parameters are the statistical weights of the various partners of the reaction. Detailed balance results from microscopic reversibility and states that, for a chemical reaction which can be decomposed into elementary reactions (i.e. states to states reaction), these are in equilibrium, which can be expressed by

$$\frac{k_f(ij \rightarrow lm)}{k_r(lm \rightarrow ij)} = \frac{g_{Cl}g_{Dm}}{g_{Ai}g_{Bj}}\left(\frac{\mu_{CD}}{\mu_{AB}}\right)^{3/2}exp^{-\frac{\Delta E}{k_BT}} \qquad (1.51)$$

with ΔE being

$$\Delta E = \Delta E^0 + E_{Cl} + E_{Dm} - E_{Ai} - E_{Bj} \qquad (1.52)$$

in which ΔE^0 results from the formation energy of the various species and where internal state energies do not include the energy of formation.

For the complete reaction mass action law, i.e. chemical equilibrium, the equation can then be derived:

$$\frac{k_f}{k_r} = \frac{Z_C Z_D}{Z_A Z_B} \left(\frac{\mu_{CD}}{\mu_{AB}} \right)^{3/2} exp^{-\frac{\Delta E^0}{k_B T}} \tag{1.53}$$

The right-hand side of this reaction is the statistical expression of the equilibrium constant for the bimolecular reaction $A + B \leftrightarrows C + D$.

The concepts discussed in the present section are not restricted to chemical reactions, but their generalization is beyond the scope of this chapter. What must be retained here is that, due to the exponential terms in the equations, most reactions, except for nearly thermoneutral, proceed only "one way" at very low temperature.

1.2.8 *Chemistry in PLTE conditions: photo chemistry and radiochemistry*

There are many natural and artificial media where the chemistry is not driven by the local temperature, as defined above. In fact, this occurs when media are only in PLTE and when radiation or electromagnetic fields disturb distribution functions and chemical equilibrium. Mostly, two kinds of situations can occur: first, energetic radiations (photons, cosmic rays etc.) penetrate the medium and interact with atoms and molecules leading to a variety of processes (excitation, dissociation, and ionization mainly but also nuclear processes) that are described in detail for the astrochemical case in Chapter 10 by E. Herbst. These processes lead to the formation of energized and active species which initiate the chemistry. Second, among these species, charged particles are very sensitive to electromagnetic fields, which consequently can disturb their distribution function leading to a specific chemistry. This is the case, for example, in drift tube reactors where an electric field heats up ions at a mean energy much higher than the thermal energy of the neutral. This ability to manipulate ions and electrons by the electromagnetic field has been widely used in experimental techniques as described in Section 1.4.

Usually, photochemistry is used to name a chemistry driven by photons, whereas radiochemistry is mostly reserved for chemistry involving nuclear processes, although it can be used for chemistry induced by very energetic particles as cosmic rays as well.

In the following section, we give a brief introduction to the case of astrochemistry, which is both photochemistry and radiochemistry driven, and is developed at length in Chapter 10.

1.3   Astrophysical Media at Low and Very Low Temperatures: The Birth of Astrochemistry

1.3.1   *Overview of such media*

For millennia, humankind has been interested in observations of stars and planets and has strongly impacted the development of mechanical sciences such as celestial mechanics over the centuries. Nowadays, the existence of stars, planets, comets, and galaxy is basic astrophysical knowledge shared by almost everyone and the man on the street is kept aware of space missions like Rosetta or of the discovery of exo-planets.

However, at the scale of human life, it is only recently (i.e. since the early seventies) that the formidable development of ground-based observatories such as the Atacama Large Millimeter/submillimeter Array (ALMA), currently the largest radio telescope in the world [26], as well as spatial missions and observations (Hubble space telescope [27], Rosetta spatial probe [28]), have revealed the existence of a very large number of molecular species in space. It must be mentioned, however, that the very first observations of simple radicals, i.e. CN, CH, and CH^+ were made in the thirties of the last century by observation of absorption spectra [29].

Of course, molecules cannot survive at the large temperatures of stars, but they are ubiquitous in a large variety of astrophysical objects. Figure 1.3 gives an overview of the present understanding of the life cycle of stars with their possible cortege of planets. It is seen that this life cycle involves a large number of cold and ultracold media such as dense interstellar clouds where the typical temperature is close to 10 K.

A non-exhaustive list of cold astrophysical media includes interstellar clouds (from dense to diffuse), planetary atmospheres, comets, and outer parts of circumstellar shells.

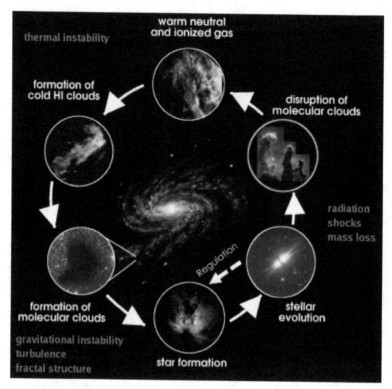

Fig. 1.3.   The star formation cycle. Copyright 2013 Steward Observatory Radio Astronomy Laboratory.

Molecules are, on the one hand, the probe of the physical conditions in all these media, but on the other hand the question of their formation and destruction, especially in such harsh environments, remains an important question that can be considered as the foundation of a branch of science known as astrochemistry. Many of these environments are at a lower temperature than our usual room temperature, i.e. well below 300 K.

Strengthened by the discovery of a large number of exo-planets, the open question of the origins of life has led a part of astrochemistry to evolve toward a specific scientific area known as astrobiology, with a key question concerning the possible formation of prebiotic molecules in cold astrophysical media [30], possibly leading to the cosmic origins of life in Earth-like planets.

1.3.2 *Non-LTE chemistry in cold astrophysical objects*

If cold astrophysical objects such as interstellar clouds were in CLTE, most of the matter there would be in solid state without an energetic radiation field and chemistry would be essentially frozen. The observation of molecules such as water or large organic molecules in the gas phase clearly shows that the initial chemistry is not driven by temperature. In fact, chemistry in space is initiated by energetic particles such as UV and VUV photons, X rays, and cosmic rays [31, 32]. Some of them come from sources outside the media, as VUV photons from stars or cosmic rays formed in supernova explosions. Others can be secondary particles, as ultraviolet photons, formed locally in the medium.

Not only are these objects not in CLTE but LTE itself is only partial (PLTE) and concerns first the kinetic energy of atoms and molecules and therefore temperature refers first to kinetic temperature. The rotational states themselves are not always in equilibrium at the kinetic temperature as it can be shown for the ortho-para ratio of molecular hydrogen [31] leading, for example, to change of chemistry in specific processes such as isotopic fractionation of H_2D^+ [31].

Besides the chemistry, the understanding of observation and of physics of these media also requires data concerning a variety of physical processes such as cross sections for rotational or spin-orbit excitation at very low energies [33].

It appears therefore that cold space is a veritable chemical factory of amazing complexity, which makes astrochemistry a very challenging science for laboratory work.

1.3.3 *The dusty universe: role of heterogeneous chemistry and its link to the gas phase*

Most cold astrophysical objects are not pure gaseous media. They can be considered as dusty media, i.e. besides their gaseous component, they contain a variety of tiny grains of solid matter. This dust component plays key roles in the physics and chemistry of the objects. For example, it shields the interior of dense interstellar clouds from the harsh UV radiation of surrounding stars. Grains also catalyze reactions such as molecular hydrogen formation and, on their surface, a rich heterogeneous chemistry can occur, as shown by laboratory experiments [34].

From the thermodynamic point of view, it is interesting to note that grains are not necessarily in temperature equilibrium with the surrounding gas, depending on the considered object [31].

From the laboratory point of view, most heterogeneous chemistry has been studied on a single surface even if a few experiments have been devised for the study of Earth atmosphere chemistry in two-phase flows [35]. To date, uniform supersonic flows have not been used with dusty flows for this purpose and while heterogeneous chemistry has to be mentioned due to its key role in cold media, it is clearly beyond the scope of this book.

1.4 Laboratory Methods Used to Study Gas Phase Kinetics at Low and Very Low Temperatures

1.4.1 *Historical perspective*

At the end of the seventies and early eighties, two different fields of laboratory work emerged independently. The first one was stimulated by the development of astrochemistry and especially by the modeling of dense interstellar cloud chemistry, which required rate coefficients and branching ratio data around 10 K, a great challenge at that time. For this purpose, a variety of techniques then emerged including CRESU, cryogenic traps, and merged beams. Some of them are true thermal techniques, using gas phase in PLTE, and others are devoted to collisions at very low KE_{cm}.

On the contrary, the second field can be characterized as a race toward ultimate cooling of atoms or molecules, i.e. at even lower temperatures (or kinetic energy) than a few Kelvin. There, the main application was toward quantum physics and some techniques of choice are laser cooling of atoms and optical traps.

These two fields have not interacted so much at the beginning of their study as shown by reference [36]. However, since the last two decades, there have been attempts to develop chemistry studies using techniques developed in this second field, but, at the present time, they have been restricted to very special and limited cases, as discussed in Section 1.4.5.

1.4.2 *Cryogenic reactors (cells, flow tubes, and traps)*

Chemical reactivity has been studied for several decades at room temperature using reaction cells [37] and flow tubes [38]. The gas can be static in a cell or flowing in a tube, the advantages in this latter case being to reduce reactions at the walls and minimize the need for initial cleaning of the apparatus. Most often, studies have been devoted to reactions between very reactive species (radicals, excited states, ions, or electrons) and a stable molecule which can be easily premixed with the buffer gas. The unstable one is usually created in situ from another stable species (the parent gas). For example, radicals can be created by pulse laser photolysis (PLP) of a convenient precursor. Nevertheless, the unstable species can also be injected into the system from outside, especially when it is a charged species (see, for example, the so-called SIFT technique [39]). Note that sometimes the buffer gas itself can be the stable reactant and that there exist a few techniques where the two reactants are unstable, for example, in the FALP-MS [40] devoted to the study of electron–ion reactions.

To extend the temperature range of these experimental techniques beyond room temperature, it has been the standard to heat up or cool down the apparatus. However, cryogenic cooling suffers from a serious drawback: condensation of reactants on the walls of the reactor. Of course, the problem worsens as the temperature is lowered: at 10 K, the vapor pressures of molecular species, excepted hydrogen, are such that they all stick as ices on the walls.

However, charged species can be easily manipulated by electromagnetic fields and therefore prevented from being lost on the walls of a cooled apparatus. This led to the development of a variety of ion traps which could be cooled down to interstellar temperatures using liquid helium as a cryogenic coolant.

It is beyond the scope of the present chapter to present a review of the various existing ion traps. The trapping of ions by radio frequency (rf) fields is a method of choice here and has been reviewed for its principle by Gerlich [41].

As an example, Fig. 1.4 presents the 22-pole rf traps developed at Freiburg University [42].

Of course, these traps have been mostly restricted to the study of ion reactions with molecular hydrogen, which is nevertheless the most abundant molecule in interstellar clouds by orders of magnitude.

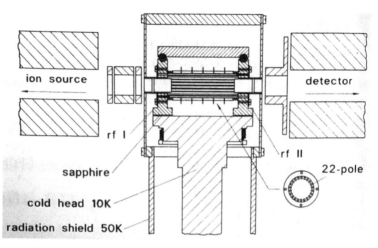

Fig. 1.4.   The 22-pole radio frequency trap developed by Gerlich *et al.* [42]. Reproduced with permission from Ref. [42], Elsevier Science B.V. Copyright 1995.

Monitoring the reactant and product ions versus time in the trap yields the rate coefficient and branching ratio, knowing the hydrogen density within the trap.

One of the serious advantages of the technique is that it allows the study of relatively slow reactions (rate coefficient $<10^{-12}$ $cm^3 s^{-1}$).

1.4.3   *Supersonic expansion (underexpanded free jet and CRESU flows)*

1.4.3.1   *Generalities*

As discussed in the introduction, a straightforward way to cool down a gas to very low temperatures and to avoid heterogeneous condensation is to expand it at supersonic velocity. The increase of fluid barycentric velocity v_b leads to an increase of its macroscopic kinetic energy at the expense of thermal energy following the first thermodynamic principle, which for conditions detailed in Section 1.5 can be written as

$$c_p T_0 = c_p T + \frac{v_b^2}{2} \tag{1.54}$$

where T_0 is the gas temperature at rest.

In practice, the gas is expanded from a large high-pressure reservoir where the gas is close to rest (temperature T_0, pressure P_0)

toward a chamber at a much lower pressure P_{ch} through a kind of tubing called nozzle. For reasons explained at length in the next paragraphs, the nozzle consists first of a converging part sometimes followed by a diverging portion. The junction between converging and diverging parts, of minimum area, is called throat. A hole in a wall can be considered as the simplest form of a nozzle.

The gas flowing beyond the nozzle exit forms a supersonic jet in the low-pressure chamber provided that the pressure P_e at the nozzle exit is sufficiently low. Details about the calculation of P_e/P_0 as a function of the Mach number are presented in Chapter 2. The jet can be considered as underexpanded, overexpanded, and perfectly expanded, respectively, for $P_e > P_{ch}$, $P_e < P_{ch}$, and $P_e = P_{ch}$. As such, the supersonic character of the flow is ultimately destroyed by phenomena such as shock waves and viscous friction with a great entropy loss and the gas in subsonic conditions then recovers a temperature very close to the reservoir temperature.

The great interest of supersonic cooling, as explained in the introduction, is the possibility to keep species in the gaseous phase in supersaturated conditions without condensation.

1.4.3.2   *The underexpanded free jet*

The simplest nozzle which can be used to generate a supersonic flow is a conical convergent or even a hole resulting in an underexpanded free jet (most often only referred to as *free jet* in physics and chemistry) of high Mach number limited by shock waves with a geometrical form looking like a barrel as illustrated in Fig. 1.5 [43]. This kind of jet has been widely used in physics, for example, in spectroscopy in order to simplify spectra by depopulating levels of various internal modes of energy. However, the underexpanded free jet suffers from serious drawbacks from the fluid mechanics and thermodynamics points of view, especially for the study of gas phase chemical reactions.

From the nozzle exit down to the Mach disk shock, pressure and density experience a very sharp drop, sometimes with a transition between the continuous and molecular free regime through the kinetic regime (see Section 1.2.3).

As the flow leaves the continuous regime, a strong thermodynamic disequilibrium arises, which means that it cannot longer be considered either in LTE or even in PLTE. The kinetic energy of atoms and molecules constituting the gas cannot be described by an

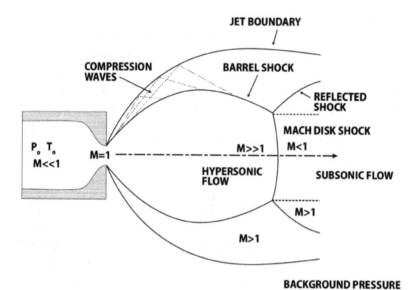

Fig. 1.5. Schematic view of a free jet expansion structure.

isotropic Maxwell-Boltzmann distribution with a true thermal temperature. At times, "parallel" and "normal temperatures" are defined and refer to distribution functions of the velocity component parallel and orthogonal to the flow, respectively. These "temperatures" may be different for different species [44].

The internal states themselves depart from a thermal distribution and a correct calculation of their populations requires data on rotational cross sections, especially at low KE_{cm}, which could be unavailable. Nowadays, calculations of underexpanded free jets of this kind are still a matter of research and publication [45].

Due to these problems, the use of such underexpanded free jets as flow reactors has been limited to the study of ion–molecule reactions with only one apparatus built at the University of Tucson by Mark Smith and his colleagues [46]. A sketch of the apparatus can be seen in Fig. 1.6. A free jet pulsed expansion of a mixture including a reactant gas, an ion precursor gas, and a suitable buffer gas is created in the experimental chamber. Then, a packet of reactant ions is selectively produced by resonantly enhanced multiphoton ionization (REMPI). These ions flow downstream and react with the reactant gas. A time

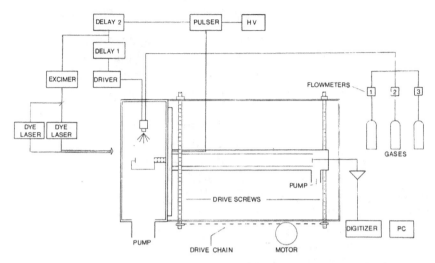

Fig. 1.6.   Sketch of the free jet apparatus developed at the University of Tucson. Reproduced with permission from Ref. [46], Elsevier B.V. Copyright 1990.

of flight mass spectrometer (TOF) allows extraction and analysis of the reactant and product ions downstream of the REMPI source. The distance between the source and the TOF can be varied and the rate coefficient can be obtained in the various ways that are standard in flow tubes (varying reaction time by varying position or varying neutral reactant concentration).

Obtaining the rate coefficients as a function of a true temperature required a complicated deconvolution [44, 46]. However, this was mainly possible for the kinetic temperature, but large uncertainties remained for the characterization of internal state populations of various molecular species such as, for example, NH_3 [47], and hence their internal temperature. In fact, it has been shown that the concept of rotational temperature itself is an ill-defined one in this case, especially for light molecules [45]. This remained a serious drawback since, for example, it is known that rotational temperature plays a key role in the kinetics of ion reactions with polar molecules [48] as shown in Chapter 3 by Biennier *et al.* To our knowledge, underexpanded free jet flows are no longer used nowadays as chemical reactors for kinetic studies.

1.4.3.3 *The perfectly expanded free jet — the CRESU flow*

For the study of supersonic and hypersonic flights, especially at high altitude, aerodynamicists faced the problem of building wind tunnels with a test section as large as possible where conditions allow reproducing the flow around an object moving at supersonic/hypersonic speed in a quiet atmosphere.

For a model at rest in the wind tunnel, the very first requirement is to obtain a uniform flow in the test section. It is clear from Section 1.4.3.2 that the simple underexpanded free jet is unusable for this purpose and then, since the early fifties, aerodynamicists have used the perfectly expanded ($P_e = P_{ch}$) free jet obtained at the exit of a well-contoured Laval nozzle.

The second requirement for aerodynamicists was to reproduce in the uniform flow the thermodynamic conditions prevailing in the Earth's (or other planets) atmosphere. If this is easily done for pressure (function of altitude) using convenient pumps, it is far more difficult to obtain atmospheric temperatures due to conservation of energy (Eq. (1.54)), which results in a very low temperature of the uniform supersonic flow in the test section. Very few wind tunnels try to overcome this problem by heating the reservoir [49] and a lot of them used room temperature reservoir (temperature T_0) and were known as "cold supersonic wind tunnels".

In the late seventies, the first author of this chapter was working as a young aerodynamicist at the "Aérothermique" fluid mechanics laboratory of the CNRS at Meudon (France). He realized then that this serious drawback for aerodynamics studies could be turned into a great advantage for chemical kinetics studies at very low temperatures. The very first apparatus based on this principle was then built in the early eighties and implemented in an existing rarefied wind tunnel [7] It was dedicated to the study of ion–molecule reactions. In order to avoid confusion with underexpanded free jets, the technique was named "CRESU" with reference made to the uniformity of the flow and not to its perfectly expanded conditions. Note that in fact the condition $P_e = P_{ch}$ is necessary but not enough to obtain the uniformity and that a well-designed Laval nozzle is required (see Chapter 2). In the early days of the CRESU project, existing N_2/O_2 Laval nozzles developed for aerodynamic studies were used but in order to expand the possibility of the technique new contoured nozzles were built, following the method of calculation explained at length in the next chapter. The lowest temperature reached with pre-cooling of the reservoir at liquid nitrogen temperature was 8 K.

Figure 1.7 shows a view of the SR3 wind tunnel inside which the original CRESU experiment was implanted, a sketch of which can be seen in the insert of this figure. Note that this wind tunnel was huge with very large pumping capacities (about 1,44,000 m^3/h) allowing one to reach exceptional conditions of low pressure and density (typically 10^{16} cm^{-3} or below for the buffer gas).

In its very first version, an ion parent gas and a reactant gas were added to the convenient buffer gas. An electron beam of 20 keV ionized mainly the buffer gas, which then reacted with the parent to yield the reactant ions. In some cases, no parent gas was added to study reactions of the buffer gas ions with the buffer gas itself. Ions were monitored by a quadrupole mass spectrometer set in housing downstream of the electron beam. They were probed by a skimmer with a small aperture of a few tenths of millimeter.

By monitoring in two locations, the decrease of reactant ions versus the neutral reactant flow rate (which converts easily into density) allowed one to straightforwardly obtain the reaction rate coefficient.

Rate coefficients could also be obtained by monitoring reactant ions versus distance to the electron beam for a given constant reactant concentration, a technique that was used to study reactions with the buffer gas itself.

Throughout this book, several CRESU apparatuses dedicated to specific kinds of studies are presented. One of the most serious drawbacks of the technique is that, in its continuous form, it requires the use of very large pumps and large mass flow rates, especially to obtain the lowest temperatures at high Mach numbers and for reasons that will be explicated in Section 1.5. Therefore, following the seminal work performed in Meudon, only a few continuous apparatuses were developed, mostly at Rennes University and at Birmingham by I. Smith and his colleagues. A smaller-sized apparatus restricted to lower Mach numbers was also built at Bordeaux University [50].

An obvious way to strongly reduce this problem is to pulse the flow. This is especially relevant when pulsed analytical tools such as pulsed lasers are used in the experiment. A first pulsed apparatus

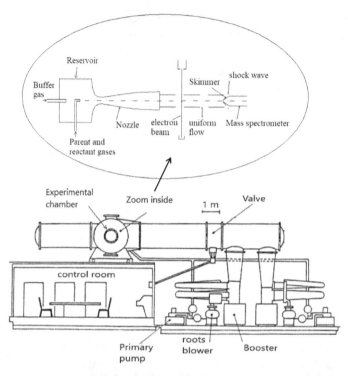

Fig. 1.7. The SR3 wind tunnel available at the *Laboratoire d'Aérothermique de Meudon* (France) in the eighties of the 20th century. A sketch of the original CRESU reactor implemented into SR3 can be seen on the top panel, whereas the viewgraph is courtesy of J.C. Lengrand.

was originally developed by M. Smith *et al.* [51] and its design reproduced by other research groups [52–56]. This apparatus used small
commercial electromagnetic valves at the entrance of a very small
nozzle reservoir. Two valves are mounted face to face in an attempt to
reduce the flow velocity to zero. This way to introduce the buffer gas
is prone to generate strong turbulences in the tiny reservoir, which
due to the values of the Reynolds number propagates downstream in
the jet. This is the reason why aerodynamicists use large reservoirs
with honeycomb in supersonic and hypersonic rarefied wind tunnels.
To overcome this serious drawback, Rowe and co-workers developed
a totally different pulsing method based on a rotating disk valve
located close to the nozzle throat and a large reservoir to avoid turbulence generation herein [57]. Following this design, a complete pulsed
and very successful version of the CRESU apparatus was then built
at the University of Castilla-La Mancha [58]. In parallel, an upgrade
of M. Smith's method was developed by A. Suits's group using faster
pulsing tools [59]. Details of the historical evolution of the technique
can be found in the next chapter.

1.4.4 *Collisions at very low KE_{cm} versus thermal experiments: merged beam apparatuses*

As shown previously in this chapter, if the collision cross section is
known at a very low center of mass kinetic energy (i.e. for KE_{cm}
corresponding to a $k_B T$ with T close to absolute zero) and for the
lowest internal energy level, then the rate coefficient can be derived
at low temperatures following Eq. (1.49).

If two mono-kinetic beams of species A and B with respective
velocities $\overrightarrow{v_A}$ and $\overrightarrow{v_B}$ and respective masses m_A and m_B are crossed
(see Fig. 1.8), the kinetic energy in the center of mass can be
written as

$$KE_{cm} = \frac{\mu}{2}\left|\overrightarrow{v_A} - \overrightarrow{v_B}\right|^2 = \frac{\mu}{2}\left(\left|\overrightarrow{v_A}\right|^2 + \left|\overrightarrow{v_B}\right|^2 - 2\left|\overrightarrow{v_A}\right|\left|\overrightarrow{v_B}\right|\cos\theta\right) \quad (1.55)$$

where θ is the angle between the two beams. To obtain very low val-
ues of KE_{cm}, the very first condition to fulfill is $\left|\overrightarrow{v_A}\right| = \left|\overrightarrow{v_B}\right|$. Then,
from Eq. (1.55) reducing KE_{cm} can be obtained either by reduction
of the velocity or by reduction of the angle θ. However, it is not

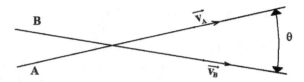

Fig. 1.8.   Crossed beams, velocity angle between the two mono kinetic beams.

possible to reduce the beam velocities so much without a considerable reduction of the beam fluxes, making this solution practically
not feasible. Therefore, the only solution is to reduce the angle θ with
the ultimate target of a zero angle.

Two overlapping beams of particles with such zero angles are
called *merged beams*. Inside their overlapping zone, inelastic and reactive collisions can occur resulting in products that can be ultimately
detected to measure cross sections.

Merging beams is a problem quite easy to solve when at least one
of the beams consists of charged particles which can be easily manipulated by electromagnetic fields. For this reason, merged beams have
been developed first for the study of either ion–molecule reactions [42]
or of dissociative recombination of ions with electrons [60] with the
ultimate lowest KE_{cm} achievable around 0.1 meV.

Merging neutral beams has been more difficult to achieve. It was
shown by the Bordeaux group that crossed beams can already result
in quite low KE_{cm} [61, 62], but it is only recently that, using Zeeman
effects with inhomogeneous fields, it became possible to merge a beam
of metastable atoms with some polar molecules [63, 64]. The apparatus used by Henson *et al.* [64] for these studies is shown in Fig. 1.9.
The main drawbacks in merged beam techniques are as follows:

- The knowledge of the internal state of the species, especially the
  rotational states.
- The detection of the products, which is required to obtain the
  cross section and which is easy only for ions or for very energetic
  particles.
- The difficulty of obtaining absolute cross sections.
- The limited versatility of the technique for non-polar neutral
  species.

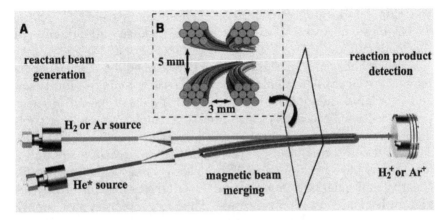

Fig. 1.9. The merged beam apparatus developed for neutral species at the Weizmann Institute of Science in Rehovot (Israel). Reproduced with permission from Ref. [64]. Copyright American Association for the Advancement of Science (Science), 2012.

1.4.5 *The race toward the lowest temperature: laser cooling and other methods*

As evoked in Sections 1.1 and 1.4.1, the end of the last century and the present one have seen an intense activity in the development of apparatuses allowing the reduction of the mean kinetic energy far below 1 K. For atoms, the main techniques use transfer of momentum between photons and atoms. They use lasers, which provide an intense radiation field. Further, the laser wavelength needs to be very close to a resonant atomic transition. For example, one of the main techniques uses the Doppler effect, which allows the atomic transition to be tuned to the laser wavelength by the atom velocity. It is a directional effect, permitting the reduction of the velocities of the atoms and their trapping thanks to several laser beams set in different directions. An excellent review of laser cooling and trapping of atoms can be found in reference [8]. As discussed in this review, the notion of "temperature" in these techniques is required to be discussed carefully as the system is in a steady state in the laser field and not in thermal equilibrium. Temperature is then defined from the mean kinetic energy and not from the MB velocity distribution function. The system of atoms cannot be considered as a closed and insulated system as it is in constant interaction with the laser field and that fluorescent light can escape from the system. Therefore, its entropy can decrease without violating the second principle of thermodynamics.

Cooling molecules by the techniques developed for atoms is not a straightforward task for molecules owing to their complex vibrational and rotational energy level structure, which makes efficient momentum transfer with laser photons impossible, and therefore laser cooling of molecules is inapplicable in the vast majority of cases [65]. This situation seems to have evolved in the very recent years and a few groups have performed work on laser cooling of molecules [66]. Clearly, molecules are fundamental for studies in chemical physics and for chemical reaction processes. Therefore, over the past twenty years, a variety of techniques have been developed to create and trap molecules at very low kinetic energy (far below "1 K"). They can be based on association of cold atoms (essentially alkali), sympathetic cooling, or field decelerator via Stark and Zeeman effects for dipolar molecules. Reviews of the progress in this field can be found in references [65–67].

However, as stated above, the versatility of these techniques for studies of chemical reactivity remains to be proved and only very specific cases of what can be considered as chemical reactions have been published. One of the fundamental reasons is probably because most of the techniques achieve very low densities for trapped or beam molecules. For example, in recent work performed on radiofrequency magneto-optical trapping of CaF [68], densities of $2.5 \ 10^5 \, \mathrm{cm}^{-3}$ are achieved for a time of 0.5 sec. For such a density and a neutral reaction, a typical "chemical reaction time" will be of several hours for reactions occurring at collisional frequency, which will make extremely difficult any reaction observation for the above conditions.

1.5 An Introduction to Compressible Flows and Supersonic Expansions

1.5.1 *General equations of reacting gas dynamics (aerothermochemistry) in the continuous regime*

The flow of a reacting gas mixture can be described by conservation equations (also called hydrodynamic equations or equations of change) as already discussed in previous sections of the present chapter. They include total mass conservation, mass balance for each chemical species, momentum conservation (i.e. Newton second law), and energy conservation. In their most general form, these equations can be written (as stated in the introduction) independent of any assumption concerning LTE. They govern the evolution of the fields of macroscopic quantities such as density, barycentric velocity, specific internal, and kinetic energy as a function of space and time. They can be obtained as such for an elementary volume followed in its movement in the so-called Lagrangian point of view, but it is easier to use an Eulerian point of view. [69] describing the velocity and thermodynamic parameter fields in geometrical space at any time. As discussed throughout this chapter, they can also be obtained directly from Boltzmann equations. Using microscopic quantities $\Psi_{Ai}(\vec{r}, t)$ defined either relative to $\overrightarrow{v^*}$ (see Eq. (1.6)) or to $\vec{V} = \vec{v} - \vec{v}_b$, it is possible to derive conservation equations relative to their mean values $\widehat{\Psi_{Ai}}$. The most general form of these equations can be found in [3, p. 32], [4] for the interested reader, but is beyond the scope of the present chapter.

Using the total derivative D/Dt defined by

$$\frac{D}{Dt} = \frac{\partial}{\partial t} + \vec{v}_b \cdot \overrightarrow{grad} \qquad (1.56)$$

the *mass conservation* equation usually referred to as the *continuity equation* is obtained by setting $\Psi(\vec{V}, A, i) = m_A$ and can be expressed in the general form

$$\frac{D\rho}{Dt} + \rho \, div(\vec{v}_b) = 0 \qquad (1.57)$$

which involves only quantities defined, respectively, by Eqs. (1.3) and (1.5) and is independent of any assumption on the distribution functions.

However, other conservation equations in their more general form include terms such as the stress tensor, heat, and mass flux that are dependent on the local thermodynamic state, i.e. of the knowledge of the velocity distribution function f and internal state distributions.

When the velocity distribution function is close to MB, defined with a local temperature T and barycentric velocity $\vec{v}_b$, it can be expressed as

$$f = f_{MB}(1 + \Phi) \tag{1.58}$$

with $\Phi \ll 1$. Then, following the Chapman-Enskog method developed in 1916–1917 [63], it is possible to solve Boltzmann equations and calculate velocity distribution functions defined by Eq. (1.58), i.e. to find Φ. A multicomponent gas mixture evolution can then be described, besides Eq. (1.57), by equations of conservation for momentum, energy and mass of chemical species, respectively.

Momentum conservation can be obtained using $\Psi(\vec{V}, A, i) = m_A \vec{V}$, and is expressed as [70, p. 170].

$$\rho \frac{\partial \vec{v_b}}{\partial t} + \rho \left(\overrightarrow{v_b} \cdot \overrightarrow{grad} \right) \overrightarrow{v_b} = \rho \vec{f} - \overrightarrow{grad}\, P$$
$$+ \mu \Delta \overrightarrow{v_b} + \left(\varsigma + \frac{\mu}{3} \right) \overrightarrow{grad}(div(\overrightarrow{v_b})) \tag{1.59}$$

where $\vec{f}$ is the body force per unit mass vector (gravitation, electrostatic if fluid is charged), P is the pressure, μ and ς are the dynamic and volume viscosity components, respectively, $\overrightarrow{grad}, div$, and Δ are the usual gradient, divergence, and Laplace operators, respectively.

Conservation of energy can be obtained using $\Psi(\vec{V}, A, i) = E_{Ai} + m_A \frac{V^2}{2}$ where E_{Ai} includes the energy of formation of species A. A general expression for the energy balance in a flow can then be represented by Eq. (1.60) [71, 72]:

$$\frac{\partial}{\partial t} \left[\rho \left(e + \frac{v_b^2}{2} \right) \right] + div \left[\rho \left(e + \frac{v_b^2}{2} \right) \overrightarrow{v_b} \right]$$
$$= -div(P\overrightarrow{v_b}) + \rho \vec{f} \cdot \overrightarrow{v_b} + div(\overline{\overline{\sigma}}_v \cdot \overrightarrow{v_b}) - divS + H_{int} \tag{1.60}$$

where e is the specific internal energy (J/m^3 units), $\overline{\overline{\sigma}}_v$ is the viscosity stress tensor, S is the flux of heat exchanged with the surroundings by conduction, and H_{int} is the internal heat source including contribution of chemical reactions.

Finally, equations governing the *change of density* of each chemical species A are given by

$$\frac{Dn_A}{Dt} + n_A div(\vec{v_b}) + div(n_A \overrightarrow{v_{DA}}) = K_R \tag{1.61}$$

with

$$\overrightarrow{v_{DA}} = \overrightarrow{v_{0A}} - \vec{v}_b \tag{1.62}$$

and $n_A \overrightarrow{v_{DA}}$ is the diffusion flux of species A.

The above equations can be considered as an extension to a multicomponent mixture of the Navier-Stokes equations that govern the movement of a fluid. They include a variety of transport laws, as the Newtonian law relating the stress tensor $\overline{\overline{\sigma}}$ to the strain rate tensor or Fick's law, which relates diffusion flux of a species to density gradient [5, 73]. These laws are valid in LTE conditions and imply transport coefficients, such as viscosity, heat conductivity, or diffusion coefficients that can be calculated provided intermolecular potentials are known.

To close the system of equations, an equation of state must also be added, i.e. the perfect gas law (Eq. (1.9)) in the case of standard gases. For a mixture of ideal gases, it is assumed that each component behaves as a perfect gas at its partial pressure and, following Dalton's law, that the total pressure of a mixture of gases is equal to the sum of the partial pressures of the individual components.

1.5.2   *Euler equations and quasi one-dimensional flow in a duct*

Simpler conservation equations than the generalized NS ones can be derived from the Boltzmann equation when it is assumed that the distribution functions are strictly of the MB kind, i.e.

$$f = f_{MB} \tag{1.63}$$

Then, a straightforward calculation [4] shows that the terms of transport (momentum, heat, and mass) vanish, the stress tensor reduces to the pressure multiplied by the unit tensor, and the flow can be considered as isentropic. This is linked to the Boltzmann H theorem, which allows deriving the form of the distribution function at equilibrium [5, p. 465].

Of course, in case of discontinuity of the distribution function, as in a shock wave, there is a less thickness on which the distribution function deviates from the MB form on a typical length of a few mean free paths (see the discussion in Section 1.2.3). Therefore, entropy is not conserved for the flow through a shock wave.

The generalized Euler equations, besides the continuity Eq. (1.57), can be written as follows:
Momentum:

$$\rho \frac{\partial \vec{v}_b}{\partial t} + \rho (\vec{v}_b \cdot \overrightarrow{grad}) \vec{v}_b = \rho \vec{f} - \overrightarrow{grad} P \qquad (1.64)$$

Energy:

$$\frac{\partial}{\partial t} \left[\rho \left(e + \frac{v_b^2}{2} \right) \right] + div \left[\rho \left(e + \frac{v_b^2}{2} \right) \vec{v}_b \right] = \rho \vec{f} \cdot \vec{v}_b - div\,(P\vec{v}_b) + H_{int}$$
$$(1.65)$$

For chemical species A, Eq. (1.61) reduces to

$$\frac{Dn_A}{Dt} + n_A div\,(\vec{v}_b) = K_R \qquad (1.66)$$

For an incompressible flow (low Mach number), the mass density ρ is constant and Eq. (1.57) reduces to $div(\vec{v}_b) = 0$, so that Eq. (1.66) becomes equivalent to Eqs. (1.37)–(1.38).

As developed in the previous paragraph, the generalized NS equations are extremely difficult to use in any way other than numerical computation. This is also the case for Euler equations of a multi-component reacting gas for a three-dimensional problem. However, following a convenient hypothesis, much simpler equations can be used in some cases.

Due to its practical importance, one of the most widely studied cases is the flow in a duct. If along the duct the variation of the sectional area is slow, it is common to make the one-dimensional assumption, namely, that the flow is uniform over this section. Then, all parameters depend only on the spatial coordinate along the duct axis. However, this situation should be named the *quasi-one-dimensional* hypothesis since it is just an approximation, except for a few cases, i.e. strictly planar, cylindrical, or spherical flows [74].

Let's assume now a steady-state flow ($\partial/\partial t = 0$) without external forces and no internal sources of energy in the flow. Strictly speaking, this latter condition is not valid for a flow with exothermic (or endothermic) reactions, but can be considered as true for reacting species in very small amount in a buffer flow. Then, conservation Eqs. (1.57), (1.64), and (1.65) yield the following simplified forms [71, p. 198]:

Mass:

$$\rho v_b A = const \tag{1.67}$$

Momentum:

$$dP = -\rho v_b dv_b \tag{1.68}$$

Energy:

$$e + \frac{P}{\rho} + \frac{v_b^2}{2} = h + \frac{v_b^2}{2} = const \tag{1.69}$$

where A is the section perpendicular to the velocity $\vec{v}_b$ through which the mass flow rate ρv_b moves. Further developments assume a constant specific heat capacity c_p for the gas, or in other words the specific enthalpy h reduces to $h = c_p T$. In the case of a perfect gas, the state law expressed by Eq. (1.10) must be taken into account in addition to these conservation equations.

Combining Eq. (1.68), the differential form of Eq. (1.67), and the speed of sound defined below (Eq. (1.72)), it is possible to obtain a relation between the area and velocity variations dA and dv_b, respectively:

$$\frac{dA}{A} = (M^2 - 1) \frac{dv_b}{v_b} \tag{1.70}$$

where M is the Mach number defined as the ratio of the flow velocity v_b to the sound velocity a:

$$M = \frac{v_b}{a} \tag{1.71}$$

Let's remember here that a is the local speed of a sound wave moving through an environment in which pressure change results from the mass density variation induced by the passage of the wave. Since these changes within the sound wave are slight, irreversible dissipative effects are negligible, implying that propagation of a sound wave is an isentropic process. The local speed of sound is expressed as

$$a^2 = \left(\frac{\partial P}{\partial \rho} \right)_S \tag{1.72}$$

Equation (1.70) is usually known as the Hugoniot theorem. It shows that the flow velocity varies in an inverse way with the section of a duct between the subsonic and supersonic case. This leads to the use of Laval nozzles to accelerate a gas flow to supersonic velocities. Such a nozzle is made of a convergent duct followed by a divergent outlet. Then, the sign of dA is negative in the subsonic ($M < 1$) convergent section, which means that the flow is accelerating. Conversely, in the supersonic ($M > 1$) divergent section, $dA > 0$ and then the flow velocity still increases. The section reaches an extremum ($dA = 0$) when the flow velocity becomes sonic ($M = 1$). This can only occur at the throat of the Laval nozzle, which corresponds to its smallest section.

Let's now consider Eq. (1.69) and apply it upstream of the Laval nozzle in the stagnant reservoir volume and downstream of the nozzle exit. Physical quantities inside the reservoir will be indexed with 0 such as P_0, T_0, and ρ_0, whereas we will use P_e, T_e, and ρ_e in the supersonic jet outside the nozzle. In the reservoir, the flow is almost at rest and then its kinetic energy is very small and can be neglected. Specific enthalpy h_0 in the reservoir is given by $c_p T_0$ since we are considering a calorically perfect gas. Similarly, h downstream of the Laval nozzle is $c_p T$. Here, however, the kinetic energy is high because the flow velocity has become supersonic and is equal to $v_b = M \times a$ according to Eq. (1.71). From this, one retrieves Eq. (1.54) or introducing the Mach number

$$c_p T_0 = c_p T + \frac{M^2 a^2}{2} \tag{1.73}$$

Defining γ as the ratio of specific heat capacities c_p/c_v, this equation can be changed to

$$\frac{T_0}{T} = \left(1 + \frac{\gamma - 1}{2} M^2 \right) \tag{1.74}$$

remembering that $c_p - c_v = r_m$ and $a^2 = \gamma r_m T$ for a perfect gas.

This equation is the cornerstone of supersonic cooling, as explained throughout this chapter. Considering the Laplace isentropic law $P\rho^{-\gamma} = const$, a variety of relationships between thermodynamic parameters and Mach number can be developed. They are made explicit in Chapter 2.

The quasi-one-dimensional results presented in this section are seminal in the understanding of compressible gas flows in a duct and many relationships can be used to yield exact results. But, it has of course some limitations, especially for the design of suitable contoured Laval nozzles for CRESU apparatuses. First, apart from the flow centerline, expansion requires some lateral motion in order to fulfill space downstream of the nozzle throat. Hence, a more rigorous treatment of the expansion through the nozzle needs to consider a flow velocity with a small radial (or perpendicular to the propagation axis for a planar nozzle) component. This is of importance in the determination of the geometric shape $A(z)$ of the divergent. The method of characteristics is a powerful tool commonly used to calculate these contours for an isentropic flow. It will be described in detail together with the introduction of the potential equation in the next chapter.

The other limitation which is presented in its principle in the next section is the non-inclusion of viscosity and heat transfer. It will also be discussed at length in Chapter 2.

1.5.3 *Reynolds number and boundary layer*

The viscous terms in the NS equations are made of second-order partial derivatives which, when ignored, lead to the Euler equations, which contain only first-order derivatives and are used in their unidimensional form in the previous section.

This has immediate implications for the boundary conditions of the flow at the walls either in an inner (duct) or outer (body in a stream) flow. In the case of the Euler equation, the velocity boundary condition is a slip condition (the velocity component normal to the wall surface is zero), whereas the second-order character of the NS equation requires a further condition, which is that the tangential velocity is also zero.

It is usual in order to evaluate the importance of the viscous term in the NS equations to use a non-dimensional velocity field where $\vec{v_b}$ is replaced by $\vec{v_b^*} = \vec{v_b}/v_\infty$ where v_∞ is chosen such that the order of magnitude of $|\vec{v_b^*}|$ is close to unity for the main flow. The spatial and temporal coordinates are also made non-dimensional with $\vec{r^*} = \vec{r}/L$ and $t^* = tv_\infty/L$, L being a characteristic length of the system under study as well as other quantities such as $\rho^* = \rho/\rho_\infty$, $P^* = P/\rho_\infty v_\infty^2$, and $\mu^* = \mu/\mu_\infty$ for volume mass, pressure, and viscosity, respectively.

Then, the momentum equation, in the absence of body forces, takes the following non-dimensional form [75, p. 76]:

$$\frac{\partial \rho^* v_{bi}^*}{\partial t^*} + \sum_j \frac{\partial \rho^* v_{bi}^* v_{bj}^*}{\partial x_j^*} = -\frac{\partial P^*}{\partial x_i^*} + \frac{1}{Re_L} \sum_j \frac{\partial \tau_{i,j}^*}{\partial x_j^*} \qquad (1.75)$$

where $i, j = x^*, y^*$, or z^* coordinates of the $\vec{r^*}$ vector and the non-dimensional stress tensor components $\tau_{i,j}^*$ are

$$\tau_{i,j}^* = \mu^* \left[\frac{\partial v_{bi}^*}{\partial x_j^*} + \frac{\partial v_{bj}^*}{\partial x_i^*} - \frac{2}{3} \delta_{i,j} \sum_l \frac{\partial v_{bl}^*}{\partial x_l^*} \right] \qquad (1.76)$$

with $\delta_{i,j} = 1$ if $i = j$ and $\delta_{i,j} = 0$ when $i \neq j$.

In this equation, the Reynolds number Re_L appears defined as

$$Re_L = \frac{\rho_\infty v_\infty L}{\mu_\infty} \qquad (1.77)$$

which gives in order of magnitude the ratio of inertia forces $\rho_\infty v_\infty^2$ to viscous forces $\mu_\infty v_\infty / L$, both per unit surface.

From this expression of the NS equations, it can be deduced that they convert to Euler equations when the Reynolds number goes to infinity. The same analysis can be made for the energy equation. However, due to the boundary conditions discussed above, the solution of NS equations converges nearly everywhere to the Euler one except at the wall. There are a variety of problems for which this problem cannot be ignored (for example, calculation of transfer at the wall) and a way to obtain a solution for flows of high Reynolds number everywhere can be found by splitting the problem using the boundary layer (hereafter BL) theory [75]. An outer flow is calculated using Euler equations: there are no transfer phenomena in this part of the stream. Then, in a zone of less thickness close to the wall, a simplified version of the NS equations is used to obtain a velocity profile linking its zero value at the wall to an outer value equal to the wall tangential component in the solution of Euler equations (v_∞). Transfer of momentum is then confined to this portion of the flow.

It is far beyond the scope of this chapter to develop at length the BL theory. Hereafter, only a short description of the seminal

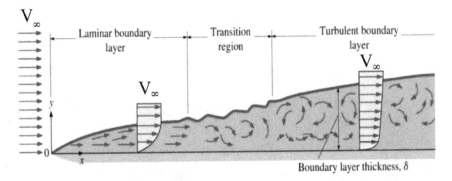

Fig. 1.10.   Development of a boundary layer along a flat plate.

treatment of the isothermal flat plate problem will be given. A much more complete development will be presented in the next chapter for the description of the Laval nozzle design. For a non-isothermal situation, the BL heat transfer must be considered, which is naturally the case for supersonic flows.

Figure 1.10 lays down the subject: a flat plate is set in a uniform flow, parallel to the stream of velocity v_∞. The problem is treated as bidimensional, i.e. the plate is assumed infinite in the dimension normal to the figure. The thickness of BL relative to the length x along the plate is enlarged for comprehension in this drawing, which also shows the Cartesian frame used. In this frame, velocity components are v_{bx} and v_{by}, respectively.

The solution of the Euler equation in this case would be $v_{bx} = v_\infty$ everywhere with a slip condition at the wall and a zero-velocity component $v_{by} = 0$ normal to the plate. Nevertheless, as discussed above, to consider the zero tangential velocity at the wall $v_{bx} = 0$, a calculation of the flow must be done taking into account its viscous character. For this purpose, a simplified version of NS equations known as BL equations can be used. This one can be derived using the same dimensional analysis as for Eq. (1.75), but introducing now a length δ, small compared to L characteristic of the length of the plate, and a non-dimensional parameter $\delta^* = \delta/L$ [75, Chapter 6]. This analysis will not be detailed here, but leads to the following dimensional form of the BL equations for the flat plate:

$$\frac{\partial v_{bx}}{\partial x} + \frac{\partial v_{by}}{\partial y} = 0 \tag{1.78}$$

$$\rho v_{bx}\frac{\partial v_{bx}}{\partial x} + \rho v_y\frac{\partial v_{bx}}{\partial y} = \mu\frac{\partial^2 v_{bx}}{\partial y^2} \tag{1.79}$$

$$0 = \frac{\partial P}{\partial y} = \frac{\partial P}{\partial x} \tag{1.80}$$

The boundary layer thickness δ can be defined rather arbitrarily as the distance to the wall for which $v_{bx} = 0.99 \times v_\infty$. It can be shown that its evolution along x, distance downstream from the plate leading edge, is given by

$$\frac{\delta}{x} = \frac{5.0}{\sqrt{Re_x}} = 5.0\sqrt{\frac{\mu_\infty r_m T_\infty}{P_\infty v_\infty x}} \tag{1.81}$$

where Re_x is defined as Eq. 1.77 with x instead of L. Therefore, small boundary layers require large Reynolds numbers. At this stage, it must be kept in mind that at a low Reynolds number, a flow cannot be considered as inviscid and that heat transfer also comes into play. **As a result, there is no way to generate a uniform and isentropic supersonic flow in this case.**

As shown in Fig. 1.10, Eq. (1.81) indicates that the boundary layer grows as the square root of distance to the leading edge of the flat plate. It is also worth mentioning that for a given position and temperature, the boundary layer diminishes when pressure is increased.

Finally, it is worth pointing out that for a mixture of two gases at low density, the Schmidt number defined by $\mu/(\rho\mathcal{D})$ (where $\mathcal{D}$ is the mass diffusion coefficient) is close to unity. This means that the analog of the Reynolds number for diffusion (i.e. Péclet number defined as $v_\infty L/\mathcal{D}$) is large in CRESU flows. Consequently, diffusion effects in the isentropic core of the flow are most often negligible. However, in some special cases, diffusion can be enhanced in CRESU flows. For example, this is the case for forced diffusion of ions due to an electric field (Chapter 3) or if some non-equilibrium effects are involved such as the formation of species following photolysis and chemical reaction in a narrow laser beam.

1.5.4   *Turbulent and laminar flow*

It is a common experience for the flow of liquid that can be seen by the naked eye that at high speed, the flow starts to look highly unsteady with eddies and instabilities even if boundary conditions are steady. This has been studied by Osborne Reynolds [76] in a famous seminal experiment on duct flows, which showed that at low mass flow rates Q_m the flow occurred without these instabilities and mixing of streamlines. Such flows are termed laminar. At higher mass flow rates, instabilities (turbulences) start to appear and become more and more invading in the duct as far as Q_m is increased.

In fact, Reynolds showed that this phenomenon is not led exactly by the mass flow rate but by the Reynolds number (now expressed respective to the duct diameter instead of L) defined above and named after his surname. It can be shown that for a duct of diameter D the Reynolds number can be written as

$$Re_D = \frac{4}{\pi}\frac{Q_m}{\mu_\infty D} \tag{1.82}$$

Therefore, for a fixed diameter and viscosity, the important parameter is the mass flow rate.

When moving from laminar to turbulent flows, the transport phenomena change in nature. In a laminar flow, momentum, heat, and mass transfers are accomplished by molecular transport linked to the small departure to the MB distribution function, as discussed above. In a turbulent flow, transport is made by macroscopic eddies, the size of which change very much from one problem to another.

As a result, transport phenomena are greatly enhanced by mixing in a turbulent flow and as shown in Fig. 1.10 this yields a large increase of the boundary layer thickness.

1.5.5   *Implications of previous paragraphs and of the useful length concept for a CRESU design*

When designing a CRESU experiment, the first target is to obtain a uniform supersonic flow using a Laval nozzle and expanding the flow from a quiet reservoir. In such a flow, the thermodynamic conditions can be considered as constant along the flow axis and in a

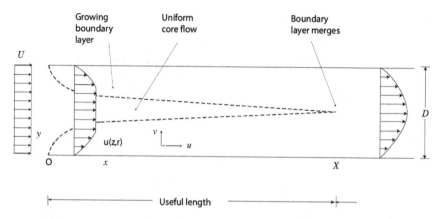

Fig. 1.11. The concept of entrance length or equivalently length of uniformity.

small kernel around it. This implies an isentropic flow without transfer phenomena: viscous flow will imply a pressure drop, and heat transfer toward the outer part of the flow will provoke an increase of temperature. Of course, the worse dissipative phenomenon would be a shock wave which would convert the flow to subsonic for a normal shock wave and will increase the temperature eventually back to a value close to the reservoir temperature.

We define the useful length L of a nozzle as the length along the flow axis downstream of the nozzle exit, which satisfies this condition of uniformity along the axis.

An order of magnitude for this useful length can be obtained in the same way as the so-called entrance length in a duct depicted in Fig. 1.11. A velocity profile nearly flat at the entrance of a duct with an isentropic core will evolve toward a parabolic profile due to the growth of the boundary layers. The distance from the entrance of the duct at which the boundary layer merges is named the entrance length. Although there is no strict equivalence with the jet without a wall after the nozzle exit (for example, in a duct the velocity in the center increases due to flow rate conservation), the calculation of the useful length can be performed in the same way, keeping in mind that it is mostly an order of magnitude.

It can be shown [77] that for a laminar flow this length can be written as

$$L \approx 0.06 \frac{Q_m}{\mu_\infty} \qquad (1.83)$$

It has to be kept in mind that Eq. (1.83) yields just an order of magnitude of the useful length, which itself depends on the stringency set for uniformity. However, the important conclusion here is that independent of nozzle size and pressure, a minimum mass flow rate has to be used in order to obtain a uniform supersonic flow on a given length. Then, this will also fix the size of the experiment for the chosen operating pressure and therefore also the pumping speed.

Nonetheless, increasing Q_m in order to increase the useful length has two limitations. The first one is a cost problem. The building and operating cost of the apparatus grow with it due to the apparatus size and buffer gas consumption, respectively. The second one is the problem of transition from laminar to turbulent flow, already discussed in Section 1.4.3.3 for pulsed CRESU flows. Even if strong precautions are taken in order to have very quiescent conditions in the reservoir (honeycomb, large size), the Reynolds number at the throat will increase with the mass flow rate leading to turbulent transition at some point, therefore reducing the useful length and the quality of the flow.

In practice, when designing an experiment and for permanent flows, cost limitations will mainly concern helium buffer since the gas cost is linked to the standard volume flow rate, i.e. much higher at an equivalent mass flow rate for helium than for gases such as nitrogen or argon. Of course, for minor gases added to the buffer, this also has to be considered for very expensive species such as some isotopes.

1.6 Conclusions

In this chapter, we have tried to give a guiding thread in a disparate patchwork of basic concepts that are used in the study of "cold" and "ultracold" molecules and of their interaction processes. Besides some fundamentals in fluid mechanics, we have emphasized the importance of the thermodynamic state in dedicated experimental techniques.

Indeed, "cold" and "ultracold" refer not only to the center of mass kinetic energy of the reactants but also to their internal states. Then, it can be distinguished mainly between techniques which seek LTE where a true thermodynamic temperature can be defined, at least for translational and rotational motion, and those which look for a very low collision energy (KE_{cm}) with specified internal energies.

A rate coefficient versus temperature $k(T)$ is obtained in the former case and a cross section versus energy $\sigma(KE_{cm})$ in the latter case. If the cross section has been measured on a sufficiently wide range of conditions, then $k(T)$ can be derived in the way presented throughout this chapter. A prototype of the first class of techniques is the CRESU device, whereas merged beams are representative of the second category.

There exist of course a variety of techniques that have been presented in Section 1.4 of this chapter; each of them has its own merits (chemical versatility, ability to measure small rate coefficients, ease of use, etc.). But, in any case, the question of thermodynamic conditions remains central when results are used for comparison to theory or use for modeling of cold natural media.

References

[1] Bondi A. Van der Waals volumes + Radii. J Phys Chem. 1964;68(3):441–51.

[2] Delcroix JL, Bers A. Physique des Plasmas Vols 1 and 2. Savoirs actuels, Inter éditions EDP Sciences/CNRS éditions, Paris; 1994.

[3] Hochstim AR. Kinetic Processes in Gases and Plasmas. Academic Press, New York and London; 1969.

[4] Rowe BR. Etude des Réactions Chimiques et Ioniques dans un Plasma en Fort Déséquilibre Thermodynamique. Cas du Mélange Azote-Argon [PhD thesis]. France: Université de Paris VI; 1975.

[5] Hirschfelder JO, Curtiss CF, Bird RB. Molecular theory of gases and liquids. New York: Wiley; 1954. p. 465.

[6] Rowe BR. Coupled relaxation in a plasma with strong thermodynamic non equilibrium. Phys Fluids. 1977;20(11):1827–35.

[7] Rowe BR, Dupeyrat G, Marquette JB, Gaucherel P. Study of the reactions $N_2^+ + 2\,N_2 \rightarrow N_4^+ + N_2$ and $O_2^+ + 2\,O_2 \rightarrow O_4^+ + O_2$ from 20–160 K by the CRESU technique. J Chem Phys. 1984;80(10): 4915–21.

[8] Metcalf HJ, van der Straten P. Laser cooling and trapping of atoms. J Opt Soc Am B–Opt Phys. 2003;20(5):887–908.

[9] van de Meerakker SYT, Vanhaecke N, Meijer G. Stark deceleration and trapping of OH radicals. Ann Rev Phys Chem. 2006;57(1):159–190.

[10] Einstein A. Quantum theory of mono-atomic ideal gas. Second paper. Sitz Preuss Akad Wiss Phys-Math Klas. 1925;3–14.

[11] Smith IWM. Kinetics and dynamics of elementary gas reactions. London: Butterworths and co.; 1980.

[12] Atkins PW. Physical chemistry. 5th ed. Oxford, Melbourne, Tokyo: Oxford University Press; 1994. p. 448.

[13] Braun S, Ronzheimer JP, Schreiber M, Hodgman SS, Rom T, Bloch I, Schneider U. Negative absolute temperature for motional degrees of freedom. Science. 2013;339(6115):52–5.

[14] Thomson W. On an absolute thermometric scale founded on Carnot's theory of the motive power of heat and calculated from Regnaut's observations. Mathematical and Physical Papers of William Thomson (Cambridge University Press, Cambridge, 1882), v. 1, pp. 100–106.

[15] NIST-JANAF thermochemical tables, Fourth Edition, Monograph 9 (Part I and Part II) (1998), 1963 pp. by M. W. Chase, Jr. J. Phys. Chem. Ref. Data Mon. Publisher: AIP Publishing. https://srd.nist.go v/JPCRD/jpcrdM9.pdf

[16] Cukrowski AS, Fritzsche S. Relaxation of translational energy in binary-mixtures of dilute gases composed of hard-spheres. Ann Phys. 1991;48(6):377–86.

[17] Marquette JB, Rebrion C, Rowe BR. Reactions of $N^+(^3P)$ ions with normal, para and deuterated hydrogens at low temperatures. J Chem Phys. 1988;89(4):2041–47.

[18] Golse F, Levermore CD. Hydrodynamic limits of kinetic models. In: Passot T, Sulem C, Sulem PL, editors. Topics in kinetic theory. Fields Institute Communication, American Mathematical Society, Providence, Rhode Island; 2005.

[19] Ivchenko IN, Loyalka SK, Tompson Jr. RV. Analytical methods for problems of molecular transport. Springer; 2007. https://www.spring er.com/gp/book/9781402058646

[20] Ross J, Mazur P. Some deductions from a formal statistical mechanical theory of chemical kinetics. J Chem Phys. 1961;35(1):19–28.

[21] Levine RD. Molecular reaction dynamics. Cambridge University Press, Cambridge (United Kingdom); 2005.

[22] Laidler KJ, King MC. The development of transition-state theory. J Phys Chem. 1983;87(15):2657–64.

[23] Truhlar DG, Garrett BC, Klippenstein SJ. Current status of transition-state theory. J Phys Chem. 1996;100(31):12771–12800.

[24] Bao JL, Truhlar DG. Variational transition state theory: theoretical framework and recent developments. Chem Soc Rev. 2017;46(24): 7548–96.

[25] Eliason MA, Hirschfelder JO. General collision theory treatment for the rate of bimolecular, gas phase reactions. J Chem Phys. 1959;30(6): 1426–36.

[26] 2020. Available from: https://www.almaobservatory.org/en/home

[27] 2020. Available from: https://www.spacetelescope.org/

[28] Taylor MGGT, Altobelli N, Buratti BJ, Choukroun M. The Rosetta mission orbiter science overview: the comet phase. Phil Trans Roy Soc A–Math Phys Eng Sci. 2017;375(2097):20160262.

[29] Herzberg G. Historical remarks on the discovery of interstellar-molecules. J Roy Astron Soc Can. 1988;82(3):115–27.

[30] Thaddeus P. The prebiotic molecules observed in the interstellar gas. Phil Trans Roy Soc B–Biol Sci. 2006;361(1474):1681–7.

[31] Millar TJ. Astrochemistry. Plasm Sour Sci Tech. 2015;24(4):43001.

[32] Herbst E, Klemperer W. The formation and depletion of molecules in dense interstellar clouds. Astrophys J. 1973;185:505–33.

[33] Roueff E, Lique F. Molecular excitation in the interstellar medium: recent advances in collisional, radiative, and chemical processes. Chem Rev. 2013;113(12):8906–38.

[34] Potapov A, Jager C, Henning T. Photodesorption of water ice from dust grains and thermal desorption of cometary ices studied by the INSIDE experiment. Astrophys J. 2019;880(1):12.

[35] Harrison RM, Collins GM. Measurements of reaction coefficients of NO_2 and HONO on aerosol particles. J Atm Chem. 1998;30(3):397–406.

[36] Smith IWM. Low temperature and cold molecules. Imperial College Press, London; 2008.

[37] Martinez E, Albaladejo J, Jimenez E, Notario A, Aranda A. Kinetics of the reaction of CH_3S with NO_2 as a function of temperature. Chem Phys Lett. 1999;308(1–2):37–44.

[38] Howard CJ. Kinetic measurements using flow tubes. J Phys Chem. 1979;83(1):3–9.

[39] Adams NG, Smith D. Selected ion flow tube (SIFT)–technique for studying ion-neutral reactions. Int J Mass Spec Ion Proc. 1976;21(3–4):349–59.

[40] Laubé S, Le Padellec A, Sidko O, Rebrion-Rowe C, Mitchell JBA, Rowe BR. New FALP-MS measurements of H_3^+, D_3^+ and HCO^+ dissociative recombination. J Phys B. 1998;31(9):2111–28.

[41] Gerlich D. Inhomogeneous Rf-fields–a versatile tool for the study of processes with slow ions. Adv Chem Phys. 1992;82:1–176.

[42] Paul W, Lucke B, Schlemmer S, Gerlich D. On the Dynamics of the Reaction of Positive Hydrogen Cluster Ions (H_5^+ to H_{23}^+) with Para and Normal Hydrogen at 10 K. Int. J. Mass Spectrom. Ion Proc. 1995;149/150:373–87.

[43] Dupeyrat G, Giat M. Experimental-study of velocities in a free jet in a rarefied atmosphere. AIAA J. 1976;14(12):1706–8.

[44] Mazely TL, Roehrig GH, Smith MA. Free jet expansions of binary atomic mixtures–a method of moments solution of the Boltzmann-equation. J Chem Phys. 1995;103(19):8638–52.

[45] Montero S. Molecular description of steady supersonic free jets. Phys Fluids. 2017;29(9):96101.

[46] Hawley M, Mazely TL, Randeniya LK, Smith RS, Zeng XK, Smith MA. A free jet flow reactor for ion molecule reaction studies at very low energies. Int J Mass Spec Ion Proc. 1990;97(1):55–86.

[47] Smith MA, Hawley M. Ion/molecule reaction-rate coefficients at translational temperatures below 5 K–selected bimolecular reactions of $C_2H_2^+$ and NH_4^+. Int J Mass Spec Ion Proc. 1995;149:199–206.

[48] Rebrion C, Marquette JB, Rowe BR, Clary DC. Low temperature reactions of He^+ and C^+ with HCl, SO_2 and H_2S. Chem Phys Lett. 1988;143(2):130–4.

[49] Sagnier P, Verant JL. Flow characterization in the ONERA F4 high-enthalpy wind tunnel. AIAA J. 1998;36(4):522–31.

[50] Daugey N, Caubet P, Retail B, Costes M, Bergeat A, Dorthe G. Kinetic measurements on methylidyne radical reactions with several hydrocarbons at low temperatures. Phys Chem Chem Phys. 2005;7(15):2921–7.

[51] Atkinson DB, Smith MA. Design and characterization of pulsed uniform supersonic expansions for chemical applications. Rev Sci Inst. 1995;66(9):4434–46.

[52] Schlappi B, Litman JH, Ferreiro JJ, Stapfer D, Signorell R. A pulsed uniform Laval expansion coupled with single photon ionization and mass spectrometric detection for the study of large molecular aggregates. Phys Chem Chem Phys. 2015;17(39):25761–71.

[53] Taylor SE, Goddard A, Blitz MA, Cleary PA, Heard DE. Pulsed Laval nozzle study of the kinetics of OH with unsaturated hydrocarbons at very low temperatures. Phys Chem Chem Phys. 2008;10(3):422–37.

[54] Lee S, Hoobler RJ, Leone SR. A pulsed Laval nozzle apparatus with laser ionization mass spectrometry for direct measurements of rate coefficients at low temperatures with condensable gases. Rev Sci Inst. 2000;71(4):1816–23.

[55] Spangenberg T, Kohler S, Hansmann B, Wachsmuth U, Abel B, Smith MA. Low-temperature reactions of OH radicals with propene and isoprene in pulsed Laval nozzle expansions. J Phys Chem A. 2004;108(37):7527–34.

[56] Sánchez-González R, Eveland WD, West NA, Mai CLN, Bowersox RDW, North SW. Low-temperature collisional quenching of NO $A^2\Sigma^+(v' = 0)$ by $NO(X^2\Pi)$ and O_2 between 34 and 109 K. J Chem Phys. 2014;141(7):74313.

[57] Morales S, Rowe BR. Hacheur aérodynamique pour la pulsation des gaz. Patent: FR 2948302—WO 2011018571–US 8.870.159; 2009.

[58] Jiménez E, Ballesteros B, Canosa A, Townsend TM, Maigler FJ, Napal V, Rowe BR, Albaladejo J. Development of a pulsed uniform supersonic gas expansion system based on an aerodynamic chopper for gas phase reaction kinetics studies at ultra-low temperatures. Rev Sci Inst. 2015;86(4):045108.

[59] Oldham JM, Abeysekera C, Joalland B, Zack LN, Prozument K, Sims IR, Park GB, Field RW, Suits AG. A chirped-pulse Fourier-transform microwave/pulsed uniform flow spectrometer. I. The low-temperature flow system. J Chem Phys. 2014;141(15):54202.

[60] Larsson M. Dissociative recombination with ion storage rings. Ann Rev Phys Chem. 1997;48:151–79.

[61] Naulin C, Costes M. Experimental search for scattering resonances in near cold molecular collisions. Int Rev Phys Chem. 2014;33(4):427–46.

[62] Berteloite C, Lara M, Bergeat A, Le Picard SD, Dayou F, Hickson KM, Canosa A, Naulin C, Launay JM, Sims IR, Costes M. Kinetics and dynamics of the $S(^1D_2) + H_2$ reaction at very low temperatures and collision energies. Phys Rev Lett. 2010;105(20):203201.

[63] Osterwalder A. Merged neutral beams. EPJ Tech Instrum. 2015;2:10.

[64] Henson AB, Gersten S, Shagam Y, Narevicius J, Narevicius E. Observation of resonances in penning ionization reactions at sub-kelvin temperatures in merged beams. Science. 2012;338(6104):234–8.

[65] Schnell M, Meijer G. Cold molecules: preparation, applications, and challenges. Ang Chem Int Ed. 2009;48(33):6010–31.

[66] Tarbutt MR. Laser cooling of molecules. Comtemp Phys. 2018;59(4):356–76.

[67] Bohn JL, Rey AM, Ye J. Cold molecules: progress in quantum engineering of chemistry and quantum matter. Science. 2017;357(6355):1002–10.

[68] Anderegg L, Augenbraun BL, Chae E, Hemmerling B, Hutzler NR, Ravi A, Collopy A, Ye J, Ketterle W, Doyle JM. Radio frequency magneto-optical trapping of CaF with high density. Phys Rev Lett. 2017;119(10):103201.

[69] Germain P. Mécanique des Milieux Continus. Masson edition, Paris; 1962.

[70] Guyon E, Hulin JP, Petit L. Hydrodynamique physique. 3rd ed. Savoirs actuels, Inter éditions EDP Sciences/CNRS éditions, Paris; 2012.

[71] Anderson JD. Modern compressible flow with historical perspectives. McGraw-Hill Series in Aeronautical and Aerospace Engineering, New York; 2004.

[72] Giovannini A, Airiau C. Aérodynamique Fondamentale. Cépaduès éditions, Toulouse (France); 2016.

[73] Bird RD, Stewart WE, Lightfoot EN. Transport phenomena. John Wiley and sons, New York, London; 1960.

[74] Ben-Artzi M, Falcovitz J. Generalized Riemann problems in computational fluid dynamics. New York: Cambridge University Press; 2003.

[75] Schlichting H, Gersten K. Boundary layer theory. 9th ed. Berlin Heidelberg: Springer-Verlag; 2017.

[76] Reynolds O. An experimental investigation of the circumstances which determine whether the motion of water shall be direct or sinuous, and of the law of resistance in parallel channels. Phil Trans Roy Soc London. 1883;174:935–82.

[77] Bejan A. Heat transfer. John Wiley and sons, New York; 1993.

Chapter 2

Laval Nozzles: Design, Characterization, and Applications

André Canosa*,§ and Elena Jiménez[†,‡,¶]

*CNRS, IPR (Institut de Physique de Rennes)-UMR 6251,
Université de Rennes, F-35000 Rennes, France
†Departamento de Química Física, Facultad de Ciencias y Tecnologías
Químicas, Universidad de Castilla-La Mancha,
Avda. Camilo José Cela 1B, 13071 Ciudad Real, Spain
‡Instituto de Investigación en Combustión y Contaminación Atmosférica,
Universidad de Castilla-La Mancha, Camino de Moledores s/n,
13071 Ciudad Real, Spain
§andre.canosa@univ-rennes1.fr
¶elena.jimenez@uclm.es

Abstract

This chapter focuses on Laval nozzles. After emphasizing their ubiquitous usage in a large variety of scientific fields, a methodology, based on the method of characteristics and Michel's integrals, is presented in detail to calculate their divergent contours taking into account the boundary layer. The main experimental techniques allowing the characterization of the flow exiting the nozzle are then reviewed. These include the Pitot tube, rotational spectroscopy, and velocimetry. Many examples of the temperature evolution downstream of Laval nozzles issued from various laboratories are depicted to illustrate the high level of uniformity that a supersonic flow expanding through these ducts can achieve. The versatility of Laval nozzles is also highlighted in detail, making them real temperature tunable tools. The chapter ends with a historical review of the development of the CRESU technique worldwide with special attention to the recent improvements for the generation of pulsed uniform supersonic flows.

Keywords: Laval nozzles; Contour design; Inviscid compressible flow; Boundary layer; Method of characteristics; Von Karman equation; Uniform supersonic flows characterization; CRESU historical perspective.

2.1   Laval Nozzles in Science

Laval nozzles were imagined at the end of the 19th century by Carl Gustaf Patrick De Laval (Fig. 2.1), a Swedish engineer who became famous as the inventor of the first high-speed centrifugal milk–cream separator and who significantly contributed to the improvement of steam turbines [1, 2]. In 1890, he presented his own machine, which was able to rotate at speeds of about 30000 rounds per minute thanks to a series of hourglass-contoured ducts. These were shooting pressurized steam, accelerated to supersonic speeds, at the turbine's blades (Fig. 2.1). The *De Laval nozzles*, as they were named later, were born (hereafter simply Laval nozzles). The present DeLaval Company, world leader in milking equipment for dairy farmers, is one of the heritages of this prolific inventor.

Fig. 2.1.   Left: Carl Gustaf Patrick De Laval (1845–1913), inventor of the De Laval nozzles. Right: Sketch of the De Laval steam turbine (from Kennedy Rankin (1903), Electrical Installations vol III, London).

Fig. 2.2. The Vulcain 2.1 engine, Europe's most powerful rocket engine, was successfully tested in January 2018 for use with the future Ariane 6 launcher in 2022. The engine is 3.7 m high, 2.5 m in diameter, and weighs about 2 tonnes. It will deliver 135 tonnes of thrust in vacuum. (From DLR and ESA: https://www.esa. int/ESA_Multimedia/Images/2018/01/Vulcain_2.1_hot_firing).

The concept of a convergent–divergent duct was then widely used in many fields of science and technology with the purpose of generating very fast dense gas flows. It has been central in the development of *rockets and missiles propulsion* [3] for which one major concern is the optimization of the thrust produced by matter ejection during the take-off phase (Fig. 2.2). Laval nozzles designed for parallel uniform exit flows can expand, accelerate, and guide large quantities of chemical propellants introduced in a high-pressure combustion chamber upstream, favoring a maximization of the thrust which is directly proportional to the mass flow rate and the velocity of the exhaust gas along the nozzle axis. In parallel, Laval nozzles triggered the development of supersonic wind tunnels in which miniaturized bodies at rest, such as shuttles or planes, face a supersonic rarefied flow with the aim to investigate the aerodynamic drag around them. Among others, this kind of experimental simulation is of major importance to understand the factors at play during the reentry of a spacecraft, although the flow temperature in cold wind tunnels is not representative of the atmospheric temperature (see Chapter 1).

Hypersonic flows are ubiquitous throughout the Universe. The so-called *astrophysical outflows* or *jets* are emitted from various objects such as young stars, compact bodies, supermassive black holes located in the center of active galactic nuclei or X-ray binary star systems. A remarkable feature of these jets is their high degree of collimation, so that Laval nozzles constitute an ideal tool for

simulating and characterizing these environments in the laboratory
[4] Exploration of the interstellar medium (ISM) has revealed an
impressive variety of environments evolving in extreme physical
conditions of pressure and temperature (see Chapter 10). Among
these, the so-called molecular clouds conceal an amazing richness of
molecules surviving at temperatures as low as 7 K in the gas phase.
In essence, a uniform supersonic expansion issued from a Laval nozzle
generates a physically well-characterized cooled (with respect to the
reservoir) environment which appears to be an ideal chemical reactor to study low-temperature *reaction kinetics* and *energy transfer
processes*, as will be discussed in detail in Chapters 3–5, 7, and 8.
As mentioned in Chapter 1, one of the main advantages of uniform
supersonic expansions is that condensable species can be maintained
in the gas phase along the flow under thermodynamic supersaturation conditions. However, if the concentration of these species is high
enough in the flow, the onset of *nucleation* can be activated and
the first steps of the process including dimerization and clustering
can be studied as well, as discussed in Chapters 6 and 9. The field
of *spectroscopy* also takes advantage of the properties of uniformity
of supersonic jets issued from Laval nozzles to, for instance, study
vibrationally hot and rotationally cold infrared spectra or determine
the density of absorbers in the infrared or microwave domain (see
Chapters 8 and 9).

Beyond the scope of this book, many other applications of Laval
nozzles can be found in the literature and a brief non-exhaustive
overview is given here. In the sixties, Laval nozzles were used in
high-power gas lasers to rapidly expand a hot equilibrium gas mixture (typically, CO_2, N_2, and H_2O or He) and generate, "gas-
dynamically", a population inversion of CO_2 states. This resulted
from the large vibrational relaxation time of the highest achievable
excited state with respect to the expansion time, whereas lower vibrational excited states, significantly more populated in the reservoir,
were quenched much faster along the expansion [5, 6]. In *laser cutting*
technologies, one of the crucial aspects is the cut edge quality. During
the laser cutting process, molten debris can be formed and cling to
the cut edge. High-pressure gas issued from Laval nozzles is currently
used to blow these away from the cut kerf thanks to their high mass
flow rate, fast velocity, and very good collimation [7, 8]. *Laser plasma
interaction* is a broad branch of matter science including areas such

as X-ray lasers, laser particle acceleration, and high harmonic generation. Experiments rely on the creation of an interaction gas media spatially well confined with a constant transverse and longitudinal high density. Hence, flat top density profiles are highly desirable; conditions that suitably designed Laval nozzles can match easily [9–11]. Surface coating is another vast field of research and industrial application in which Laval nozzles play a major role in techniques like the *cold gas dynamic spray* (CGDS). In this process, sprayed micro- or nanometric powdered particles are seeded within a propellant gas either in the high-pressure (∼20 bars) hot (∼800 K) reservoir of a Laval nozzle or in the diverging section of the nozzle. Particles are accelerated at a sufficiently high supersonic speed to surpass a critical velocity beyond which, when impacting a substrate, they undergo plastic deformation and bond to the surface without melting [12, 13]. Beyond transportation of very-small-sized particles, Laval nozzles play a major role in the *fabrication of metal powders* as well. In gas atomization processes, they are used to generate a high-velocity gas flow which impinges on a liquid metal flow. As a consequence, the latter disintegrates into mini droplets which subsequently cool and solidify instantaneously to form metal powders the size of which is inversely proportional to the supersonic speed of the gas and is typically in the micrometric range. In addition, the uniformity of the gas flow improves the narrowness of the size distribution of the droplets, and hence that of the eventually formed particles [14, 15]. Production of clean energy is one of the key issues of modern society. In this frame, *production of molecular hydrogen* from plasma-assisted reformed ethanol using the Laval nozzle gliding arc discharge technique can be a promising source of energy. In this method, a plasma is generated during the expansion of an ethanol-water-air mixture through a Laval nozzle by applying a voltage between its body and a stainless steel rod set coaxial to the Laval nozzle. The supersonic gas flow, including H_2 products, is then trapped into a condenser and mass analyzed via gas chromatography [16]. Innovative developments for the *production of super-thin fibers* of micrometric size can take advantage of the qualities of supersonic flows issued from Laval Nozzles. Here, a melted polymer is pumped through a die orifice and dragged toward the nozzle throat by a room-temperature air flow which draws and aligns the extruded polymer stream in the Laval nozzle axis direction. Permanent acceleration of cooling air down to the nozzle exit and well-defined orientation beyond allow one to generate continuous fibers without filaments melting together as can occur using other techniques [17].

From these various examples of Laval nozzle applications, one of the striking aspects is the diversity of physical conditions for which they are employed. For instance, pressure inside the reservoir can be of several tens of bars or a few mbars, whereas temperature in this reservoir can reach several thousands of degrees or be even that of liquid nitrogen.

The geometrical size can be very different as well, varying from several meters in diameter for rocket thrusters down to millimeter or even submillimeter dimensions for laser cutting applications, for example. Usually, Laval nozzles are axisymmetric ducts, but rectangular or even annular sections are also designed for some applications. When the nozzle contour is known, it is possible to characterize the flow expanding through the nozzle and downstream using computational fluid dynamics (CFD) methods. This is the so-called *direct method*. A more common situation, however, consists in designing geometry able to deliver a uniform supersonic flow with defined characteristics within the supersonic jet downstream of the nozzle. This is the so-called *inverse problem* which is very difficult to handle with modern CFD to our knowledge.

The following two Sections 2.2 and 2.3 will concentrate on these geometrical aspects and more specifically on how the walls of the diverging part of a Laval nozzle can be properly shaped to generate a uniform long enough parallel supersonic flow downstream of the nozzle exit with well-defined physical conditions of velocity, temperature, pressure, and density.

2.2   Designing Inviscid Laval Nozzles

2.2.1   *Philosophy and strategy*

Most chemical physics or CRESU applications of uniform supersonic jets deal with low-pressure steady flows passing through a Laval nozzle. In this situation, viscosity effects and heat transfer cannot be neglected. In essence, these flows are compressible which means that the mass density ρ varies with space. They are supposed to be ideal gases with specific heat capacities c_p (constant pressure) and c_v (constant volume) independent of the temperature.

The determination of a suitable Laval nozzle profile should necessitate resolving the Navier-Stokes equation (see Chapter 1) for the *inverse problem* together with the complete equation of heat transfer. As this is a quite hard numerical task because of its high nonlinear nature, other approximated approaches are commonly considered. As mentioned in Chapter 1, uniform supersonic flows are made up of an inviscid core surrounded by a viscous boundary layer separating the inner part of the jet from the environment (wall of a nozzle, chamber, atmosphere, etc.). Within this inviscid core, the supersonic flow is isentropic (adiabatic and reversible), in thermal equilibrium, and is not subject to heat conduction and body forces such as gravitational and electromagnetic forces. Downstream of the nozzle exit, the gas velocity is oriented along the flow axis z, $\vec{v} = (v_z, 0, 0)$. Figure 2.3 summarizes the main flow parameters which will be used in this chapter.

In the present section, a method will be described to obtain the inviscid contour A_z for a Laval nozzle allowing expansion of a given gas to form a uniform supersonic stream at a given Mach

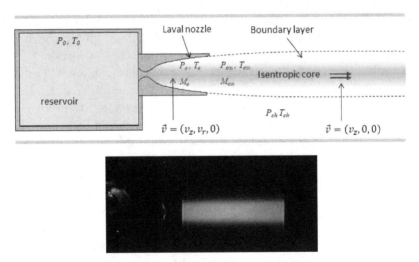

Fig. 2.3. Schematic view of a uniform supersonic expansion issued from a Laval nozzle as well as a flow visualization issued from the SR3 wind tunnel in the early days of the CRESU developments at the *Laboratoire d'Aérothermique in Meudon* (Nov. 1981).

number $M_{e\infty}$. The following Section 2.3 will explain how the boundary layer thickness (or better said the *displacement thickness*) can be calculated and combined with the inviscid contour to form the final shape of the Laval nozzle. Since these sections contain a great deal of mathematics, the most significant equations have been set in boxes.

As developed in Chapter 1, the Euler equations (Eqs. (1.64) and (1.68)) prevail within an isentropic inviscid flow. Together with the mass conservation equation (Eq. (1.57)), it can deliver the so-called *potential equation* which we will establish in Section 2.2.3. This latter equation, which controls the spatial velocity coordinates v_z and v_r evolution of the expanding flow, remains insufficient by itself, however. The fundamentals of the strategy to determine the isentropic frontier are then to choose a series of additional constraints that the flow must respect. In other words, the nozzle contour will apply to a certain category of flows only. Doing so, several new equations can be obtained allowing the design of the inviscid contour eventually.

The first assumption is to consider that the flow is one dimensional along the centerline of the system ($v_z(r=0)=0$) leading to keystone relations within temperature, pressure, and density upstream and downstream of the Laval nozzle (Section 2.2.2). Still along the centerline, the Mach number evolution will be imposed to reach the expected $M_{e\infty}$ value at a given position $(z_b, 0, 0)$ within the inviscid contour and representing the top of a Mach cone (Section 2.2.6). Beyond this cone delimitation, the flow becomes uniform. Outside the centerline, the evolution of the radial coordinate r along z and the evolution of the radial velocity v_r with respect to the longitudinal velocity v_z will be depicted by two new relations (Section 2.2.4) obtained from specific assumptions, basements of the *Method of Characteristics* (MOC hereafter). The $r(z)$ curves will define the so-called *characteristic lines* along which the flow properties can be calculated. Provided that these properties are known for a few points within the flow (for instance, close to the nozzle throat, Section 2.2.5; or at the Mach cone line, Section 2.2.7), a grid of points where the flow properties can be calculated iteratively can be constructed using the characteristic lines. Propagation of the grid of points will terminate when the radial coordinate r will be large enough so that the mass flow rate passing through the nozzle throat is conserved along a given characteristic line (Section 2.2.7).

2.2.2 *One-dimensional compressible, isentropic, inviscid supersonic flows*

Conservation laws established in Chapter 1 for a quasi-one-dimensional flow can be applied *within the isentropic kernel*. They led to the important energy conservation relation that we repeat here for convenience:

$$c_p T_0 = c_p T_e + \frac{M_e^2 a_e^2}{2} \tag{2.1}$$

where index 0 represents the reservoir location where the gas flow is almost at rest and index e is for the isentropic core. M_e is the Mach number and a_e the speed of sound at temperature T_e. Defining γ as the ratio of specific heat capacities c_p/c_v, this equation can be changed to

$$\frac{T_0}{T_e} = \left(1 + \frac{\gamma-1}{2} M_e^2\right) \tag{2.2}$$

Similar relations can be obtained for pressure and mass density if we apply the isentropic assumption to the flow $P \times \rho^{-\gamma} = const$ or equivalently $P^{1-\gamma} \times T^\gamma = const$:

$$\frac{P_0}{P_e} = \left(1 + \frac{\gamma-1}{2} M_e^2\right)^{\gamma/(\gamma-1)} \tag{2.3}$$

$$\frac{\rho_0}{\rho_e} = \frac{n_0}{n_e} = \left(1 + \frac{\gamma-1}{2} M_e^2\right)^{1/(\gamma-1)} \tag{2.4}$$

Here, n_0 and n_e are the particle (molecules or atoms) density.

Since the nature of the buffer gas is known and the reservoir conditions are fixed, Eqs. (2.2)–(2.4) indicate that the supersonic flow characteristics are determined for every Mach number. These equations are valid all along the expansion of any isentropic flow. An interesting situation is the throat location where $M_e = 1$. Very simple relations can then be extracted for the pressure, temperature, and density at the throat ($P_{th}, T_{th},$ and ρ_{th}, respectively):

$$T_{th} = \frac{2}{\gamma+1} T_0; \quad P_{th} = \left(\frac{2}{\gamma+1}\right)^{\gamma/(\gamma-1)} P_0; \quad \rho_{th} = \left(\frac{2}{\gamma+1}\right)^{1/(\gamma-1)} \rho_0 \tag{2.5}$$

For a monoatomic gas such as helium or argon ($\gamma = 5/3$), we find

$$T_{th} = 0.750 \times T_0; \;\; P_{th} = 0.487 \times P_0; \;\; \rho_{th} = 0.650 \times \rho_0 \qquad (2.6)$$

For a diatomic gas such as N_2 or O_2 ($\gamma = 7/5$), Eq. (2.5) gives

$$T_{th} = 0.833 \times T_0; \;\; P_{th} = 0.528 \times P_0; \;\; \rho_{th} = 0.634 \times \rho_0 \qquad (2.7)$$

Applying the energy conservation between the throat and any other station, one can derive a relation between the local sonic velocity a_e and the sonic velocity at the throat a_{th}. By replacing the left-hand side term in Eq. (2.1) with $c_p T_{th} + a_{th}^2/2$ and expressing c_p as a function of r_m (the specific gas constant defined in Chapter 1) and γ, one obtains [2].

$$a_e^2 = \frac{\gamma + 1}{2} a_{th}^2 - \frac{\gamma - 1}{2} v_z^2 \qquad (2.8)$$

It is also worth noting that Eq. (2.1) says that the flow velocity v_z will attain an upper limit v_{max} when enthalpy is completely converted into kinetic energy. This is accomplished when $T_e = 0$ leading to the expression for v_{max}:

$$v_{max} = \sqrt{\frac{2\gamma}{\gamma - 1} r_m T_0} \qquad (2.9)$$

When the Mach number increases, the flow velocity tends to v_{max} asymptotically. As a matter of example, the limit velocity is given in Table 2.1 for helium, argon, and nitrogen when the reservoir temperature is set to 295 K. The Mach number is also indicated for a flow velocity reaching 90%, 95%, and 99% of the limit velocity.

The one-dimensional model presented in this section is exact for any uniform supersonic flow, hence especially downstream of the nozzle exit. Inside the diverging contour, it is also strictly correct

Table 2.1. Limit velocity v_{max} for typical buffer gases and Mach conditions corresponding to a supersonic velocity reaching 90%, 95%, and 99% of v_{max}.

	v_{max} (m/s)	$M_{90\%}$	$M_{95\%}$	$M_{99\%}$
He	1751	3.6	5.2	12.0
Ar	554	3.6	5.2	12.0
N_2	783	4.6	6.8	15.5

along the centerline and beyond the Mach cone (see Section 2.2.6). Apart from the centerline, expansion requires, however, some lateral motion in order to fulfill space downstream of the nozzle throat. Hence, a more rigorous treatment of the expansion through the nozzle needs consider a flow velocity with a small radial (or perpendicular to the z axis for a plane section) component v_r. This is of importance in the determination of the geometric shape A_z of the divergent. As mentioned earlier, in practice, A_z has to be initially calculated for the isentropic core of the nozzle, that is, in the absence of any viscosity contribution. The MOC is a powerful tool commonly used to calculate these isentropic contours. Before describing the main features of this technique, let us take a moment in order to derive the *potential equation* which is one of the starting points of the MOC.

2.2.3 *The potential equation*

This equation is obtained by combining the mass conservation equation with the Euler equation for an isentropic, irrotational $(rot(\vec{v}) = 0)$ flow representative of an expansion through a Laval nozzle with velocity vector $\vec{v} = (v_z, v_r, 0)$. As shown in Chapter 1, the Euler equation can be written as

$$dP_e = -\rho_e v dv = -\frac{\rho_e}{2} d(v_z^2 + v_r^2) \qquad (2.10)$$

whereas the mass conservation equation reduces to

$$\frac{\partial}{\partial z}(\rho_e v_z) + \frac{\partial}{\partial r}(\rho_e v_r) + \varepsilon \frac{\rho_e v_r}{r} = 0 \qquad (2.11)$$

where $\varepsilon = 1$ for an axisymmetric flow and $\varepsilon = 0$ for a rectangular section (in this case, r would assimilate to the usual y coordinate).

Remembering the definition of the speed of sound given in Section 1.5.2 of Chapter 1, the total differential of ρ_e can be obtained from Eq. (2.10) and the partial differential with respect to z and r as well:

$$d\rho_e = -\frac{\rho_e}{2a_e^2} d\left(v_r^2 v_z^2 + v_r^2\right) = -\frac{\rho_e}{a_e^2}(v_z dv_z + v_r dv_r) \qquad (2.12)$$

$$\frac{\partial \rho_e}{\partial z} = -\frac{\rho_e}{a_e^2}\left(v_z \frac{\partial v_z}{\partial z} + v_r \frac{\partial v_r}{\partial z}\right) \qquad \frac{\partial \rho_e}{\partial r} = -\frac{\rho_e}{a_e^2}\left(v_z \frac{\partial v_z}{\partial r} + v_r \frac{\partial v_r}{\partial r}\right)$$
$$(2.13)$$

Developing the mass conservation Eq. (2.11) and substituting the partial differentials of ρ_e by the expressions given in Eq. (2.13), one obtains the so-called *potential equation* [2, 18]:

$$(a_e^2 - v_z^2)\frac{\partial v_z}{\partial z} + (a_e^2 - v_r^2)\frac{\partial v_r}{\partial r} - 2v_z v_r \frac{\partial v_r}{\partial z} = -\varepsilon \frac{a_e^2 v_r}{r} \qquad (2.14)$$

The factor 2 in the equation takes into account that the flow is irrotational and then $\partial v_r/\partial z = \partial v_z/\partial r$. The irrotational character of the flow involves that velocity can be expressed as the gradient of a function usually called *potential velocity* Ø (since $\overrightarrow{rot(grad(\psi))} = 0$ for any scalar function ψ) giving its name to Eq. (2.14). This equation represents the velocity components' evolution with space for either an axisymmetric or a planar steady compressible flow. It is a highly nonlinear equation with partial derivatives, which is hard to solve and for which no exact analytical solutions exist. However, it is a keystone in the determination of flow properties in the supersonic and transonic regime.

2.2.4 *The method of characteristics (MOC)*

This method is an exact numerical approach for the resolution of Eq. (2.14). It is based on the calculation of flow-field properties at discrete points in a non-rectangular grid. By definition, a characteristic line is a line in which angle α with a streamline of property u (u can be any flow variables: $\vec{v}$, P, ρ, or T) is such that $\sin \alpha = 1/M_e$. For such a line, the derivative of u with respect to the z coordinate is indeterminate and may sometimes be discontinuous.

As we will see below, this condition will allow us to derive two new equations permitting determination of the properties evolution along the flow. In the following, we will consider u as the velocity vector $(v_z, v_r, 0)$. Considering Eq. (2.14) and the total differentials of v_z and v_r

$$dv_z = \frac{\partial v_z}{\partial z}dz + \frac{\partial v_r}{\partial z}dr \quad dv_r = \frac{\partial v_r}{\partial z}dz + \frac{\partial v_r}{\partial r}dr \qquad (2.15)$$

one ends up with a system of three equations with three variables that can be written in a matrix format as

$$\left[\begin{array}{ccc} \left(1-\dfrac{v_z^2}{a_e^2}\right) & \dfrac{-2v_z v_r}{a_e^2} & \left(1-\dfrac{v_r^2}{a_e^2}\right) \\ dz & dr & 0 \\ 0 & dz & dr \end{array}\right] \left[\begin{array}{c} \dfrac{\partial v_z}{\partial z} \\ \dfrac{\partial v_r}{\partial z} \\ \dfrac{\partial v_r}{\partial r} \end{array}\right] = \left[\begin{array}{c} -\varepsilon\dfrac{v_r}{r} \\ dv_z \\ dv_r \end{array}\right] \qquad (2.16)$$

This system can be solved for $\partial v_r/\partial z$ using the determinant formalism as

$$\frac{\partial v_r}{\partial z} = \frac{det\left[\begin{array}{ccc} \left(1-\dfrac{v_z^2}{a_e^2}\right) & -\varepsilon\dfrac{v_r}{r} & \left(1-\dfrac{v_r^2}{a_e^2}\right) \\ dz & dv_z & 0 \\ 0 & dv_r & dr \end{array}\right]}{det\left[\begin{array}{ccc} \left(1-\dfrac{v_z^2}{a_e^2}\right) & \dfrac{-2v_z v_r}{a_e^2} & \left(1-\dfrac{v_r^2}{a_e^2}\right) \\ dz & dr & 0 \\ 0 & dz & dr \end{array}\right]} \qquad (2.17)$$

At a characteristic line, this derivative is indeterminate by definition. This can only occur when the numerator and denominator determinants are null simultaneously. Let us consider first the denominator of Eq. (2.17). At the characteristic line, it leads to the following relation:

$$\left(1-\frac{v_z^2}{a_e^2}\right)\left(\frac{dr}{dz}\right)^2 + \frac{2v_z v_r}{a_e^2}\frac{dr}{dz} + \left(1-\frac{v_r^2}{a_e^2}\right) = 0 \qquad (2.18)$$

which constitutes a second-order expression as a function of variable dr/dz. Hence, for a given point in space (z,r) having a velocity (v_z, v_r), the motion along the r axis and a characteristic line must respect the following relation for a known displacement dz along the z axis:

$$\frac{dr}{dz} = \frac{-\dfrac{v_z v_r}{a_e^2} \pm \sqrt{M_e^2 - 1}}{\left(1-\dfrac{v_z^2}{a_e^2}\right)} \qquad (2.19)$$

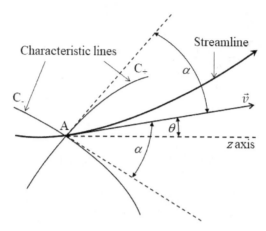

Fig. 2.4.　Illustration of ascending (left running) C_+ and descending (right running) C_- characteristic lines at point A moving along a streamline with velocity $\vec{v} = (v_z = v \times \cos\theta, v_r = v \times \sin\theta, 0)$ and a Mach number such as $M_e = 1/\sin\alpha$.

where M_e is the Mach number at position (z, r). This result says that there are two real characteristic lines C_+ and C_- through each point of the supersonic part of the flow, whereas in the sonic region, at the throat, only one characteristic exists. Interestingly, for subsonic flows, solutions are imaginary and the MOC is not applicable in this situation. Expressing v_z and v_r in polar coordinates $v_z = v \times \cos\theta$ and $v_r = v \times \sin\theta$ leads to a transformation of Eq. (2.19) into a very simple formula

$$\frac{dr}{dz} = \tan(\theta \mp \alpha) \tag{2.20}$$

illustrated in Fig. 2.4 Equation (2.20) indicates that at the (z, r) station A, the characteristic lines make an angle α with the velocity vector (v_r, v_z). Since α is another expression of the Mach number through $\sin\alpha = 1/M_e$ as mentioned earlier, these lines are also called *Mach lines* at point A.

To go further, one needs now to evaluate how the velocity properties evolve along the characteristic lines when one moves from (z, r) to $(z + dz, r + d)$. For this, let us now focus on the numerator of Eq. (2.17) which must be zero as well along the characteristic line.

This leads to the new expression:

$$\frac{dv_r}{dv_z} = -\frac{\left(1 - \frac{v_z^2}{a_e^2}\right)}{\left(1 - \frac{v_r^2}{a_e^2}\right)}\frac{dr}{dz} - \varepsilon\frac{\frac{v_r}{r}\frac{dr}{dv_z}}{\left(1 - \frac{v_r^2}{a_e^2}\right)} \tag{2.21}$$

Along the characteristic line, remember that dr/dz is given by Eq. (2.19). Using the same transformation of the velocity coordinates as before, one can obtain after some algebraic and trigonometric manipulation

$$d\theta = \mp\sqrt{M_e^2 - 1}\frac{dv}{v} \pm \varepsilon\frac{1}{\sqrt{M_e^2 - 1} \mp \cot\theta}\frac{dr}{r} \tag{2.22}$$

This expression gives the angular evolution of the velocity vector $\vec{v} = (v_z, v_r, 0)$ for a given displacement dr and a modification of the velocity module dv. As the Mach number is defined as v/a_e, the term dv/v can be written as a function of M_e. Expressing Eq. (2.2) in terms of the sound velocity ratio a_0/a_e, and differentiating, one can eventually obtain the following relation [2]:

$$\frac{dv}{v} = \frac{1}{1 + \frac{\gamma - 1}{2}M_e^2}\frac{dM_e}{M_e} \tag{2.23}$$

Introducing the $\nu(M_e)$ function known as the Prandtl-Meyer function and defined as

$$\nu(M_e) = \int \frac{\sqrt{M_e^2 - 1}}{1 + \frac{\gamma - 1}{2}M_e^2}\frac{dM_e}{M_e}$$

$$= \sqrt{\frac{\gamma + 1}{\gamma - 1}}\arctan\left(\sqrt{\frac{\gamma - 1}{\gamma + 1}(M_e^2 - 1)}\right) - \arctan\left(\sqrt{M_e^2 - 1}\right) \tag{2.24}$$

Equation (2.22) ends up as

$$d(\theta \pm \nu(M_e)) = \pm\varepsilon\frac{1}{\sqrt{M_e^2 - 1} \mp \cot\theta}\frac{dr}{r} \tag{2.25}$$

Equation (2.25) is known in the literature as the *compatibility equation* for an irrotational isentropic flow. It reflects the angular evolution of $\vec{v}$ for a given displacement dr from point A (z, r). The positive sign in the differential left-hand side term corresponds to the evolution along the C_- characteristic line, whereas the negative sign is for the C_+ one. It is interesting to note that for a two-dimensional flow, that is a rectangular section ($\varepsilon = 0$), the right-hand side term cancels and Eq. (2.25) becomes algebraic with $\theta \pm \nu(M_e) = const$ along the characteristic lines C_- and C_+, respectively. For an axisymmetric flow, Eq. (2.25) is differential and is usually linearized and discretized in numerical computations [18]. It is solved in a step-by-step process in conjunction with Eq. (2.20), which defines the characteristic lines' curves, i.e. the grid of points. In order to apply linearization of the curved characteristic lines, this grid is always chosen so that the flow field properties are slightly modified from one point to its closest neighbor. Contrary to other methods for which the grid of calculation is defined once and for all, the methods of characteristics build a network of points simultaneously with the properties determination since Eq. (2.20) says that the motion from one point to another one along the grid depends on the local flow properties.

The method can only be used in the supersonic zone, hence in the divergent part of the Laval nozzle. It also necessitates initial conditions consisting in a specific characteristic line (see further). Propagation of the characteristic net can then proceed by intersecting the C_+ (ascending line) and C_- (descending line) characteristic lines issued from two points 1 and 2 for which all properties are known. Hence, the properties of that intersecting point must respect laws provided by the two characteristics C_+ and C_-. Supposing that point 1 (z_1, r_1) and 2 (z_2, r_2) are known as shown in Fig. 2.5, the intersecting point 3 (z_3, r_3) is geometrically easily obtained as [19].

$$r_3 = r_1 + (z_3 - z_1) \tan(\theta_1 + \alpha_1) = r_2 + (z_2 - z_1) \tan(\theta_2 - \alpha_2) \tag{2.26}$$

and

$$z_3 = \frac{r_2 - r_1 - z_2 \tan(\theta_2 - \alpha_2) + z_1 \tan(\theta_1 + \alpha_1)}{\tan(\theta_1 + \alpha_1) - \tan(\theta_2 - \alpha_2)} \tag{2.27}$$

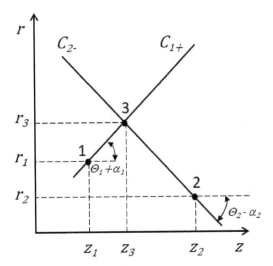

Fig. 2.5. Construction of the characteristic net from two known initial points 1 and 2 and the associated characteristic lines C_{1+} and C_{2-}, respectively.

Having r_3, the dr components in Eq. (2.22) (or equivalently (2.25)) are known and one is left with a system of two equations with two unknowns: v_3 and θ_3. Linearizing Eq. (2.22), i.e. setting $d\theta = \theta_3 - \theta_1$ (resp. $\theta_3 - \theta_2$), $dv = dv_3 - dv_1$ (resp. $dv_3 - dv_2$) and $dr = r_3 - r_1$ (resp. $r_3 - r_2$), the velocity module v_3 at point 3 can be determined by subtracting equations one from the other. The velocity angle θ_3 at point 3 is then straightforwardly deduced. Note that it may be convenient to use the dimensionless velocity term defined as $\acute{v} = v/v_{max}$ instead of v in Eq. (2.22) in order to determine the Mach number M_e from the relation

$$M_e^2 = \frac{2}{\gamma - 1} \frac{\acute{v}^2}{1 - \acute{v}^2} \tag{2.28}$$

This equation is easily derived from the v_{max} expression 2.9, Eq. (2.2), and the Mach number definition. Once M_{e3} is known, all physical parameters can be inferred at point 3.

Finally, a particular situation must be stressed here. In the event that one initial point lies in the centerline of the flow ($r_i = 0$ with $i = 1$ or 2), its velocity is parallel to the axis meaning $\theta_i = 0$. This leads to an indetermination of the second term of the right-hand side part of Eq. (2.22):

$$\frac{1}{\sqrt{M_{ei}^2 - 1 \mp \cot \theta_i}} \frac{dr}{r_i} = \frac{\sin \theta_i \sin \alpha_i}{\sin(\theta_i \mp \alpha_i)} \frac{r_3 - r_i}{r_i} \approx$$

$$\mp r_3 \lim_{\substack{\theta_i \to 0 \\ r_i \to 0}} \frac{\sin \theta_i}{r_i} = \mp r_3 \lim_{\substack{\theta_i \to 0 \\ r_i \to 0}} \left(\frac{d\theta}{dr}\right)_i = \mp r_3 \frac{\theta_3}{r_3} = \mp \theta_3 \quad (2.29)$$

As mentioned earlier, in order to propagate the grid construction, it is worth knowing a series of points or in other words an initial characteristic line which can be determined easily. A peculiar situation is found close to the nozzle throat where the flow characteristics must be obtained using a different treatment that will be described in the next section. Another special characteristic line is the one defined by the Mach cone that will be commented upon in Section 2.2.7.

Note that the MOC technique can be used either to obtain the flow field properties for a given existing contour or, as it is our concern here, to derive the diverging shape of the isentropic core of a nozzle, which would deliver a flow with initially specified qualities such as the gas nature and the Mach number outside of the nozzle.

2.2.5   *Flow properties at the throat*

Let us turn back to the potential Eq. (2.14). In the throat zone, several techniques can be applied in order to determine the flow field properties by approached methods [18]. Here, we will present the one used by Owen and Sherman [20, 21] which is inspired by a method developed by Sauer [19]. In the throat region, sections A_z vary smoothly so that the flow velocity remains very close to the sonic value at the throat a_{th} (with $a_{th}^2 = \gamma r_m T_{th}$). Hence, a perturbation method can be used for the velocity coordinates' v_z and v_r evolution in order to solve for Eq. (2.14). Using dimensionless velocity components, one writes

$$V_z = \frac{v_z}{a_{th}} \qquad V_r = \frac{v_r}{a_{th}} \qquad (2.30)$$

and then close to the throat

$$V_z = 1 + \tilde{v}_z \qquad V_r = \tilde{v}_r \qquad (2.31)$$

with $\tilde{v}_z, \tilde{v}_r \ll 1$. These expressions are only valid close to the nozzle throat that is for the region with $z \leq \chi \ll L$ where L is the length of the diverging section and χ the longitudinal limit for the validity of Eq. (2.31) (see below). Replacing the velocity components v_z and v_r by their new formulations in Eq. (2.14) and considering Eq. (2.8) leads, after some algebraic gymnastics, to

$$(\gamma + 1)\,\tilde{v}_z \frac{\partial \tilde{v}_z}{\partial z} - \frac{\partial \tilde{v}_r}{\partial r} + 2\tilde{v}_r \frac{\partial \tilde{v}_z}{\partial r} - \varepsilon \frac{\tilde{v}_r}{r} = 0 \qquad (2.32)$$

where terms of order $n + 1$ have been neglected with respect to terms of order n (i.e. $\tilde{v}_z^2, \tilde{v}_r^2 \ll \tilde{v}_z, \tilde{v}_r \ll 1$). Note that in the original treatment by Sauer, the third term was also neglected [18]. In order to obtain $\tilde{v}_z$ and $\tilde{v}_r$, a perturbation potential function is defined as

$$\emptyset(z, r) = \sum_{n=0}^{\infty} f_{2n}(z) r^{2n} \sim f_0(z) + r^2 f_2(z) + r^4 f_4(z) \qquad (2.33)$$

with

$$\tilde{v}_z = \frac{\partial \emptyset(z, r)}{\partial z} \qquad \tilde{v}_r = \frac{\partial \emptyset(z, r)}{\partial r} \qquad (2.34)$$

$$\frac{\partial \tilde{v}_z}{\partial z} = \frac{\partial^2 \emptyset(z, r)}{\partial z^2} \qquad \frac{\partial \tilde{v}_r}{\partial r} = \frac{\partial^2 \emptyset(z, r)}{\partial r^2} \qquad \frac{\partial \tilde{v}_z}{\partial r} = \frac{\partial \emptyset(z, r)}{\partial r \partial z} \qquad (2.35)$$

Function $\emptyset(z, r)$ does not contain odd powers because the geometry (planar or cylindrical) of the flow is symmetric in r. At the centerline of the flow close to the throat it, can also be assumed that the velocity perturbation $\tilde{v}_z(z, 0)$ evolves linearly with z : $\tilde{v}_z(z, 0) = K \times z = f_0'(z)$ (where f_0' is the derivative of f_0). Inserting these new expressions for $\tilde{v}_z$ and $\tilde{v}_r$ in Eq. (2.32), one ends up with a polynomial in r^n which must be valid for every z position (with $z \leq \chi$). Thus, all r^n coefficients must be set to zero which allows one to obtain expressions of the unknown $f_2(z)$ and $f_4(z)$ functions

$$f_2(z) = \frac{\gamma + 1}{2} \frac{f_0'(z) f_0''(z)}{1 + \varepsilon} = \frac{\gamma + 1}{2} \frac{K^2 z}{1 + \varepsilon} \qquad (2.36)$$

$$f_4(z) = \frac{\gamma + 1}{4} \frac{f_0'(z)f_2''(z) + f_0''(z)f_2'(z)}{3 + \varepsilon} + \frac{2f_2(z)f_2'(z)}{3 + \varepsilon}$$

$$= \frac{(\gamma + 1)^2 K^3}{8(3 + \varepsilon)(1 + \varepsilon)} + \frac{(\gamma + 1)^2 K^4}{2(3 + \varepsilon)(1 + \varepsilon)^2}z \qquad (2.37)$$

and consequently the V_z and V_r expressions

$$V_z = 1 + Kz + \frac{\gamma + 1}{2}\frac{K^2}{1 + \varepsilon}r^2 + \frac{(\gamma + 1)^2 K^4}{2(3 + \varepsilon)(1 + \varepsilon)^2}r^4 \qquad (2.38)$$

$$V_r = \frac{\gamma + 1}{1 + \varepsilon}K^2 zr + \frac{2(\gamma + 1)^2 r^3}{(3 + \varepsilon)(1 + \varepsilon)}\left[\frac{K^3}{4} + \frac{K^4 z}{1 + \varepsilon}\right] \qquad (2.39)$$

At this point, we are left with only one unknown K before being able to quantify the velocity components for every point in space (z, r) close to the nozzle throat. Determination of K is not straightforward and needs several intermediate steps that we will summarize below. Let us first consider the locus of points for which $V_r \equiv 0$. This physically includes the nozzle wall at the throat. Naming r_{th} the radius (or half height for a planar section) of the nozzle throat, we can express the position (z_{th}, r_{th}) which satisfies the preceding statement (note that the origin of coordinates $(z = 0, r = 0)$ corresponds to a point where the Mach number is 1).

$$z_{th} = -\frac{\gamma + 1}{2}\frac{(1 + \varepsilon)Kr_{th}^2}{(3 + \varepsilon)(1 + \varepsilon) + 2(\gamma + 1)K^2 r_{th}^2} \qquad (2.40)$$

The streamline passing through (z_{th}, r_{th}) is at the nozzle wall and parallel to the flow axis z. Hence, its radius of curvature $\hat{R}$ at that point can be expressed by

$$\frac{1}{\hat{R}} = \left(\frac{\frac{\partial \tilde{v}_r}{\partial z}}{1 + \tilde{v}_z}\right)_{(z_{th}, r_{th})}$$

$$= \frac{\frac{\gamma + 1}{1 + \varepsilon}K^2 r_{th} + \frac{2(\gamma + 1)^2 r_{th}^3 K^4}{(3 + \varepsilon)(1 + \varepsilon)^2}}{1 + Kz_{th} + \frac{\gamma + 1}{2}\frac{K^2}{1 + \varepsilon}r_{th}^2 + \frac{(\gamma + 1)^2 K^4}{2(3 + \varepsilon)(1 + \varepsilon)^2}r_{th}^4} \qquad (2.41)$$

Defining Γ as

$$\Gamma = (\gamma + 1)\, K^2 r_{th}^2 \tag{2.42}$$

one can obtain the following relation between the radius at the throat r_{th} and the radius of curvature $\hat{R}$ of the streamline at the throat wall (z_{th}, r_{th}):

$$\frac{r_{th}}{\hat{R}} = \frac{2(3+\varepsilon)\Gamma\left[1+\frac{2\Gamma}{(3+\varepsilon)(1+\varepsilon)}\right]^2}{2(3+\varepsilon)(1+\varepsilon)+6\Gamma+\frac{3}{1+\varepsilon}\Gamma^2+\frac{2}{(3+\varepsilon)(1+\varepsilon)^2}\Gamma^3} \tag{2.43}$$

which can be reformulated as a polynomial of the third order in Γ as

$$\frac{1}{(3+\varepsilon)\,(1+\varepsilon)^2}\left[4-\frac{r_{th}}{\hat{R}}\right]\Gamma^3 + \left[\frac{4}{1+\varepsilon} - \frac{3}{2\,(1+\varepsilon)}\frac{r_{th}}{\hat{R}}\right]\Gamma^2$$

$$+ \left[3+\varepsilon - 3\frac{r_{th}}{\hat{R}}\right]\Gamma - (3+\varepsilon)\,(1+\varepsilon)\,\frac{r_{th}}{\hat{R}} = 0 \tag{2.44}$$

The ratio $r_{th}/\hat{R}$ is an input of the problem. In practice, it has been observed that $\hat{R} \geq 4 \times r_{th}$ is a necessary criteria for the validity of the method developed above [18]. In usual calculations, we personally use $r_{th}/\hat{R} = 1/5$ [21]. From this, the gamma value, which is a real positive number, is unique and independent of the nozzle properties excepting for the planar or revolution symmetry. For a ratio of 0.2, one finds $\Gamma = 0.385655$ for a symmetry of revolution and $\Gamma = 0.1904415$ for a planar configuration.

From Eq. (2.42), the K factor can then be obtained provided that the nozzle throat radius is known and subsequently the position (z_{th}, r_{th}) (Eq. (2.40)) and the velocity perturbation terms (Eq. (2.38) and (2.39)) can be determined as well. As a result, at the throat position (z_{th}, r_{th}), V_z is only a function of Γ and ε and consequently is independent of the gas nature.

It is also interesting to note that the locus of points for which the flow velocity is sonic $(V_z^2 + V_r^2 = 1)$ is not a simple straight line perpendicular to the nozzle axis z at the throat position. As shown

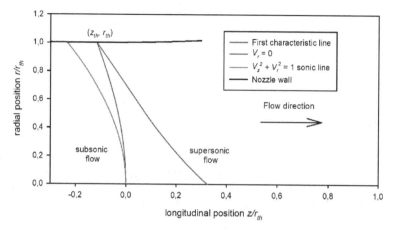

Fig. 2.6. Sonic curve at the throat vicinity for the expansion of a monoatomic gas through an axisymmetric nozzle.

in Fig. 2.6, it is a curve which intersects the nozzle wall upstream and the centerline downstream of the geometrical throat.

As a consequence, the point (z_{th}, r_{th}) is already in the supersonic zone of the flow and it can be used as a starting point in order to design the characteristics lattice. In practice, the C_{-0} characteristic curve joining the throat point (z_{th}, r_{th}) to the axis of the flow $(z_{C-0}, 0)$ is calculated step by step for a few tens of points using Eqs. (2.20), (2.38), and (2.39) with $\theta = 0$ at (z_{th}, r_{th}) and at $(z_{C-0}, 0)$ as well.

In order to design the isentropic core of the nozzle, two additional steps must be considered to propagate the initial characteristic line along the future nozzle. First, one needs to build the second characteristic line downstream C_{-0} by moving from $(z_{C-0}, 0)$ to $(z_{C-1}, 0)$ where $z_{C-1} - z_{C-0} = \delta z$ is a small elementary step chosen by the operator. Determining the properties at $(z_{C-1}, 0)$ requires one to propose an empirical law for the centerline Mach number taking into account some boundary conditions. Once this is done, the series of characteristics can be designed, but one needs to know at which distance from the centerline calculations must be stopped. For this, conservation of the mass flow rate is considered using as a reference the mass flow passing through the throat. These two points will be discussed in the next sections.

2.2.6 Choice of the Mach number distribution along the centerline

Within the small region $z \leq \chi$, the local Mach number can be obtained easily along the centerline along which the velocity vector is parallel to the axis $\vec{v} = (v_z, 0, 0)$. Defining $\zeta = z/r_{th}$ and using Eqs. (2.8) and (2.38),

$$M_e^2 = \frac{v_z^2}{a_e^2} = \frac{a_{th}^2(1 + Kz)^2}{\frac{\gamma+1}{2}a_{th}^2 - \frac{\gamma-1}{2}a_{th}^2(1 + Kz)^2}$$

$$= \frac{2(1 + Kr_{th}\zeta)^2}{\gamma+1-(\gamma - 1)(1 + Kr_{th}\zeta)^2} \qquad (2.45)$$

and its derivative is given by

$$\frac{dM_e}{d\zeta} = \frac{\sqrt{2}(\gamma + 1)Kr_{th}}{[\gamma+1-(\gamma - 1)(1 + Kr_{th}\zeta)^2]^{3/2}} \qquad (2.46)$$

Note here that Kr_{th} is a constant independent of the nozzle throat radius from Eq. (2.42). In addition, we wish that the searched Mach number expression reaches a constant value $M_{e\infty}$ at a position $\zeta = b = z_b/r_{th} \gg \chi/r_{th} (\equiv \zeta_\chi$ hereafter) so that beyond b the flow reaches its expected final velocity. This way, at $\zeta = b$, the derivative $dM_e/d\zeta$ of the Mach number must be zero. Note that b does not represent the length of the diverging section of the nozzle, but it is closely related to it. *It is a free parameter that the operator can choose* with some limitations detailed further. Finally, we also need the Mach number formulation to agree with Eqs. (2.45) and (2.46) at $\zeta = \zeta_\chi$.

The following expression was proposed by Owen and Sherman [20, 21] in the range $\zeta_\chi \leq \zeta \leq b$:

$$M_e\left(\zeta\right) = C_0 \left(C_1 - exp \left(\frac{b - \zeta}{C_2} \right)^2 \right) \quad \text{with}$$

$$\frac{dM_e(\zeta)}{d\zeta} = 2\frac{C_0}{C_2}\frac{(b - \zeta)}{C_2}exp \left(\frac{b - \zeta}{C_2} \right)^2 \qquad (2.47)$$

where C_0, C_1, and C_2 are constants that can be obtained from Eqs. (2.45) and (2.46) and the condition $M_e(\zeta = b) = M_{e\infty}$. By construction, this function respects $dM_e/d\zeta = 0$ at $\zeta = b$. Determination of the C_n coefficients is not trivial, however, and only the main steps will be explicated here. Playing with the three equations, Eqs. (2.45) and (2.46) and $M_e(b) = M_{e\infty}$, one can show that C_0 can be expressed as a function of C_2 (or better say x as defined below) and C_1 as a function of C_0. Hence, the determination of C_2 becomes the major step to be completed. It appears that one has to resolve an equation for C_2 of the form

$$\frac{x}{1-e^{-x}} = U \text{ with } x = \left(\frac{b - \zeta_\chi}{C_2}\right)^2 \tag{2.48}$$

U is a complicated expression which only depends on known parameters such as K, r_{th}, ζ_χ, γ, and $M_{e\infty}$.

We can express it as

$$U = \frac{1}{4} \frac{S_k^3 (\gamma + 1)(b - \zeta_\chi) K r_{th}}{M_{e\infty} - S_k (1 + K r_{th}\zeta_\chi)} \text{ with}$$

$$S_k = \frac{\sqrt{2}}{[\gamma + 1 - (\gamma - 1)(1 + K r_{th}\zeta_\chi)^2]^{1/2}} \tag{2.49}$$

Several comments are worth making here. First, Eq. (2.48) has a solution for x only when $U > 1$. This solution is unique when it exists. Although C_2 can be positive or negative, its sign does not matter as it affects neither Eq. (2.47) nor the determination of C_0. Parameter S_k can be easily evaluated because, as mentioned earlier, $K r_{th}$ is known from Eq. (2.42). For a symmetry of revolution, we have $K r_{th} = 0.38$ for a monoatomic gas and $K r_{th} = 0.40$ for a diatomic gas. For a rectangular section, $K r_{th}$ is even smaller, close to 0.27. Parameter ζ_χ is usually very small since it also represents the elementary step for all calculations. *Typically,* $\zeta_\chi = 0.02$ (or less). With this in mind, it appears that $K r_{th} \zeta_\chi \ll 1$ and consequently $S_k \approx 1$. Hence, the denominator for the U expression is very close to $M_{e\infty} - 1$. It can also be observed that U is directly proportional to the b parameter. This is an important point because the constraint $U > 1$ involves that b cannot be as small as one would like. In other words, the diverging section of the nozzle has a minimum length below which the

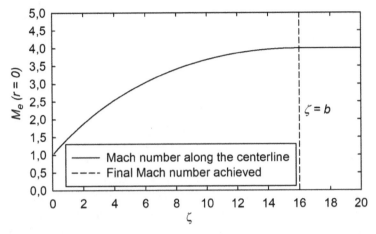

Fig. 2.7. Centerline Mach evolution for a diatomic gas generating a Mach 4 supersonic flow ($b = 16, \zeta\chi = 0.1$).

Mach number representation from Eq. (2.47) does not hold anymore. Furthermore, the higher the Mach $M_{e\infty}$, the higher b must be.

Taking into account all these considerations, x and then C_2 can be obtained resolving Eq. (2.46). Eventually, C_1 and C_0 are obtained from

$$C_0 = \frac{M_{e\infty} - S_k\left(1 + Kr_{th}\zeta\chi\right)}{e^x - 1} \text{ and } C_1 = 1 + \frac{M_{e\infty}}{C_0} \qquad (2.50)$$

The Mach number evolution (Fig. 2.7) along the centerline of the nozzle is now perfectly defined and depends only on four input parameters, namely, $\zeta\chi$, b, γ, and $M_{e\infty}$.

2.2.7 Obtaining the final isentropic contour

Propagation of the grid of points constructed with the MOC method needs be ended at some station taking into account conservation of the desired mass flow rate to pass through the nozzle. Here, we limit the calculation to a cylindrical symmetry. The mass flow rate $\dot{m}$ through an area of radius r_i making an angle $\Pi/2 - \theta$ with the velocity vector $\vec{v} = (\nu_z = \nu\cos\theta, \nu_r = \nu\sin\theta, 0)$ of the fluid can be expressed as

$$\dot{m} = 2\pi \int_0^{r_i} \rho\nu\cos\theta r\,dr \qquad (2.51)$$

Taking into account Eqs. (2.4) and (2.8) and the expression of the velocity close to the throat (Eqs. (2.30) and (2.31)), the mass flow rate at the throat can be expressed as a function of the reservoir conditions as

$$\dot{m}_{th} = 2\pi r_{th}^2 \rho_0 \sqrt{\frac{2\gamma}{\gamma-1} r_m T_0} \sqrt{\frac{\gamma-1}{\gamma+1}} \int_0^1 (1+\tilde{v}_z)$$

$$\times \left[1 - \frac{\gamma-1}{\gamma+1}\left[(1+\tilde{v}_z)^2 + \tilde{v}_r^2\right]\right]^{1/(\gamma-1)} r'dr' \qquad (2.52)$$

r' is a dimensionless variable defined as $r' = r/r_{th}$; $\tilde{v}_z$ and $\tilde{v}_r$ are given by Eqs. (2.31), (2.38), and (2.39) as a function of r (or r') fixing $z = z_{th}$. Apart from the nature of the gas (ρ_0, γ) and the geometrical factor r_{th}, the mass flow rate appears to be a function of $r_{th}/\hat{R}$ only, which we have fixed to 0.2 previously.

Beyond $z = \chi$, the flow velocity cannot be expressed by Eqs. (2.30) and (2.31) anymore. Then, the mass flow rate needs be determined differently. Using $\acute{v} = v/v_{max}$, the mass density ρ can be expressed as a function of $\acute{v}$ from Eqs. (2.1) and (2.9):

$$\rho = \rho_0 \left(1 - \acute{v}^2\right)^{1/(\gamma-1)} \qquad (2.53)$$

A general expression for the mass flow rate at any (z_i, r_i) station within the divergent of the nozzle can be deduced then:

$$\dot{m}_i = 2\pi\rho_0 \sqrt{\frac{2\gamma}{\gamma-1} r_m T_0} \int_0^{r_i} \acute{v} \left[1 - \acute{v}^2\right]^{1/(\gamma-1)} \cos\theta \, rdr \qquad (2.54)$$

A peculiar situation is when one reaches the exit plane of the nozzle. Here, the flow has constant parameters corresponding to the final Mach number $M_{e\infty}$. Hence, velocity is parallel to the z axis meaning $\cos\theta = 1$ and independent of radial position r. The mass flow rate at the exit of the nozzle is then

$$\dot{m}_{exit} = 2\pi\rho_0 \sqrt{\frac{2\gamma}{\gamma-1} r_m T_0} \acute{v}_\infty \left[1 - \acute{v}_\infty^2\right]^{1/(\gamma-1)} \frac{r_{exit}^2}{2} = \dot{m}_{th} \qquad (2.55)$$

As all parameters are known, excepting the radius at the nozzle exit r_{exit}, and the mass flow rate is conserved, it is now possible

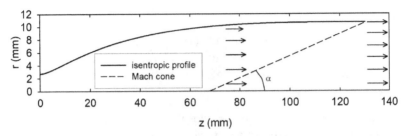

Fig. 2.8. Mach cone at the exit of the diverging section for a helium Laval nozzle calculated to generate a uniform supersonic flow at 23 K downstream of the nozzle exit. Arrows indicate the flow velocity perfectly constant within the Mach cone and slightly decreasing to the walls outside the Mach cone.

to determine the isentropic radius at the nozzle exit as a function of r_{th}. Subsequently, this allows us to calculate the total length L of the diverging section. Indeed, we said previously that at point $(z_b, 0)$ the final Mach number has to be reached, and beyond, the flow must be uniform with a constant velocity parallel to the z axis. Hence, the characteristic curve issued from that point must have an angle $\alpha = \arcsin(1/M_{e\infty})$ with the z axis and all points along this characteristic will have the same properties, meaning that the angle with the horizontal axis will be kept constant and equal to α. This characteristic curve is then a straight line defining the so-called *Mach cone* (see Fig. 2.8). Eventually, this line intercepts the nozzle wall at r_{exit}. The length separating point $(z_b, 0)$ from point (z_{exit}, r_{exit}) is then given by $r_{exit}/\tan\alpha$. The total length L of the divergent is then given by

$$L = br_{th} + r_{exit}\sqrt{M_{e\infty}^2 - 1} \qquad (2.56)$$

Along the preceding discussion, we said that the characteristic closest to the nozzle throat could be used as the starting point of the characteristic mesh propagation in a forward direction. However, it may also be convenient to use the characteristic limiting the Mach cone as the initial characteristic since properties along it are constant and known very simply. Propagation will then occur in a backward direction from the exit to the throat. Note, however, that the length of this Mach cone characteristic will be limited by r_{exit} which can only be determined once flow properties close to the throat are known (Sections 2.2.5 and 2.2.7).

2.2.8   *Summary*

The strategy to construct the inviscid contour of a Laval nozzle able
to generate a uniform supersonic flow at a given exit Mach number
$M_{e\infty}$ for a specified mass flow rate of a buffer gas can be summarized
in five steps as follows:

- Determine the flow properties close to the nozzle throat.
- Calculate the starting characteristic either close to the throat or
  at the Mach cone.
- Choose a Mach number evolution along the centerline compatible
  with the conditions reigning close to the throat on the one hand
  and downstream of the Mach cone on the other hand. Choose the
  b (typ. 10–40) and $\zeta\chi$ (typ. 0.01–0.1) parameters as well.
- Calculate the mass flow rate at the throat which must be kept
  constant along the expansion.
- Using the initial characteristic and the properties along the centerline, the characteristic grid can be propagated taking into account
  mass conservation in order to determine the limit of each characteristic, hence the isentropic contour (z, r_e).

The mechanism is illustrated in Fig. 2.9 when the initial characteristic is the one closest to the throat. The mesh of the grid is
chosen quite large in order to clearly illustrate the process of propagation. The initial characteristic C_{-0} intersects the axis at point
A_0 $(z_{C-0}, 0)$, simply quoted as 00 in Fig. 2.9. The operator must
choose a nonlinear step to define the horizontal grid of points at the
centerline for which the Mach number evolution has been imposed
(Section 2.2.6). Then, from points 00 and 10, point 11 can be obtained

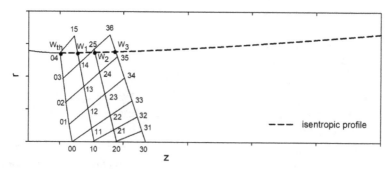

Fig. 2.9.   Design of the isentropic border from a characteristics network.

as explained in Section 2.2.4. From 01 and 11, one obtains 12 and so on until the second characteristic is completed. This is achieved when the mass flow rate along the characteristic exceeds the reference mass flow rate at the throat allowing bracketing of the true isentropic wall W1 which is then deduced by interpolation. The process is then propagated to other characteristics until abscissa b is reached (Mach cone).

2.3   The Boundary Layer

2.3.1   *General aspects*

The notion of boundary layer for flows with a high Reynolds number has been introduced in Chapter 1. Here, we will comment in more detail about this concept and explain how the boundary layer can be taken into account in the Laval nozzle design.

Beyond inertia and pressure forces mentioned in the preceding section, a real flow is also submitted to friction forces which arise between layers of the fluid as well as between the fluid and the walls of the chamber in which it is propagating or an obstacle it can encounter along its displacement. These are forces tangential to the motion which slow down the fluid leading to a reduction of its velocity until it cancels at the wall where the fluid adheres. In the approximation of a one-dimension stream, the non-dimensional stress tensor $\tau_{i,j}^*$, described in Section 1.5.3 of Chapter 1, reduces to one component. In dimensional coordinates, this component represents the friction force per surface unit τ, also called *shear stress*. It has been observed to be proportional to the flow velocity gradient normal to the wall:

$$\tau = \mu \frac{\partial v_z}{\partial r} \tag{2.57}$$

The proportionality factor μ is the so-called *dynamic viscosity coefficient* which reflects the momentum transport perpendicular to the flow axis. In practice, it has been shown that the effect of viscosity on the fluid only occurs within a thin space separating the wall from an outer flow where the friction forces are negligible compared to inertia and pressure forces. The space separating the wall from the outer flow was called *boundary layer* by L. Prandtl in 1904 [22]. The size of this layer is not precisely defined; however, conventionally it

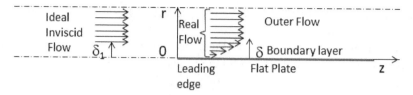

Fig. 2.10.   Boundary layer δ and displacement thickness δ_1 resulting from the presence of a flat plate.

is considered that the boundary layer ends at a distance δ (defined as the boundary layer thickness) from the wall where the longitudinal flow velocity $v_z(\delta)$ reaches 99% of the flow velocity in the outer region (v_{ze} hereafter). In the boundary layer area, as the velocity is slowed down, the temperature and density are also affected and change continuously from the outer flow values to the ones prevailing at the wall.

As a consequence of the velocity drop, the geometry of the ideal inviscid flow described in the previous section is modified and the dimension of the entire real flow needs to be increased in order to take into account the flow rate reduction within the border of the flow. This increase of diameter can be viewed as a displacement of the isentropic wall obtained in Section 2.2 and is usually referred to as the *displacement thickness* δ_1. The objective of this section is to explain how δ_1 can be determined so that the final geometry of a Laval nozzle can be obtained by addition of the isentropic radius and the displacement thickness for every longitudinal position z.

The displacement and boundary layer thicknesses are intimately linked by a simple formula that can be obtained considering the situation of a two-dimensional (rectangular) inviscid flow encountering a flat plate (see Fig. 2.10). Considering the leading edge of the plate as the origin of distances, if r is the length between the wall and the axes of the system by definition of the displacement thickness, the height of the ideal inviscid flow is $r - \delta_1$. Hence, mass conservation leads to

$$\int_{\delta_1}^{r} \rho v_z dr = \int_0^{\delta} \rho v_z dr + \int_{\delta}^{r} \rho v_z dr \qquad (2.58)$$

where the left member of the equation represents the mass flow rate per unit length transported by the ideal inviscid flow in the absence of the flat plate, while the right member of the equation represents the sum of the mass flow rates per unit length issued from the boundary layer on the one hand and from the inviscid outer flow on the other hand.

Within the ideal inviscid flow and the outer flow, the mass density ρ and the flow velocity v_z are constant by definition and equal to ρ_e and v_{ze}, respectively. Hence, the preceding equality becomes

$$\delta - \delta_1 = \int_0^\delta \frac{\rho v_z}{\rho_e v_{ze}} dr \text{ or equivalently } \delta_1 = \int_0^\delta \left(1 - \frac{\rho v_z}{\rho_e v_{ze}}\right) dr \quad (2.59)$$

These expressions immediately indicate that the boundary layer is always greater than the displacement thickness or in other words that the presence of the wall of the flat plate induces a smaller dimension of the outer flow compared to the ideal inviscid flow, illustrating the partial mass loss in the outer flow compared to the ideal one.

Calculating the flow characteristics within the boundary layer focused attention of a large community of researchers during the last century especially just after the Second World War. Dedicated textbooks have been written and updated even until recently [23–29] as well as a huge number of technical reports. The field remains still very active nowadays. According to the physical conditions at play, the boundary layer can be either laminar or turbulent and transition between these two regimes is also a matter of interest. They can develop for compressible as well as incompressible flows. The present discussion will be restricted to *laminar boundary layers in compressible axisymmetric or rectangular steady flows of a perfect gas with a specific heat capacity independent of temperature.*

2.3.2   *The Navier-Stokes equation within the boundary layer*

In order to characterize a flow within the boundary layer, one is led to resolve the Navier-Stokes equation within this area. In the present situation, however, simplification of the Navier-Stokes equation can arise from the assumption that the boundary layer thickness δ is very small compared to characteristic lengths of the problem such as the throat diameter, the length of the nozzle, or the curvature radius of its wall. Let us first consider a plane flow (Fig. 2.11) approaching at

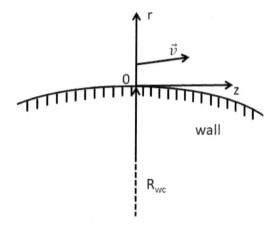

Fig. 2.11. Coordinate axis for a plane flow approaching a wall.

velocity $\vec{v} = (v_z, v_r)$ a wall having a large contour curvature R_{wc} with respect to δ(*a flat plate being the ultimate limit with* $R_{wc} = \infty$).

Ignoring body forces, the Navier-Stokes equations projected along the longitudinal axis z and the normal axis r of the wall are [23, 26]

$$\rho v_z \frac{\partial v_z}{\partial z} + \rho v_r \frac{\partial v_z}{\partial r} = -\frac{\partial P}{\partial z} + \frac{\partial \tau_{zr}}{\partial r} + \frac{\partial \tau_{zz}}{\partial z} \qquad (2.60)$$

$$\rho v_z \frac{\partial v_r}{\partial z} + \rho v_r \frac{\partial v_r}{\partial r} = -\frac{\partial P}{\partial r} + \frac{\partial \tau_{zr}}{\partial z} + \frac{\partial \tau_{rr}}{\partial r} \qquad (2.61)$$

with

$$\tau_{zr} = \mu \frac{\partial v_z}{\partial r} + \mu \frac{\partial v_r}{\partial z}$$

$$\tau_{zz} = \frac{2\mu}{3} \left(2 \frac{\partial v_z}{\partial z} - \frac{\partial v_r}{\partial r} \right)$$

$$\tau_{rr} = \frac{2\mu}{3} \left(2 \frac{\partial v_r}{\partial r} - \frac{\partial v_z}{\partial z} \right) \qquad (2.62)$$

Within the boundary layer, orders of magnitude considerations show that the shear stress derivatives $\partial \tau_{zr}/\partial z$ and $\partial \tau_{rr}/\partial r$ are comparable but remain very small, of the order of δ/z for the $\partial \tau_{zr}/\partial r$ value. Similarly, the $\partial \tau_{zz}/\partial z$ derivative is of the order of $(\delta/z)^2$ for the $\partial \tau_{zr}/\partial r$ value. Furthermore, let us remember that the radial component of the velocity v_r is very small with respect to the longitudinal one v_z. Hence, its longitudinal derivative is also negligible with respect to the radial derivative of v_z, and τ_{zr} reduces to τ defined earlier (Eq. (2.57)). In the radial projection of the momentum equation, the term $\rho v_r \frac{\partial v_r}{\partial r}$ can also be disregarded. However, the first term of the left side of the equation can be estimated as $\rho v_z \frac{\partial v_r}{\partial z} \sim \rho \frac{v_z^2}{R_{wc}}$ indicating that its contribution is directly dependent on the wall curvature. This means that the radial dependence of pressure is driven by the wall curvature. In the situation of our interest, however, R_{wc} is large enough so that this contribution can be eliminated as well. In the frame of these assumptions, the two momentum equations reduce to

$$\rho v_z \frac{\partial v_z}{\partial z} + \rho v_r \frac{\partial v_z}{\partial r} = -\frac{\partial P}{\partial z} + \frac{\partial}{\partial r}\left(\mu \frac{\partial v_z}{\partial r}\right) \qquad (2.63)$$

$$0 = -\frac{\partial P}{\partial r} \qquad (2.64)$$

The result of Eq. (2.64) is of great importance because it will allow determining the longitudinal pressure gradient within the boundary layer by applying Eq. (2.63) to the junction between the boundary layer and the isentropic flow. At this location, the radial derivative of v_z vanishes because the longitudinal velocity reaches the constant isentropic v_{ze} value. It comes immediately that

$$-\frac{\partial P}{\partial z}\bigg|_{r=\delta} = \rho_e v_{ze}\frac{\partial v_{ze}}{\partial z} = -\frac{\partial P}{\partial z}\bigg|_{allr} \qquad (2.65)$$

A corollary of the non-radial dependence of pressure is that for a perfect gas the mass density ρ varies as the inverse of temperature within the boundary layer.

As already specified, the preceding developments hold for a plane flow. However, within the boundary layer, Eqs. (2.63)–(2.65) are also valid for a flow with symmetry of revolution. Strictly speaking, the friction term in Eq. (2.63) should be $\frac{1}{r}\frac{\partial(r\tau)}{\partial r}$ for an axisymmetric flow [23]; nonetheless, within the boundary layer, r always stands very close to r_e, the inviscid wall radius coordinate obtained using the methodology explained in Section 2.2. Hence, r can be reasonably approximated to r_e and can be withdrawn from the derivation, leading to the expression of the friction term of Eq. (2.63).

As a conclusion, for a two-dimensional flow, either plane or axisymmetric, we are left with the following local momentum equation to solve:

$$\boxed{\rho v_z \frac{\partial v_z}{\partial z} + \rho v_r \frac{\partial v_z}{\partial r} = \rho_e v_{ze}\frac{\partial v_{ze}}{\partial z} + \frac{\partial \tau}{\partial r}} \qquad (2.66)$$

Let us remember that this final equation includes the radial independence of pressure (Eq. (2.64)) by using the longitudinal pressure dependence obtained in Eq. (2.65).

It is worth remembering at this point that at the wall ($r = 0$) the no-slip condition implies $v_z(0) = 0$. Further, we will consider a non-porous wall: $v_r(0) = 0$; in other words, no suction nor blowing is allowed. At the junction between the outer flow and the boundary layer ($r = \delta$), by definition, the longitudinal velocity v_z reaches the external constant value: $v_z(\delta) = v_{ze}$. Here, the shear stress is null $\tau(\delta) = 0$, while the wall shear stress $\tau(0)$ will be denoted τ_w hereafter:

$$\tau_w = -\int_0^\delta \frac{\partial \tau}{\partial r} dr \tag{2.67}$$

2.3.3   *Mass and energy conservation*

Mass conservation (Eq. (2.11)) is also useful for our purpose and within the boundary layer it can be expressed as

$$\boxed{\frac{\partial}{\partial z}(r_e \rho v_z) + \frac{\partial}{\partial r}(r_e \rho v_r) = 0} \tag{2.68}$$

a formulation that holds for planar flows ($r_e = 1$) or axisymmetric geometries (r_e is the inviscid contour). Here again, we take advantage of the possibility to approximate the radial coordinate r to the inviscid radius r_e everywhere within the boundary layer.

Within the boundary layer, energy conservation, shown in Chapter 1, becomes

$$\rho v_z \frac{\partial h_0}{\partial z} + \rho v_r \frac{\partial h_0}{\partial r} = \frac{\partial(v_z \tau - \mathcal{S})}{\partial r} \tag{2.69}$$

with

$$h_0 = h + \frac{v_z^2}{2} \quad \text{and} \quad \mathcal{S} = \lambda \frac{\partial T}{\partial r} \tag{2.70}$$

where λ is the thermal conductivity and h_0 is the stagnation enthalpy.

2.3.4   *The Von Karman equation*

Resolution of Eqs. (2.66) and (2.68) would allow one to obtain local characterization of the velocity everywhere inside the boundary layer.

This can be done exactly only for a few specific situations such as a flow along a thin flat plate [23, 25, 30]. Several numerical techniques can be applied as explained elsewhere [23, 24]. Nonetheless, in the present situation, it is not necessary to know the details of the velocity field everywhere inside the boundary layer; rather, we are interested in determining a "macroscopic" parameter, namely, the displacement thickness δ_1 in order to add it to the previously calculated inviscid flow. From its definition (Eq. (2.59)), the displacement thickness is not dependent on position r inside the boundary layer as it is a global value for every longitudinal position z. Hence, integration of Eq. (2.66) with respect to the radial coordinate r over the boundary layer will allow us to go one stage further in order to obtain δ_1.

Before this, it is worth outlining several important quantities that will appear in the following developments. In a similar way from which the displacement thickness was introduced earlier, a *momentum thickness* δ_2 can be defined as

$$\delta_2 = \int_0^\delta \frac{\rho v_z}{\rho_e v_{ze}} \left(1 - \frac{v_z}{v_{ze}}\right) dr \text{ with } \delta_1 + \delta_2 = \int_0^\delta \left(1 - \frac{\rho v_z^2}{\rho_e v_{ze}^2}\right) dr$$

$$(2.71)$$

This quantity characterizes the decrement of momentum within the boundary layer compared to the outer flow. From this, a *shape factor H* is set as

$$H = \frac{\delta_1}{\delta_2} \qquad (2.72)$$

Taking into account the continuity Eq. (2.68), the left-hand part of Eq. (2.66) can be rewritten. From this, Eq. (2.68) can be transformed as

$$\frac{1}{r_e}\left[\frac{\partial}{\partial z}\left(\rho v_z^2 r_e - \rho_e v_{ze}^2 r_e\right)\right] + \frac{v_{ze}}{r_e}\frac{\partial}{\partial z}\left(\rho_e v_{ze} r_e\right) + \frac{1}{r_e}\frac{\partial}{\partial r}\left(\rho v_r v_z r_e\right) = \frac{\partial \tau}{\partial r}$$

$$(2.73)$$

This expression can now be integrated over the boundary layer and the radial component v_r can be removed using the integration of the continuity equation (remember here that $v_r(0) = 0$). One obtains

$$\frac{1}{\rho_e v_{ze}^2 r_e} \int_0^\delta \frac{\partial}{\partial z} \left[\rho_e v_{ze}^2 r_e \left(1 - \frac{\rho v_z^2}{\rho_e v_{ze}^2} \right) \right] dr$$

$$- \frac{v_{ze}}{\rho_e v_{ze}^2 r_e} \int_0^\delta \frac{\partial}{\partial z} \left[\rho_e v_{ze} r_e \left(1 - \frac{\rho v_z}{\rho_e v_{ze}} \right) \right] dr = \frac{\tau_w}{\rho_e v_{ze}^2} \qquad (2.74)$$

Because the two derivative functions are null at the boundary limit position δ, the integral and the partial derivative signs can be inversed and the partial derivative becomes a total derivative [25]. Then, the displacement and momentum thicknesses appear naturally considering that ρ_e, v_{ze}, and r_e are independent of the radial coordinate. Finally, after some additional algebraic effort, we obtain the so-called *Von Karman equation* [24, 25, 31]:

$$\boxed{\frac{d\delta_2}{dz} + \delta_2 \left[\frac{H + 2}{v_{ze}} \frac{dv_{ze}}{dz} + \frac{1}{\rho_e} \frac{d\rho_e}{dz} + \frac{1}{r_e} \frac{dr_e}{dz} \right] = \frac{\tau_w}{\rho_e v_{ze}^2} = \frac{C_f}{2}} \qquad (2.75)$$

where C_f is defined as the local skin friction coefficient. It can be worth pointing out that the term involving the mass density ρ_e can be expressed as a function of v_{ze} and the corresponding Mach number M_e using Eq. (2.12) where the radial component of the velocity is ignored. The Von Karman equation has been obtained using dimensional parameters; however, some authors prefer writing it using dimensionless variables after more or less sophisticated transformations of the initial set of coordinates (r, z) [24, 25, 30].

There are several ways to deal with the Von Karman equation and it will be too long here to go into details for all techniques that have been applied [23, 32]. In the following, we will restrict the discussion to the method developed by R. Michel [31, 33, 34] and which was implemented in the early eighties by B.R. Rowe and J.B. Marquette in their pioneering work aimed at designing uniform supersonic flows suitable for the study of ion–molecule reactions [33]. The method can be applied to laminar or turbulent boundary layers for a two-dimensional flow with either a plane or axisymmetric geometry. Although it is an approximate method, Michel's integral technique has proven its great efficiency in designing a great deal of Laval nozzles within the last 40 years with a high level of aerodynamic quality (see examples in Section 2.4).

The theory of the flat plate allows one to obtain a relation between enthalpy h and velocity v_z by combining Eqs. (2.66) and (2.70) along a streamline at constant velocity [31, 36]. Note here that in the flat plate approximation, pressure is independent of the longitudinal position z and hence v_{ze} so that the first term of the right-hand side of Eq. (2.66) cancels. The enthalpy–velocity relation is given here without further demonstration:

$$\frac{h(\overline{v_z})}{h_e} = \frac{h_w}{h_e} + \left(1 - \frac{h_w}{h_e}\right)\frac{I(\overline{v_z})}{I_e} + \frac{v_{ze}^2}{h_e}\left[\frac{I(\overline{v_z})}{I_e}J_e - J(\overline{v_z})\right] \qquad (2.84)$$

where h_w is the enthalpy at the wall and I and J are defined as

$$I(\overline{v_z}) = \int_0^{\overline{v_z}} \mathcal{P}_r exp - \left\{\int_0^{\overline{v_2}} \frac{1-\mathcal{P}_r}{\overline{\tau}}\frac{d\overline{\tau}}{d\,\overline{v_1}}d\,\overline{v_1}\right\}d\,\overline{v_2} \qquad (2.85)$$

$$J(\overline{v_z}) = \int_0^{\overline{v_z}} G(\overline{v_2})\frac{dI(\overline{v_2})}{d\overline{v_2}}d\overline{v_2} \qquad (2.86)$$

with

$$G(\overline{v_2}) = \int_0^{\overline{v_2}} exp\left\{\int_0^{\overline{v_3}} \frac{1-\mathcal{P}_r}{\overline{\tau}}\frac{d\overline{\tau}}{d\overline{v_1}}d\overline{v_1}\right\}d\overline{v_3} \text{ and } \mathcal{P}_r = \frac{\mu c_p}{\lambda} \qquad (2.87)$$

$\mathcal{P}_r$ is the so-called Prandtl number and I_e and J_e are obtained setting $\overline{v_z} = 1$ (i.e. $v_z = v_{ze}$) in the preceding formulas for $I(\overline{v_z})$ and $J(\overline{v_z})$.

Provided that the friction coefficient and the Prandtl number dependences with velocity are known within the boundary layer, these integrals can be numerically determined. Obtaining Eq. (2.84) requires several additional important assumptions, however: the wall temperature must be constant, i.e. independent of position z; the expression is valid at a given position z only and extension to other positions requires considering that the Prandtl number and the friction coefficient are not dependent on the longitudinal coordinate.

Considering Eq. (2.84) and defining h_f, the recovery enthalpy is

$$h_f = h_e + 2J_e\frac{v_{ze}^2}{2} \qquad (2.88)$$

The displacement thickness takes its final form

$$\delta_1 = \frac{\mu_e v_{ze}}{\tau_w} g \left[\int_0^1 (1 - \overline{v_z}) \frac{d\overline{v_z}}{\overline{\tau}} + \frac{h_w - h_f}{h_e} \int_0^1 \left(1 - \frac{I(\overline{v_z})}{I_e} \right) \frac{d\overline{v_z}}{\overline{\tau}} \right.$$

$$\left. + \frac{v_{ze}^2}{h_e} \int_0^1 (J_e - J(\overline{v_z})) \frac{d\overline{v_z}}{\overline{\tau}} \right] \tag{2.89}$$

The momentum thickness δ_2 can also be expressed as a function of $\overline{v_z}$ and $\overline{\tau}$ by modifying Eq. (2.71) using the same methodology as for the transformed formulation of δ_1. The result is straightforward and gives

$$\delta_2 = \frac{\mu_e v_{ze}}{\tau_w} \int_0^1 \frac{\rho\mu}{\rho_e \mu_e} (1 - \overline{v_z}) \overline{v_z} \frac{d\overline{v_z}}{\overline{\tau}} = \frac{\mu_e v_{ze}}{\tau_w} g \int_0^1 (1 - \overline{v_z}) \overline{v_z} \frac{d\overline{v_z}}{\overline{\tau}} \tag{2.90}$$

Using these two new formulations for δ_1 and δ_2, the shape factor H defined in Eq. (2.72) can now be expressed as

$$H = \frac{\int_0^1 (1 - \overline{v_z}) \frac{d\overline{v_z}}{\overline{\tau}}}{\int_0^1 (1 - \overline{v_z}) \overline{v_z} \frac{d\overline{v_z}}{\overline{\tau}}} + \frac{h_w - h_f}{h_e} \frac{\int_0^1 \left(1 - \frac{I(\overline{v_z})}{I_e} \right) \frac{d\overline{v_z}}{\overline{\tau}}}{\int_0^1 (1 - \overline{v_z}) \overline{v_z} \frac{d\overline{v_z}}{\overline{\tau}}}$$

$$+ \frac{v_{ze}^2}{h_e} \frac{\int_0^1 (J_e - J(\overline{v_z})) \frac{d\overline{v_z}}{\overline{\tau}}}{\int_0^1 (1 - \overline{v_z}) \overline{v_z} \frac{d\overline{v_z}}{\overline{\tau}}} \tag{2.91}$$

The first term on the right-hand side of the preceding equation represents the δ_1/δ_2 ratio for an incompressible flow ($\rho = cte$ and $\mu = cte$). It is usually denoted as H_i. All integrals can be calculated provided that the friction law as a function of velocity within the boundary layer is known. Historically, the friction profile was calculated by Blasius in 1908 [37] for an incompressible flow along a flat plate. Then, Crocco [38] observed that in the case of a compressible flow, the enthalpy distribution $h(\overline{v_z})$ was hardly affected by the compressibility factor ($\rho\mu/\rho_e\mu_e$) so that any viscosity law could be used without significant changes. For an incompressible flow, the compressibility factor is clearly 1 (equivalent to a linear viscosity law

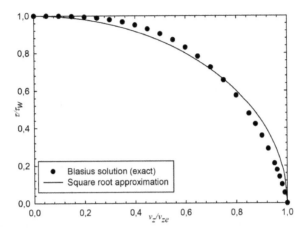

Fig. 2.12.   The numeric Blasius friction profile from [39] and an approximate analytical representation.

with temperature); the Blasius profile obtained in this situation can then also be used for compressible flows. The Blasius numeric profile is reproduced in Fig. 2.12 from data gathered by Van Driest [39]. The dimensionless friction coefficient $\bar{\tau}$ must respect three boundary conditions: at the wall, $r = 0, \nu_z = 0$, and $\tau = \tau_w$ hence $\bar{\tau}(0) = 1$; at the outer edge of the boundary layer, $r = \delta, \nu_z = \nu_{ze}$, and $\tau = 0$ hence $\bar{\tau}(\delta) = 0$; finally, the friction coefficient must be at its maximum at the surface of the flat plate $r = 0 : d\bar{\tau}/d\,\overline{v_z} = 0$. From this, it is possible to reasonably approximate the Blasius profile by the analytic formula $\bar{\tau} = \sqrt{1 - \bar{v}_z^2}$, which is also depicted in Fig. 2.12 for comparison with the exact Blasius profile.

The $I/\mathcal{P}_r$, $I_e/\mathcal{P}_r$, $J/\mathcal{P}_r$, and $J_e/\mathcal{P}_r$ functions are numerically given by Van Driest for constant Prandtl numbers varying from 0.5–2 so that the H factor can take the final form

$$H = H_i + \alpha M_e^2 + \beta \frac{T_w - T_f}{T_e} \tag{2.92}$$

with [33, 34]

$$H_i = 2.591 \tag{2.93}$$

$$\alpha = 2.053(\gamma - 1)\mathcal{P}_r^{0.638} \tag{2.94}$$

$$\beta = -0.711\mathcal{P}_r + 3.416 \tag{2.95}$$

comparison possible. For the present purpose, it was opted for the Monaghan expression based on literature recommendations [25, 31]:

$$T^* = T_e + 0.54(T_w - T_e) + 0.16(T_f - T_e) \qquad (2.99)$$

As the common buffer gases used in uniform supersonic expansions have Prandtl numbers that are quite similar (see Table 2.2), this approximation was found to be perfectly acceptable. Let us finally mention that the reference temperature T^* was interpreted by Monaghan as an average temperature within the boundary layer and its value was obtained by integrating the temperature profile as a function of $\bar{v}_z$.

The g factor can then be calculated provided that the temperature dependence of the viscosity is known. A Sutherland-type law is presently used:

$$\mu(T) = \mu_{ref} \left(\frac{T}{T_{ref}} \right)^n \frac{1 + T_1/T_{ref}}{1 + T_1/T} \qquad (2.100)$$

Since the mass density is inversely proportional to the temperature within the boundary layer, the g factor can be finally expressed as

$$g = \left(\frac{T^*}{T_e} \right)^n \frac{T_e + T_1}{T^* + T_1} \qquad (2.101)$$

Table 2.3 summarizes the various constants used for N_2, O_2, Ar, He, and H_2 and the g factor is calculated for a Mach number of 4

Table 2.3. Parameters for the viscosity law and value of the g factor for a Mach number of 4 and a wall at 295 K with the exception of H_2 for which $T_0 = T_w = 77$ K.

Gas	μ_{ref} $(10^{-6} Pa.s)$	$T_{ref}(K)$	n	$T_1(K)$	$T_e(K)$	$T_f(K)$	$T^*(K)$	g
N_2	5.33	80.6	0.50	114	70.2	262.3	22.3	0.975
O_2	6.10	80.4	0.66	53.2	70.2	259.6	221.9	0.959
Ar	21.16	273.0	0.93	0	46.6	248.4	213.0	0.899
He	18.87	273.0	0.647	0	46.6	248.4	213.0	0.585
H_2	2.65	59.75	0.575	25.8	12.2	64.8	55.6	1.118

and a wall temperature of 295 K (excepting H_2 for which $T_w = 77\,K$, see Section 2.5.5).

Since the Mach number is a function of the longitudinal coordinate z (via Eq. (2.47) at the centerline, for instance), it is important to remember here that T_e, T_f, T^*, and g are z dependent as well.

2.3.7 Solution of the Von Karman equation

From the expression of H obtained in Section 2.3.5 (Eqs. (2.92)–(2.95)), it is now possible to calculate the integral of the function $P(z')$ defined in Section 2.3.4 taking into account that the Mach number can be expressed as a function of $d\rho_e/\rho_e$ from Eq. (2.10) and the speed of sound definition (see Chapter 1). Hence, we have

$$P(z')dz' = 2\left[(H_i + 2)\frac{dv_{ze}}{v_{ze}} + (1 - \alpha)\frac{d\rho_e}{\rho_e} + \frac{dr_e}{r_e}\right] + 2\beta\frac{T_w - T_f}{T_e}\frac{dv_{ze}}{v_{ze}} \tag{2.102}$$

and

$$exp\int_{z_1}^{z} P(z')dz' = \frac{M(z)N(z)r_e^2(z)}{M(z_1)N(z_1)r_e^2(z_1)} \tag{2.103}$$

defining the $M(z)$ and $N(z)$ functions as

$$M(z) = \left[v_{ze}^{H_i+2}\,\rho_e^{1-\alpha}\right]^2 \tag{2.104}$$

and

$$N(z) = exp\left[2\beta\int_{z_1}^{z}\frac{T_w - T_f}{T_e}\frac{dv_{ze}}{v_{ze}}\right] \text{ with } N(z_1) = 1 \tag{2.105}$$

From this, the solution of Eq. (2.77) can be written as

$$\delta_2^2(z) = \delta_2^2(z_1)\frac{M(z_1)N(z_1)r_e^2(z_1)}{M(z)N(z)r_e^2(z)}$$

$$+ \frac{0.441}{M(z)N(z)r_e^2(z)}\int_{z_1}^{z}\frac{\mu_e g}{\rho_e v_{ze}}M(z'')N(z'')r_e^2(z'')dz'' \tag{2.106}$$

Once z_1 has been chosen, (usually the nozzle throat position z_{th}), the momentum thickness $\delta_2(z)$ can be obtained at every z position along the flow since the z dependence of all functions with subscript e

has been obtained by the determination of the inviscid flow characteristics. The wall temperature T_w is set by the operator; $T_f(z)$ is calculated from Eq. (2.96) and the compressibility factor g is obtained from Eqs. (2.99) and (2.101). The momentum thickness at the origin point z_1 is usually estimated by the user fixing first the displacement thickness $\delta_1(z_1)$ and then using the relation $\delta_2(z_1) = \delta_1(z_1)/H(z_1)$. The displacement thickness at the nozzle throat is usually a few percent of the inviscid radius r_{th}.

Eventually, the z dependence of the displacement thickness $\delta_1(z)$ is obtained from the definition of the shape factor H (Eq. (2.72)):

$$\delta_1(z) = \delta_2(z) \times H(z) \tag{2.107}$$

The final wall geometry of the Laval nozzle $r_w(z)$ is then obtained by summation of the inviscid radius $r_e(z)$ and the displacement thickness $\delta_1(z)$:

$$r_w(z) = r_e(z) + \delta_1(z) \tag{2.108}$$

The boundary layer $\delta(z)$ is not explicitly used in the wall calculation of a Laval nozzle. However its knowledge is of interest in order to evaluate the size of the free inviscid supersonic flow at the exit of the Laval nozzle (see Fig. 2.13). For this, the solution of Eq. (2.97) for the flat plate leads to the following expression for the ratio $\delta(z)/\delta_1(z)$ [31]:

$$\frac{\delta(z)}{\delta_1(z)} - 1 = \left[\frac{\delta(z)}{\delta_1(z)} - 1\right]_i \frac{H_i}{H(z)} \tag{2.109}$$

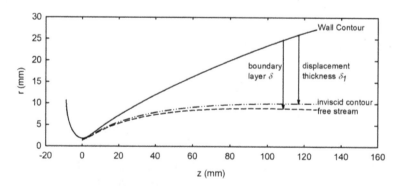

Fig. 2.13.   Wall contour for a Mach 8 helium Laval nozzle.

where the subscript i stands for *incompressible*. The ratio $\delta(z)/\delta_1(z)$ for the incompressible situation can be very well approximated using a polynomial of 6th degree for the velocity profile [25, 31, 46] with respect to the r/δ variable given by

$$\frac{v_z}{v_{ze}} = 2\left(\frac{r}{\delta}\right) - 5\left(\frac{r}{\delta}\right)^4 + 6\left(\frac{r}{\delta}\right)^5 - 2\left(\frac{r}{\delta}\right)^6 \qquad (2.110)$$

Using Eq. (2.59) with $\rho/\rho_e = 1$, it is then easy to find that the ratio $[\delta(z)/\delta_1(z)]_i$ is 3.5. Hence, the boundary layer within the Laval nozzle can be determined from

$$\frac{\delta(z)}{\delta_1(z)} = 1 + \frac{5}{2}\frac{H_i}{H(z)} \qquad (2.111)$$

This expression shows on the one hand that the boundary layer is always greater than the displacement thickness as commented earlier and on the other hand that the ratio $\delta(z)/\delta_1(z)$ is marked out. In effect, from Eq. (2.92) and Table 2.2, it can be shown that the shape factor H is of the order of the incompressible shape factor H_i when the Mach number approaches 1. Hence, $\delta(z)/\delta_1(z)$ is always lower than 7/2. Conversely, when the Mach number rises to a hypersonic situation, the shape factor H increases significantly and consequently the boundary layer tends to the displacement thickness. As a matter of example, for a Mach 5 helium nozzle, the ratio $\delta(z)/\delta_1(z)$ is about 1.2, whereas for a Mach 9, it is lowered down to 1.06.

Figure 2.13 shows an example of a Laval nozzle profile including the inviscid contour (Section 2.2), the displacement thickness contribution, the boundary layer contour, and the free stream.

2.3.8 *The Laval nozzle converging portion*

The design of the convergent section is quite empirical and based on a kind of rule of thumb. The major concern is to elaborate a smooth geometry that will not generate any detachment of the flow when crossing the nozzle throat. A quarter of a circle tangent to the nozzle throat usually matches this constraint very well. Its radius r_{conv} is related to the nozzle throat radius via the empirical relation

$$r_{conv} = 5(r_{th} + \delta_1(z_{th})) - \delta(z_{th}) \qquad (2.112)$$

The factor of 5 was chosen to match with the curvature at the throat $\hat{R}$, which in the present description respects the relation $\hat{R} = 5r_{th}$ (see Section 2.2.5). It is worth pointing out, however, that the factor 5 in Eq. (2.112) can be increased somewhat without any damage on the final quality of the uniform flow. Likewise, an arc of a circle shorter than a quarter can sometimes be preferred. These small adjustments with the usual rule can be worthwhile when space limitations are encountered within the experimental chamber.

2.3.9 *Summary*

In the present Section 2.3, we have learnt how to determine the real wall contour of a Laval nozzle taking into account the viscous nature of a gaseous flow. The main steps for this objective can be summed up as follows:

- Assume that the flat plate model can be applied within the boundary layer at every station z along the flow in the diverging section of the nozzle
- Resolve the Von Karman momentum-integral equation (Eq. (2.75)) with the momentum thickness δ_2 taken as the unknown variable and the skin friction coefficient C_f derived from the flat plate model. This requires the knowledge of the following:
  - The recovery temperature T_f (Eq. (2.96))
  - The wall temperature T_w (experimentally fixed)
  - The reference temperature T^* (Eq. (2.99))
  - The viscosity law for the buffer gas of interest (Eq. (2.100) and Table 2.3)
  - The compressibility factor g (Eq. (2.101))
  - The shape factor H (Eqs. (2.92)–(2.95))
  - The centerline Mach number longitudinal profile $M_e(z)$ within the isentropic core (Eqs. (2.47)–(2.50)), which by extension allows one to determine the $\nu_{ze}(z)$ and $\rho_e(z)$ functions
- Derive the displacement thickness δ_1 (Eq. (2.72))
- Derive the wall radius at each station z via summation of the inviscid contour $r_e(z)$ and the displacement thickness $\delta_1(z)$ (Eq. (2.108))

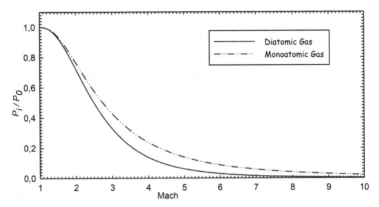

Fig. 2.15. Impact pressure to reservoir pressure ratio as a function of the Mach number M_e.

whereas for a diatomic gas we obtain

$$\frac{P_i}{P_0} = \left(\frac{6M_e^2}{M_e^2 + 5} \right)^{3.5} \left(\frac{6}{7M_e^2 - 1} \right)^{2.5} \qquad (2.119)$$

From this, the supersonic flow behind the Laval nozzle can be explored longitudinally provided that the Pitot tube can be moved along the flow axis if the Laval nozzle is at a fixed station or vice versa displacing the nozzle if the Pitot tube is immobile. Within the uniform part of the supersonic flow, time and space are equivalent since the flow velocity is constant. The typical hydrodynamic time corresponding to the optimal length of uniformity is usually found in the range 100–1000 μs. A perpendicular exploration of the flow can also be made provided that the Pitot tube is mounted on a two-dimensional moving platform. The impact pressure evolution along the flow leads to the determination of the Mach profile $M_e(z)$ or $M_e(z, r)$. A consequence of Eqs. (2.117)–(2.119) is the extreme sensitivity of the impact pressure P_i to the Mach number. Therefore a good uniformity of $P_i(z)$ is a very accurate check of the quality of the nozzle. Since inside the reservoir, temperature T_0 and pressure P_0 can be measured and mass density ρ_0 and molecular density n_0 can be deduced from the law of perfect gases, the supersonic flow characteristics T_e, P_e, ρ_e, and n_e can be calculated using Eqs. (2.2)–(2.4).

2.4.1.2   *Pulsed flow regime*

The previous discussion was implicitly considering a continuous isentropic expansion through a Laval nozzle. However, as commented in Chapter 1 and later on in this chapter, there are several advantages in pulsing a uniform supersonic flow. From now on, for simplicity of discussion, we will assume that the flow is pulsed at 10 Hz, that is, every 100 ms, a burst of gas is expanded through the Laval nozzle say for about 10 ms. This is a dramatic difference with respect to a continuous flow in terms of pressure measurements because it is now required to get the correct pressures within a few ms only, whereas in a continuous regime one could wait for several tens of seconds, if needed, before validating the measurement. Consequently, fast response pressure gauges are mandatory to accurately measure the impact pressure and this requires a reduction of the length of the Pitot tube to the minimum. In other words, to suppress all useless volumes that would induce very long times to be filled and stabilized, it is necessary to mount a mini pressure sensor at the mouth of the Pitot tube. An illustration of this kind of Pitot tube developed by the authors and B.R. Rowe can be viewed in Fig. 2.16.

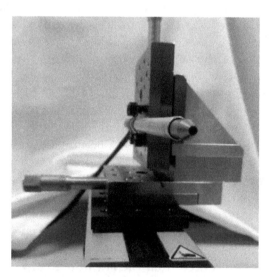

Fig. 2.16.   A fast response Pitot tube mounted on a remote control movable rail in use at the University of Castilla-La Mancha (UCLM, Ciudad Real, Spain).

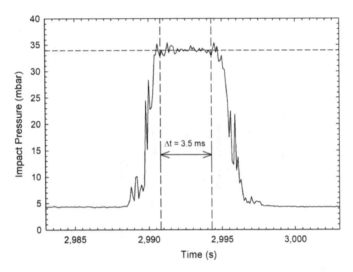

Fig. 2.17.   Structure of a typical impact pressure pulse registered from the fast response pressure sensor shown in Fig. 2.16. During the useful time Δt, the flow is stable and the average impact pressure is (33.924 ± 0.573) mbar over Δt, whereas the average impact pressure obtained over a dozen pulses is 34.148 mbar. The background pressure of about 4 mbar corresponds to the pressure in the chamber P_{ch} optimized to obtain the longest uniformity for these flow conditions (from an argon flow operating at the UCLM at 95.4 K; 10 Hz, 10 cm downstream of the nozzle exit).

The impact pressure measurement consists in the acquisition of a train of pulses, each one corresponding to a burst of expanded gas. Each pulse presents a plateau as illustrated in Fig. 2.17.

This plateau can last for several milliseconds during which the stream has the same characteristics as if it were continuous. This time is usually much longer than the aerodynamic time of the uniform flow (typically 0.1–1 ms) so that during one pulse a time-resolved experiment can probe the entire length of uniformity of the supersonic flow. Provided that the volume of the chamber is large enough, the chamber pressure remains constant all time. In such a situation, the Pitot measurement gives the chamber pressure P_{ch} in the absence of flow as a background pressure as shown in Fig. 2.17. A train of a few pulses can be seen in Fig. 2.18.

The impact pressure can be measured within the plateau of every pulse and then a mean value can be extracted from statistics, including a few tens of pulses. Once the impact pressure is obtained, the

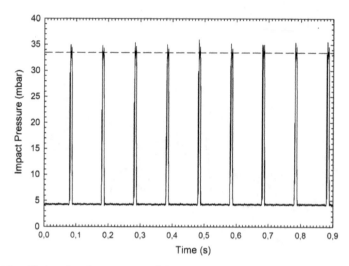

Fig. 2.18.   Train of pulses generated by a 10-Hz rotary disk chopping the flow. The dashed horizontal line represents the average impact pressure obtained over 12 individual pulses (from an argon flow operating at the UCLM at 95.4 K at the nozzle exit).

flow characteristics can be determined using Eqs. (2.117) and (2.2)–(2.4) like for the continuous regime. According to the method used to pulse the gas, it is important to note that the reservoir pressure could be time dependent and then a fast response pressure gauge could be necessary as well for an accurate measurement of P_0. More details about this will be given in Section 2.6. Finally, it is worth mentioning that the characteristics of an expansion through a given Laval geometry are identical for a continuous or pulsed stream within the experimental uncertainties.

2.4.1.3   *Temperature cartographies*

In practice, to characterize the supersonic flow issued from a newly made Laval nozzle, one begins with the theoretical conditions chosen for the nozzle geometry design. Experimentally speaking, these are the gas nature, its standard flow rate, and the residual pressure within the main chamber P_{ch}. This one is expected to be equal to the pressure within the isentropic core P_e since as explained in Chapter 1 the uniform supersonic flow is perfectly expanded. Alternatively to the standard flow rate, the reservoir pressure P_0 can be used as a reference input because it only depends on the nozzle throat diameter and the standard flow rate. Since mass flow controllers can sometime deviate from their factory calibration, P_0 results in a better indicator in order to adjust the theoretical conditions for the trial session.

Nonetheless, the theoretical conditions are not necessarily the best inputs to obtain the uniform flow with the lowest temperature fluctuation along the expansion since, as explained in Sections 2.2 and 2.3, the design of Laval nozzle contours is based on several approximations. Optimization of the flow characteristics can be carried out by adjusting the residual pressure P_{ch} to minimize aerodynamic oscillations. An example of such a situation is given in Fig. 2.19. In this case, a Laval nozzle was calculated for a nitrogen supersonic flow with $T_e = 140$ K and $P_e = 6.38$ mbar. A first Pitot exploration was then performed adjusting the chamber pressure P_{ch} to P_e (red curve). A uniform flow was obtained within 17 cm with a mean temperature along the flow of 137.1 K ($\sim$2% lower than the theoretical value) and fluctuations of 1.8 K estimated using the standard deviation of the temperature dataset obtained within the 17 cm length of uniformity. By reducing the chamber pressure down to 6 mbar (about 6% less),

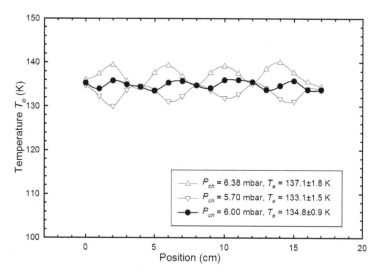

Fig. 2.19.   Temperature fluctuations along a nitrogen supersonic flow for different residual pressures P_{ch} set into the chamber for a flow with theoretical conditions: $P_e = 6.38$ mbar, $T_e = 140$ K.

the mean temperature was slightly reduced (134.8 K) and more interestingly the temperature fluctuation was reduced by a factor of two
(0.9 K, black curve). Still reducing the chamber pressure by about
the same amount, hence down to 5.7 mbar, leads to a colder jet
(133.1 K) with, however, worse temperature fluctuations (1.5 K, blue
curve). For these three examples, the resulting supersonic flows are
all acceptable since the temperature fluctuation remains small with
respect to the mean temperature. Nevertheless, the profile obtained
for a chamber pressure of 6 mbar is significantly better than the other
two. Reduction of the mean temperature with the chamber pressure
can be easily understood keeping in mind that for this series of tests,
the reservoir pressure has been maintained unchanged (same mass
flow rate). Hence, the ratio of pressures upstream and downstream
of the Laval nozzle is increased indicating that the Mach number is
also increased (Eq. (2.3)), causing a cooling of the flow (Eq. (2.2)).

As a matter of further illustration, some optimized temperature
profiles are given in Fig. 2.20 using different buffer gases and various
Laval nozzles. These profiles have been obtained for the centerline
position $r = 0$ such as those shown in the preceding graph. Figure 2.20
demonstrates large lengths of uniformity L_h for the three supersonic
conditions.

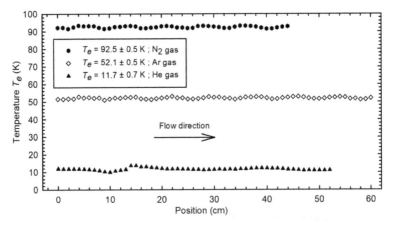

Fig. 2.20.   Temperature profile T_e at $r = 0$ for three pulsed supersonic expansions
issued from different Laval nozzles available at the UCLM (Spain). Position 0
corresponds to the Laval nozzle exit. The temperature fluctuation represents a
1σ standard deviation due to small aerodynamic cells that develop along the flow
progression.

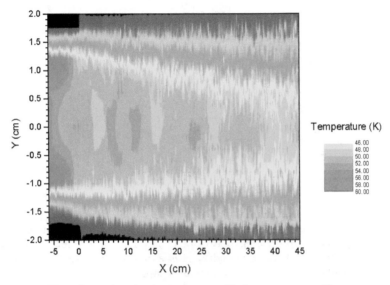

Fig. 2.21. Two-dimensional temperature profile for an argon uniform supersonic flow at 52 K obtained at the University of Rennes 1 (France).

Additionally, it is of interest to explore radially a section of the flow with the purpose of determining the diameter of the free isentropic stream and the boundary layer thickness. An example of a two-dimensional temperature profile is presented in Fig. 2.21 for a continuous expansion of argon delivering a uniform supersonic flow at say 52 K within about 40 cm. This plot clearly shows that the isentropic core (orange and red colors) is about 1 to 2 cm in diameter and this diminishes as the flow propagates downstream of the nozzle exit. This reduction results from the boundary layer growth that eventually totally destroys the supersonic core. A warning is required here with regard to the color scale: the boundary layer is depicted in green and blue colors which could make one think that the temperature is lower than in the isentropic core. This is obviously impossible and these colors are an artifact issued from the use of Eq. (2.117) to obtain the Mach number. Indeed, Eq. (2.117) assumes isentropicity behind the shock wave, a condition that no more holds when the Pitot tube starts to probe the boundary layer. However, the limit of the isentropic core remains correct in the presented figure.

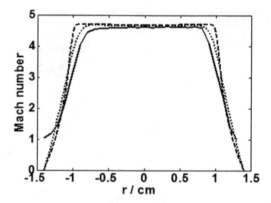

Fig. 2.22. Radial Mach number profile for a Mach 4.6 Laval nozzle (solid line) obtained at the Texas A&M University (USA). Reproduced with permission from Ref. [47] (by courtesy of Prof. S.W. North). The medium dashed line represents a CFD simulation for a laminar boundary layer situation, whereas the dotted line is a CFD simulation for a turbulent boundary layer.

Figure 2.22 shows an example of a Mach number radial profile clearly illustrating the very good radial uniformity of a nitrogen flow within the isentropic core.

2.4.2   *The rotational spectroscopy technique*

The Pitot tube technique is very economical, simple, and easy to set up especially for probing the aerodynamics of continuous supersonic flows. Yet, other methods can be applied in order to characterize supersonic flows. One of them is based on the induced fluorescence of a molecular radical seeded in the flow. This approach was used by those research groups using laser-induced fluorescence (LIF) to explore reaction kinetics within the supersonic expansion. Very briefly, a radical molecule is produced along the flow from the laser photolysis of a suitable precursor inseminated within the reservoir together with the main buffer gas. The radicals are cooled down and quickly thermalized to the buffer gas temperature in the uniform flow. Then, they can be electronically excited using a second laser and their fluorescence can be collected thanks to an optical device such as a photomultiplier (PMT). Complementary details can be found in Chapters 5 and 7.

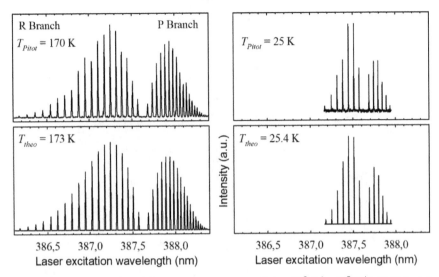

Fig. 2.23. Laser-induced fluorescence spectra of the $(B^2\Sigma^+-X^2\Sigma^+)$ (0,0) band of the CN radical from two different Laval nozzles compared to the corresponding synthetic spectra.

As mentioned in Chapter 1, one of the strengths of uniform supersonic expansions is that they are dense enough so that thermalization occurs rapidly between all the components of the stream. In particular, rotational relaxation processes are very fast and therefore the rotational temperature T_{rot} derived from many molecular species can be considered as representative of the translational temperature T_e in the flow. In other words, the rotational spectrum will present a Boltzmann distribution of relative intensities. Several radicals have been used to probe uniform supersonic flows including CN, OH, and NO. Among these, CN makes an ideal thermometer because the rotational level spacing of the $X^2\Sigma^+$ state is quite small, 3.78 cm^{-1} ($B_{CN} = 1.899$ cm^{-1}), equivalent to 5.4 K. Hence, small temperature differences can be tackled quite easily. Figure 2.23 shows some rotational spectra corresponding to the electronic transition $(B^2\Sigma^+-X^2\Sigma^+)$ (0,0) band taken for different CRESU flows at the University of Rennes 1 (France) as well as the corresponding synthetic spectra. An excellent agreement is found between the temperature obtained using the aerodynamic method T_{Pitot} and the one deduced from the synthetic spectra T_{theo} calculated by Sims *et al.* [48].

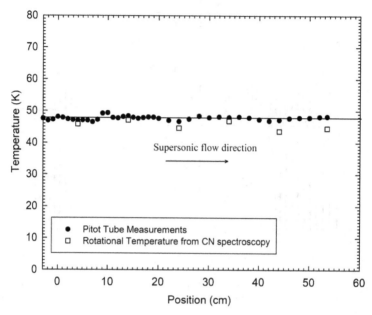

Fig. 2.24. Temperature along a nitrogen uniform supersonic flow calculated from impact pressure measurements (black circles, $(T = 47.7 \pm 0.6 \text{ K})$). The open squares represent estimates of the temperature $(T_{rot} = 45.4 \pm 1.4 \text{ K})$ from relative line intensities in the LIF spectrum of the CN $(B^2\Sigma^+ - X^2\Sigma^+)$ (0,0) band.

The preceding spectra were obtained at a given position within the two supersonic flows. Figure 2.24 shows how rotational temperatures obtained at different positions within a nitrogen flow at about 48 K compare to the complete temperature profile calculated from impact pressure measurements.

The same kind of optical measurements of the flow temperature can also be undertaken using the $A^2\Sigma^+ - X^2\Pi(1,0)$ electronic transition band of the OH radical. Nevertheless, since the rotational constant of the $X^2\Pi$ state is much higher than the one of CN ($B_{OH} = 18.9 \text{ cm}^{-1}$), small changes in temperature are more difficult to appreciate. An illustration of experimental spectra obtained at 11.7 K and 89 K at the University of Castilla-La Mancha (Spain), hereafter UCLM, can be viewed in Fig. 2.25. These are compared to simulated spectra calculated at the experimental temperature T_{Pitot} using the LIFBASE software [49].

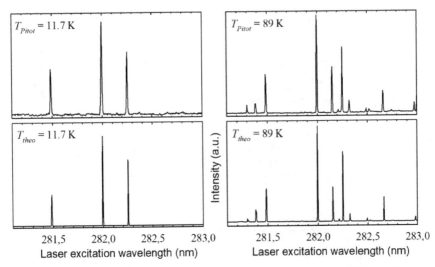

Fig. 2.25. Laser-induced fluorescence spectra of the $(A^2\Sigma^+ - X^2\Pi)$ $(1,0)$ band of the OH radical from two different Laval nozzles compared to the corresponding synthetic spectra (from UCLM, Spain).

The LIF technique was alternatively employed to determine the rotational temperature of nitric oxide NO [50]. The electronic transition $A^2\Sigma^+(\nu'=0 - X^2\Pi(\nu'=0$ at *ca.* 226 nm was considered for this purpose. Figure 2.26 shows a comparison of the translational temperature profile deduced from impact pressure measurements to the rotational temperature determined from a Boltzmann plot issued from a LIF spectrum as exemplified in the bottom panel. Complementary details can be found in [50].

Recently, an alternative method captured NO fluorescence images using an intensified charge coupled device (ICCD) camera instead of a PMT [47]. For this, a 226-nm laser beam was geometrically transformed into a very thin 20-mm-wide sheet using a lens system to probe a broad area of a nitrogen supersonic stream seeded with 1% of NO. This technique is usually known as Planar Laser-Induced Fluorescence (PLIF). Two-dimensional temperature maps can then be derived comparing images issued from the rotational transitions $R_1 + Q_{21}(1.5)$ and $R_1 + Q_{21}(8.5)$.

A good agreement was found between these measurements and aerodynamic temperatures determined with a Pitot tube. For instance, in the centerline upstream of a Mach 4.6 nozzle, the

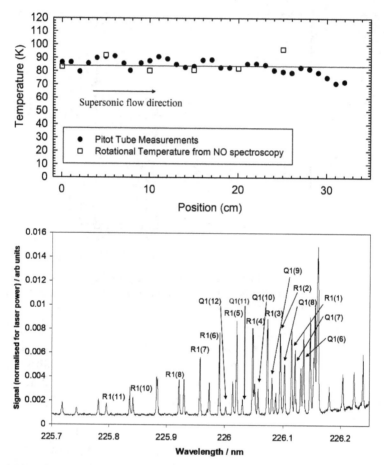

Fig. 2.26. Temperature profiles obtained for a Mach 3 Laval nozzle working with nitrogen either from (top panel) impact pressure measurements ($T = 84 \pm 5$ K) or LIF relative intensity lines of NO spectra ($T_{rot} = 86 \pm 3$ K) such as the one presented in the bottom panel (by courtesy of Prof. D.E. Heard).

aerodynamic temperature was found to be (55.9 ± 0.6) K, whereas the rotational temperature was (56.2 ± 1.2) K. More details can be found in [47].

2.4.3   Velocimetry diagnostics

The PLIF technique, mentioned in the preceding section, can also be coupled to the Molecular Tagging Velocimetry (MTV) approach to obtain a velocity mapping within the supersonic flow [47]. Here, the 226-nm laser sheet has to be directed through an array of micro-cylindrical lenses to form a comb of fluence-enhanced narrow beams before crossing the supersonic flow perpendicular to the expansion. By acquiring two time-delayed images of the NO fluorescence with the ICCD camera within the fluorescence lifetime, it is possible to determine the streamwise velocity by measuring the displacement of the fluorescence lines between the two images since the delay time can be perfectly controlled. This technique, however, requires a sufficiently long fluorescence lifetime. This is the case using NO as a probe of a nitrogen flow. For instance, the displacement of a flow moving at 750 m/s would represent a distance of 450 μm within 600 ns. The method leads to consistent results compared to velocities deduced from impact pressure measurements.

Although nitric oxide can be directly introduced in a chamber from a manufactured cylinder, it can also be produced by photodissociation of NO_2 at 355 nm [51]. This allows an extension of the method discussed in the previous paragraph. In this new configuration, a frequency-tripled Nd-YAG laser (355 nm) crosses the uniform supersonic flow perpendicularly after a geometrical transformation to a series of flat narrow beams as explained above. In this situation, NO is produced in the ground and first vibrational states within the periodic modulated pattern generated by the photolysis laser beam. Electronic transitions $A^2\Sigma^+(v'=1-X^2\Pi(v''=1))$ and $A^2\Sigma^+(v'=0-X^2\Pi(v''=0))$ can be used for its detection still at ca. 226 nm using a dye laser at 600–630 nm frequency mixed with the 355-nm residual beam from an Nd-YAG laser. This laser beam is geometrically modified through sheeting optics to a thin sheet which crosses the uniform flow within the same plane as the photolysis beam with an angle of 70 degrees from the stream direction. With such a design, velocity measurements can be carried out by first imaging the position of the NO lines in the absence of flow (static environment), which fixes the initial time and second by probing and imaging the uniform flow after a controlled delay time between the probe dye laser pulse and the photolysis pulse. Like in the previous paragraph, the displacement of lines comparing the two images gives the velocity. The advantage of this method with respect to the previous one is that the delay time is no more limited to the fluorescence lifetime of NO transitions. Larger displacements can then be observed, which increases the accuracy of the velocity measurement. As explained in [51], the method can permit determination of the NO rotational temperature as well. Eventually, simultaneous velocimetry and thermometry measurements can be undertaken through the VENOM (Vibrationally Excited Nitric Oxide Monitoring) technique [51, 52].

Other time-of-flight methods can be employed to determine the streamwise velocity. For instance, the supersonic flow can be locally ionized by means of an energetic pulsed electron gun perpendicular to the flow direction. The burst of created electrons is eventually collected with a movable Langmuir probe facing the supersonic stream and held at a constant voltage. The current is monitored as a function of time and for different positions of the Langmuir probe with respect to the electron beam trace through the flow, leading to a series of pulses shifted one to the other according to that distance. The method was successfully used by Mostefaoui *et al.* [53] who measured flow velocities in excellent agreement with those deduced from impact pressure measurements. More details of the technique can be found in [53].

2.5   Versatility of Laval Nozzle Contours

From Sections 2.2 and 2.3, one could be left with the impression that for every desired flow condition a specific nozzle contour must be designed. This is basically true, but once a nozzle has been constructed there are several means for using it in a variety of physical conditions making these tools very versatile. The aim of this section is to briefly overview the different methods that one can handle to extend the range of available physical conditions accessible from a given Laval geometry.

2.5.1   *The similarity method*

Let us imagine that we have designed a Laval nozzle from a series of conditions A: gas nature (i.e. γ_A number and molecular mass m_A), mass flow rate q_A, Mach number M_{eA} (with reservoir temperature T_{0A} and flow temperature T_{eA}), and flow pressure downstream of the nozzle exit P_{eA} (hence reservoir pressure P_{0A} easily be deduced from

Eq. (2.3)). As commented in Section 2.2, the isentropic contour is calculated for a given Mach number which means for a given temperature ratio T_0/T_e as expressed by Eq. (2.2). What would occur if we would like to change the reservoir temperature, for instance? Equation (2.2) says that the flow temperature T_e will be modified in the same proportion. Would that be the only modification in the flow conditions? What about the flow pressure P_e and the mass flow rate q? Since the Mach number is maintained constant, the pressure ratio upstream and downstream of the Laval nozzle will be kept constant through Eq. (2.3). However, can we choose any pressure value P_e and calculate the corresponding P_0 using Eq. (2.3)? This is what we will investigate right now. Let us consider a panel of conditions B: q_B, m_B, T_{0B}, T_{eB}, P_{eB}, and P_{0B} keeping in mind that $\gamma_A = \gamma_B$ and $M_{eA} = M_{eB}$. We will assume additionally that the wall temperature of the Laval nozzle is that of the reservoir $(T_0 = T_w)$ for both conditions A and B. First, we will express the ratio of mass flow rates q_B/q_A through the section A_z within the isentropic core in the divergent of the nozzle. More precisely, we will consider the exit station (subscript e). As usual, when we name ρ the mass density and v_z the flow velocity indexed with A or B accordingly, it comes immediately that

$$\frac{q_B}{q_A} = \frac{\rho_{eB} v_{zeB} A_{ze}}{\rho_{eA} v_{zeA} A_{ze}} = \frac{P_{eB}}{P_{eA}} \sqrt{\frac{m_B}{m_A}} \sqrt{\frac{T_{eA}}{T_{eB}}} = \frac{P_{0B}}{P_{0A}} \sqrt{\frac{m_B}{m_A}} \sqrt{\frac{T_{0A}}{T_{0B}}} \quad (2.120)$$

where we have expressed velocities v_z as a function of the sound velocity and used the law of ideal gases for the mass density ρ. Furthermore, Eq. (2.120) is valid for the flow pressure and the reservoir pressure because both conditions A and B respect Eq. (2.3) with $\gamma_A = \gamma_B$ and $M_{eA} = M_{eB}$.

From a practical point of view, it is often easier to use flow rates (Q) in standard conditions of pressure and temperature, 1 atm and 273 K, instead of mass flow rates q because Q is the value usually delivered by mass flow controllers. The ratio of mass flow rates q is simply coupled to the ratio of standard flow rates Q via

$$\frac{q_B}{q_A} = \frac{m_B Q_B}{m_A Q_A} \quad (2.121)$$

from which we can deduce the relation

$$\frac{Q_B}{Q_A} = \frac{P_{eB}}{P_{eA}} \sqrt{\frac{m_A}{m_B}} \sqrt{\frac{T_{eA}}{T_{eB}}} = \frac{P_{0B}}{P_{0A}} \sqrt{\frac{m_A}{m_B}} \sqrt{\frac{T_{0A}}{T_{0B}}} \qquad (2.122)$$

At this point, the new standard flow rate and flow pressure are simply related. Their ratio is known provided that the new flow temperature has been defined. Hence, this equation alone proposes infinity of solutions for the pair (Q_B, P_{eB}). In order to univocally determine this pair, another constraint can be now imposed considering that the boundary layer at any position within the divergent section of the nozzle must be kept constant for both conditions $(\delta_A = \delta_B)$. We have seen in Chapter 1 that the boundary layer δ is directly linked to the Reynolds number Re_z (see Eqs. (1.81) and (1.77) for the definition of Re_z). Writing the ratio of boundary layers for conditions B and A leads to

$$\frac{\delta_B}{\delta_A} = \frac{z_B}{z_A} \sqrt{\frac{\rho_A v_{zA} z_A \mu_B}{\rho_B v_{zB} z_B \mu_A}} \qquad (2.123)$$

Still reasoning at the nozzle exit (subscript e), hence at the same characteristic length for conditions A and B $(z_A = z_B)$, one obtains

$$\frac{\delta_{eB}}{\delta_{eA}} = \sqrt{\frac{\mu_{eB}}{\mu_{eA}} \frac{P_{eA}}{P_{eB}}} \sqrt{\frac{T_{eB}}{T_{eA}} \frac{m_A}{m_B}} \qquad (2.124)$$

Equaling boundary layers of conditions A and B as mentioned above leads to a simple relation for the new flow pressure P_{eB} which depends only on condition A as soon as the flow temperature T_{eB} has been fixed (hence μ_B as well):

$$\boxed{\frac{P_{eB}}{P_{eA}} = \frac{\mu_{eB}}{\mu_{eA}} \sqrt{\frac{T_{eB}}{T_{eA}} \frac{m_A}{m_B}}} \qquad (2.125)$$

Introducing this expression in Eqs. (2.120) and (2.122) gives the very simple relation for the mass and standard flow rates ratio:

$$\boxed{\frac{q_B}{q_A} = \frac{\mu_{eB}}{\mu_{eA}} \quad \text{and} \quad \frac{Q_B}{Q_A} = \frac{m_A \mu_{eB}}{m_B \mu_{eA}}} \qquad (2.126)$$

Let us then write how the ratio Q/P changes from condition A to condition B using Eq. (2.122). This ratio is essential because it corresponds to the required pumping capacity for ensuring a correct uniform expansion.

$$\boxed{\frac{Q_B}{P_{eB}} = \frac{Q_A}{P_{eA}} \sqrt{\frac{T_{eA}\, m_A}{T_{eB}\, m_B}}} \qquad (2.127)$$

Interestingly, this relation is independent of the viscosity law. Using the temperature dependence given in Eq. (2.100) for the viscosity $\mu_e(T_e)$, Eqs. (2.125) and (2.126) can be expressed as a function of temperature only. When the buffer gas remains <u>unchanged</u> ($m_A = m_B$), we can then propose the following formulas for the flow pressure and the mass flow rate:

$$\frac{P_{eB}}{P_{eA}} = \left(\frac{T_{eB}}{T_{eA}}\right)^{n+3/2} \frac{T_{eA} + T_1}{T_{eB} + T_1} \qquad (2.128)$$

$$\frac{q_B}{q_A} = \frac{Q_B}{Q_A} = \left(\frac{T_{eB}}{T_{eA}}\right)^{n+1} \frac{T_{eA} + T_1}{T_{eB} + T_1} \qquad (2.129)$$

where T_1 and n are given in Table 2.3 for O_2, N_2, Ar, He, and H_2. Several practical comments are worth mentioning at this point.

Let us keep the first mass unchanged. If $T_{eB} < T_{eA}$ (hence $T_{0B} < T_{0A}$), meaning that the reservoir is cooled down, Eq. (2.126) immediately shows that the mass flow rate in condition B will be smaller than in condition A. This mass flow reduction comes from the fact that the viscosity is lower when the temperature is decreased; in other words, one needs a smaller mass flow rate to compensate for friction effects. At first sight, cooling down the reservoir appears to be, economically speaking, more advantageous, consequently. Nonetheless, the flow pressure has to be diminished following Eq. (2.128), but in a stronger manner with respect to the mass flow rate because the temperature law has a higher power. Hence, the required pumping speed increases when one wants to cool down the reservoir of the apparatus. In practice, this is clearly a crucial point since one must first ensure that he can afford the necessary pumping capacities before designing a machine allowing temperature changes of the nozzle walls and reservoir! As a matter of example, let us assume that we wish to cool down the reservoir from room

Table 2.4. Flow conditions obtained by the method of similarities from a Laval nozzle initially designed to generate a supersonic helium flow at 36 K with a reservoir maintained at room temperature.

T_0(K)	T_e(K)	Mach M_e	P_0 (mbar)	P_e (mbar)	Q (l/min.atm = slm)	Q/P_e (m^3/h)
470	57.4	4.644	49.2	0.256	65	16800
295	36.0	4.646	28.8	0.150	48	21200
77	9.4	4.645	6.2	0.032	20	41400

temperature (say 295 K) to liquid nitrogen temperature ($T = 77$ K). The ratio of temperatures T_{eA}/T_{eB} at the exit of the nozzle will follow the same reduction. Hence, whatever the buffer gas is, the pumping speed must be augmented by a factor of about 2 (1.957 precisely). This is clearly significant especially if the initial condition A already requires a pumping speed of say 20,000 m^3/h (see Table 2.4 for an example)!

Conversely, if one wants to heat up the reservoir, this would result in an increase of the mass flow rate and the flow pressure opposite to a reduction of the pumping speed. Heating up the reservoir may appear somewhat orthogonal to the objective of the CRESU apparatus which was essentially designed to generate very cold gas reactors. However, this can be of great interest in a situation where one partner of a chemical process has a very low vapor pressure at room temperature or even is just about solid. A practical example of this situation was the studies carried out by Goulay *et al.* [54, 55] on the reaction of anthracene with the OH or CH radicals. Even though anthracene is solid at room temperature, it has been possible to study its gas phase reaction at temperatures down to 58 K! In Table 2.4, we have gathered a series of flow conditions for a helium nozzle initially designed for a reservoir left at room temperature. This was later used to generate the 58-K flow just mentioned above after heating the reservoir up to 470 K using oil as a fluid circulating inside the double jacket of the reservoir and nozzle.

Another interesting situation to be explored is the effect of changing the buffer gas (keeping obviously the same gamma). Let us maintain the reservoir temperature at the usual room temperature for the following discussion. The first interesting case is using molecular oxygen instead of nitrogen. Let us then imagine that we have designed a nitrogen Laval nozzle (condition A). Assuming that the conditions within the reservoir are kept constant, Eq. (2.122) leads to

$$\frac{Q_{O_2}}{Q_{N_2}} = \sqrt{\frac{m_{N_2}}{m_{O_2}}} = 0.935 \qquad (2.130)$$

Let us have a look now at Eq. (2.126) which is a consequence of conservation of the boundary layer. In the situation of oxygen and nitrogen, viscosity coefficients are quite close to each other and more importantly the ratio of viscosity μ_{O_2}/μ_{N_2} is almost constant below room temperature with a typical value of 1.13 ± 0.02 (Table 2.3). This leads to a ratio of standard flow rates of 0.99 quite close to the value calculated above. This small difference between the two calculations simply tells us that the boundary layer thickness is not strictly constant. However, it is only slightly affected when nitrogen is substituted by oxygen. This change can be calculated from Eq. (2.124) where we keep pressure and temperature constant. Using the good approximation for the viscosity ratio of 1.13, we find $\delta_{O_2}/\delta_{N_2} = 1.028$, which confirms the previous assertion. Hence, nitrogen and oxygen are perfectly equivalent provided that the oxygen flow rate is decreased by about 7% with respect to the nitrogen conditions. As P_0 and T_0 are not changed, the flow characteristics P_e and T_e are not changed either.

Can this sort of interchange be possible between two monoatomic gases such as helium and argon? There are two major differences here with respect to the previous situation. First, the mass ratio becomes much bigger and, second, the viscosity ratio μ_{He}/μ_{Ar} is temperature dependent. From 180–40 K, μ_{He}/μ_{Ar} increases from about 1 to about 1.5. From these two considerations, we can estimate a value of 2.2 for the boundary layer ratio δ_{He}/δ_{Ar} in the worse situation at 40 K. Even in a more favorable situation at 180 K, the boundary layer ratio would still be large (1.8), essentially because of the too high mass difference. Since the boundary layer of helium appears significantly larger than that of argon, this generates an increase of the nozzle section consequently. As a conclusion, it is not possible to use the same nozzle for argon and helium keeping the same flow conditions P_e, T_e. We will see in Section 2.5.4, however, that good flows can be obtained sometimes interchanging these two gases, modifying the pressure and temperature conditions as well as the standard flow rate.

2.5.2   *Slight contour distortions*

Beyond similarity methods that are essentially interesting for a given
gas, experience demonstrates that Laval nozzle contours designed for
one specific buffer could be used with other gases in differing standard
flow rate, temperature, and pressure conditions, however. Further,
solutions are not unique when changing from a gas A to a gas B
and no mathematical support is available to determine these new
conditions partly because of this multiplicity of solutions and also
because many parameters interplay in the construction of a divergent
contour. Then, we are usually left with choosing a set of reasonable
parameters (gas, Q, P_e, T_e, T_0) which will lead to the design of a
nozzle geometry similar to the original one.

The length of the divergent section is the easiest parameter to
adjust because it is user controlled in the computing through the
b parameter mentioned in Section 2.2.6. It is quite simple as well
to match the throat diameter because it is directly linked to the
standard flow rate and the reservoir pressure. The exit diameter
is somewhat trickier because its size depends on two contributions
as explained in Sections 2.2 and 2.3: the isentropic core radius
and the boundary layer thickness, the latter qualitatively increasing
with viscosity (and then temperature) and decreasing when pressure
increases (see Eq. (1.81)). Nevertheless, a series of successive tries
of the set of parameters mentioned above allows one to obtain these
three mandatory dimensions: length, throat, and exit diameters.

This may not be sufficient, however, because a large number of
contours having different curvatures can join the throat to the exit.
Therefore, a satisfactory set of parameters for the extremities may
not be convenient within the divergent section itself. Fortunately, as
discussed by Canosa *et al.* [56], experience shows that some flexibil-
ity is acceptable concerning the curvature shape within the divergent section, a difference of a few hundred of microns being usually
tolerable. Compared to the usual diameter of the divergent section
(typically a few centimeters at the control station), this alteration
remains very small. Hence, a series of more or less similar contours
can be obtained for different sets of parameters and these new conditions can then be tested experimentally via Pitot characterizations
for instance. Here, the mass flow rate plays a crucial role. As a matter
of illustration, a series of temperature profiles obtained for various

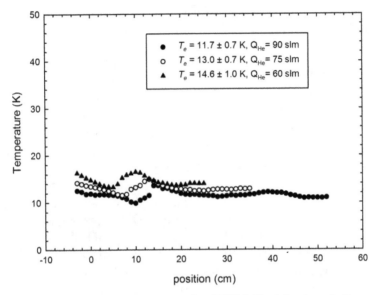

Fig. 2.27.   Temperature profiles obtained at UCLM (Spain) using a helium nozzle with different standard flow rates.

helium flow conditions introduced through a nozzle, initially designed to generate a helium jet at 15 K, is shown in Fig. 2.27.

Using this nozzle, the standard flow rate was changed from 60 slm (the original value for the contour calculation) up to 90 slm. Standard flow rates are given here for a continuous regime, although this nozzle was characterized with the pulsed CRESU available at UCLM (Spain). Since the optimal flow pressure P_e was found to be very similar for these three conditions (0.11–0.12 mbar), the improvement in quality required a pumping speed increase. Note that for other tested nozzles, it could be necessary to increase P_e more significantly when the standard flow rate was augmented. Figure 2.27 represents a good example of the significant role of the mass flow rate on the length of uniformity of the flow as discussed in Chapter 1 (Eq. (1.83)). An increase of the standard flow rate by 50% increases the length of uniformity by a factor of about two in the present situation. Increasing the standard flow rate is, however, limited because it simultaneously increases the Reynolds number. Thus, at one point, the flow will transit from the laminar to the turbulent regime and the length of uniformity will first stagnate and eventually shorten. Such flow rate changes have also been successfully tested for nitrogen or argon nozzles and additional examples can be found in [56] and [57, Fig. S2].

2.5.3 *Gas mixtures*

Testing a nozzle with a different gas than the one for which it has been designed offers interesting opportunities. A special situation is when the two gases have different heat capacity ratios. In this case, one of the most remarkable results of flow characterizations is the significant difference between the optimal flow temperatures obtained for one gas compared to the other one. Many times indeed, the temperature change is found to reach a factor close to 2. This immediately gives the chance to test binary mixtures A and B, the proportion of which is varied from the two extremities 0–100% to 100–0%. This makes Laval nozzle temperature-tunable flow generators and considerably enhances their versatility. Combinations including a monoatomic gas (helium or argon) and a diatomic one (nitrogen) were found to give very good flows in perfect agreement with the set of input parameters determined to mimic the available Laval contour [56]. For this, several considerations must be taken into account during the contour design.

First, the heat capacity of a mixture of a monoatomic and a diatomic gas depends on the proportion of the mixture and then the heat capacity ratio γ_{mix} changes from 5/3 for a monoatomic gas to 7/5 for a diatomic one. For a given mixture of volume proportion x_A of a diatomic gas and $x_B = 1 - x_A$ of a monoatomic gas, γ_{mix} can be simply expressed as

$$\gamma_{mix} = 1 + \frac{2}{3 + 2\,x_A} \qquad (2.131)$$

Second, the viscosity of a mixture cannot be obtained by simply combining linearly the individual weighted viscosities. This is due to interactions prevailing between the partners of the mixture. A simple approach is detailed in [58] in order to calculate the viscosity of a gas mixture based on the knowledge of the transport integral $\Omega^{(2,2)}$ for viscosity. For a unique gas A, a good approximation of the viscosity $\mu_A(T)$ (g cm^{-1}s^{-1} units) is given as

$$\mu_A(T) = 266.93 \times 10^{-7} \frac{\sqrt{M_A T}}{\sigma_A^2 \Omega_A^{(2,2)}(T_A^*)} \qquad (2.132)$$

where M_A is the molar mass (g mol^{-1} units), σ_A is the Lennard-Jones radius (Å units), T_A^* is a reduced temperature expressed as $k_B T / \varepsilon_A$ (with ε_A / k_B being the Lennard-Jones energy expressed in Kelvin), and $\Omega_A^{(2,2)}(T_A^*)$ is the tabulated transport integral, a function of T_A^* (see Annex 1). An additional viscosity term $\mu_{AB}(T)$ must be taken into account because of interactions between the two partners A and B:

$$\mu_{AB}(T) = 266.93 \times 10^{-7} \frac{\sqrt{M_{AB}T}}{\sigma_{AB}^2 \Omega_{AB}^{(2,2)}(T_{AB}^*)} \text{ with } M_{AB} = \frac{2M_A M_B}{M_A + M_B}$$

(2.133)

where T_{AB}^* is the interaction reduced temperature expressed as $k_B T / \varepsilon_{AB}$. The Lennard-Jones interaction force constants σ_{AB} and ε_{AB} are derived from the individual force constants as

$$\sigma_{AB} = \frac{\sigma_A + \sigma_B}{2} \text{ and } \varepsilon_{AB} = \sqrt{\varepsilon_A \varepsilon_B}$$

(2.134)

and $\Omega_{AB}^{(2,2)}(T_{AB}^*)$ is again the tabulated transport integral for viscosity depending on T_{AB}^*. Values for the Lennard-Jones coefficients are given for the usual gases in Table 2.5.

The viscosity of the mixture $\mu_{mix}(T)$ is then given by the following formula [58]:

$$\mu_{mix}(T) = \frac{1+Z}{X+Y}$$

(2.135)

with

$$X = \frac{x_A^2}{\mu_A(T)} + \frac{2x_A x_B}{\mu_{AB}(T)} + \frac{x_B^2}{\mu_B(T)}$$

(2.136)

$$Y = \frac{3}{5}A_{AB}^* \left\{ \frac{x_A^2}{\mu_A(T)}\left(\frac{M_A}{M_B}\right) + \frac{2x_A x_B}{\mu_{AB}(T)}\frac{(M_A + M_B)^2}{4M_A M_B}\frac{\mu_{AB}^2(T)}{\mu_A(T)\mu_B(T)} \right.$$

$$\left. + \frac{x_B^2}{\mu_B(T)}\left(\frac{M_B}{M_A}\right) \right\}$$

(2.137)

Table 2.5.   Lennard-Jones parameters for usual carrier gases in CRESU reactors from [59] (He) [58], (H₂), and [60] (Ar, N₂, O₂).

	Ar	**He**	**N₂**	**O₂**	**H₂**
$\sigma(\text{Å})$	3.35	2.556	3.656	3.428	2.959
$\varepsilon/k_B(\text{K})$	143.2	10.22	98.94	118.5	36.5

$$
Z = \frac{3}{5} A^*_{AB} \left\{ x_A^2 \left(\frac{M_A}{M_B} \right) \right.
$$

$$
+ 2 x_A x_B \left[\frac{(M_A + M_B)^2}{4 M_A M_B} \left(\frac{\mu_{AB}(T)}{\mu_A(T)} + \frac{\mu_{AB}(T)}{\mu_B(T)} \right) - 1 \right]
$$

$$
\left. + x_B^2 \left(\frac{M_B}{M_A} \right) \right\} \tag{2.138}
$$

where A^*_{AB} is another tabulated parameter function of T^*_{AB} (see Annex 1).

Finally, the Prandtl number $\mathcal{P}_{rmix}$ of a binary mixture is correctly approximated by [58]

$$
\mathcal{P}_{rmix} = c_{p,mix} \left(\frac{x_A^2}{\alpha_A} + \frac{2 x_A x_B}{\sqrt{\alpha_A \alpha_B}} + \frac{x_B^2}{\alpha_B} \right) \text{ with } \alpha_{A,B} = \frac{c_{pA,B}}{\mathcal{P}_{rA,B}} \tag{2.139}
$$

where $\mathcal{P}_{rA}$ and $\mathcal{P}_{rB}$ are the Prandtl numbers of the individual species (see Table 2.2) and $c_{pA,B,mix}$ the specific heat capacities at constant pressure for species A, B, or the mixture, respectively.

Taking this into account, Laval nozzle contours can be calculated for any mixture of the common carrier gases in CRESU reactors: He-N₂; He-O₂; Ar-N₂; and Ar-O₂.

An extensive evaluation of the method was performed recently by Canosa *et al.* [56] and demonstrated the very good agreement between the input set of parameters used for the nozzle design and the physical flow conditions deduced from Pitot measurements. An example of flow temperature profiles is given in Fig. 2.28 corresponding to a nozzle initially designed to produce a uniform supersonic flow in argon at 50 K and later characterized for a series of argon–nitrogen mixtures. The accessible temperature ranges in between 52

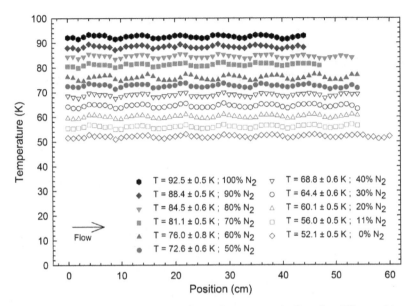

Fig. 2.28. Temperature evolution along the supersonic flow for different binary mixtures Ar–N$_2$ expanding through a nozzle initially designed for a pure argon stream expanding at 50 K. Reproduced with permission from Ref. [56], Copyright Springer-Verlag Berlin Heidelberg 2016.

and 92 K in this specific case. Finally, it is important to remember again that in the flow optimization process for each gas mixture, the flow pressure P_e must be set at the value corresponding to the nozzle geometry. Changing the gas mixture proportions delivers a series of uniform streams, each of them characterized by a unique combination (T_e, P_e). Such binary mixtures involving nitrogen are presently routinely used at UCLM [57, 61–65]. A combination of argon and oxygen was also used earlier by the Rennes group in order to evaluate the stability of HO$_3$ during the OH + O$_2$ + M association reaction [66–68].

To conclude with gas mixtures, it is worth mentioning that some nozzles were also characterized using mixtures of nitrogen and SF$_6$ [69]. Since SF$_6$ has a heat capacity ratio of 4/3, a mixture of N$_2$/SF$_6$ allows one to reduce γ_{mix} and then obtain higher temperatures than if nitrogen is introduced alone. A nozzle with molar fractions 40% N$_2$ and 60% SF$_6$ was validated at the University of Bordeaux (France), leading to a flow temperature $T_e = 241$ K [69, 70].

2.5.4   *Interchanging argon and helium*

In Section 2.5.1, we have seen that interchanging argon and helium maintaining the flow conditions constant was not feasible. Releasing this constraint makes it possible and, for some nozzles, argon could be used instead of helium and vice versa for different standard flow rates, temperatures, and pressures. Since the two gases are monoatomic, a mandatory constraint can be found by expressing the standard flow rate passing through the throat of the nozzle for both gases. By writing the obvious equality of sections at the throat, one can obtain the following guiding relation between standard flow rates, flow pressures, and flow temperatures of argon and helium:

$$\frac{Q_{He}}{P_{eHe}} = \sqrt{\frac{m_{Ar}}{m_{He}}} \left(\frac{T_{eAr}}{T_{eHe}}\right)^{\gamma/\gamma-1} \frac{Q_{Ar}}{P_{eAr}} = \sqrt{10} \left(\frac{T_{eAr}}{T_{eHe}}\right)^{5/2} \frac{Q_{Ar}}{P_{eAr}} \quad (2.140)$$

As commented above, since the ratio of viscosity μ_{He}/μ_{Ar} is temperature dependent and the mass ratio is significant, the boundary layers will differ between the two gases and it is not possible to define a second relation as for the similarity method. Nevertheless, Eq. (2.140) can bring interesting guidelines when searching for new convenient conditions. Let us suppose that an argon nozzle is available and characterized. Provided that the temperature ratio is known, Eq. (2.140) gives a direct simple relation between the necessary pumping speed for helium and the one known for argon. It also says that if one wants to keep the same temperature for the supersonic flow $T_{eAr} = T_{eHe}$ the pumping speed must be increased by a factor of 3.16. Can this be affordable with your CRESU apparatus? If this appears not possible, one is left to increase the helium flow temperature until it matches the capacity of the installation. Thus, Eq. (2.140) can simply give the minimum temperature that the operator can attain when changing argon to helium. Once the helium flow target temperature has been chosen, one has to find the combination Q_{He}, P_{eHe} complying with Eq. (2.140) and reproducing closely the existing nozzle contour. If needed, a temperature readjustment can be considered and a new search for Q_{He}, P_{eHe} can be carried out.

As a matter of illustration, an argon nozzle, the contour of which was calculated in order to deliver a uniform supersonic flow at 50 K, 1.3 mbar for a 60 slm standard flow rate, has been tested using helium as the carrier gas. The nozzle calculation with helium succeeded in replicating the existing contour with a huge standard flow rate of

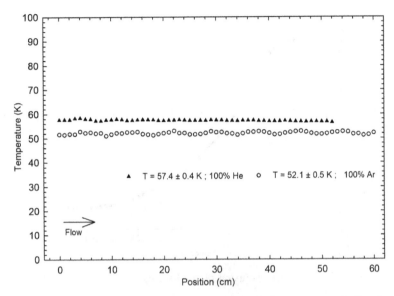

Fig. 2.29. Temperature profile for an argon and a helium supersonic flow issued from a Laval nozzle originally calculated for an argon expansion at 50 K (UCLM, Spain).

300 slm expanding at $T_e = 56$ K and $P_e = 2.73$ mbar, complying with Eq. (2.140). Although this set of parameters looks prohibitive in a continuous regime, it was possible to handle it in the pulsed CRESU reactor available at UCLM (Spain). The obtained temperature profile is compared with the one characterized for argon in Fig. 2.29.

2.5.5 *The peculiar case of H_2*

The most usual carrier gases in CRESU reactors are nitrogen, argon, and helium. Due to its ubiquity in the interstellar medium and more specifically in regions dense enough to be shielded from the ultraviolet light issued from neighboring stars, dihydrogen H_2 appears like a key player on the interstellar chessboard. Consequently, generating uniform supersonic flows of H_2 should be of great interest. Nevertheless, H_2 has a major drawback since its heat capacity ratio γ_{H2} is temperature dependent in the usual temperature range of CRESU reactors. This results from its high rotational constant (87.6 K). Further, internal relaxation within the ortho (or para) manifold of rotational states is relatively slow. As mentioned earlier in Sections 2.2

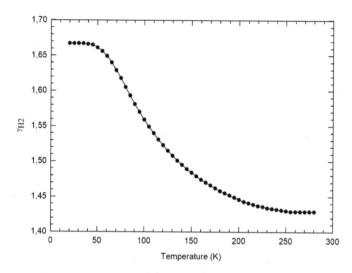

Fig. 2.30.   Temperature evolution of the heat capacity ratio for normal H_2.

and 2.3, the Laval nozzle design model assumes a temperature-independent γ along the expansion from the throat to the nozzle exit. In addition, no energy transfer kinetic equations are considered within the expansion, an aspect which would be hard to numerically handle.

The temperature dependence of γ_{H2} for *normal* H_2 is plotted in Fig. 2.30. Interestingly, γ_{H2} approaches the monoatomic value of 5/3 at low temperatures, especially below the liquid nitrogen temperature (77 K). At this temperature, essentially all of the H_2 is in the two lowest rotational states $j = 0$ and $j = 1$. Since relaxation of the ortho/para ratio can only occur through a nuclear spin exchange which is extremely slow in the gas phase, the ratio of 3:1 prevailing at 300 K is preserved at 77 K.

From this discussion, it is then conceivable to cool the reservoir and the nozzle down to 77 K, provided that they are both equipped with a double-walled envelope within which liquid nitrogen can be introduced. Doing this, H_2 can be considered like a monoatomic gas of mass 2. As seen in Fig. 2.30, $\gamma_{H2}(77$ K$)$ remains lower than 5/3, but this approximation was validated a posteriori by the characterization of specific molecular hydrogen nozzles which lead to excellent uniform flows. Figure 2.31 shows a series of three temperature profiles obtained for three specially designed pre-cooled nozzles using H_2

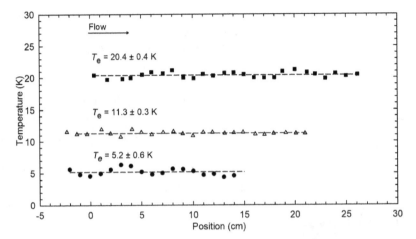

Fig. 2.31. Temperature profiles for three different hydrogen supersonic flows issued from three specifically designed Laval nozzles pre-cooled with liquid nitrogen (Rennes, France).

as the carrier gas at 20, 10, and 5 K. Agreement with the input flow temperature is quite satisfactory. Two out of these three nozzles were used more particularly in the kinetic study of the reaction $F + H_2 \rightarrow HF + H$ [71]. Further, if a pre-cooled nozzle contoured for helium expansions is available, substitution of He by H_2 is also likely, taking advantage of the considerations explained in Section 2.5.4 for Ar and He. It is interesting to note that the hydrogen viscosity coefficient is significantly lower than the one of helium: $\mu_{H2}/\mu_{He} = 0.25$ at 20 K and 0.1 at 5 K. This means that a reduced flow rate of H_2 with respect to helium can be introduced in order to compensate friction forces. For supersonic flows having the same temperature, Eq. (2.140) applied to He and H_2 tells us that the required pumping speed will be about 40% greater for hydrogen expansions.

2.6 The CRESU Machines

2.6.1 *Overview of the CRESU technique*

The CRESU technique has been widely described in the literature (see for example [72–76]. This technique combines a time (or space)-resolved kinetic method with the uniform supersonic expansion of a gas mixture through a Laval nozzle from a high-pressure zone P_0 to a very-low-pressure region P_{Ch} in a vacuum chamber. Therefore, in a simplistic manner, a CRESU machine is composed by the following:

- A **pre-expansion chamber or reservoir**: this chamber can have an external jacket to cool down the gas before the expansion in order to get lower temperatures. The reservoir is usually placed inside a vacuum chamber.
- A **Laval nozzle**: especially designed to achieve the desired jet temperature as explained in Sections 2.2 and 2.3 of this chapter.
- An **expansion (or vacuum) chamber**, where the cold uniform supersonic jet propagates after exiting from the nozzle.

For the characteristics of the kinetic experiments, all CRESU systems have in common that the reservoir (or the whole chamber, depending on the design) is mounted on a translation system to vary the downstream distance from the nozzle to the detection region along the gas propagation axis, i.e. allowing a change in the available timescale for performing a kinetic experiment.

The two operational (continuous and pulsed) modes of the CRESU apparatuses will be described in Section 2.6.3. In summary, in the continuous CRESU machines, the gas is constantly flowing and expanding through the Laval nozzle, getting a continuous uniform cold jet (as schematically shown in Fig. 2.13). In contrast, in the pulsed mode, the gas mixture is pulsed by using various methods such as *solenoid valves* or *stacked piezoelectric valves* or a *rotary disk*.

In addition to the main components to get ultralow temperatures, a CRESU setup needs specific instrumentation to generate and monitor the desired reacting species. For example, in the gas phase kinetic studies on reactions of either charged species (see Chapters 3 and 4) or radicals with molecules (see Chapters 5, 7 and 8), laser techniques, ionizing sources or electron beams are needed. The CRESU technique has also been applied to nucleation and energy transfer processes (Chapters 6 and 7). More recently, it has been coupled to powerful detection techniques to probe reaction products and to determine branching ratios (see Chapter 8).

2.6.2 *History of the CRESU development*

In Table 2.6, a summary of the past and present CRESU machines around the world is listed in chronological order of their development or construction.

The first *continuous CRESU machine* (*CC1*) was developed in the 1980s by B.R. Rowe and co-workers to study ion–molecule reactions in the *Laboratoire d'Aérothermique* of the *Centre National de la Recherche Scientifique* (CNRS) at Meudon (France) [77]. But, in the early nineties, Rowe moved to the University of Rennes 1 and

Table 2.6. CRESU machines in chronological order of development/construction.

Code[a]	First Publication	Institution	Ref.
CC1	1984	CNRS–Aérothermique de Meudon (France)	[77]
CC2a	1994	*CNRS–University of Rennes 1* (France)	[48]
PC1	1995	University of Arizona (Tucson, USA)	[85]
CC2b	1995	*CNRS–University of Rennes 1* (France)	[79, 80]
CC3	1997	University of Birmingham (United Kingdom)[b]	[78]
PC2	2000	University of Colorado (Boulder, USA)[c]	[86]
PC3	2004	University of Göttingen (Germany)	[87]
CC4	2005	*University of Bordeaux* (France)	[81]
PC4	2008	*University of Leeds* (United Kingdom)	[50]
PC5a	2009	CNRS–University of Rennes 1 (France)	[88, 89]
CC2c	2010	*CNRS–University of Rennes 1* (France)	[83]
PC6	2012	*Texas A&M University* (College Station, USA)	[47]
PC7a	2014	*Wayne State University* (Detroit, USA)[d]	[90]
PC8	2015	*University of Castilla-La Mancha* (Ciudad Real, Spain)	[91]
PC9	2015	*ETH Zurich* (Switzerland)	[92]
PC7b	2019	*University of Missouri* (Columbia, USA)	[93]
CC2d	2020	*CNRS–University of Rennes 1* (France)	[84]
PC5b	2020	*CNRS–University of Rennes 1* (France)	[94]

Note: [a]*CC*: Continuous CRESU; *PC*: Pulsed CRESU; [b]moved to the University of Rennes 1 in 2003, [c]now at the University of Berkeley (USA); [d]now at the University of Missouri (USA). Institutions indicated in italics are still active.

built a new continuous CRESU apparatus for the express purpose of studying the kinetics of neutral–neutral reactions (*CC2a*) in 1992. The project issued from a nascent collaboration between the Rennes team and I.W.M. Smith's group at the University of Birmingham (UK). As a result of this fruitful collaboration, a third continuous CRESU apparatus (*CC3*), replicating *CC2a*, was developed in 1994 in Birmingham by I.W.M. Smith and I.R. Sims to study neutral–neutral reactions and energy transfer processes [78]. The Rennes-Birmingham collaboration demonstrated that many neutral–neutral reactions remain rapid at temperature conditions under which they had been assumed to be irrelevant. In the mid-nineties, a new CRESU machine (*CC2b*) was constructed in Rennes and dedicated to the study of ion–molecule reactions and electron attachment [79, 80] since the activity in Meudon was definitely stopped in the early nineties. At present, this is the only apparatus worldwide designed for investigating kinetics of charged species. In the early 2000s a miniaturized continuous CRESU apparatus (*CC4*) was built at the University of Bordeaux. Due to the small size of this system compared to previous ones, the minimum temperature achievable was initially 77 K in Ar [81], and later it was brought down to 50 K [82]. In parallel, Rowe's group developed a third *CC2* apparatus (*CC2c*) adapted for the study of the kinetics of dimerization of condensable species such as pyrene ($C_{16}H_{10}$). In this case, the chamber was coupled to a time-of-flight mass spectrometer and the *reservoir* was heated by silicone oil in order to increase the vapor pressure of solid pyrene [83]. Currently, a new CRESU apparatus has been developed in Rennes (*CC2d*) under the CRESUSOL project (see Chapter 8) for performing kinetics and branching ratio studies of collisional processes at very low temperatures (down to 20 K). This transportable apparatus, to which a time-of flight mass spectrometer is attached, is routinely moved and coupled to the VUV Desir line at the SOLEIL synchrotron employed as a source of photoionization at threshold for the detection of reaction products [84]. This CRESU reactor is presently operating in a continuous mode, but a pulsed device is planned to be implemented in the future. Currently, there are six continuous CRESU apparatuses available in Rennes, including the one initially developed and built at Birmingham (*CC3*) and moved to Rennes in 2003 when I.R. Sims joined the Breton team and the CRESUCHIRP (*PC5b*) apparatus mentioned below. In addition, another chamber is also available for infrared spectroscopy studies within a uniform supersonic flow (see Chapter 9). Besides the Rennes group, the only CRESU machine working constantly in a continuous mode is the one established in Bordeaux (*CC4*).

In continuous CRESUs, large pumping capacities (typically, 5000 L/s) are required to achieve *temperatures lower than 50 K* and long timescales for performing a kinetic study (see Section 2.6.3 for details). This requirement, which is only accomplished by the various *CC2* apparatuses in Rennes, makes the CRESU technique rather expensive and voluminous. For that reason, the development of **pulsed CRESU apparatuses**, where the gas is pulsed by mechanical systems, allowed a significant reduction of the apparatus size (since a lower pumping speed is needed) and costs (i.e. gas consumption).

Historically, the first pulsed CRESU (*PC1*), developed in the early nineties by M.A. Smith and collaborators, employed a pulsed valve for the gas injection in the pre-expansion chamber [85]. The affordability of this pulsed technology promoted the appearance of new pulsed CRESU machines in academic laboratories, such as at the University of Boulder (USA) in work conducted by S. R. Leone's group (*PC2*) [86], at the University of Göttingen (Germany) by B. Abel's group (*PC3*) [87], at the University of Leeds (UK) by D.E. Heard's group (*PC4*) [50], and more recently at the Texas A&M University (USA) by S.W. North's group (*PC6*) [47] and at the ETH Zurich (Switzerland) by R. Signorell's group (*PC9*) [92]. The *PC2-PC4*, *PC6*, and *PC9* apparatuses are all based on the pioneering pulsed version implemented by M.A. Smith and co-workers (*PC1*) that uses electromechanical valves, and thus, they present similarities in the design. In 2003, the *PC3* apparatus was moved to the University of Berkeley (USA).

In 2009, S. Morales and B.R. Rowe patented a prototype pulsing method (*PC5a)* in Rennes [88]. In this prototype, the gas was pulsed by means of an *aerodynamic chopper* (or rotary disk). A few years later, the pulsing method was improved in new CRESU devices, one built at Wayne State University (USA) (*PC7a*) and another one at the UCLM (Spain) (*PC8*) using different technical approaches. The improvement in *PC7a*, developed by A.G. Suits' group [90], was achieved using a stacked piezoelectric valve instead of a solenoid valve to pulse the gas (see Section 2.6.3.2). This system, called CPUF for chirped pulse in uniform supersonic flow, combines chirped pulse Fourier Transform millimeter wave detection with the CRESU flow. It was recently moved to the University of Missouri where a second reactor (**PC7b**) coupled to a Cavity Ring Down Spectrometer was newly constructed [93]. The Spanish pulsed CRESU **PC8** [91] constituted an upgrade of the patented prototype **PC5a.** It was developed by E. Jiménez's group in 2011 as a result of a narrow collaboration with A. Canosa and B.R. Rowe. This latter apparatus is nowadays the most powerful and versatile pulsed CRESU machine worldwide in terms of accessible supersonic flow temperature T_e. Further, it produces excellent supersonic flows such as those examples shown in Section 2.5. The improvements are described in Section 2.6.3.2. A third version of the rotary disk technique is actually being developed by the Rennes group as part of a new CRESU apparatus (**PC5b**) aimed at measuring product branching ratios (see Chapter 8) using the chirped pulse millimeter wave detection method. Among those pulsed CRESU machines, only **PC5b** and **PC8** can work in the continuous mode as well. Most of these **PC** devices are dedicated to the kinetics of neutral–neutral processes either reactive or inelastic.

2.6.3   *Specifications*

2.6.3.1   *Continuous gas expansion*

In a continuous CRESU apparatus, if a P_{Ch} of 10^{-2} Torr is required, huge pumping capacities are needed to satisfy the necessity of sufficiently small boundary layers with respect to the isentropic kernel. For that reason, the first CRESU apparatus (**CC1**) was developed by B.R. Rowe in the Laboratoire d'Aérothermique at Meudon, where a large rarefied atmosphere wind tunnel SR3 was already available (see Fig. 1.7). This facility offered a pumping capacity up to 144,000 m^3/h for $P_{Ch}{\sim}10^{-2}$ Torr. These very low pressures were needed to avoid clustering problems in ion–molecule reactions investigations. However, the whole pumping capacity was not used and the volume flow rate ranged from 13,500 to 109,000 m^3/h. After B.R. Rowe moved to the University of Rennes 1, a new CRESU apparatus was built for studying the kinetics of neutral–neutral reactions (**CC2**). This version was designed to use a somewhat higher operating pressure

Table 2.7.   Characteristics of the continuous apparatuses used worldwide.

CRESU	V_{res} (L)	Buffer Gas	T_0 (K)	T_{min} (K)	Working P_{Ch} (mbar)	Pumping Speed (m^3/h)	Ref.
CC1	50	Ar;He;O$_2$;N$_2$	77–295	8	0.007–0.2	144,000	[77]
CC2a	22.5	He;N$_2$;Ar	77–295	13	0.1–6	22,500	[48]
CC2b	22.5	He;N$_2$;Ar	295	23	0.1–6	22,500	[80]
CC2c	10	He;N$_2$;Ar	295–470	14	0.3–6	14,000	[55]
CC2d	2.4	N$_2$;Ar	295	52	0.3–1.1	2,200	[84]
CC3*	22.5	Ar;He;O$_2$;N$_2$;H$_2$	77–295	6	0.1–6	33,100	[78]
CC4	0.1	Ar;N$_2$;N$_2$/SF$_6$	295	50	1.5–2.1	1,400	[82]

Note: *CC3 is at the University of Rennes 1 since 2003.

(see Table 2.7) than that at Meudon to reduce the cost of the experiment, since it is linked to the pumping speed, which is inversely proportional to the square of the pressure. Excepting the *CC4* at Bordeaux, the rest of the continuous CRESUs have a large reservoir volume V_{res} of typically 10–20 liters according to the apparatus. In Table 2.7, a comparison of continuous CRESU machines is presented.

2.6.3.2   *Pulsed gas expansion*

As stated above, the main differences between the pulsed CRESU systems are related to the way the gas is pulsed, which determined the size of the pre-expansion chamber or reservoir. Table 2.8 summarizes some technical data of all pulsed CRESUs.

In CRESU apparatuses based on *solenoid and stacked piezoelectric valves*, the gas pulsation is accomplished by means of an electromagnetic valve. *Solenoid valves* are electromechanical devices in which an electric current is used to generate a magnetic field that exerts a force on a ferromagnetic plunger. In rest position, the gas flow through a small orifice (usually 1–3 mm) can be closed off by the plunger. A *stacked piezo actuator* is a layer of piezo ceramic elements that expand or contract when an electrical charge is applied, generating linear movement and force on a plunger. The schematics of typical configurations of the electromagnetic valves are shown in Fig. 2.32. The gas is pulsed before entering the reservoir. *PC1-PC4*, *PC6*, and *PC9* apparatuses employ small *reservoir* volumes (~1 cm^3). In order for the gas to pass through the nozzle almost instantaneously, the valves must be capable of rapid loading and may be held open

Table 2.8.   Comparison among pulsed CRESU systems. The lowest temperature (T_{min}) achieved is also presented, when T_0 is *ca.* 300 K.

CRESU	T_{min}(K)	V_{res}(cm^3)	Pulsed System	Pumping Speed(m^3/h)	Ref.
PC1	92	1	Solenoid valve	170	[85, 95]
PC2	70	1	Solenoid valve	216	[86]
PC3	58	<1	Solenoid valve	400	[87]
PC4	31	0.78	Solenoid valve	210	[50, 96]
PC5a	21.8	22,500	Rotary disk	22500	[88, 89]
PC6	34	2.25	Solenoid valve	252	[97]
PC7a	22	15	Piezoelectric stack actuator	3960	[90]
PC7b	68	20	Piezoelectric stack actuator	8600	[93]
PC8	11.7	12,000	Rotary disk	3800	[57, 91]
PC9	29	0.9	Solenoid valve	1900	[92, 98]

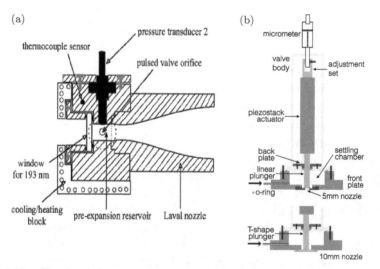

Fig. 2.32.   Sketch of the gas pulsing system based on (a) commercial solenoid valves. Reproduced with permission from Ref. [86], Copyright AIP Publishing 2000. (b) Piezoelectric stack valves. Reproduced with permission from Ref. [90], Copyright AIP Publishing 2014.

for comparatively long times (>10 ms) while a steady state builds throughout the setup.

However, as explained by Jiménez *et al.* [91], the pulsed valves have a small diameter and therefore high Reynolds numbers which are susceptible to generate turbulences since the *fill-and-empty* cycles do not allow obtaining a gas at rest inside the mini reservoir. In *PC7* setups, the volume reservoir is a bit larger, but still small ($\sim$15 cm^3). The reservoir is repetitively filled and rapidly loaded upstream from the Laval nozzle. This procedure must be quick enough to achieve the gas throughput requirements needed in large nozzles at a high stagnation pressure. This is the main limitation of the solenoid valves, which was solved by using high-force piezoelectric stack actuators by A. G. Suits' group [90]. These actuators can open much faster and allow larger flow rates to penetrate into the reservoir (up to 250 slm [93]. Oldham *et al.* [90] reported that the reservoir was initially overpressured with respect to the nominal working conditions to ensure a satisfactory uniformity of the flow, getting hydrodynamic times of $200-500$ μs for helium and argon, respectively. Further, as the reservoir volume V_{res} in *PC7a,b* is significantly larger than that used in *PC* reactors using solenoid valves, this ensures a negligible mean velocity of the gas inside the reservoir and limits the effects of turbulence.

In the *PC5a* and *PC8* versions, the large reservoir volume ($\sim$10–20 L) is continuously filled since the transmission of gas through the Laval nozzle is pulsed by the use of a stainless steel rotary disk or chopper sealed in the divergent part of the nozzle (Fig. 2.33). Opposite to what occurs with the electromagnetic valve system for which the reservoir is constantly filled and emptied, here the reservoir pressure is kept constant. During the periodic expansion, P_0 fluctuates by about 0.2% with respect to the optimal value due to the large volume employed. Hence, in the reservoir, the expanding gas is maintained at rest and at a constant pressure P_0 equal to that of the continuous regime during the short time the disk is open. This significantly improves the quality of the flow and consequently increases its length of uniformity. The rotary disk is 2 mm thick and 240 mm in diameter and possesses one or two symmetrical apertures. When the disk rotates, its

(a)

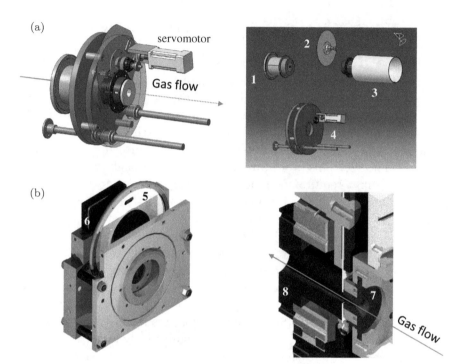

Fig. 2.33.   Sketch of the gas pulsing system based on a rotary disk. (a) **1** convergent part of the Laval nozzle; **2** disk with an aperture and transmission; **3** divergent part of the Laval nozzle; **4** mounting and servomotor. Adapted from Ref. [89] and (b) **5** disk with two symmetrical apertures; **6** optocoupler for timing synchronization; sectional view of the convergent **7** and divergent **8** part of the nozzle. Adapted from Ref. [91].

orifice periodically coincides with the nozzle section, allowing the passage of the fluid or/and any light source from upstream to downstream. When the aperture is out of the flow axis, the disk acts as a shutter or a closed valve. The movement is transmitted to the disk by a servomotor either attached to the mounting in *PC5a* (Fig. 2.33(a)) or placed outside of the vacuum chamber in *PC8*. Although setting a servomotor under vacuum shortens transmission of the motion, it must be cooled down with water, which is unnecessary when it is set outside of the chamber. Further, in *PC8*, the disk is sandwiched in between two stainless steel rings which rigidify it considerably avoiding buckling, which had to be considered in *PC5a*.

Compared with a continuous CRESU, the design and the small pumping capacities of several pulsed CRESU machines restrict the lowest achievable temperature, T_{min}. These are listed in Table 2.8 for a reservoir left at room temperature. With the exception of the first pulsed CRESU apparatus, **PC1**, all others work with a reservoir maintained at about 300 K. M.A. Smith's device was designed such that the Laval nozzles could be cooled down to 250 K and warmed up to 400 K [95]. The lowest temperature, 53 ± 4 K, reached by the **PC1** machine was then achieved by pre-cooling the reservoir at 263 K [99, 100]. This record for an electromechanical pulsed system lasted for about 10 years until the TAMU team used a uniform supersonic flow at 34 K for the investigation of the collisional quenching of $NO(A^2\Sigma^+, v' = 0)$ by $NO(X^2\Pi)$ and O_2 [97]. More recently, the Zurich group lowered this mark down to 29 K, designing a Mach 5 argon nozzle employed during the characterization of CO_2 clusters formed in their flows [98].

For all the **PC** machines based on pulsed solenoid valves, the usual length of uniformity is typically around 10 cm and buffer gases are essentially nitrogen and argon. Consequently, the hydrodynamic time is confined in the range of $100-200$ μs. To achieve lower temperatures, however, argon and nitrogen are usually not suitable because of their propensity to dimerize or form aggregates with the reactants seeded in the flow. Nevertheless, if the flow pressure is sufficiently low, this can be significantly limited, and nitrogen or oxygen pre-cooled flows were obtained in the early CRESU apparatus **CC1** at 20 K and a flow density of 3×10^{15} cm^{-3} [77, 101]. This density is about a factor of 10–30 lower than the typical ones found in the pulsed CRESU systems and required a prohibitive 73,000 m^3/h pumping speed in a continuous regime. Even for pulsed CRESU machines, this remains a considerable challenge. Helium is then mandatory for generating pulsed supersonic flows at the lowest temperatures of particular interest for the interstellar medium. Further, as helium is light, the velocity of the jet is consequently fast and a 10-cm-long uniform flow will only result in a 60 μs available time for kinetic experiments. In practice, this reduces the kinetic field to the fastest neutral–neutral processes or ion–molecule reactions. Hence, longer uniform flows are required at the lowest temperatures to extend the domain of reaction kinetics to *not so fast* reactions: $\sim 10^{-11}$ cm^{3}molecular.$^{-1}$s^{-1}. Obtaining long useful lengths requires designing nozzles for large helium flow rates of typically ~ 100 slm (in the continuous regime).

The **PC7a** apparatus was able to achieve lower temperatures, down to 22 K using helium, because the associated pumping capacities are significantly higher than in most of the traditional **PC** arrangements and the reservoir volume is also larger, thus reducing the undesired effects of turbulence. A 10 cm uniform helium flow was obtained with this configuration [90]. Such a temperature was also attained by the rotary disk technique (**PC5a** and later **PC8**) in a first step and it was recently lowered down to (11.7 ± 0.7) K by the **PC8** apparatus in the frame of the study of the reactivity of OH radicals with methanol [57]. The temperature profile for this helium flow has already been displayed in Fig. 2.27, showing a length of uniformity of 50 cm resulting from the large helium flow rate introduced through the Laval nozzle. This temperature of 11.7 K represents the lowest temperature ever achieved by a pulsed CRESU apparatus as well as the lowest temperature ever reached for any CRESU either pulsed or continuous when the reservoir is maintained at room temperature. The present absolute published record, 5.8 ± 0.8 K, has been obtained for a continuous helium flow, pre-cooled at liquid nitrogen temperature, at the University of Rennes 1 during the study of the S(^{1}D) + H$_2$ reaction [102], although a slightly cooler flow was characterized in molecular hydrogen (Fig. 2.31). In terms of gas consumption, **PC8** employs standard flow rates which are 10–20 times smaller than in the corresponding continuous regime, whereas the piezoelectric stack valve used in **PC7a,b** allows a reduction of about two orders of magnitude [93].

2.6.3.3   *CRESU gallery*

A series of viewgraphs of the principal CRESU reactors active nowadays is presented here including two chambers available at the University of Rennes 1 and the CRESU machines from the Universities of Bordeaux, Leeds, Castilla-La Mancha, Zurich, Missouri, and TAMU.

Fig. 2.34. The CRESU chamber dedicated to the study of ion–molecule reactions at the University of Rennes 1 (13–300 K).

Fig. 2.35. The CRESU chamber dedicated to the study of neutral–neutral reactions and energy transfer processes originally built at the University of Birmingham and moved to the University of Rennes 1 in 2003 (6–300 K).

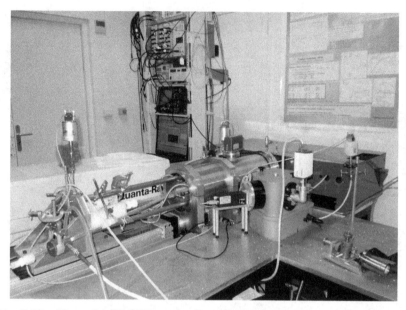

Fig. 2.36.   The mini CRESU at the University of Bordeaux dedicated to the study of neutral–neutral reactions (50–300 K).

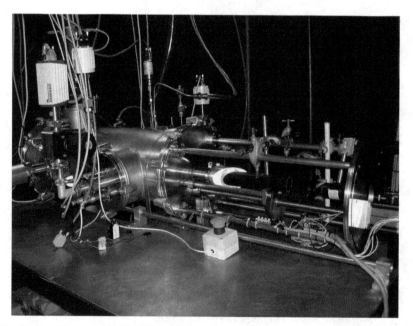

Fig. 2.37.   The pulsed CRESU at the University of Leeds dedicated to the study of neutral–neutral reactions (31–148 K).

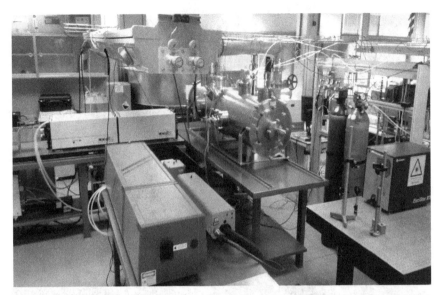

Fig. 2.38. The pulsed CRESU at the University of Castilla-La Mancha in Ciudad Real dedicated to the study of neutral–neutral reactions (12–178 K).

Fig. 2.39. The pulsed CRESU machine at the Eidgenössische Technische Hochschule (ETH) in Zurich dedicated to the study of nucleation processes (29–100 K).

Fig. 2.40.   The pulsed CRESU machine at the Texas A&M University (TAMU) in College Station dedicated to the characterization of hypersonic flows via laser diagnostics and the study of energy transfer ($T > 34$ K).

2.7   Digest

As the heart of the CRESU technique, the Laval nozzle has been placed at the center of the present Chapter. Beyond its use in reaction kinetics, we have briefly shown that convergent–divergent nozzles are ubiquitous devices in Science and Technology. We also gave a highly detailed explanation of a proven methodology to design Laval nozzle contours with a great deal of practical situations, demonstrating the versatility of this instrument. Characterization techniques have also been reviewed since they are essential in obtaining the best uniform supersonic flows, with special attention given to the Pitot tube method. We have stressed the importance of the competition between inertia forces and friction ones on the quality of the uniformity of supersonic flows exemplified by the major role of the mass flow rate

Fig. 2.41. The CPUF (Chirped Pulse in Uniform supersonic Flow) machine at the University of Missouri in Columbia (USA) coupling a pulsed CRESU reactor to a chirped pulse microwave spectrometer (T $\geq$ 22 K).

upon the length of uniformity. The various CRESU apparatuses available worldwide have been briefly described, and placed in a historical context. Pulsed and continuous regimes have been commented and compared focusing on recent technological breakthroughs that improve the quality of pulsed supersonic expansions. Advantages and drawbacks of both regimes have also been highlighted.

In the following chapters, attention will concentrate on specific applications of the CRESU technique in the field of reaction kinetics in the gas phase, and nucleation or aggregate characterization. Coupling of optical diagnostics or mass spectrometry techniques to the uniform supersonic flows will be of fundamental importance for these investigations among which the most prominent results will be emphasized.

Glossary of Symbols

Arabic symbols

a	speed of sound
A_z	section of the divergent nozzle contour at station z
A^*_{AB}	tabulated parameter available in Annex 1 (Eq. (2.138))
b	Mach cone tip longitudinal dimensionless position normalized to the throat radius
c_p	specific heat capacity at constant pressure
C_+, C_-	left running and right running characteristic lines, respectively
C_0, C_1, C_2	coefficients of the centerline Mach number function
C_f	local skin friction coefficient
f_0, f_2, f_4	zero-, second-, and fourth-order coefficients of the potential velocity polynomial expansion
F	function solution of the momentum Eq. (2.97)
F_i	normalized function F
g	compressibility factor for the Monaghan reference temperature
h	enthalpy
H	shape factor
$I(\overline{v_z}), J(\overline{v_z}), G(\overline{v_z})$	functions involved in the calculation of H (Eqs. (2.85)–(2.87))
k_B	Boltzmann constant
K	proportionality factor of $\tilde{v}_z$ with respect to position $z(\tilde{v}_z(z,0) = K \times z)$
L	length of the nozzle diverging section
m	molecular masse
$\dot{m}_i$	mass flow rate through an area of radius r_i
M	Mach number
M_A	molar mass of particle A
$M(z), N(z)$	functions defined in Eqs. (2.104) and (2.105)
n	particle density
P	pressure
P_i	impact pressure

$\mathcal{P}_r$	Prandtl number
q	mass flow rate
Q	standard flow rate
$Q(z)$, $P(z)$	intermediate functions in the Von Karman equation resolution (Eqs. (2.77)–(2.79))
r	radial position
r'	dimensionless radial position related to the throat radius
r_m	ideal gas constant per mole
Re_n	Reynolds number for length n
R_{wc}	radius of curvature at any location of the wall contour
$\hat{R}$	radius of curvature at the nozzle throat station (z_{th}, r_{th})
S_k	intermediate function defined in Eq. (2.49)
T	temperature
T_A^*	reduced temperature of particle A
U	intermediate function defined in Eq. (2.48)
$\vec{v}$	velocity vector with coordinates $(v_z, v_r, 0)$
v	velocity module
$\acute{v}$	ratio of the velocity module to the maximum velocity $(= v/v_{max})$
$\tilde{\nu}$	infinitesimal dimensionless velocity close to the throat
$\overline{v_z}$	dimensionless longitudinal velocity normalized to the isentropic longitudinal velocity
V	velocity normalized to the speed of sound at the throat
V_{res}	reservoir volume
x_A	volume proportion of particle A
X, Y, Z	expressions used in the calculation of the viscosity of a gas mixture (Eqs. (2.135)–(2.138))
$Y(z)$	solution function of the Von Karman equation
z	longitudinal position

Greek symbols

α	Mach angle; alternatively, function of the Prandtl number (Eq. (2.94))
α_A	ratio of the specific heat capacity at constant pressure to the Prandtl number of particle A
β	function of the Prandtl number (Eq. (2.95))
χ	longitudinal boundary station within which a perturbation method is applied to the velocity evolution
δ	boundary layer thickness
δ_1	displacement thickness
δ_2	momentum thickness
δz	elementary longitudinal step
ε	geometrical factor equal to 1 for an axisymmetric section and 0 for a rectangular one
ε_A	Lennard-Jones energy of particle A
ε_{AB}	interaction Lennard-Jones energy of particles A and B
$\varnothing$	potential velocity
γ	ratio of specific heat capacities
Γ	constant obtained from the resolution of Eq. (2.44)
λ	thermal conductivity
μ	dynamic viscosity coefficient
ν	Prandtl-Meyer function
θ	velocity angle with the longitudinal axis z
ρ	mass density
σ_A	Lennard-Jones radius of particle A
σ_{AB}	interaction Lennard-Jones radius of particles A and B
τ	shear stress
τ_w	wall shear stress
$\bar{\tau}:$	dimensionless shear stress normalized to the wall shear stress
$\tau_{i,j}$	stress tensor components
$\tau_{i,j}^*:$	non-dimensional stress tensor components
$\Omega^{(2,2)}$	transport integral
ζ	longitudinal dimensionless station related to the throat radius $(= z/r_{th})$

Superscript

*	Monaghan reference quantities

Subscripts

0	reservoir location
∞	isentropic core downstream of the Mach cone
ch	chamber (surrounding the uniform supersonic jet downstream of the Laval nozzle exit)
e	isentropic core
f	stands for film or recovery
max	stands for maximum
r	radial coordinate
th	Laval nozzle throat
w	Laval nozzle wall contour
z	longitudinal coordinate
χ	longitudinal boundary station χ

Annex 1

Table 2.A1. Transport parameters used in the calculation of the viscosity of a gas mixture (from [58]).

T^*	$\Omega(2,2)$	A_{AB}^*	T^*	$\Omega(2,2)$	A_{AB}^*	T^*	$\Omega(2,2)$	A_{AB}^*
0.30	2.785	1.046	1.70	1.248	1.095	4.20	0.9600	1.098
0.35	2.628	1.062	1.75	1.234	1.094	4.30	0.9553	1.099
0.40	2.492	1.075	1.80	1.221	1.094	4.40	0.9507	1.099
0.45	2.368	1.084	1.85	1.209	1.094	4.50	0.9464	1.099
0.50	2.257	1.093	1.90	1.197	1.094	4.60	0.9422	1.100
0.55	2.156	1.097	1.95	1.186	1.094	4.70	0.9382	1.100
0.60	2.065	1.101	2.00	1.175	1.094	4.80	0.9343	1.100
0.65	1.982	1.102	2.10	1.156	1.094	4.90	0.9305	1.101
0.70	1.908	1.104	2.20	1.138	1.094	5.00	0.9269	1.101
0.75	1.841	1.105	2.30	1.122	1.094	6.00	0.8963	1.103

(Continued)

Table 2.A1.   (*Continued*)

T^*	$\Omega(2,2)$	A^*_{AB}	T^*	$\Omega(2,2)$	A^*_{AB}	T^*	$\Omega(2,2)$	A^*_{AB}
0.80	1.780	1.105	2.40	1.107	1.094	7.00	0.8727	1.105
0.85	1.725	1.105	2.50	1.093	1.094	8.00	0.8538	1.107
0.90	1.675	1.104	2.60	1.081	1.094	9.00	0.8379	1.109
0.95	1.629	1.103	2.70	1.069	1.094	10.00	0.8242	1.110
1.00	1.587	1.103	2.80	1.058	1.094	20.00	0.7432	1.119
1.05	1.549	1.102	2.90	1.048	1.095	30.00	0.7005	1.124
1.10	1.514	1.102	3.00	1.039	1.095	40.00	0.6718	1.127
1.15	1.482	1.101	3.10	1.030	1.095	50.00	0.6504	1.130
1.20	1.452	1.100	3.20	1.022	1.096	60.00	0.6335	1.132
1.25	1.424	1.099	3.30	1.014	1.096	70.00	0.6194	1.134
1.30	1.399	1.099	3.40	1.007	1.096	80.00	0.6076	1.135
1.35	1.375	1.098	3.50	0.9999	1.097	90.00	0.5973	1.137
1.40	1.353	1.097	3.60	0.9932	1.097	100.00	0.5882	1.138
1.45	1.333	1.097	3.70	0.9870	1.097	200.00	0.5320	1.146
1.50	1.314	1.097	3.80	0.9811	1.097	300.00	0.5016	1.151
1.55	1.296	1.096	3.90	0.9755	1.097	400.00	0.4811	1.154
1.60	1.279	1.096	4.00	0.9700	1.098			
1.65	1.264	1.096	4.10	0.9649	1.098			

References

[1] Dikinson HW. A short history of the steam engine. Cambridge University Press, New York Macmillan Co.; 1939.

[2] Anderson JD. Modern compressible flow with historical perspectives. McGraw-Hill Series in Aeronautical and Aerospace Engineering, Singapore; 2004.

[3] Sutton GP, Biblarz O. Rocket propulsion elements. 9th ed. John Wiley and Sons Inc., Hoboken, New Jersey; 2016.

[4] Tordella D, Belan M, Massaglia S, De Ponte S, Mignone A, Bodenschatz E, Ferrari A. Astrophysical jets: insights into long-term hydrodynamics. New J Phys. 2011;13:043011.

[5] Anderson JD. Gasdynamics lasers: an introduction. New York, San Francisco, London: Academic Press Inc.; 1976.

[6] Gerry ET. Gasdynamic lasers. IEEE Spectr. 1970;7(11):51–58.

[7] Marimuthu S, Nath AK, Dey PK, Misra D, Bandyopadhyay DK, Chaudhuri SP. Design and evaluation of high-pressure nozzle assembly for laser cutting of thick carbon steel. Int J Adv Man Tech. 2017;92(1–4):15–24.

[8] Man HC, Duan J, Yue TM. Design and characteristic analysis of supersonic nozzles for high gas pressure laser cutting. J Mat Proc Tech. 1997;63(1–3):217–22.

[9] Lemos N, Lopes N, Dias JM, Viola F. Design and characterization of supersonic nozzles for wide focus laser-plasma interactions. Rev Sci Inst. 2009;80(10):103301.

[10] Lemos N, Cardoso L, Geada J, Figueira G, Albert F, Dia JM. Guiding of laser pulses in plasma waveguides created by linearly–polarized femtosecond laser pulses. Sci Rep. 2018;8(3165):1–9.

[11] Schmid K, Veisz L. Supersonic gas jets for laser-plasma experiments. Rev Sci Inst. 2012;83(5):53304.

[12] Moridi A, Hassani-Gangaraj SM, Guagliano M, Dao M. Cold spray coating: review of material systems and future perspectives. Surf Eng. 2014;36(6):369–95.

[13] Zhu L, Jen TC, Pan YT, Chen HS. Particle bonding mechanism in cold gas dynamic spray: a three-dimensional approach. J Therm Spray Tech. 2017;26(8):1859–73.

[14] Allimant A, Planche MP, Bailly Y, Dembinski L, Coddet C. Progress in gas atomization of liquid metals by means of a De Laval nozzle. Powder Tech. 2009;190(1–2):79–83.

[15] Si CR, Zhang XJ, Wang JB, Li YJ. Design and evaluation of a Laval-type supersonic atomizer for low-pressure gas atomization of molten metals. Int J Min Metal Mat. 2014;21(6):627–35.

[16] Du CM, Li HX, Zhang L, Wang J, Huang DW, Xiao MD, Cai JW, Chen YB, Yan HL, Xiong Y, Xiong Y. Hydrogen production by steam-oxidative reforming of bio-ethanol assisted by Laval nozzle arc discharge. Int J Hyd En. 2012;37(10):8318–29.

[17] Jarecki L, Blonski S, Zachara A. Modeling of pneumatic melt drawing of poly-L-lactide fibers in the Laval nozzle. Ind Eng Chem Res. 2015;54(43):10796–810.

[18] Délery J. Traité d'aérodynamique compressible–Vol III Applications de la théorie des caractéristiques et écoulements transsoniques. Mécanique des fluides series, Hermés–Lavoisier; 2008.

[19] Sauer R. General characteristics of the flow through nozzles at near critical speeds. NACA Tech Mem. 1947;1147.

[20] Owen JM, Sherman FS. Design and testing of a Mach 4 axially symmetric nozzle for rarefied gas flows. Engineering project report: he 150–104. University of California Berkeley; 1952.

[21] Maslach GJ, Sherman FS. Design and testing of an axisymetric hypersonic nozzle for a low density wind tunnel. Tech Rept He. University of California Berkeley; 1956. pp. 150–134.

[22] Prandtl L. Uber Flüssigkeitsbewegungen bei sehr kleiner Reihung. Heidelberg: Verhandlung III Intern. Math. Kongr; 1904. p. 484–491.

[23] Schlichting H, Gersten K. Boundary layer theory. 9th ed. Berlin Heidelberg: Springer-Verlag; 2017.

[24] Cebeci T. Convective heat transfer. 2nd ed. Berlin: Springer-Verlag, Berlin and Heidelberg Gmbh & Co.; 2002.

[25] Cousteix J. Aérodynamique: Couche Limite Laminaire. Cepaduès éditions, France; 1988.

[26] Anderson JD. Fundamentals of aerodymamics. Aeronautical and Aerospace Engineering series. 5th ed. McGraw-Hill; 2011.

[27] Bejan A. Convection heat transfer. 4th ed. John Wiley & Sons, Inc., Hoboken, New Jersey; 2013.

[28] Brun EA, Martinot-Lagarde A, Mathieu J. Mécanique des fluides, Vol III "Exemples de phénomènes instationnaires, couche limite et écoulements visqueux". Dunod, Paris; 1970.

[29] Stewartson K. The theory of laminar boundary layers in compressible fluids. London: Oxford Mathematical Monographs series, Oxford University Press; 1964.

[30] Cohen CB, Reshotko E. The compressible laminar boundary layer with heat transfer and arbitrary pressure gradient. NACA Rept. 1956;1294.

[31] Michel R. Aérodynamique: Couches Limites, Frottement et Transfert de Chaleur. Paris: ENS; 1963.

[32] Walz A. Boundary layers of flow and temperature. Cambridge, Massachusetts: The Massachusetts Institute of Technology Press; 1969.

[33] Dupeyrat G, Marquette JB, Rowe BR. Design and testing of axisymmetric nozzles for ion molecule reaction studies between 20 K and 160 K. Phys Fl. 1985;28(5):1273–9.

[34] Marquette JB. Etudes des Réactions Ion-Molécule à très Basse Température: Méthode Expérimentale, Application à la Formation de N_4^+ et O_4^+, 2 Octobre 1983. France: Université de Paris 7; 1983.

[35] White FM. Viscous fluid flow. Mechanical Engineering series. 3rd ed. MacGraw-Hill, Singapore; 2006.

[36] Lengrand JC. Couche Limite. Ecole Centrale des Arts et Manufactures, Paris; 1981.

[37] Blasius H. Grenzschichten in Flussigkeiten mit kleiner Reibung. (Available in translation as NACA TM 1256, 1950). Zeitschr Math und Phys. 1908;56(1):1–37.

[38] Crocco L. Lo Strato Limite Laminare nei Gas (available in translation in North American Aviation Aerophysics Lab., Rep. AL- 684, July 15, 1948). Roma: Monografie Scientifiche di Aeronautica No.3, Ministero della Difesa Aeronautica; 1946.

[39] Van Driest ER. Investigation of laminar boundary layer in compressible fluids using the crocco method. NACA Tech Note. 1952;2597.

[40] Wilson RE. Handbook of supersonic aerodynamics sections 13 and 14 viscosity and heat transfer effects. NAVORD Rept 1488. 1966; p. 5.

[41] Michel R, Kretzschmar G. Détermination théorique d'une enthalpie de référence pour la couche limite laminaire de la plaque plane. Rech Aeros. 94, 1963; p. 3–7.

[42] Meador WE, Smart MK. Reference enthalpy method developed from solutions of the boundary-layer equations. AIAA J. 2005;43(1): 135–9.

[43] Rubesin MW, Johnson HA. A critical review of skin-friction and heat-transfer solutions of the laminar boundary layer of a flat plate. Trans ASME. 1949;71(4):383–388.

[44] Eckert ERG. Survey of heat transfer at high speeds. ARL 189; 1961.

[45] Monaghan RJ. On the behaviour of boundary layers at supersonic speeds. Los Angeles, California: Fifth International Aeronautical Conference Institute of the Aeronautical Sciences, Inc.; 1955. pp. 277–315.

[46] Schlichting H, Ulrich A. Zur Berechnung des Umschlages laminar-turbulent. Jahrbuch d dt Luftfahrtforschung. 1942;I:8–35.

[47] Sanchez-Gonzalez R, Srinivasan R, Hofferth J, Kim DY, Tindall AJ, Bowersox RDW, North SW. Repetitively pulsed hypersonic flow apparatus for diagnostic development. AIAA J. 2012;50(3): 691–7.

[48] Sims IR, Queffelec JL, Defrance A, Rebrion-Rowe C, Travers D, Bocherel P, Rowe BR, Smith IWM. Ultra-low temperature kinetics of neutral-neutral reactions: the technique, and results for the reactions $CN + O_2$ down to 13 K and $CN + NH_3$ down to 25 K. J Chem Phys. 1994;100(6):4229–41.

[49] Luque J, Crosley DR. LIFBASE: database and spectral simulation program for diatomic molecules, version 2.1.1. SRI International Report MP-99-009;1999.

[50] Taylor SE, Goddard A, Blitz MA, Cleary PA, Heard DE. Pulsed Laval nozzle study of the kinetics of OH with unsaturated hydrocarbons at very low temperatures. Phys Chem Chem Phys. 2008;10(3): 422–37.

[51] Sánchez-González R, Bowersox RDW, North SW. Simultaneous velocity and temperature measurements in gaseous flowfields using

the vibrationally excited nitric oxide monitoring technique: a comprehensive study. Appl Opt. 2012;51(9):1216–28.

[52] Sánchez-González R, Srinivasan R, Bowersox RDW, North SW. Simultaneous velocity and temperature measurements in gaseous flow fields using the VENOM technique. Opt Lett. 2011;36(2): 196–8.

[53] Mostefaoui T, Rebrion-Rowe C, Travers D, Rowe BR. A comparison of flow velocities measured using an impact-pressure probe and electron time of flight in a supersonic flow. Implications for electron thermalization. Meas Sci Tech. 2000;11(4):425–9.

[54] Goulay F, Rebrion-Rowe C, Biennier L, Le Picard SD, Canosa A, Rowe BR. The reaction of anthracene with CH radicals: an experimental study of the kinetics between 58 K and 470 K. J Phys Chem A. 2006;110(9):3132–7.

[55] Goulay F, Rebrion-Rowe C, Le Garrec JL, Le Picard SD, Canosa A, Rowe BR. The reaction of anthracene with OH radicals: an experimental study of the kinetics between 58 K and 470 K. J Chem Phys. 2005;122(10):104308.

[56] Canosa A, Ocaña AJ, Antiñolo M, Ballesteros B, Jiménez E, Albaladejo J. Design and testing of temperature tunable de Laval nozzles for applications in gas-phase reaction kinetics. Exp Fluids. 2016;57(9):152.

[57] Ocaña AJ, Blázquez S, Potapov A, Ballesteros B, Canosa A, Antiñolo M, Vereecken L, Albaladejo J, Jiménez E. Gas phase reactivity of CH_3OH toward OH at interstellar temperatures (11.7–177.5 K): experimental and theoretical study. Phys Chem Chem Phys. 2019;21(13):6942–57.

[58] Hirschfelder JO, Curtiss CF, Bird RB. Molecular theory of gases and liquids. New York: Wiley; 1954.

[59] Mason EA, Monchich L. Transport properties of polar gas mixtures. CM-1009. The Johns Hopkins University Applied Physics Laboratory; 1962.

[60] Lemmon EW, Jacobsen RT. Viscosity and thermal conductivity equations for nitrogen, oxygen, argon, and air. Int J Thermophys. 2004;25(1):21–69.

[61] Blázquez S, González D, García-Sáez A, Antiñolo M, Bergeat A, Caralp F, Mereau R, Canosa A, Ballesteros B, Albaladejo J, Jiménez E. Experimental and theoretical investigation on the $OH + CH_3C(O)CH_3$ reaction at interstellar temperatures (T = 11.7–64.4 K). Earth Spac Chem. 2019;3(9):1873–83.

[62] Ocaña AJ, Blázquez S, Ballesteros B, Canosa A, Antiñolo M, Albaladejo J, Jiménez E. Gas phase kinetics of the $OH + CH_3CH_2OH$

reaction at temperatures of the interstellar medium (T = 21–107 K). Phys Chem Chem Phys. 2018;20(8):5865–73.

[63] Ocaña AJ, Jiménez E, Ballesteros B, Canosa A, Antiñolo M, Albaladejo J, Agúndez M, Cernicharo J, Zanchet A, Del Mazo P, Roncero O, Aguado A. Is the gas-phase OH + H_2CO reaction a source of HCO in interstellar cold dark clouds? A kinetic, dynamic and modelling study. Astrophys J. 2017;850(1):28.

[64] Antiñolo M, Agúndez M, Jiménez E, Ballesteros B, Canosa A, El Dib G, Albaladejo J, Cernicharo J. Reactivity of OH and CH_3OH between 22 and 64 K: modelling the gas phase production of CH_3O in Barnard 1B. Astrophys J. 2016;823(1):25.

[65] Jiménez E, Antiñolo M, Ballesteros B, Canosa A, Albaladejo J. First evidence of the dramatic enhancement of the reactivity of methyl formate (HC(O)OCH$_3$) with OH at temperatures of the interstellar medium: a gas-phase kinetic study between 22 K and 64 K. Phys Chem Chem Phys. 2016;18(3):2183–91.

[66] Tizniti M, Le Picard SD, Canosa A, Sims IR, Smith IWM. Low temperature kinetics; the association of OH radicals with O_2. Phys Chem Chem Phys. 2010;12(39):12702–12710.

[67] Smith IWM, Le Picard SD, Tizniti M, Canosa A, Sims IR. The quest for the hydroxyl-peroxy radical. Z Phys Chem. 2010;224(7–8): 949–65.

[68] Le Picard SD, Tizniti M, Canosa A, Sims IR, Smith IWM. The thermodynamics of the elusive HO_3 radical. Science. 2010;328(5983): 1258–62.

[69] Hickson KM, Caubet P, Loison JC. Unusual low-temperature reactivity of water: the CH + H_2O reaction as a source of interstellar formaldehyde? J Phys Chem Lett. 2013;4(17):2843–46.

[70] Daranlot J. Réactivité de l'azote atomique et du radical OH à basse température par la technique CRESU: Réactions d'intérêt pour l'astrochimie. France (phD dissertation): Université de Bordeaux 1;2012.

[71] Tizniti M, Le Picard SD, Lique F, Berteloite C, Canosa A, Alexander MH, Sims IR. The F + H_2 reaction at very low temperatures. Nat Chem. 2014;6(2):141–5.

[72] Cooke IR, Sims IR. Experimental studies of gas-phase reactivity in relation to complex organic molecules in star-forming regions. Earth Spac Chem. 2019;3(7):1109–34.

[73] Potapov A, Canosa A, Jiménez E, Rowe BR. Uniform supersonic chemical reactors: 30 years of astrochemical history and future challenges. Ang Chem Int Ed. 2017;56(30):8618–40.

[74] Canosa A, Goulay F, Sims IR, Rowe BR. Gas phase reactive collisions at very low temperature: recent experimental advances and perspectives. In: Low Temperatures and Cold Molecules. ed. Imperial College Press; 2008. pp. 55–120.

[75] Smith IWM. Reactions at very low temperatures: gas kinetics at a new frontier. Ang Chem Int Ed. 2006;45(18):2842–61.

[76] Smith IWM, Rowe BR. Reaction kinetics at very low temperatures: laboratory studies and interstellar chemistry. Account Chem Res. 2000;33(5):261–268.

[77] Rowe BR, Dupeyrat G, Marquette JB, Gaucherel P. Study of the reactions $N_2^+ + 2\ N_2 \rightarrow N_4^+ + N_2$ and $O_2^+ + 2\ O_2 \rightarrow O_4^+ + O_2$ from 20–160 K by the CRESU technique. J Chem Phys. 1984;80(10): 4915–21.

[78] James PL, Sims IR, Smith IWM. Rate coefficients for the vibrational self-relaxation of $NO(X^2\Pi, v = 3)$ at temperatures down to 7 K. Chem Phys Lett. 1997;276(5–6):423–9.

[79] Le Page V. Conception et Mise au Point d'un Moyen d'Essai CRESU (Cinétique de Réaction en Ecoulement Supersonique Uniforme) pour l'Etude des Réactions Ion-Molécule et Application aux Températures Ultra-Basses (10 K) (phD dissertation). France: Université de Rennes 1; 1995.

[80] Le Garrec JL, Mitchell JBA, Rowe BR. CRESU studies of electron attachment and penning ionization at temperatures down to 48 K. Act Phys Com. 1996;37:15–26.

[81] Daugey N, Caubet P, Retail B, Costes M, Bergeat A, Dorthe G. Kinetic measurements on methylidyne radical reactions with several hydrocarbons at low temperatures. Phys Chem Chem Phys. 2005;7(15):2921–7.

[82] Daranlot J, Bergeat A, Caralp F, Caubet P, Costes M, Forst W, Loison JC, Hickson KM. Gas-phase kinetics of hydroxyl radical reactions with alkenes: experiment and theory. Chemphyschem. 2010;11(18):4002–10.

[83] Sabbah H, Biennier L, Sims IR, Rowe BR, Klippenstein SJ. Exploring the role of PAHs in the formation of soot: pyrene dimerization. J Phys Chem Lett. 2010;1(17):2962–7.

[84] Durif O, Capron M, Messinger J, Benidar A, Biennier L, Bourgalais J, Canosa A, Courbe J, Garcia GA, Gil JF, Nahon L, Okumura M, Rutkowski L, Sims IR, Thiévin J, Le Picard SD. A new instrument for kinetics and branching ratio studies of gas phase collision processes at very low temperatures. Rev Sci Inst. 2021;92(1): 014102.

[85] Atkinson DB, Smith MA. Design and characterization of pulsed uniform supersonic expansions for chemical applications. Rev Sci Inst.
1995;66(9):4434–46.

[86] Lee S, Hoobler RJ, Leone SR. A pulsed Laval nozzle apparatus with
laser ionization mass spectrometry for direct measurements of rate
coefficients at low temperatures with condensable gases. Rev Sci Inst.
2000;71(4):1816–23.

[87] Spangenberg T, Kohler S, Hansmann B, Wachsmuth U, Abel B,
Smith MA. Low-temperature reactions of OH radicals with propene
and isoprene in pulsed Laval nozzle expansions. J Phys Chem A.
2004;108(37):7527–34.

[88] Morales S, Rowe BR. Hacheur aérodynamique pour la pulsation des
gaz. Patent: FR 2948302—WO 2011018571–US 8.870.159; 2009.

[89] Morales S. Le Hacheur Aérodynamique: un nouvel instrument dédié
aux processus réactionnels à ultra-basse température (phD dissertation). France: Université de Rennes 1, 2009; pp. 209.

[90] Oldham JM, Abeysekera C, Joalland B, Zack LN, Prozument K, Sims
IR, Park GB, Field RW, Suits AG. A chirped-pulse Fourier-transform
microwave/pulsed uniform flow spectrometer. I. The low-temperature
flow system. J Chem Phys. 2014;141(15):54202.

[91] Jiménez E, Ballesteros B, Canosa A, Townsend TM, Maigler FJ,
Napal V, Rowe BR, Albaladejo J. Development of a pulsed uniform
supersonic gas expansion system based on an aerodynamic chopper
for gas phase reaction kinetics studies at ultra-low temperatures. Rev
Sci Inst. 2015;86(4):045108.

[92] Schlappi B, Litman JH, Ferreiro JJ, Stapfer D, Signorell R. A pulsed
uniform Laval expansion coupled with single photon ionization and
mass spectrometric detection for the study of large molecular aggregates. Phys Chem Chem Phys. 2015;17(39):25761–71.

[93] Suas-David N, Thawoos S, Suits AG. A uniform flow-cavity ring-
down spectrometer (UF-CRDS): a new setup for spectroscopy and
kinetics at low temperature. J Chem Phys. 2019;151(24):244202.

[94] Hearne TS, Abdelkader-Khedaoui O, Hays BM, Guillaume T, Sims
IR. A novel Ka-band chirped-pulse spectrometer used in the determi-
nation of pressure broadening coefficients of astrochemical molecules.
J Chem Phys. 2020;153(8):084201.

[95] Atkinson DB, Smith MA. Radical-molecule kinetics in pulsed uniform
supersonic flows–termolecular association of OH + NO between 90 K
and 220 K. J Phys Chem. 1994;98(23):5797–800.

[96] West NA, Millar TJ, Van de Sande M, Rutter E, Blitz MA, Decin L,
Heard DE. Measurements of low temperature rate coefficients for the
reaction of CH with CH_2O and application to dark cloud and AGB
stellar wind models. Astrophys J. 2019;885(2):134.

[97] Sánchez-González R, Eveland WD, West NA, Mai CLN, Bowersox RDW, North SW. Low-temperature collisional quenching of NO $A^2\Sigma^+(v'=0)$ by NO(X $^2\Pi$) and O_2 between 34 and 109 K. J Chem Phys. 2014;141(7):74313.

[98] Lippe M, Szczepaniak U, Hou GL, Chakrabarty S, Ferreiro JJ, Chasovskikh E, Signorell R. Infrared spectroscopy and mass spectrometry of CO_2 clusters during nucleation and growth. J Phys Chem A. 2019;123(12):2426–37.

[99] Mullen C, Smith MA. Low temperature NH(X $^3\Sigma^-$) radical reactions with NO, saturated, and unsaturated hydrocarbons studied in a pulsed supersonic Laval nozzle flow reactor between 53 and 188 K. J Phys Chem A. 2005;109(7):1391–9.

[100] Mullen C, Smith MA. Temperature dependence and kinetic isotope effects for the OH + HBr reaction and H/D isotopic variants at low temperatures (53–135 K) measured using a pulsed supersonic Laval nozzle flow reactor. J Phys Chem A. 2005;109(17):3893–902.

[101] Rowe BR, Dupeyrat G, Marquette JB, Smith D, Adams NG, Ferguson EE. The reaction $O_2^+ + CH_4 \rightarrow CH_3O_2^+ + H$ studied from 20–560 K in a supersonic jet and in a SIFT. J Chem Phys. 1984;80(1):241–5.

[102] Berteloite C, Lara M, Bergeat A, Le Picard SD, Dayou F, Hickson KM, Canosa A, Naulin C, Launay JM, Sims IR, Costes M. Kinetics and dynamics of the $S(^1D_2) + H_2$ reaction at very low temperatures and collision energies. Phys Rev Lett. 2010;105(20):203201.

https://doi.org/10.1142/9781800610996_0003

Chapter 3

Ion Chemistry in Uniform Supersonic Flows

Ludovic Biennier[*,§], Sophie Carles[*], François Lique[*,†] and
James Brian Mitchell[‡]

*CNRS, IPR (Institut de Physique de Rennes)-UMR 6251,
Université de Rennes, F-35000 Rennes, France
†Laboratoire Ondes et Milieux Complexes, UMR CNRS 6294, Université
du Havre, 25 rue Philippe Lebon, BP 1123, 76063 Le Havre Cedex, France
‡Merl-Consulting, 21 Rue Sergent Guihard, 35000 Rennes, France
§ludovic.biennier@univ-rennes1.fr

Abstract

Ion processes are a key driver of chemistry in a wide variety of natural
dilute environments such as the ionosphere of Earth, the atmosphere
of solar system planets, and satellites and interstellar clouds. They also
play a role in low-temperature plasmas that are commonly encountered
in industrial settings. Uniform supersonic flows have proved to be pivotal
in gaining insights into the kinetics of ionic processes. In more than
three decades, a variety of ion–molecule reactions has been investigated
with the CRESU (French acronym standing for Kinetics of Reactions
with Uniform Supersonic Flows) method. They include bimolecular and
termolecular reactions and often depart from predictions made by simple
empirical models.

Neglected in the 90s for the benefit of the study of less predictable
radical–neutral reactions, the exploration of ion–molecule reactive collisions has seen a recent revival. It has been in particular stimulated
by the discovery of molecular anions in astrophysical environments for
which the formation and destruction processes remain elusive, mostly
due to the lack of kinetic and branching ratio data.

Despite the growing sophistication of competing methods which include crossed beams or cooled ion traps, uniform supersonic flows continue to be a method of choice, one of the rare techniques to provide rate coefficients obtained under well-controlled thermalized conditions over a wide range of low temperatures. Uniform supersonic flows turn out to be well suited for heavy neutral co-reactants.

Some challenges remain and call for new directions of action. In a last section, the most promising routes and the limits of the method are outlined. The extension of the method to the study of dissociative recombination sounds perilous, and state-selective chemistry with ions could only marginally benefit from the CRESU. From a technical point of view, the implementation of isomer-specific detection schemes could, however, greatly expand the scope of ion–molecule reaction studies. In tight connection with planetary sciences, ion-induced nucleation appears to be a topic within reach today using the CRESU technique. Further development of the approach will contribute to validate approximate treatments and push further the knowledge of ion–molecule reactions at low temperatures.

Keywords: Ion processes; Ion–molecule reactions; Chemical kinetics.

3.1   Context

The interaction of photons or high-energy particles with gaseous matter readily leads to the emergence of atomic and molecular ions. Ion chemistry plays an important role in dilute natural environments such as interstellar clouds, star-forming regions, circumstellar shells of dying stars, and closer to us, comae of comets, planetary atmospheres, and the Earth's ionosphere. If we restrict ourselves to low-temperature processes, charged species are also key constituents of very cold plasmas that are generated in low-pressure gases and that are widely used in industry for chemical-vapor deposition, plasma etching, and other treatments of solid surfaces. Among all the ionic processes involved, ion–molecule reactions play a pivotal role in many environments. This has fostered a strong and lasting interest in understanding and predicting their behavior.

3.1.1   *Fundamental aspects*

Schematically, the dynamics of gas phase reactive collisions is governed by two mechanisms (see Fig. 3.1): abstraction in direct

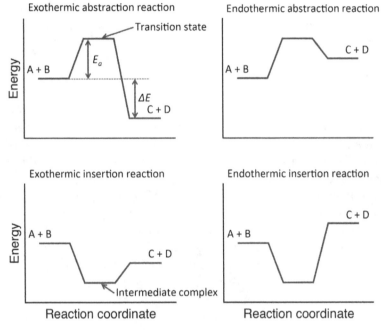

Fig. 3.1. Different kinds of chemical reactions.

reactions or insertion with the formation of a stable intermediate complex in indirect reactions. Both kinds of reactions can be either endoergic or exoergic.

Ion–molecule reactions generally belong to the latter case. They are usually fast exoergic reactions that can be reasonably modeled using simple capture approaches [1]. However, especially at low temperatures, several ion–molecule reactions do not follow the so-called "Langevin behavior" (Chapter 11 and Section 3.4) predicted by basic capture theory (e.g. constant rate coefficients). Sometimes, the rate coefficients can even increase with increasing temperatures, whereas simple capture models predict the opposite for ion–polar neutral reactions.

Several reasons can explain the deviation between simple models and experimental measurements. While most of the time, there are no energetic barriers to reaction, reactive trajectories typically form van der Waals intermediates and must overcome submerged barriers to form products. If van der Waals complexes cannot be formed efficiently, this leads to a lowering of the reaction rate coefficients, especially at low temperatures. Also, for some ion–molecule reactions such as nucleophilic substitution reactions, the reaction path does not correspond to a direct rebound only and alternative reaction paths play a significant role [2]. As a consequence, such reaction processes are far from simple to model since knowledge about stationary energy points is not sufficient to predict the dynamics.

Accurate modeling of ion–molecule reactions at low temperatures would then require the use of quantum time-independent approaches. The main difficulty with such systems is the number of dimensions to consider and also the excessively large number of angular couplings between the reactants, which make full quantum calculations prohibitively expensive, both in terms of computer time and memory requirements. Quantum time-dependent approaches can be an alternative, but they are difficult to implement not only because of the number of dimensions but also because of the difficulty to converge calculations at low collisional energies. As a consequence, with the exception of some prototypical tri- or tetra-atomic ion–molecule reactions, most calculations beyond simple capture approaches have been performed from (quasi-) classical trajectories or using reduced dimension approaches. Hence, the validation of theoretical developments strongly relies on detailed and extensive comparisons with experimental results, especially at low temperature where quantum effects can predominate.

There are, however, few experimental methods for studying ion–molecule reactions in the absence of solvent or environmental effects. As a result, if we put calculations of capture rate coefficients aside, ion–molecule reaction mechanisms are still largely unexplained at low temperatures despite significant progress in the experimental techniques to study them. Among the experimental approaches available, the use of uniform supersonic flows is of particular interest since it provides both the temperature dependence and the absolute value of the rate coefficients over a wide range of temperatures including the lowest ones. Such an approach produces valuable data for benchmarking theoretical models and quantum-mechanical calculations.

3.1.2 *Interstellar cold ion chemistry*

The generation of positive ions in the interstellar medium is initiated by UV/X irradiation, cosmic rays, and shock waves. The respective contribution of these mechanisms is closely linked to the environment [3]. In diffuse clouds, characterized by temperatures T around 100 K and densities n of 100 molecule cm^{-3}, the main processes are the photoionization of carbon atoms by stellar UV photons which generates C^+, and the ionization of atomic and molecular hydrogen by galactic cosmic rays leading to H^+ and H_2^+. In dense clouds ($T \sim 10\,K, n \sim 10^5$ molecule cm^{-3}), only cosmic rays penetrate and ionize H_2 and He producing H^+, H_2^+, and He^+ as primary ions. The ion fraction, which varies greatly from 10^{-8} of the abundance of [H] in dense clouds to 10^{-4} in diffuse clouds [4], traces to non-local thermodynamic equilibrium (Chapter 1), and is an indicator of the role of ions in the physics and chemistry of dilute astrophysical environments. In diffuse clouds, for instance, ionized carbon is considered as the main gas phase reservoir of carbon. Atomic hydrogen with an ionization potential of 13.6 eV absorbs all the incoming photons at energies greater than or equal to this, but lower-energy photons can pass right through these clouds. Thus, carbon with an ionization potential of 11.26 eV is ionized throughout and 90% exists as C^+. Once formed, the primary ions trigger a rich chemical network that contributes to the formation of a variety of molecules. In the cold interstellar medium (ISM), the chemistry is initiated by the H_3^+ ion that has been formed through the $H_2 + H_2^+$ chemical reaction.

More than 250 molecules have been discovered in the interstellar medium so far, from diatomics to large, complex molecules (Chapter 10), with ions representing about one sixth of the detections [5]. Most of them are positively charged. With the exception of CH^+, H_3^+, and C_{60}^+, molecular ions have all been identified in the sub-mm to mm wavelength range through their rotational emission lines employing ground-based or space telescopes. However, other spectral regions, such as the infrared window, may also conceal ions, possibly heavy ones. The mid-infrared emission features at 3.3, 6.2, 7.7, 8.6, 11.2, 12.7, and 16.4 μm which dominate the IR spectra of almost all objects with associated gas and dust and are illuminated by UV photons may partly result from ionized polycyclic aromatic hydrocarbons (PAHs) [6]. The near-UV to near-IR spectral range, over which diffuse clouds — in which the interstellar matter is highly ionized — are probed, should also not be overlooked. Only a few of the 500+ diffuse interstellar bands (DIBs), observed in absorption in reddened lines of sight crossing diffuse clouds, have thus far been assigned.

They were all attributed to the fullerene C_{60}^+ [7]. Before this long awaited first success, the spotlight was set on PAH cations [8], but no carrier was decisively identified and some potential candidates, such as HC_4H^+ [9] or C_7^- [10], were refuted.

Overall, the presence of molecular ions is found almost all along the evolutionary cycle of matter starting with the outer layers of dying stars, continuing with diffuse clouds, then dense cold clouds, and terminating with star-forming regions.

In terms of formation routes for positive ions, in addition to UV irradiation, cosmic ray bombardment, and shockwave processing, reactions between ions and neutrals are key recycling pathways. These processes can promote a rapid ion growth. Governed by long-range forces, ion–molecule reactions are most often faster than neutral–neutral reactions, even the indirect ones with formation of a stable intermediate complex. Hence, even if molecular cations are generally present only as trace species (e.g. at the particles per billion level in dense clouds), they play a crucial role in the chemistry of astrophysical environments [4, 11]. Reactions of ions have been invoked as intermediate steps in the build-up of complex organic molecules [12].

Among the destruction mechanisms of cations, dissociative recombination ($AB^+ + e^- \rightarrow A + B$) is a major route. In highly ionized media, it is a dominant pathway for the formation of neutrals. In dense clouds, the low electron abundance is compensated by the high rate coefficient for the process, and so dissociative recombination plays some important role there too.

The detection of negative ions in space more than a decade ago altered our view of interstellar ion chemistry. Negative ions were already recognized to play a key role in the early universe through the associative detachment of H^- with H to form H_2, the first molecular coolant available. However, the proof of their presence remained elusive. It came only in the 2000s with the identification of charged linear polyynes, C_4H^- [13] C_6H^- [14, 15], C_8H^- [16, 17], and cyanopolyynes, CN^- [18], C_3N^- [19], C_5N^- [20], in a range of sources including dark clouds, prestellar cores, protostellar envelopes, and circumstellar envelopes of carbon stars. Photochemical models that have followed suggest that anions have a larger abundance than free electrons in some sources, such as the outer layers of the circumstellar envelope of the carbon star IRC + 10216, and that they play an important role in the synthesis of very large hydrocarbon species [21]. In the regions shielded from UV photons, dust building blocks, such as PAHs, may be negatively charged and therefore play some role in regulating the interaction of magnetic fields with the gas.

Dissociative attachment and radiative association are considered as major pathways for anion production [11]. In the mechanism of dissociative attachment, a free electron attaches to a neutral forming a highly excited intermediate that subsequently fragments into a neutral and an anion following the process

$$AB + e^- \rightarrow (AB^-)^* \rightarrow A^- + B \qquad (3.1)$$

A few of these types of reactions are exoergic and therefore possible in cold astrophysical environments. In particular, CN^- could be formed in the circumstellar envelopes by dissociative attachment to MgCN or MgNC, which have both been reported [22] The other competitive route is radiative electronic attachment.

$$A + e^- \rightarrow (A^-)^* \rightarrow A^- + h\upsilon \qquad (3.2)$$

Until recently, most of the anions detected so far were believed to be formed through the mechanism of radiative electron attachment initially proposed by Herbst [23]. However, several theoretical studies have cast some doubts on the prominence of the process [24, 25]. The topic remains controversial, notably because theoretical calculations disagree with sophisticated experiments on radiative electron attachment of C_6 based on a detailed balance approach [26].

Anions can be destroyed by processes such as photo-detachment, electron-impact detachment, associative detachment, collisional detachment, and mutual neutralization with cations [21]. Photo-detachment by interstellar UV photons appears to be the dominant destruction channel for anions in particular CN^- and C_3N^- in the circumstellar envelope of the carbon-rich star IRC+10216 [27]. According to Millar [21], the relevance of electron-impact detachment and collisional detachment is unclear for cold astrophysical environments.

By contrast, ion–molecule reactions are considered to contribute strongly to the anion life cycle. Such pathways could provide valid alternative routes to radiative electron attachment for the production of some carbon-rich anions. The efficiency of radiative electron attachment, formerly presumed to be the process at the origin of anions in the ISM, is today contested [21].

3.1.3 *Ion processes in comets and planetary atmospheres*

Beyond the interstellar medium, ions have also been discovered in the coma of comets. The outgassing from the comet, a large fraction of which is composed of volatile compounds, generates neutral species such as H_2O, CO, and CO_2 that once liberated, undergo a variety of ionic processes. Among them are photoionization by VUV solar photons, charge exchange with solar wind ions, electron impact ionization from solar wind, and suprathermal cometary electrons [28]. The Rosetta encounter with comet 67P/Churyumov-Gerasimenko provided a unique opportunity for an in situ, up-close investigation of ion–neutral chemistry in the coma of a weakly outgassing comet far from the Sun. The low-energy ion composition in the coma was measured by mass spectrometry and revealed the presence of light organic ions, mainly from water (O^+, OH^+, H_2O^+, H_3O^+) but also CH_x^+ ($x = 1$–4), $C_2H_x^+$ ($x = 2, 3$), or NH_x^+ ($x = 2$–4), with likely contributions from higher mass carbon molecules and their breakup products [29]. The transient nature of the cometary atmosphere strongly drives the coma out of equilibrium. Thus, the rate coefficients of various reactions are critically important in determining the ion species that are produced and temporarily reside in the coma. Negative ions have been detected in surprisingly large abundances during spacecraft flybys of solar system bodies such as the coma of comet 1P/Halley [30].

Ions are also found in the atmosphere of rocky and gaseous planets and their satellites. For planets with no or weak magnetospheres like Venus and Mars, ionic processes contribute, along with some other mechanisms, to the atmospheric escape process. Dissociative recombination of O_2^+ with electrons, for instance, can produce two O atoms with enough energy to exceed the escape velocity of Mars.

Ion processes are also pivotal in Earth's atmosphere, where large cluster ions are found up to the D layer of our ionosphere [31] and whose possible role in the formation of aerosols, which have a vast influence on the climate, is still the subject of intensive discussion. More on this topic is addressed in Section 3.5.5.

One of the most captivating illustrations of the role of ionic processes can be found in the atmosphere of Titan, Saturn's largest satellite. The chemistry of Titan's atmosphere is initiated by the dissociation and ionization of nitrogen and methane, its main constituents, by solar UV photons and solar wind particles. The absence of magnetosphere does not stop the penetration of these particles, solar wind, or galactic cosmic rays into the low layers of Titan's dense and cold atmosphere. The primary ions N_2^+ and CH_4^+ trigger a rich chemical network [32, 33] that rapidly leads to the formation of a great diversity of ionic and neutral species. Measurements performed by the Ion and Neutral Mass Spectrometer (INMS), embarked on the Cassini spacecraft during its flybys, revealed the presence of hydrocarbons and nitriles at an altitude of $\sim$1000 km [34]. Good correspondence is observed between the groups of peaks of the ion and neutral mass spectra that cover the 1–100 amu/q range. The Cassini Plasma Spectrometer measurements demonstrated that the production of large ions extends even further in size as positive ions with m/z up to 350 amu/q were detected. Ion–molecule reactions contribute strongly to the growth of positive ions, while the main loss process is radiative or dissociative electron recombination. The abundance of some prominent ions such as $HCNH^+$ and $C_2H_5^+$ remains overestimated by chemical models, suggesting some gaps in our understanding of the photochemical network [33]. One of the most surprising results obtained by Cassini remains, however, the detection of heavy negative ions in its upper atmosphere with the Electron Spectrometer, one of the three sensors embarked on the spacecraft [35]. The derived low-resolution mass spectra highlighted the presence of ions with m/z up to 13,800 [36] with some broad features tentatively assigned to CN^-, C_3N^-/C_4H^-, and C_5N^-/C_6H^- in line with the findings of a negative ion photochemical model [37]. Although abundant, their density does not exceed $\sim$1% of the positive ion density. Production mechanisms for negative ions include photoionization (ion-pair formation) as well as dissociative and radiative electron attachment, while loss mechanisms include photodetachment, associative detachment with neutrals, and mutual neutralization [33]. Observations made by the instruments on board the Cassini orbiter show that molecular growth detected in Titan's ionosphere leads to the formation of aerosols. Recent studies of this layer of the atmosphere have established that the positive and negative heavy ions detected by the Cassini mass spectrometers are two sides of the same coin: aerosol embryos. A recent model combining photochemistry and micro-physics [38] shows that the macromolecules produced by ionic chemistry capture free electrons and, once charged, recombine easily with positive ions. Depending on the exit channels, this interaction may lead to a rapid mass gain in macromolecules, thus generating aerosol embryos in this region of the atmosphere, which is very tenuous.

3.1.4   *Cold plasmas*

More marginally, the understanding of very cold plasmas can also benefit from the investigation of ion chemistry conducted in uniform supersonic flows. It is widely recognized that ions and radicals play an important role in low-temperature plasma processes encountered in the industry such as etching, film deposition, and surface modification. Ions are involved in the overall plasma process, both in the gas phase and at the plasma–surface interface [39]. However, the exact role of gas phase ion–molecule reactions is still unclear in both plasma etching and plasma polymerization processes. Several studies suggest that gas phase ion–molecule reactions that lead to large polyatomic ions represent key steps in the overall polymerization process. Ion–molecule reactions may also give access to products that cannot be directly produced by electron impact ionization.

The reaction rate coefficients and energetics for the species of interest in plasma deposition and etching [40] are required to improve the models of these dilute environments. Most of the relevant data can be obtained with room-temperature flow tube studies. Experiments at lower temperatures may nevertheless help to constrain the theoretical models and explore further a potential temperature dependence of the rate coefficients and of the branching between the exit channels. Note that non-thermal experiments with state-selected ions, mostly out of reach of uniform supersonic flow methods, can also provide a deep insight into the plasma characteristics.

A majority of "cold plasma" studies actually are performed at or near room temperatures (as opposed to hot plasmas such as in thermonuclear fusion studies). Indeed, an interesting domain to study would be ultracold plasmas as there the chemistry could be quite different. In particular, electron molecular ion recombination generally scales as $T_e^{-0.5}$, while electron attachment has a rather flat temperature dependence, except when the molecules are rovibrationally excited. As seen in Chapter 4, internal excitation of the attaching molecules can have dramatic effects on this process. Thus, this could play an important role in attaching plasmas and possibly have industrially important consequences, in particular, as regards the formation of dust in plasmas. This is a process that is triggered by negative ion formation as these ions are trapped in the positive plasma potential of plasma reactors, leading to nucleation and growth phenomena. The CRESU technique offers the possibility of studying temperature effects related to internal energies of the attaching molecules and such studies could point the way to subsequent ultracold plasma research. With regard to dissociative recombination, as will be shown in Section 3.5.4, this is a topic that has not been studied using the CRESU method and the reasons for this and possible future approaches will be discussed in that section.

Interestingly, cold plasmas can in turn be employed to explore the chemistry of planetary atmospheres. For instance, plasma inductively coupled with a radio frequency power source can be used to generate analogs of Titan's aerosols, also known as tholins, starting with, e.g. a N_2/CH_4 gas mixture [41, 42]. A better understanding of the ion chemistry processes in play is vital for a meaningful comparison with *in situ* measurements by space probes and astronomical observations.

3.2 Ion Production and Detection in Uniform Supersonic Flows

Specific methods have been developed to explore the reactivity of ions, first to ensure their efficient production, e.g. with the help of discharges and second to allow their detection [43]. The abundance of ions in experimental setups is commonly a few orders of magnitude lower than radicals. Consequently, the simple adaptation of techniques employed for radical–neutral reactions is hindered by the detection sensitivity required to probe ions that remains out of reach of many diagnostics used for radicals, such as absorption methods (whether direct or enhanced). The methodology adopted usually rests on the use of electrostatic and radio-frequency fields to efficiently mass select, guide, and manipulate the charged species and allow the reactions to be studied over a wide range of well-defined collision energies. The detection of ions is facilitated by the use of charge-sensitive devices such as electron multipliers.

We can identify a few broad families of methods (see Chapter 1) to study ion–neutral reactions: crossed and merged molecular beams, ion traps, and flow reactors. The frontier between these approaches remains subjective and follows unremitting technical evolutions. In this section, we focus on the use of uniform supersonic flows that belong to the family of flow reactors. Alternative methods are briefly evoked further below (Section 3.3).

Uniform flows are well adapted to quantitatively explore the reactivity of ions with large neutral molecules. Several ion production methods, explored since the design and construction of the first CRESU, are detailed hereafter. Uniform flow-based methods overcome some of the limits underlined for the techniques of crossed beams, ion traps, and subsonic flows.

3.2.1   *Electron beam ionization*

The first CRESU machine was specifically developed in the 80s in Meudon (Chapters 1 and 2) initially to study ion–molecule reactions. At that time, reactions involving charged species, including dissociative recombination, were thought to rule unchallenged over the gas phase chemistry of the ISM. The experimental investigation of low-temperature kinetics of reactions with charged species appeared necessary to assess more accurately the role played by ionic processes. In a seminal experiment in 1984, Rowe *et al.* implemented a high-energy (10–20 keV) electron gun on the CRESU reactor to produce an electron beam transversally crossing the flow [44]. This configuration led to the generation of cations either directly through electron impact ionization or indirectly by charge transfer or other processes with the ionized carrier gas [44]. Ion number densities typically fell in the 10^8-10^9 molecule cm^{-3} range. These values were estimated from Langmuir probe measurements from which the electron density could be accurately determined. The electron beam also produced some metastable excited species that could ionize the parent gas through the so-called Penning effect $(A^* + B \rightarrow A + B^+ + e^-)$.

This, of course, is detrimental for quantitative measurements as only ionic species can be monitored by mass spectrometry. One way to overcome this issue is to add a quencher to rapidly convert, i.e. a few mm downstream of the electron beam, the metastable species into ions. Because of collisions, the ions rapidly thermalize their rotation. By contrast, vibrational excited species may not have sufficient time to relax to their ground states given the short hydrodynamic time and number of collisions in typical uniform flows and the usual low rate coefficients for such process. The mode of ion production, mostly through dissociative charge transfer or proton transfer between the ionized carrier gas and the parent molecule, may lead to the generation of several competing ions. In that case, the difficulty resides in the determination of branching ratios when some of the exit channels for the reactions of the ions with the neutral gas overlap. The ion content of the uniform flow is determined by a movable quadrupole mass spectrometer equipped with a sampling truncated cone. As the speed of the flow is well known and constant (typically 1.6 kms^{-1} for helium), ion spatial measurements are easily converted into ion time profiles. The ion time profiles are then measured for a range of initial neutral co-reactant number densities (see Fig. 3.2) maintained in great excess (in the $10^{11}-10^{13}$ molecule cm^{-3} range) to ensure pseudo-first-order conditions. By plotting the loss rate (in units of s^{-1}) of the reacting ion as a function of the neutral number density, one can easily derive the rate coefficient of the reaction at a given temperature, $k(T)$. Alternatively, the decrease of the reactant ions versus reactant neutral concentration is measured at two locations, which allows in the same way a straightforward determination of the rate coefficient. With this well-established procedure, knowledge of the *absolute* ion number density is not required.

As stated earlier, the apparatus was first employed to measure the rate coefficients for the termolecular reactions $N_2^+ + N_2(+N_2)$ and $O_2^+ + O_2(+O_2)$ [44] and the bimolecular reaction $O_2^+ + CH_4$ [46] down to 20 K. The latter one demonstrated an unexpected 100-fold increase of the rate coefficient between 300 and 20 K. The measurements were quickly extended down to 8 K thanks to the use of helium as a carrier gas and the exceptional pumping capacity of Meudon's facilities topping 1,44,000 m^3 hr^{-1} (see also Chapter 2) [47]. Reaction rate coefficients of He$^+$ ions with N_2, O_2, and CO, and N$^+$ ions with O_2, CO, and CH$_4$ at 8 K were determined. With the

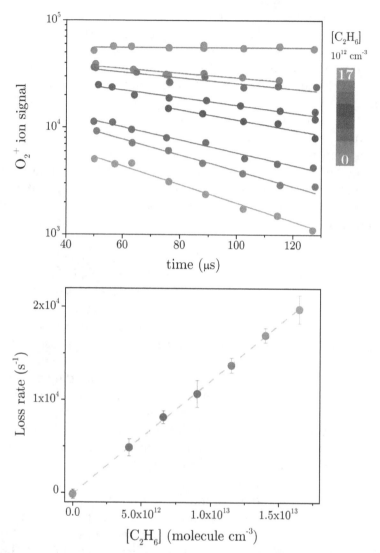

Fig. 3.2. Example of kinetic rate coefficient derivation for the $O_2^+ + C_2H_6$ reaction at 49 K. Top: ion counts of O_2^+ (log scale) as a function of flow time. Bottom: corresponding loss rates versus initial neutral co-reactant density C_2H_6. Reproduced with permission from Ref. [45], Copyright 2019 AIP Publishing.

exception of $N^+ + CO$, the reactions were found to be all fast and temperature independent by comparison with measurements performed at 300 K.

This arrangement can also be used to examine the reactivity of anions. Anions can be produced by dissociative attachment of secondary cold electrons onto a neutral precursor AP following the mechanism, $AP + e^- \rightarrow A^- + P$. For instance, the generation of C_3N^- anions can be achieved by the dissociative electron attachment onto BrC_3N injected in small amounts into the flow [48]. The precursor BrC_3N was selected for its efficiency of attaching cold electrons and giving predominantly C_3N^- fragments. One of the challenges is to identify precursors with the proper characteristics: high vapor pressure, chemical stability, exoergicity of the exit channel leading to the desired ion, and favorable branching.

3.2.2 *Selected ion nozzle*

Inspired by the success and the versatility of the SIFT (Selected Ion Flow Tube) method (Section 3.3.4) [49], Rowe and coworkers undertook in the late 80s the development of a CRESU device with ion selection (CRESUS) [50]. Under this configuration displayed Fig. 3.3, ions were produced by a low-pressure ($\leq 10^{-4}$ mbar) source, mass selected, and then injected into the flow through a 1-mm-diameter exit orifice located on the divergent part of the nozzle wall. The ions were then drifted in the nozzle using an electric field. Similarly to the SIFT technique, only the primary ions of interest and the neutral coreagents injected into the reservoir were present in the reaction zone. A nozzle-wall potential of $\sim$10 V was found to be adequate for an efficient drifting of ions. The nozzle was designed for helium as a carrier gas and for operation at flow temperatures of 20 and 70 K with a chamber pressure kept below 100 μbar. The uniformity of ions in the flow was validated by measurements of the radial and axial density profiles of the injected ions. The electric field was kept low enough to ensure that under the density conditions of the flow, the energy acquired by the ions was insufficient to populate electronic or vibrational energy levels. The primary ion count rate was monitored versus the position of the movable mass spectrometer and neutral reactant flow density to determine the rate coefficient.

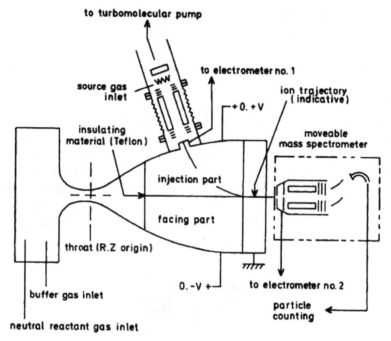

Fig. 3.3. Schematic view of the CRESUS machine (Cinétique de Réactions en Ecoulement Supersonique Uniforme avec Sélection). Reproduced with permission from Ref. [50], Copyright 1989 Royal Society of Chemistry (RSC).

The approach was validated through the measurement of a series of reactions ($He^+ + N_2$, $O_2^+ + CH_4$, and $N^+ + H_2$), some of which were known to be strongly influenced by the temperature and that were already studied using the CRESU technique [51] The drawback of the device lies in its limited temperature coverage as the ion source is incorporated in the Laval nozzle. Its complex operation also restricted its deployment.

3.2.3   *Pulsed injection of ions*

An alternative configuration was experimented later on by the Rennes group. In that version developed in the early 2000s, the ions are produced externally in a source and then drifted into the supersonic flow core [52]. This pulsed source was designed with the aim of overcoming the issue encountered with the CRESU's electron gun where several species are present in the plasma (e.g. He^+, He_2^+, and

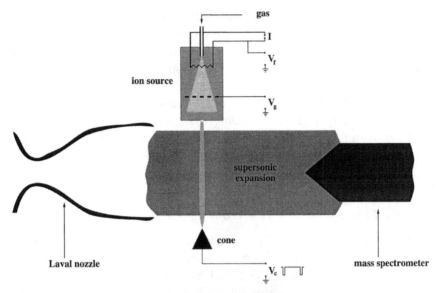

Fig. 3.4. Schematic view of the pulsed ion source implemented on the CRESU. Reproduced with permission from Ref. [52], Copyright 2001 Elsevier Science B.V.

metastable He), and may react with other molecules present in the flow and affect the measurements.

In the pulsed ion source (Fig. 3.4), which is located at the nozzle exit perpendicular to the flow, ions are produced by electron impact of the injected gas in the source and are then accelerated (by a typical potential of 110 V) toward the exit of the source. To drift the ions into the center of the supersonic jet, the potential of a cone (typically −120 V) that is mounted on the opposite side of the flow is turned on and off (during 250 μs) periodically (4 kHz) in order to release the ions from the source and allow them to propagate and thermalize within the uniform supersonic flow. The aim of a pulsed injection is to avoid heating of the ions by electric fields.

The injection of ions is similar to the motion of a motorboat in a fast river. The motorboat (ion) needs some significant velocity to reach the center of the river (supersonic flow) from the riverside where it departs. Once in the middle (center), its motor is stopped (electric field is shut down) and the boat (ion) then follows the streamline [53]. After a short travel in the flow, the ions pass through a quadrupole filter coupled with a channeltron to be detected during a given time gate. With the help of this experimental technique, the kinetics of several ion–molecule reactions (binary and ternary processes) was investigated to examine the relaxation of the primary ion. The targets were the reactions of Ar^+ with CH_4, C_2H_6, and N_2, $O_2^+ + CH_4$, $CH_3^+ + H_2(+He)$ and $CH^+ + CO$ which were investigated over the 23–300 K temperature range. The authors concluded that the primary ions are efficiently rotationally and translationally relaxed in the examined cases. However, the drawback of this simple experimental technique is the absence of mass selection in the ion injection.

3.2.4 *Mass-selective ion transfer line*

Inspired by the selected ion nozzle, the Rennes group in collaboration with the Fasmatech company recently designed and built a mass-selective ion transfer line that was implemented on a CRESU reacting chamber [45] In this setup (Fig. 3.5), the ions are first produced externally in a hollow cathode discharge ($P{\sim}1$ mbar), although

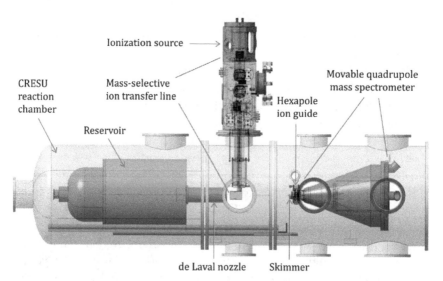

Fig. 3.5. Mass-selective ion transfer line implemented on a dedicated CRESU reacting chamber. A movable quadrupole mass spectrometer upgraded with a skimmer/hexapole ion guide combination is used for monitoring ions in the uniform supersonic flow. Reproduced with permission from Ref. [45], Copyright 2019 AIP Publishing.

a variety of ionization sources can be implemented. Ions are then transferred into a low-pressure ion guide through a radiofrequency ion funnel.

They are subsequently introduced into a quadrupole mass filter operated at the pressure of 10^{-5} mbar. The selected ions are transferred into a second radiofrequency ion guide and finally injected into the reaction chamber through a 2-mm aperture — the exit lens — located near the tip of the Laval nozzle. The voltage applied to a bottom deflector plate, which is positioned 5 cm from the exit lens, is adjusted to seed the isentropic core of the transverse uniform supersonic flow with the mass-selected ions. The mass-selective transfer line has been designed with the objective of maximizing ion transmission, ensuring the prompt thermalization of the mass-selected ions, and directing the ion beam into the isentropic core of the uniform supersonic flow. The high pressure in the source module at the starting point of the line (Fig. 3.6) ensures the initial cooling of all degrees of freedom of the ions including vibration. The heating of the ions

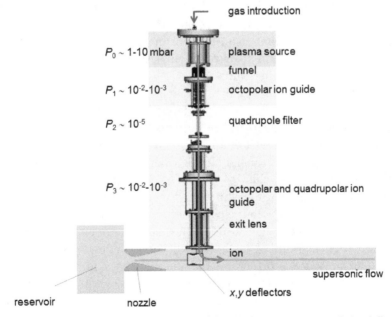

Fig. 3.6. Schematic view of the setup which emphasizes the role of the different components of the mass-selective transfer line associated with different pressure regimes.

during their transport from the source to the uniform flow is minimized by the use of moderate electric fields and by collisions with the neutral background gas.

Contrary to the external pulsed injector (Section 3.2.3), the device is continuously operated. The attractive potential applied to the deflector is adjusted (typ. 30 V) so that the ions are smoothly guided toward the center of the flow. Simulations of ion trajectories and properties — from the last section of the ion guide to the core of the uniform flow — show that thermalization (within a few K) of the ions in the reactor chamber takes place within a few centimeters downstream of their injection where the electric field E has dropped down to a few tens of V/m. The operation at lower densities n will mechanically raise the ion drift velocity v_d that linearly depends on the reduced field E/n. In that case, one mitigation strategy may be to suppress residual electric fields E by, e.g. adding a ring electrode set to the ground downstream of the deflectors. The preservation of a continuous operation guarantees a 100% duty cycle.

The approach has been validated in particular by the study of the $Ar^+ + N_2$ reaction which exhibits some marked temperature dependence over the 36–300 K range [45]. The reaction of O_2^+ with C_2H_6 was also investigated and the branching between the different exit channels was found to agree well with previous measurements reported at room temperature.

It is well known that quadrupole mass analyzers exhibit an overall ion transmission with an m/z dependence. This effect must be taken into account to ensure an accurate determination of branching ratios between the exit channels of the reaction. To calibrate the ion transmission, the transfer line is operated to select a single ion with a given mass to charge ratio. The ion current in the uniform supersonic flow is then measured on the tip of the weakly polarized skimmer with the help of an electrometer. Ions, that are sampled and transmitted through the hexapole ion guide (in which the pressure is maintained at 10^{-4} mbar to limit additional chemical reactions) and the quadrupole mass filter ($P \sim 10^{-5}$ mbar), are counted at the end of the chain by an electron counter under the exact same conditions. The procedure is repeated for a few other selected ions with different masses (e.g. by seeding the plasma with different atomic or molecular gases) and over the whole mass range.

3.3 Concurrent Methods to Investigate Ion–Molecule Reactions

3.3.1 *Crossed and merged beams*

Ion–molecule reactions can be probed under single collision conditions employing the crossed beam scattering technique developed in the 60s (see [54] and references therein). Crossed beam studies can give access not only to the identity of the products but also their velocity, angular distribution, and possibly their internal energy. Ion–neutral reactions can be explored over a wide range of collision energies from a few tens of meV to many eV, providing some insights into the influence of translational energy on the chemical processes. Crossed beam scattering experiments with ions have also benefited from the development of velocity map imaging (VMI) [55]. VMI allows the detection of reaction products irrespective of their scattering angle or velocity magnitude. The strength of the method lies in its ability to provide simultaneously and with high resolution details on velocity, angular distribution, and mass of charged particles [56, 57]. Crossed beams experiments represent a method of choice to investigate in details ion–molecule reaction dynamics and to test theory (Potential Energy Surface (PES) and quantum calculations) [2].

The derivation of the rate coefficient $k(T)$ at low temperatures from crossed beam measurements requires (1) a good determination of the cross section $\sigma(E)$ at low energies, which is generally challenging to obtain, (2) a high-energy resolution, and (3) a calibration of the cross-section measurements. This appears to restrain the capacity of the method in the determination of rate coefficients.

In practice, most apparatuses operate with an intersection angle of 90°, hence limiting the potential of the technique. Low relative collision energies also remain demanding to obtain, notably because of the energy resolution of the ion beam. In this regard, merged beams represent an attractive technique to reach very low relative energies (see Chapter 1 for an introduction to concepts). As for crossed beams devices, the neutral beam is produced by dual skimming of a supersonic jet generated by a pulsed valve. For the ion beam, ions are first prepared in an ion source, mass selected, and thermalized in a variable-temperature ion trap, and then transferred and merged with the neutral beam by magnetic deflection. Reactant and product ions are detected by mass spectrometry. The method leads to the measurement of integral reactive cross sections down to very low kinetic energies of 1 meV [58] (Chapter 1).

3.3.2   *Ion traps*

Ion trapping is another state-of-the-art method to study ion–molecule reactions at temperatures as low as 10 K. The technique was pioneered by Dunn in the early 80s [59, 60] and popularized by Gerlich with the iconic 22-pole trap [61]. In the later approach, the cryogenically cooled device is made of multiple electrodes (up to 22) on which non-homogeneous radio frequency fields are applied in order to produce an effective attractive potential to trap the ions [62]. In such multi-electrode traps with large field-free central regions, the radiofrequency heating is kept small. After each filling, an unreactive buffer gas is injected to cool down the ions through collisions. During their trapping time which usually varies between milliseconds and minutes, ions are then given the possibility to react with neutral gases introduced in a controllable amount into the trap. The extended range of storage times offers the possibility to measure both fast and slow ($< 10^{-12}$ cm^3 molecule^{-1} s^{-1}) bimolecular reactions. After a defined interval, ions are extracted and analyzed using a mass spectrometer. Since ion trapping rests on cryogenic cooling to obtain low temperatures, reactants, other than H_2 or He, partly condense onto the walls of the chamber. As a consequence, the derivation of kinetic rate coefficients at low temperatures is generally limited to reactions with light neutrals [63, 64], whose absolute concentration in the trap can be accurately determined.

3.3.3   *Hybrid traps*

In the last decade, studies with ions have been extended to temperatures well below 10 K as a result of the development of hybrid methods able to trap both ions and neutrals. Initially limited to atomic ions and neutrals, the techniques are progressively entering the field of molecular physics [65]. Such an approach can be employed to explore both the spectroscopy and reaction kinetics of ions. Ions are usually confined in electrodynamic radiofrequency traps or in combined electric and magnetic traps. Cooling the sample further in the millikelvin range or lower requires Doppler laser cooling. Molecular ions can be efficiently cooled sympathetically by the interaction with trapped, laser-cooled atomic ions. The cold trapped ions form ordered structures in the trap, usually referred to as Coulomb crystals. For neutral molecules, Stark deceleration of molecular beams can be employed to control the internal states and external motion with high precision [66]. This methodology able to generate cold molecular ions and cold neutral molecules opens the way to the study of ion–neutral reactions, energy transfer, and chemical dynamics in a much colder regime [67]. Sympathetic cooling by collisions with ultracold atoms in hybrid traps could also be applied to molecular anions [68]. In practice, however, the experiments done so far on molecular systems are generally limited by the residual collision energy between the cold trapped molecular ions and the incoming molecular neutrals (a few meV). Similar to cryogenic radiofrequency ion traps, the derivation of quantitative reaction rate coefficients can turn out to be challenging in the case of large neutral co-reactants (Chapter 1).

3.3.4 *Subsonic flow tubes*

Methods based on subsonic flows have been extensively used to investigate ion–molecule reactions. Pioneering studies were conducted using flowing afterglows developed for ionospheric chemistry by Ferguson and co-workers [69]. One of the key features of these methods is to provide kinetic rate coefficients under thermal conditions. In the 70s, Adams and Smith [49] achieved a breakthrough with the development of the selected ion flow tube (SIFT) technique, hence simplifying the chemistry while giving access to a wide variety of molecular ions. One of the strengths of subsonic flow methods, including SIFT, is to decouple production, reaction, and detection zones. Under the latter approach, the ion of interest is generated in a remote source (microwave or continuous discharge, electron gun, etc.), mass selected, and injected into a subsonic flow in which it rapidly thermalizes through collisions with the buffer gas. Ions can then react with the neutral co-reactant gas introduced in small amounts directly into the flow. The absence of the ion precursor gas in the flow avoids side reactions which could complexify the chemistry. Primary and product ions are then sampled downstream by mass spectrometry. Kinetic rate coefficients are derived by measuring the ion loss rate for varying initial density of the neutral reactant. The flow temperature can be cryogenically controlled down to typically 80 K with liquid nitrogen. Condensation onto the walls of the flow reactor restricts the method to high vapor pressure neutral reactants.

3.3.5   *Underexpanded free jets*

With the aim of reaching very low temperatures, Mark Smith and co-workers proposed in the 90s the use of a free jet reactor to measure ion–molecule reactions down to 1 K [70]. In this approach, ions are generated by resonantly enhanced multi-photon ionization (REMPI) of a precursor in an underexpanded free jet (Chapter 1). The precursor, along with the co-reactant, is diluted in an inert gas and introduced into the reservoir, onto which the nozzle is mounted. The ionization takes place a few mm away from the exit of the pulsed nozzle. The ion packet then travels downstream in a field-free environment while reacting with neutral collisional partners. The ion population is monitored by time-of-flight mass spectrometry, giving access to branching ratios between the exit channels. The kinetic analysis turns out to be arduous as it needs to take into account the free jet dynamics, i.e. constantly changing temperature and pressure. Free jets in essence, generate a non-equilibrium medium and vibrational, rotational, and translational (perpendicular and parallel) temperatures end up decoupled, sometimes very strongly. Insights into the concept of temperature can be found in Chapter 1. A theoretical framework was developed to allow the deconvolution of the data obtained to provide meaningful rate coefficients both for bimolecular and termolecular reactions. The expansion of the method was, however, hampered by the complicated processing required and by the difficulty to assign a temperature to the derived rate.

3.4   Low-Temperature Reactivity of Ions

The quest for understanding the reactivity of ions at low temperatures has been largely driven by the need for reliable data to feed photochemical models. It benefited from a variety of dedicated methods with specific advantages (Sections 3.2 and 3.3). Results obtained with the CRESU played a significant part in this pursuit. In the last three decades, over eighty five chemical reactions were studied using the CRESU apparatus. The temperature range covered, routinely down to 20 K, extended to 8 K with the Meudon original setup. The full list of ion–molecule reactions examined is listed in Tables 3.1 and 3.2.

Table 3.1. Bimolecular ion–molecule reactions studied with CRESU.

Reactions with Atomic Ions				
Ion	**Neutral**	$T(\mathbf{K})$	**Reference**	**Method**
Ar^+	H_2	20; 30; 70	[50, 76]	CRESU[a], CRESUS[b]
Ar^+	N_2	20–300[e]	[45, 50, 52, 76]	CRESU, CRESUS, CRESU-PI[c], CRESU-SIS[d]
Ar^+	O_2	20; 30; 70	[50, 76]	CRESU, CRESUS
Ar^+	CO	20; 30; 70	[50, 76]	CRESU, CRESUS
Ar^+	N_2O	36; 49; 72	[45]	CRESU-SIS
Ar^+	CH_4	23; 36; 71	[52]	CRESU-PI
Ar^+	C_2H_6	23–71	[45, 52]	CRESU-PI, CRESU-SIS
$Ar^{2+}(^3P)$	Ar	30	[130]	CRESU
$Ar^{2+}(^3P)$	$He(^1S)$	30	[130]	CRESU
$Ar^{2+}(^3P)$	H_2	30	[130]	CRESU
$Ar^{2+}(^3P)$	N_2	30	[130]	CRESU
$Ar^{2+}(^3P)$	O_2	30	[130]	CRESU
$Ar^{2+}(^3P)$	CO_2	30	[130]	CRESU
C^+	HCl	27; 68	[77, 78]	CRESU
C^+	H_2S	27; 68	[77, 78]	CRESU
C^+	H_2O	27; 68	[86]	CRESU
C^+	NH_3	27; 68	[107]	CRESU

(*Continued*)

Table 3.1   (*Continued*)

Reactions with Atomic Ions				
Ion	**Neutral**	$T(K)$	**Reference**	**Method**
C^+	SO_2	27; 68	[77, 78, 131]	CRESU
C^+	C_6F_6	27; 68	[132]	CRESU
C^+	$c\text{-}C_6H_{12}$	27; 68	[132]	CRESU
Cl^-	CH_3Br	23–158	[133]	CRESU
He^+	N_2	8–70	[50, 51, 71, 134]	CRESU, CRESUS
He^+	O_2	8; 20; 50	[51, 71, 134]	CRESU
He^+	CO	8; 20	[51, 134]	CRESU
He^+	HCl	27; 68	[77, 78]	CRESU
He^+	H_2S	27; 67; 68	[77, 78]	CRESU
He^+	CO_2	70	[50]	CRESUS
He^+	SO_2	27; 67; 68	[77, 78]	CRESU
He^+	NH_3	27; 68	[107]	CRESU
He^+	H_2O	27; 68	[107]	CRESU
He^+	CH_4	23	[71]	CRESU
He^+	C_6F_6	27; 68	[132]	CRESU
He^+	$c\text{-}C_6H_{12}$	27; 68	[132]	CRESU
N^+	H_2	8–163	[51, 82, 86, 134]	CRESU, CRESUS
N^+	$p\text{-}H_2$	20–163	[82]	CRESU
N^+	HD	20	[82]	CRESU
N^+	D_2	45; 68; 163	[82]	CRESU
N^+	O_2	8	[51, 134]	CRESU
N^+	CO	8	[51, 134]	CRESU
N^+	H_2O	27; 68; 163	[107]	CRESU
N^+	NH_3	23–170	[107, 112, 114]	CRESU
N^+	CH_4	8; 163	[50, 51, 134]	CRESU
N^+	$1\text{-}1\text{-}C_2H_2Cl_2$	27; 68; 163	[135]	CRESU

Table 3.1   (*Continued*)

Reactions with Atomic Ions				
Ion	**Neutral**	$T(K)$	**Reference**	**Method**
N^+	cis-1,2-$C_2H_2Cl_2$	27; 68; 163	[135]	CRESU
N^+	trans-1,2-$C_2H_2Cl_2$	27; 68; 163	[135]	CRESU
N^+	C_6F_6	27; 68	[132]	CRESU
N^+	c-C_6H_{12}	27; 68	[132]	CRESU
O^+	NO	23–168	[136]	CRESU
O^+	O_2	20	[134]	CRESU
O^+	N_2	23–222	[137]	CRESU, CRESU-PI

Reactions with Molecular Ions				
Ion	**Neutral**	$T(K)$	**Reference**	**Method**
CH^+	CO	23–71	[52]	CRESU-PI
CN^-	HC_3N	49–294[e]	[118]	CRESU
CN^-	HCOOH	36–158	[119]	CRESU
N_2^+	H_2	20; 70	[50]	CRESUS
N_2^+	O_2	8–163	[51, 138]	CRESU
N_2^+	CO_2	70	[50]	CRESUS
N_2^+	CH_4	70	[50]	CRESUS
O_2^+	CH_4	20–163	[46, 50, 52]	CRESU, CRESUS, CRESU-PI
O_2^+	CD_4	20; 45; 68	[139]	CRESU
O_2^+	CH_3D	20–163	[115]	CRESU
O_2^+	CH_2D_2	20–163	[139]	CRESU
O_2^+	CHD_3	20–163	[139]	CRESU
O_2^+	C_2H_6	49	[45]	CRESU-SIS
H_3^+	CH_4	30	[72]	CRESU
H_3^+	CO	30	[72]	CRESU

(*Continued*)

Table 3.1   (*Continued*)

Reactions with Molecular Ions				
Ion	Neutral	$T(K)$	Reference	Method
H_3^+	N_2	30	[72]	CRESU
H_3^+	NH_3	30	[72]	CRESU
H_3^+	H_2S	30	[72]	CRESU
H_3^+	SO_2	30	[72]	CRESU
$C_3^{15}N^-$	$HC_3^{14}N$	49–294[e]	[48]	CRESU
C_3N^-	HC_3N	49–294[e]	[48]	CRESU
C_3N^-	$HCOOH$	36–158	[119]	CRESU
NH_3^+	H_2	8; 20	[134]	CRESU

Notes:
[a]CRESU with electron beam ionization (Section 3.2.1).
[b]CRESUS: "Cinétique de Réaction en Ecoulement Uniforme Supersonique avec Sélection" (Section 3.2.2).
[c]PI: external Pulsed Injection of ions (Section 3.2.3).
[d]SIS: external and continuous mass-Selective Ion Source (Section 3.2.4).
[e]Results obtained with a sub-sonic Laval nozzle at room temperature.

Attempts to enable the emergence of a global picture for bimolecular reactions have been sketched. It was found that for the majority of exoergic ion–molecule reactions, the reaction probability is close to unity, yielding a **fast rate coefficient** reaching the capture limit k_c. The reaction is then dominated by long-range interactions. Some exceptions to this general rule exist, corresponding to an exoergic reaction with a probability lower than 1. This can be due to the domination of short-range interactions leading to a rate coefficient lower than k_c. It can also be caused by the existence of several potential energy surfaces correlated to the reactants which may have energy barriers leading to a rate coefficient lower than k_c at low temperature.

For **non-polar neutral partners**, the rate coefficient is predicted by the Langevin theory to be temperature independent. Reactions of Ar^+, He^+, N^+, N_2^+, and H_3^+ with CH_4 investigated with the CRESU all fall under this category and show no temperature evolution given the uncertainty of the measures [45, 48, 50, 69, 70].

Table 3.2. Ternary ion–molecule reactions studied with CRESU methods. M is the third body.

Ion	Neutral	M	$T(K)$	Ref.	Methods
Ar^+	Ar	Ar	27–141	[140]	CRESU[a]
C^+	H_2	He	20; 82	[50]	CRESUS[b]
N_2^+	N_2	N_2	20–163	[44, 51]	CRESU
O_2^+	O_2	O_2	20–163	[46]	CRESU
CH_3^+	H_2	He	20–82	[50, 52]	CRESUS, CRESU-PI[c]
CH_3^+	CO	He	20; 70; 82	[50]	CRESUS
CH_3^+	N_2	He	20; 70; 82	[50]	CRESUS
NH_4^+	NH_3	He	15–123	[141]	CRESUS
NH_4^+	NH_3	Ar	45; 52	[141]	CRESU
NH_4^+	NH_3	N_2	48–170	[141]	CRESU
H_3O^+	H_2O	He	23–158	[125]	CRESU
H_3O^+	H_2O	N_2	48–170	[125]	CRESU

Notes:
[a]CRESU with electron beam ionization.
[b]CRESUS means CRESU method with a Selective ion source coupled with the Laval nozzle.
[c]PI: external Pulsed Injection of ions.

As displayed in Table 3.1 and illustrated in Fig. 3.7 for reactions of ions with CH_4, it was found that the measured rate coefficients of these reactions amount to 60–100% of the calculated Langevin rates. For proton transfer reactions $AH^+ + B \rightarrow A + BH^+$ with high exoergicity, the rate coefficient is commonly high at room temperature and generally remains high at low temperature. In this kind of reaction, one condition for having an effective proton transfer is that the proton affinity of the neutral reactant B is greater than the neutral product A as illustrated by the $H_3^+ + CH_4$ reaction.

Usually, for an exoergic reaction involving a polar neutral partner, the rate coefficient is fast and increases when the temperature

$$\text{ion}^+ + \text{CH}_4$$

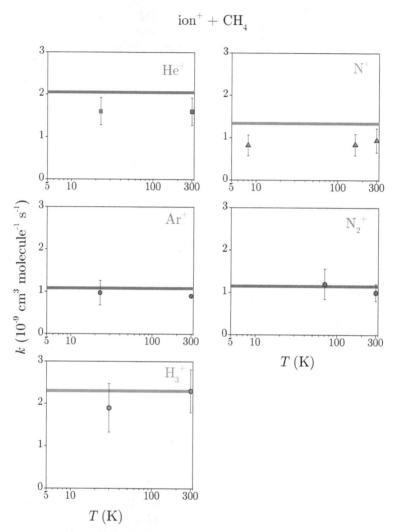

Fig. 3.7.   Rate coefficients measured with the CRESU method over the 8–163 K range for several reactions of atomic and molecular ions with methane CH_4. Room-temperature measurements were determined with an SIFT apparatus. Thick lines are calculated Langevin values with the polarizability of CH_4 taken as 2.448 Å³ [63].

decreases. When the collision energy gets lower, capture and formation of an intermediate complex are more effective. Several theories can estimate the temperature dependence of the rate coefficient considering additional electrostatic interactions such as charge–dipole (Chapter 11).

The most renowned theories for ions are the Average Dipole Orientation (ADO) theory developed by Su & Bowers [73], the ADO theory with conservation of the angular momentum (AADO) further improved by Su and collaborators [74], the adiabatic capture centrifugal sudden approximation (ACCSA) elaborated by Clary [75], and the Statistical Adiabatic Channel Model (SACM) proposed by Troe that considers some long-range additional interactions like the ion–dipole one. The $N^+ + H_2O$ and $N^+ + NH_3$ exoergic reactions, described further below, belong to the class of fast reactions with a polar partner. However, the kinetics of ion–molecule reactions does not always follow the predictions made by simple models (Section 3.1.1). In that case, experimental data can be employed to benchmark more sophisticated theoretical models.

For **slow reactions** at room temperature, i.e. with a reaction probability much lower than k_c, there is no general trend to predict the temperature dependence of the rate coefficient. Several reactions show a negative temperature dependence of the rate coefficient with, sometimes, an extrapolated limit corresponding to k_c at 0 K. Note that the increase of the rate coefficient at low temperature does not systematically correspond to a larger reaction probability because of the joint increase of the rate coefficient and k_c. For example, it can be the consequence of the formation of a long-lived intermediate complex. The $Ar^+ + N_2$ charge transfer reaction [45, 50, 52, 76], which is detailed below (Section 3.4.1.2), highlights this. This can also be explained by the open shell nature of the reactant as in the $C^+ + HCl$ reaction [77–79]. Sometimes, a reaction remains slow between room temperature and down to the lowest temperatures such as $He^+ + H_2$ for which the rate coefficient is about a few 10^{-13} cm^3 molecule^{-1}s^{-1} [80]. In some cases, the rate coefficient decreases with temperature as for the reaction $N^+ + H_2$, which is detailed below (Section 3.4.1.1).

We review below a selection of reactions explored using CRESU techniques, underlining the role of specific interactions. We consider separately the case of bimolecular ion–non-polar neutral molecule and ion–polar neutral molecule collisions. Finally, we point out an example of termolecular association, which is a pressure- and temperature-dependent process, with the $H_3O^+ + H_2O$ reaction (see Section 3.4.3).

3.4.1 *Non-polar reactions*

The long-range potential that describes the interaction between an ion and a non-polar neutral is essentially governed by the charge-induced dipole interaction [81]. This interaction depends on the inter-molecular distance as $\alpha/2R^4$ with R being the distance between the reactant center of mass and α, the polarizability of the neutral partner. This leads one to derive a cross section as

$$\sigma(E) = \pi\sqrt{(2\alpha/E)} \qquad (3.3)$$

and a Langevin rate coefficient independent of temperature as

$$k_L = 2\pi e\sqrt{\alpha}/m \qquad (3.4)$$

where m is the reduced mass and e the elementary charge (in cgs-esu units). The simple Langevin description is not always sufficient to account for the underlying mechanisms of the reaction. Examples include the $N^+ + H_2$ reaction studied by Marquette *et al.* [50, 82] and the $Ar^+ + N_2$ charge transfer reaction [45, 50, 52, 76]. They are detailed below.

3.4.1.1 *The $N^+ + H_2$ reaction*

Ammonia NH_3 was the first polyatomic molecule detected in the ISM in the direction of the Galactic Center, by means of microwave emission spectra [83]. The relative abundances of NH_3 compared with H_2 can vary from 10^{-8} in dark clouds to 10^{-5} in dense cores, such as the Orion molecular cloud [84, 85] The understanding of the elementary physical and chemical processes leading to the formation of NH_3 is then an important issue in the field of astrochemistry. Commonly, in dense dark clouds, the ion–molecule reactions prevail in forming NH_3 in the gas phase. The generation of NH_3 proceeds via a succession of H abstraction reactions, the first key step being

$$N^+ + H_2 \rightarrow NH^+ + H \qquad (3.5)$$

This reaction was explored by Rowe and co-workers between 8 and 67 K [51, 86]. In interstellar molecular clouds, it is usually admitted that H_2 is formed on dust grains with an ortho-to-para ratio (OPR) of 3:1. Several works consider that in the ISM, the OPR conversion can occur through interactions with the dust surface or through the hydrogen exchange reaction with H and H^+ (and possibly with H_3^+) in the gas phase [87, 88]. In the photodissociation regions where the observations of the H_2 emission lines give an OPR far below the standard 3:1, these processes can be very efficient [89]. In dark clouds and cold regions, H_2 is not directly observable and the OPR remains elusive even if it is usually assumed that most of H_2 is in its para form. The relative abundance of NH_3 is very sensitive to the H_2 OPR in the ISM and then to the rate coefficients of the reaction (3.5) with both H_2 isotopomers. A follow-up study was conducted with the CRESU between 8 and 163 K to investigate the reactions [82]:

$$N^+ + n\text{-}H_2 \rightarrow NH^+ + H \tag{3.6}$$

$$N^+ + p\text{-}H_2 \rightarrow NH^+ + H \tag{3.7}$$

with $p\text{-}H_2$ and $n\text{-}H_2$ standing for para-H_2 and normal-H_2, respectively. A second objective in that study was to explore the reactions

$$N^+ + HD \rightarrow ND^+ + H$$

$$NH^+ + D \tag{3.8}$$

$$N^+ + D_2 \rightarrow ND^+ + D \tag{3.9}$$

The products $NH^+ + D$ from the reaction $N^+ + HD$ were not observed due to the excessive endoergicity of the reaction. In these studies, the electron beam indirectly led to the generation of $N^+(^3P)$ ions in the ground state with an excess of kinetic energy of about 0.14 eV, from the reaction $He^+ + N_2 \rightarrow N^+(^3P) + N + He$. Because of the high density of the uniform flow, N^+ ions relax in a few microseconds to reach the temperature of the bath gas. In the ISM, the mean free path is huge and this additional kinetic energy is enough to overcome the small endoergicity of the $N^+(^3P) + H_2 \rightarrow NH^+ + H$ reaction derived from CRESU measurements ($\Delta H^0 = 18 \pm 2\,\text{meV}$) [86, 90]. Between 8 and 163 K, the general trend for $n\text{-}H_2$ and $p\text{-}H_2$ is an increase of the experimental rate coefficients with the temperature, but the reaction with $n\text{-}H_2$ is more efficient (see Fig. 3.8).

Actually, because of the very slow interconversion, $p\text{-}H_2$ and $o\text{-}H_2$ can be considered as two distinct species in supersonic flows.

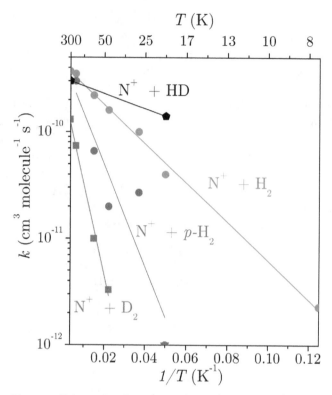

Fig. 3.8. Rate coefficients for the $N^+ + n\text{–}H_2$, $N^+ + p\text{–}H_2$, $N^+ + HD$, $N^+ + D_2$ reactions versus inverse temperature [82]. The room-temperature data are taken from Ref. [91].

The authors concluded that in addition to the kinetic energy, the rotational and spin-orbit states are essential to explain the greater efficiency of the $N^+(^3P) + n\text{–}H_2$ reaction. Indeed, even at low temperature, both excited rotational states of H_2 and spin-orbit states of N^+ can be significantly populated and will contribute to the overall reactivity.

As the reactivity of excited spin-orbit states of N^+ is much greater than that of the fundamental state, the fast rise of the reaction rate coefficients with increasing temperature is explained by the enhanced role of these states. Such conclusion was confirmed later theoretically by Grozdanov *et al.* [92]. These authors used a statistical treatment to predict state-selective reactive collisions involving N^+ ions in all fine structures (spin-orbit level) with H_2 or HD molecules in different rotational levels. Their results allowed for the determination of reaction rate coefficients for any given distribution of specific fine structure and rotational state populations of the reactants and allowed for a detailed analysis of the CRESU measurement.

3.4.1.2 *The $Ar^+ + N_2$ reaction*

Another reaction of interest is $Ar^+ + N_2$ [45, 50, 52, 76] where the possibility of forming excited vibrational states of the product plays a fundamental role in this charge transfer (CT) reaction. The CT reactions $(A^+ + B \to A + B^+)$ are very common ion–molecule reactions. From an astrochemical point of view, the CT plays a key role in the hydrogen chemistry in the Early Universe [93] but also in the planetary atmospheres of Mars [94], Venus [95], the Jovian satellite Europa [96] or Saturn's biggest satellite, Titan [33], and in the ISM [4, 11].

The CT reaction is associated to non-adiabatic processes involving $(A^+ + B)$ and $(A + B^+)$ interacting PESs which correspond to a dynamical coupling between internuclear and electronic motion (i.e. a non-adiabatic process). Usually, these reactions are explained by an electron jump at large intermolecular distance or by some "intimate" collision mechanisms which correspond to a low-energy charge transfer. The first case is characterized by a large and weakly energy-dependent cross section and no close interaction between the reactant and products states. The second case corresponds to a strong interaction (either on the initial adiabatic PES or after a non-adiabatic change on the other PES) and an efficient exchange between translational and internal energy.

When a molecule is involved, the number of its degrees of freedom complicates the description of the process. Indeed, in asymmetrical charge transfer reactions involving a molecule, the strong reaction probability can be explained by a close resonance (small energy defect) and favorable Franck-Condon factors between the reactant and product states. Then, a set of reactant internal states leads to several charge transfer reactions distinguished by their energy defect and their Franck-Condon factors.

The charge transfer reaction $Ar^+ + N_2$ has an unusual rate coefficient dependence with temperature which has been often theoretically and experimentally examined [45, 76, 99–101]. Figure 3.9 shows

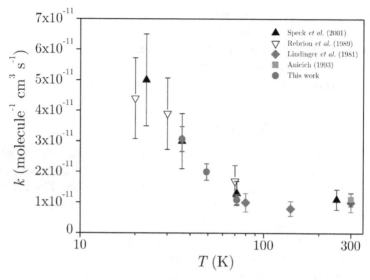

Fig. 3.9. Rate coefficients for the $Ar^+ + N_2$ reaction as a function of temperature from Joalland *et al.* [45] Blue circles: experimental data with the new CRESU-SIS apparatus. Triangles: previous CRESU measurements [52, 76]. Green diamonds: SIFT measurements [97]. Red square: average room temperature value taken from the review of Anicich [98].

the minimum of the rate coefficient around 140 K which has been observed by several studies. This general trend is explained by a competition between the endoergic (0.09 eV) $Ar + N_2^+(v=1)$ and the exoergic (–0.18 eV) $Ar + N_2^+(v=0)$ exit channels of the reaction. At high temperature, the endoergic channel dominates because of its closest energy to the reactant state, violating the Franck-Condon principle. At the opposite, at very low temperature, the $Ar + N_2^+(v=1)$ exit channel is too endoergic to be observed, then only the exoergic channel is possible. The causes behind the rise of the rate coefficient at low temperatures remain nonetheless elusive.

From these two examples, one can see that the temperature variation of ion–non-polar molecule reactions sometimes deviates from capture (Langevin) theory. Indeed, several quantum effects such as non-adiabatic effects or resonances can dominate the reaction process. Such effects are expected to play a possible stronger role in ion–polar molecule reactions, and in the next section we describe some interesting features of some well-studied collisional systems.

3.4.2 *Ion–polar neutral reactions*

Several exoergic ion–molecule reactions are found very efficient at room temperature (typically $k \geq 10^{-9}$ cm^3 molecule^{-1} s^{-1}) well beyond their charge–polarizability interaction. This is the case for reactions where the neutral partner has a permanent dipole moment which leads to a charge–dipole interaction so that a corresponding additional term to the Langevin rate coefficient must be taken into account to model the interaction potential. As a consequence, the rate coefficients usually increase with decreasing temperature and several models have been proposed. A selection of them is briefly exposed below.

In the Locked Dipole model [102, 103], the orientation of the permanent dipole μ of the neutral partner is considered. If the dipole orientation is not aligned with the incoming charge, the charge–dipole interaction is negligible and the rate coefficient is the Langevin one. If not, the charge–locked-in dipole interaction corresponds to the attractive potential term $-\frac{\mu}{r^2}$ and the rate coefficient becomes (cm^3 molecule^{-1} s^{-1} units)

$$k_{LD} = k_L + 2\pi\mu\sqrt{\frac{2}{m\pi k_B T}} \tag{3.10}$$

where T is the temperature and k_B the Boltzmann constant. Because this corresponds to an "ideal" situation, the dipole effect is frequently overestimated when compared to reality.

The Average Dipole Orientation model (ADO) [74, 104] considers the average orientation of the permanent dipole moment μ by considering the average angle $\bar{\theta}$ between the dipole and the reaction axis in the attractive term $-\frac{\mu}{r^2}\cos(\bar{\theta})$. Later, Bass *et al.* proposed to use the $\overline{\cos(\theta)}$ term instead [105]. In this model, the rate coefficient can be obtained with the formula (in cm^3 molecule^{-1} s^{-1})

$$k_{ADO} = k_L + 2\pi e\mu C\sqrt{\frac{2}{m\pi k_B T}} \tag{3.11}$$

where C is the empirical dipole locking constant obtained from a fit of several experimental rate coefficients [73, 104]. In this approach, there is no net angular momentum between the rotating molecule and the ion–molecule orbital motion.

An improvement to the ADO model was introduced by Su and Chesnavich [106]. The Angular Momentum Conserved model AADO was explicitly developed for linear rotors considering the coupling between the rotational angular momentum of the dipole (J) and the orbital angular momentum of the system (L). The rate coefficient becomes then

$$k_{AADO} = k_L \times k_{cap}(x) \tag{3.12}$$

where the dimensionless parameter x is equal to $\mu/\sqrt{(2\alpha k_B T)}$. The empirical parameterization gives

$$k_{cap}(x) = 0.9754 + \frac{(x + 0.509)^2}{10.526} \text{ for } x < 2 \tag{3.13}$$

and

$$k_{cap}(x) = 0.6200 + x \times 0.4767 \text{ for } x \geq 2 \tag{3.14}$$

There is a net angular momentum transfer between the rotating molecule and the ion–molecule orbital motion. This model works pretty well for linear rotors but usually underestimates the rate coefficient at the lowest temperatures [107, 108].

Alternative models to capture theory, such as adiabatic capture centrifugal sudden approximation (ACCSA) proposed by D. Clary [75], were also historically used to model ion–molecule reactions. Such methods (described in detail in Chapter 11) lead to a good agreement with kinetic measurements at low temperature as in the case of proton transfer reactions of H_3^+ with molecular neutrals [72]. A similar approach, the statistical adiabatic channel theory (SACM) by Quack & Troe developed in the mid-seventies [109, 110], recently revisited by Loreau *et al.* [111], was also used to determine cross sections and rate constants of ion–molecule reactions. At low temperature, this approach, based on a statistical description of the collisions, provides accurate results in good agreement with more accurate quantum calculations [111].

Below, we describe several ion–polar molecule reactions.

3.4.2.1 $N^+ + H_2O$ and $N^+ + NH_3$ reactions

The $N^+ + H_2O$ and the $N^+ + NH_3$ reactions illustrate pretty well the importance of the charge–dipole interaction [107, 112]. The dipole of

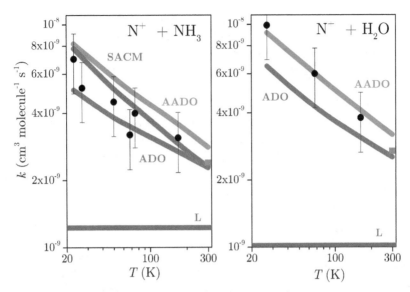

Fig. 3.10. Rate coefficients for the $N^+ + H_2O$ and $N^+ + NH_3$ reactions versus temperature. The experimental data have been obtained with the CRESU [107, 112, 114] and SIFT [115] methods. The rate coefficients k_L, k_{ADO}, and k_{AADO} are calculated with the Langevin, ADO (the C locking dipole constant is extracted from the fit of the experimental values [104]), and AADO models, respectively. The polarizabilities used to calculate the rate coefficients come from KIDA and NIST databases (https://kida.astrochem-tools.org/ and cccbdb.nist.gov/pollistx .asp).

H_2O and NH_3 is, respectively, 1.85 and 1.47 D and the measured rate coefficients are displayed in Fig. 3.10. The effect of the charge–dipole interaction is truly significant at low temperatures and thus results in negative temperature dependence well above the Langevin value.

The calculated ADO and AADO rate coefficients for the $N^+ + H_2O$ reaction are in good agreement with the experimental ones over all the temperature ranges considered. From this comparison, it is difficult to identify the most adapted model. For the $N^+ + NH_3$ reaction, available SACM calculations as well as ADO and AADO estimates all reproduce satisfactorily the experimental results. Clary *et al.* [113] have demonstrated that the strong negative temperature dependence of the measured rate coefficient is related to the sensitivity of the rotationally selected rate coefficients $k_{jk}(T)$ to the initial rotational state j, where j and k are the molecular angular momentum and its projection along the molecular axis of symmetry, respectively. At low temperature, the contribution of $k_{0k}(T)$ to the measured rate coefficient is important contrary to high temperatures where excited rotational states start being significantly populated.

CRESU studies have clearly exemplified the influence of the dipole moment on the temperature dependence of the reaction rate coefficient. In addition, various reactions studied with the CRESU method have demonstrated that the electrical quadrupole moment of the neutral partner has no significant influence on the rate coefficient even at low temperatures [78]. In recent years, particular efforts have been undertaken to explore reactions involving negative molecular ions, as shown below.

3.4.2.2 $C_xN^- + HC_xN$ reactions

A significant advantage of the CRESU is the capability to study molecular ion–heavy molecule reactions. This is very useful to understand the formation of complex organic molecules in the ISM — more than 6 atoms as defined by Herbst and Van Dishoeck [116] — and to explain the formation of the observed abundant heavy molecular systems in molecular clouds. It is also helpful to understand the complex chemistry found in planetary atmospheres such as the one of Titan.

In Titan's atmosphere, ion–molecule chemistry, neutral chemistry, and aggregation play a crucial role in the formation of aerosols, but the elementary mechanisms are still not well understood [38]. As mentioned in Section 3.1.3, the cyanide anion CN^- is likely present in the upper atmosphere of Titan [37]. Larger nitriles, such the ubiquitous cyanoacetylene HC_3N, have been detected in the mid-stratosphere of this satellite [117]. The chemical pathways toward heavier C_xN^- species may hinge on successive nucleophilic substitutions starting with the $CN^- + HC_3N$ reaction. This hypothesis has motivated recent studies of the $C_{x=1,3}N^- + HC_3N$ reactions with the CRESU between 49 and 294 K. For the $CN^- + HC_3N$ reaction, the only observed exit channel is the proton exchange leading to the C_3N^- formation, while for the $C_3N^- + HC_3N$ reaction [118], an additional exit channel corresponding to a reactive detachment process has been observed [48].

$$
\begin{aligned}
CN^- + HC_3N \quad &\rightarrow HCN + C_3N^- & (3.15)\\
C_3N^- + HC_3N \quad &\rightarrow HC_3N + C_3N^- &\\
&\rightarrow \text{neutral products} + e^- & (3.16)
\end{aligned}
$$

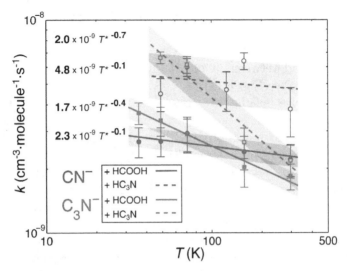

Fig. 3.11. Temperature dependence of the rate coefficients of the $C_{2n+1}N^-$ reactions with HCOOH and HC_3N measured with the CRESU technique. Experimental data are fitted with the power law $k(T) = \alpha(T^*)^{\beta}$ with $T^* = T/298$. Reproduced with permission from Ref. [119], Copyright 2016 ACS.

In both cases, the rate coefficient presents negative temperature dependence as expected, but stronger for the $C_3N^- + HC_3N$ reaction than for the $CN^- + HC_3N$ reaction (see Fig. 3.11).

In this kind of reactive collisions, i.e. S_N^2 nucleophilic substitution of general form $X^- + RY \rightarrow XR + Y^-$, as with all ion–molecule reactions, the ion-induced dipole interaction dominates at long intermolecular distance. If a polar molecule is involved, its free rotation generates a time-averaged permanent dipole moment. When the energy corresponding to a "locked" dipole is comparable to the ion-induced dipole interaction, then the influence of the permanent dipole moment is non-negligible. In the single step of the reaction, the $(X\text{-}RY)^-$ complex is formed and the charge moves from X to Y *via* a transition state, before the $XR\text{-}Y^-$ separation. In summary, the interaction energy has two minima with the reaction coordinate separated by a submerged energy barrier (i.e. much below the dissociation limit). If this energy is high, the reaction is slow.

For the $C_{x=1,3}N^- + HC_3N$ reactions, the experimental results suggest a barrierless process or a very weak submerged barrier. The difference between the temperature dependence of the rate coefficients is essentially thought to be due to the dipole–dipole interaction.

These two cyano ions indeed exhibit very different dipole moments: $\mu_{CN-} = 0.6$ D and $\mu_{C3N-} = 3.1$ D [120]. This hypothesis was reinforced by the study of the $C_{x=1,3}N^- + HCOOH$ reactions with the CRESU method between 36 and 158 K for which the same trend was observed [119] (see Fig. 3.11).

To gain more insight into these experimental findings, Joalland *et al.* [119] performed quantum mechanical calculations based on a reduced dimensional approach to simulate the $CN^- + HCOOH \rightarrow HCN + HCOO^-$ reaction dynamics. As quantum mechanical calculations that consider all the dimensions of the problem were not applicable for such a large system, the authors computed a global two-dimensional (2D) interaction potential that correctly describes the reaction path and used it to solve the nuclei motion. They employed a 2D quantum reactive dynamics code based on the finite element method in order to determine the reactive cross sections and rate coefficients. The agreement between the theoretical and experimental results was found to be rather good on an absolute scale, especially considering the reduced dimensional approximation used. The weak temperature dependence of the ion–molecule reaction involving CN^- was theoretically confirmed (see dashed line in Fig. 3.12). The calculations overestimate the rate coefficients at room temperature by a factor 2–3 because of the validity limit of the reduced dimension approach. More generally, this combined theoretical and experimental study confirmed that the dipole–dipole interactions are important when considering ion–molecule reactions involving molecules with non-zero dipole moment.

3.4.3 *Association reactions*

Association reactions of ions with neutral molecules play an essential role in rarefied environments, such as the interstellar medium, through radiative association described by

$$A^{+/-} + B \xrightarrow{k_{rad}} AB^{+/-} + h\nu \qquad (3.17)$$

with a rate coefficient k_{rad} in cm^3 molecule^{-1} s^{-1}. Traps in which ions can be stored under ultrahigh vacuum have been employed to investigate such processes. In denser media, such as planetary atmospheres

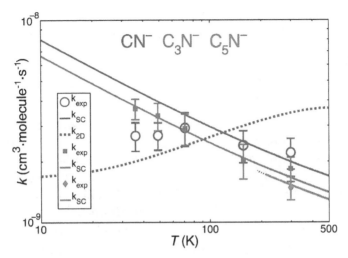

Fig. 3.12. Rate coefficients for the $C_{2n+1}N^- +$ HCOOH reactions measured experimentally from room temperature down to 36 K (k_{exp}) and estimated from the Su and Chesnavich capture model (k_{SC}, continuous line). The result of our two-dimensional quantum modeling of the $CN^- +$ HCOOH reaction is also shown (k_{2D}, dashed line). Reproduced with permission from Ref. [119], Copyright 2016 ACS.

and comae of comets, termolecular association indisputably dominates. The overall process can be summarized as

$$A^{+/-} + B + M \xrightarrow{k_t} AB^{+/-} + M \qquad (3.18)$$

and characterized by a termolecular reaction rate coefficient k_t given in units of cm^6 molecule^{-2} s^{-1}. The CRESU method is well suited to explore termolecular association given the achievable range of pressures in the laboratory.

Association and the reverse thermal dissociation are well represented within the framework of unimolecular reaction rate theory. The effectiveness of association reactions depends on the temperature and the transferred energy but also on the concentration and the nature (linearity, symmetry, electronic attraction and repulsion ...) of the third body M that stabilizes the ionic adduct product $AB^{+/-}$ taking away the excess energy of the reaction. The more atoms M has, the more effective the relaxation of $AB^{*+/-}$ is.

The RRKM theory is frequently used to interpret this chemical process through the formation of an intermediate complex AB*+/− such as

$$\text{A}^{+/-} + \text{B} \underset{k_d}{\overset{k_a}{\rightleftharpoons}} \text{AB}^{*+/-} + \text{M} \xrightarrow{k_s} \text{AB}^{+/-} + \text{M} \qquad (3.19)$$

In the first step, there is a competition between association leading to the formation of a vibrational excited ion AB*+/−, characterized by an association rate coefficient k_a, and the dissociation, characterized by k_d. If the lifetime of AB*+/− is sufficient, then a second step becomes possible with a rate coefficient k_s leading to the formation of a stable ionic adduct AB+/−. Two pressure regimes can be distinguished. In the low-pressure regime, the rate coefficient linearly depends on the third body density [M]. In the high-pressure limit, the rate coefficient is constant. In between these two regimes lies the fall-off region whose range will depend on the system of interest.

The measured rate coefficient is a bimolecular apparent rate coefficient k_{ba} which is proportional to the third body density [M] and the termolecular rate coefficient k_t.

$$k_{ba} = k_t[\text{M}] \qquad (3.20)$$

For bath gas densities [M] of $10^{11} - 10^{12}$ molecule cm^{-3}, the $k_s[\text{M}]$ term is close to the rate coefficient of the radiative association counterpart reaction k_{rad} [121]. In uniform supersonic flows, the buffer gas density (which plays the role of the third body) is about 10^{16} molecule cm^{-3}. As a consequence, $k_s[\text{M}]$ is larger than k_{rad} and the bimolecular apparent rate coefficient k_{ba} is given by

$$k_{ba} = \frac{k_s k_a}{k_d + k_s[M]}[M] \qquad (3.21)$$

Table 3.2 gathers all the ternary reactions investigated with the CRESU. Among all reactions studied, one interesting example is the termolecular reaction $\text{H}_3\text{O}^+ + \text{H}_2\text{O}(+\text{M})$. The hydronium ion H_3O^+ is abundant in comets and in dense molecular clouds [122, 123]. In dilute interstellar clouds, the gas phase proton exchange reaction $\text{H}_3\text{O}^+ + \text{H}_2\text{O} \rightarrow \text{H}_2\text{O} + \text{H}_3\text{O}^+$ (with possible ortho-para conversion of both H_3O^+ and H_2O) and the electron recombination reaction

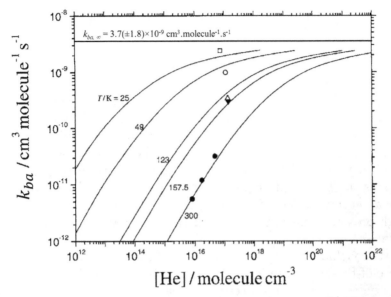

Fig. 3.13.　Experimental values ($\square$, $\circ$, $\triangle$, $\blacktriangledown$) obtained with the CRESU method and theoretical curves (full lines) for the bimolecular apparent rate coefficients k_{ba} for the $H_3O^+ + H_2O$ (+He) $\rightarrow H_2O_5^+$ (+He) reaction [125]. Experimental data at 300 K measured with the SIFT method ($\bullet$) are taken and reproduced with permission from Ref. [126]. The $k_{ba,\infty}$ is the high-pressure limit value of the bimolecular apparent rate coefficient of this reaction.

$H_3O^+ + e^-$ dominate and hence are incorporated in chemical networks [124]. The association reaction

$$H_3O^+ + H_2O + M \rightarrow H_5O_2^+ + M \tag{3.22}$$

was examined in the late nineties by Hamon *et al.* [125] using N_2 and He as buffer gases ($M = $ He, N_2 as the third body) between 23 and 170 K and at pressures between 0.16 and 3.1 mbar (see Fig. 3.13).

The standard unimolecular rate theory based on the Lindemann and Hinshelwood model [127–129] reproduces well the bimolecular apparent rate coefficient k_{ba} behavior for the low-pressure limit regime where the association process is governed by the energy transfer mechanism. In this regime, the fit of the higher temperatures in helium buffer gas gives

$$k_{ba} = [\text{He}]1.05(\pm 0.1) \times 10^{-27} \left(\frac{T}{300K}\right)^{-4.0} \text{cm}^6 \text{ molecule}^{-2} \text{s}^{-1}$$

Over the 23–123 K temperature range, a high-pressure limit $k_{ba\infty} = 3.7(\pm 1.8) \times 10^{-9}$ cm^3 molecule^{-1} s^{-1} was derived, lower than the calculated AADO capture rate coefficient value of 10^{-8} cm^3 molecule^{-1} s^{-1}. The authors suggested that the likely explanation is a PES of the ion product $H_5O_2^+$, slightly different from a long-range ion–dipolar asymmetric-top rotor one due to a non-negligible valence contribution. The interaction is stronger than for an ion–dipole interaction and the potential anisotropy is also different. Such deviation invalidates the use of capture models based on a well-known long range of the PES that can be obtained from simple multipolar expansion. This study highlights the obvious need for multidimensional *ab initio* calculations to explain the association process in the high-pressure limit situation when the ion–dipole interaction is not sufficient.

To conclude, it turns out that the CRESU method, combined with state-of-the-art theoretical calculations, is particularly well adapted to study association processes. Laval nozzles have been specifically designed to generate high-density supersonic flows (typically 10^{16} molecule cm^{-3}) with several buffer gases (essentially He, but also N_2, Ar, and H_2) to cover a large range of temperatures ($\sim$200 K down to 20 K). An undeniable advantage of the CRESU is to offer the possibility to measure the influence of pressure as well as the influence of the nature of the third body (i.e. the buffer gas) on the association reaction rate coefficient in order to probe the involved fundamental mechanisms in the high- and low-pressure regimes and the fall-off region, at least partially.

3.4.4 *List of bimolecular and termolecular ion–molecule reactions studied with CRESU*

Eighty-five ion–molecule reactions were explored with the CRESU method over more than three decades. A total of 73 bimolecular ion–molecule reactions have been studied (Table 3.1). Fifty of the reactions examined involve atomic ions: Ar$^+$, C$^+$, He$^+$, N$^+$, O$^+$, and Cl$^-$. We count 23 reactions with molecular ions, several of them organic, the largest ones having four atoms: Ar^{2+}, CH$^+$, N$_2^+$, O$_2^+$, H$_3^+$, NH$_3^+$, and CN$^-$, C$_3$N$^-$. Twelve termolecular reactions, gathered in Table 3.2, were also investigated with the ions Ar$^+$, C$^+$, CH$_3^+$, O$_2^+$, N$_2^+$, H$_3$O$^+$, and NH$_4^+$.

3.4.5 *Concluding remarks*

The previous sections have illustrated the results obtained with the various CRESU configurations (Section 3.2). While many bimolecular ion–molecule reactions follow some rule of thumb in terms of kinetics, a number of them clearly depart from predictions by simple models. This demonstrates the need for experimental data in order to both validate theoretical models and provide low-temperature data for astrophysical/atmospheric applications.

Although the CRESU track record for ion–molecule reactions is less impressive than for concurrent techniques exposed above, its merits in exploring a very-low-temperature range are well recognized and the method established.

By contrast with radical–neutral uniform supersonic flow reactors (Chapter 5) or nucleation uniform supersonic flow chambers (Chapter 6), the CRESU machine dedicated to ion–molecule reactions has never so far been replicated outside the laboratory where it was designed (if we except the move of the specific ion–molecule part of the apparatus from Meudon to Rennes). One of the causes lies certainly in the large pumping capacities required to operate the CRESU at the lowest possible pressures, which in turn are needed to avoid the competition between termolecular and bimolecular processes. The unavailability in other groups of a wind blower such as the one in Meudon likely prevented the diffusion of the method. There are a number of alternative techniques such as ion traps and guided ion beams — with their own advantages and drawbacks (see Section 3.3) — which utilize more tractable machines.

One of the strengths of the CRESU setup remains its capacity to explore the reactivity on an absolute scale of large ions over a wide range of temperatures and pressures. The implementation of a new pumping system in Rennes in 2020, with a combined pumping capacity of $72\,000$ $m^3\,hr^{-1}$, will extend further the accessible pressure range and allow a swifter identification of the nature of the reaction (bimolecular or termolecular).

The tight link maintained with the astrophysics and planetary sciences communities nevertheless continues to offer some bright perspectives to the CRESU method, notably for ion-induced nucleation studies and for the investigation of ion reactivity with heavy neutral partners.

3.5   Perspectives

Considerable work has been achieved in the investigation of the kinetics and product determination of ion–molecule reactions, in particular, at room temperature using, e.g. the well-established SIFT
method. However, the field of cold reactive collisions with molecular ions has been unevenly explored. While reactions with cations
first appear to have been examined in a systematic manner, the
branching ratios have been determined most of the time only at room
temperature and there is oftentimes some large discrepancy between
the reported branching ratios at the same temperature. Temperature dependence of the branching ratios is often missing. Moreover,
we count only a handful of experiments on the reactive collisions of
anions with neutrals below 200 K. The situation gets even worse if
we consider anions of molecular nature only. There is, therefore, a
clear avenue open for the kinetic study of ion–molecule reactions at
low temperatures, particularly of astrophysical relevance.

Methods to investigate ion–molecule reactions at low temperatures or low energies have grown in sophistication. The most
advanced techniques include crossed molecular beams, well adapted
to study the reaction dynamics, and cryogenic ion traps, tailored
to examine the reaction kinetics of ions with light neutrals over a
wide dynamic range. For their part, uniform supersonic flows turn
out to be well suited for probing the reactivity of ions with heavy
neutral co-reactants. Above all, uniform supersonic flows continue to
be a method of choice because it is one of the rare techniques to
provide rate coefficients obtained under well-controlled thermalized
conditions (with the exception of vibrational and electronic states,
see Chapter 1) over a wide range of low temperatures.

3.5.1   *Toward systematic product determination*

Of particular interest from both the fundamental and applicative
points of view is the determination of the branching ratios, especially at very low temperatures (<10 K) where quantum effects are
revealed. Indeed, as discussed above, low-temperature measurements
of reactive rate coefficients sometimes escape to theoretical predictions from basic capture models since quantum effects are dominant.

Hence, the branching ratio of the reaction can also be quite different from what is expected from statistical models despite the formation of a stable intermediate complex. Such behavior can be found when a significant submerged barrier prevents the capture. For such reactions, theoretical methods are presently developed. Reduced dimensional approaches and approximate quantum approaches are used and provide generally good agreement with experimental data since part of the quantum effects is taken into account. However, they are usually very CPU time-consuming. It is, however, obvious that development of experimental techniques is mandatory in order to validate approximate treatments and to push further the knowledge of ion–molecule reactions at low temperatures. Prototypical small-size systems, as those involving atomic or diatomic ions reacting with H_2, would be systems of choice for benchmarking the new theoretical development since these light systems could also be studied using quantum mechanical approaches.

By contrast with neutral–radical reactions, the quantitative determination of the reaction products is an intrinsic feature of the methods employed to probe charged species. Ionic products are indeed easily detected by charge-sensitive sensors. Two shortcomings are nevertheless recognized. The first one, associated with anion–molecule reactions, is reactive detachment following the reaction

$$A + B^- \rightarrow products + e^- \tag{3.23}$$

In that case, the free and fast moving electron is usually undetectable. Langmuir probes employed to determine the electron density cannot be operated at the low level of currents typically observed (<hundred picoamps). We miss here some direct way to probe that exit channel. Past studies have relied on a detailed balance of the charged ions to lead to the quantification of the associated loss channel. It is noteworthy that the measurement of the neutral products, which would be even more desirable, remains difficult too as their abundance is small for methods designed for sensing neutrals. The second deficiency is the identification of the isomer. The current mass spectrometry methods employed to probe uniform flows have not been designed to recognize isomers. This aspect is treated in the following section.

3.5.2 *Toward isomer-specific detection*

The latest instrumental developments for the study of radical–neutral reactions, based on mass spectrometry with tunable photoionization (Chapter 8), or microwave spectroscopy [142], opened the way to the (quasi-)systematic identification of isomers generated in uniform supersonic flows. Photo-ionization efficiency curves and rotational spectra are indeed both isomer specific. Such methods remain, however, unfit for ions, the main obstacle being the low abundance of ions compared to their neutral counterparts. One way around it could be provided by ion mobility mass spectrometry (IMMS), a versatile two-dimensional analytical technique for rapid separation and simultaneous detection of compounds of interest by conducting a gas phase separation prior to mass analysis [143]. The method can separate gas phase ions based on their charge, size, and shape, according to their differential mobility through a buffer gas in an electric field. The drift region should be operated in the low-field regime such that the mobility is independent of field strength and the field-induced collision energy is much less than the thermal energy. Ions would pass from the uniform supersonic flow to the drift region via a shutter grid, which has to be pulsed open to allow a finite number of ions to enter the drift region. Many challenges await the implementation of IMMS on the CRESU. The first one concerns the low pressure ($\sim$0.1 mbar) imposed on the drift tube. Although the pressure drop facilitates the introduction of gases into the drift region, the low pressure affects the drift time resolution that scales with the inverse of the density. The second one relates to the pulsed introduction of ions. Under the current configuration (see above), ions are continuously sampled. With ion mobility mass spectrometry, ions are periodically introduced into the drift chamber, with a frequency limited by the drift time (the gate is only opened when the last ions have reached the detector). As a consequence, a large drop in the duty cycle is anticipated. As there is no obvious way to compensate for that, experiments could be then conducted following two modes. The first mode could focus on kinetic measurements with "regular" mass spectrometry, while the second one could be devoted to distinguishing potential isomers with ion mobility measurements. Finally, it is to be noted that to support the identification of the isomers, quantum chemistry calculations appear essential.

3.5.3   *State-selective chemistry*

As already mentioned, the reactivity of ions significantly depends on the form under which the energy is stored (i.e. translational, rotational, vibrational, or electronic). A variety of experimental techniques have accordingly been developed that enable the study of the kinetic, vibrational, and rotational energy dependence of reactions. State-of-the-art experiments control the initial states of the reactants and detect the quantum states of the products. State-specific methods include guided ion beam mass spectrometry, resonance-enhanced multiphoton ionization (REMPI), photoelectron-photoion coincidence (PEPICO), pulsed field ionization (PFI) or zero kinetic energy (ZEKE) photoelectron spectroscopy, and mass analyzed threshold ionization (MATI) [144]. These techniques usually provide single collision conditions.

In some instances, a thermalized environment can be sought to, e.g. mimic a planetary atmosphere. In such a setting, ions can be directly produced in excited states by solar UV radiation. Uniform flows can then offer an alternative to examine the state-selective reactivity. Whereas rotational states quickly reach equilibrium in uniform supersonic flows, vibrational relaxation, which requires numerous collisions, is much less efficient within the typical flow hydrodynamic time ($\sim 0.1-1$ ms), especially for high-lying states. Therefore, one can in principle probe the reactivity of the vibrationally excited species — unaffected by collisional quenching by the bath gas. Experimental studies performed with the CRESU have demonstrated the strong effect that internal vibrational energy of the molecules has on the attachment process [133] in agreement with flowing afterglow measurements. This is also very well described in the next section devoted to dissociative recombination.

Tunable continuous VUV radiation provided by, e.g. synchrotron radiation could offer a route to the generation of vibrationally selected ions. In terms of configuration, the beam would have to transversely cross the uniform flow to form photoions upstream. Ions would then be monitored at different locations in the flow by mass spectrometry and their loss rate measured. The advantage of the method would be to provide quantitative kinetic rate coefficients for the selected excited vibrational states. It is to be noted that promising chirped-pulsed microwave methods cannot be applied due to their lack of sensitivity to probe state-specific ionic reagents or products present in the uniform flow only in very low amounts (10^6–10^7 molecule cm^{-3}).

3.5.4　*Dissociative recombination*

As can be seen by reading through the chapters of this book, the CRESU technique provides a remarkable means of studying chemistry at low temperatures and the supersonic expansion is an excellent means of achieving rotational cooling in molecular species. Vibrational cooling is slower and may not be complete depending on the molecule under study. This fact was beautifully demonstrated in an experiment involving the dissociative attachment (DA) of electrons to CF_3Br and CCl_2F_2 molecules [133] where it was found that as the temperature of the flow was lowered using different nozzles and carrier gases, the rate coefficients for this process decreased as the vibrational populations of these molecules decreased (see Chapter 4 for additional details). Internal rovibrational excitation plays an important role in determining the process of the related process of dissociative recombination (DR) of electrons with molecular ions:

$$e^- + ABC^+ \quad \rightarrow AB + C \qquad (3.24)$$
$$\rightarrow A + BC$$
$$\rightarrow A + B + C$$

where ABC^+ is a polyatomic ion and A, B, and C are atomic or molecular fragments. The question arises therefore if it is possible to apply the CRESU technique to DR.

The experimental study of dissociative recombination began using first stationary and then flowing afterglow techniques and the history of this process has been extensively reviewed [125, 126]. While the molecular ions formed in these experiments have generally (though not always) been relaxed to vibrational ground states, a majority of measurements have been made at room temperature thus reflecting the equilibrium rotational population of these species.

Dissociative recombination is a process of particular importance in interstellar clouds where the molecular ion species are rovibrationally cold and temperatures are in the 10–100 K range and so the temperature dependence of these processes is of particular interest to the astrochemical community. Unfortunately, it is not easy to perform stationary or flowing afterglow measurements over a wide temperature range and certainly not in the range of astrophysical interest. The advent of merged electron–ion beam experiments, first in single-pass and then in multi-pass versions at heavy ion storage rings, opened up access to this range and in addition, with these experiments, it was possible to study the branching ratios for the dissociation product channels. These experiments involve colliding an electron beam of known energy with a beam of ions of known energy, both traveling in the same direction so that very low center-of-mass collision energies can be obtained. The advantage of the multi-pass techniques over the single-pass method is that by storing the ions for several seconds, in the storage ring, they have time to vibrationally relax radiatively, prior to performing the experiments.

In experiments performed at the Max-Planck Institute for Nuclear Physics in Heidelberg [147–149], it was found that the method typically used in these experiments yields molecular ions that were rotationally very hot ($\approx$1000 K), excited radiatively from the room-temperature walls of the vacuum chamber during the storage period, but also by the repeated passage through the electron beam, used to translationally cool the ions. Modern theoretical studies of recombination [150–152] have shown that rotation can play an important role in the process and so this represented a serious issue in the interpretation of the experiments. This aspect has been addressed at Heidelberg by the construction of a new cryogenically cooled storage ring in addition to ion source technology, aimed at injecting ions that are already rovibrationally cold. It has to be said that while very elegant theoretically, it is a major undertaking to cool an entire storage ring down to liquid helium temperatures. The application of this technique has been recently demonstrated by the measurement of the dissociative recombination of rovibrationally cold HeH^+ ions [153] that have shown that the cross sections obtained are very significantly different from those previously obtained in storage ring experiments where the ions were rotationally hot. (Previous single-pass experiments using ions prepared in a storage ion trap had already shown evidence of this effect [154]).

The question therefore is if the CRESU technique can be used for measurements of rovibrationally cold molecular ions? The concept is very attractive when one can approach 10 K in a CRESU experiment. However, there are two major drawbacks. The first one is the fact that it is difficult to achieve a high electron density using the high-energy electron gun method described in Chapter 4 concerning electron attachment methods. Typical electron densities in those experiments were 10^8–10^9 molecule cm^{-3}, while in afterglow experiments, the density is 10^9–10^{10} molecule cm^{-3}. Electron attachment is much easier to achieve since the target number density of the reacting gas in the CRESU is typically 10^{12} cm^{-3}, while the ion density will be equal to the electron density assuming that there is no competing attachment for the molecular species under study. Thus, the signal produced is too low in DR to be measured using the Langmuir probe technique and since the products are neutral, they cannot be detected using a mass spectrometer, as is done in the DA case. Thus, the electron beam ionization technique, used for the electron attachment experiments, will not work for DR.

There is a second more insidious reason, however, why that method will not work, and that is the fact that the electron beam producing the electrons can also excite the buffer gas producing the expansion, whether it be metastable helium or argon atoms or vibrationally excited nitrogen. As discussed extensively in Chapter 4, this leads to re-excitation of the electron population via super-elastic collisions, resulting in an electron temperature typically of between 100 K and 400 K, with considerable uncertainty in the exact distribution. This is often not so serious for electron attachment for which the attachment coefficient is rather insensitive to electron temperature for S-wave attachment. Thus, it is possible to obtain useful data concerning the temperature dependence of the process for many cases (Chapter 4). Electron ion recombination coefficients typically follow a $T_e^{-0.5}$ temperature dependence and so accurate knowledge of the electron temperature is essential for producing meaningful results.

Thus, any method that could be used to perform rotationally cold molecular ion DR measurements would have to be such as not to influence the excitation state of the buffer gas. This implies that the production of ions should be performed prior to the stabilized flow achieved at the exit to the Laval nozzle. Also, the electrons needed for the recombination should be created in such a way as not to excite the buffer gas.

Attempts have been made to use photoionization of gases with low ionization potentials to generate threshold (cold) electrons, but it was found that in practice, electron number densities achieved were too low to perform recombination experiments [155] where one expects a "typical" rate coefficient of $10^{-7}\,\mathrm{cm^3\,molecule^{-1}s^{-1}}$ at temperatures close to ambient (300 K). It is found, however, that polyatomic hydrocarbon ions have rates on the order of several $10^{-6}\,\mathrm{cm^3 molecule^{-1}\,s^{-1}}$ [145, 146] at 300 K and with the negative temperature dependence, it could be expected that these may increase as we go to the low temperatures able to be achieved by the CRESU technique. Such a temperature dependence is not guaranteed, however, given the fact that rotational excitation of the ions does indeed play a role in the recombination process, a fact that was dismissed for many years in the community, prior to the experiments on H_3^+ in Heidelberg [147–149] and later theoretical work [150–152].

Another possibility is to create the ions and electrons in the Laval nozzle itself. In fact, in the early days of CRESU development, an attempt was made to fit a glass nozzle with a microwave cavity, much as is used in the Flowing Afterglow method. It was found, however, that this method did not work as the plasma, generated by the microwaves, attached itself asymmetrically to the walls of the nozzle so that the spatial distribution of the plasma downstream was not uniform [53]. At that time, there were many other applications of the CRESU that were more easily and profitably achieved and so this investigation was abandoned. It could, however, be interesting to pursue this approach, possibly by attaching radiofrequency loop antennas to a quartz nozzle rather than the microwave cavity approach. This ionization method is used in standard radiofrequency ion sources, commonly employed in Van de Graaff generators.[a] The microwave cavity functions by producing a standing wave in the resonant cavity and this perhaps introduces undesirable structure in the plasma that interacts with the expanding flow conditions. The ion generation in the radiofrequency method is longitudinal and linear with respect to the flow and perhaps this would avoid this problem. This is perhaps a subject for further investigation. Again, however, the problem of excitation of the buffer gas could call the energy of the

---

[a]Quartz is preferred to glass as, for this application, this material responds to plasma production in a uniform fashion. Glass has a tendency to form charged islands on its surface and this may have influenced the asymmetric character in the plasma observed in Ref. [53].

electrons into question and whereas in some cases, electron attachment rate coefficients are independent of T_e, electron–ion recombination is not.

There is the additional limitation with this technique in that as the dissociation products are neutral, it is not possible to be able to do full measurements of product branching ratios as can be done using beam techniques.

In summary then, despite the attractiveness of being able to produce species with low internal energy, and to cool them to the low temperatures characteristic of interstellar chemistry, the practical limitations seem to preclude measurements of dissociative recombination, barring a technical breakthrough.

3.5.5 *Ion-induced nucleation*

In 1998, H. Svensmark published an article in Physical Review Letters [156] in which he put forward evidence linking Earth's temperature cycles with solar activity resulting in the modulation of the cosmic ray flux entering the atmosphere. Cosmic rays are the major source of ion formation in the upper atmosphere and it was argued that this resulted in the modulation of cloud formation. Greater cloud cover would increase the Earth's albedo so that more sunlight would be reflected back into space. Svensmark has pursued this subject and presented evidence for the enhancement of cloud condensation nuclei (CCN) formation by ions [157]. The role of ions in the formation of aerosols in Earth's atmosphere is, however, debated. Conflicting data exist as to whether or not ion-induced nucleation is important on a global scale. In Earth's atmosphere, only a minor role is attributed to ions, while sulfuric acid is essential to initiate most particle formation. Recent experiments performed at the CERN with the CLOUD (Cosmics Leaving Outdoor Droplets) chamber showed that ion-induced nucleation of pure organic particles constitutes a potentially widespread source of aerosol particles in terrestrial environments with low sulfuric acid pollution [158]. Beyond Earth, ion-induced nucleation may play a key role in the atmosphere of planets or their moons. In Titan's upper atmosphere, heavy ions appear to act as seeds of aerosols, [38] which in turn dramatically affect its climate.

Uniform supersonic flows are unique tools to generate and study the formation kinetics of molecular clusters (Chapter 6). Such studies could be advantageously extended to ion-induced nucleation to further assess the role of this process. Only minor modifications to the dedicated CRESU reactor equipped with a mass-selective ion transfer line are needed. They basically consist in the implementation of the existing quadrupole mass spectrometer of an ionization source to measure accurately the neutral cluster distribution in the absence of ions. In addition (as described in Chapter 6), light scattering techniques can be used to track the growth of particles by nucleation along the flow. Thus, the influence of the density and nature of ions in the flow can be examined.

Ion-induced clustering is an important early-stage process in nucleation and this is something that has already been studied in the CRESU apparatus in Rennes [159]. In these experiments, a mixture of water vapor and selected hydrocarbons was introduced into the upstream reservoir, and electron attachment to clusters downstream of the electron beam production region (Section 3.2.1) was observed. This experiment was particularly significant as it was found that, for example, benzene, that does not attach electrons at thermal energies, did readily attach when clustered with water molecules. Thus, it would be very interesting to extend this study to characterize the growth of clusters as a function of experimental parameters (temperature, parent molecule density, electron density, ion density, and composition) and to explore this topic that is so crucial to many areas where the presence of ions can have a profound effect on particle growth in plasmas or cloud condensation nuclei in the troposphere.

References

[1] Faure A, Vuitton V, Thissen R, Wiesenfeld L, Dutuit O. Fast ion–molecule reactions in planetary atmospheres: a semiempirical capture approach. Faraday Discuss. 2010;147:337–48.

[2] Meyer J, Wester R. Ion–molecule reaction dynamics. Annu Rev Phys Chem. 2017;68:333–53.

[3] Petrie S, Bohme DK. Ions in space. Mass Spectrom Rev. 2007;26:258–80.

[4] Snow TP, Bierbaum VM. Ion chemistry in the interstellar medium. Ann Rev Anal Chem. 2008;1:229–59.

[5] McGuire BA. 2018 census of interstellar, circumstellar, extragalactic, protoplanetary disk, and exoplanetary molecules. Astrophys J Suppl Ser. 2018;239:17.

[6] Tielens AGGM. Interstellar polycyclic aromatic hydrocarbon molecules. Ann Rev Astron Astrophys. 2008;46:289–337.

[7] Campbell EK, Holz M, Gerlich D, Maier JP. Laboratory confirmation of C_{60}^+ as the carrier of two diffuse interstellar bands. Nature. 2015;523:322–3.

[8] Salama F, Galazutdinov GA, Krelowski J, Biennier L, Beletsky Y, Song I-O. Polycyclic aromatic hydrocarbons and the diffuse interstellar bands: a survey. Astrophys J. 2011;728:154.

[9] Krełowski J, Beletsky Y, Galazutdinov GA, Koł os R, Gronowski M, LoCurto G. Evidence for diacetylene cation as the carrier of a diffuse interstellar band. Astrophys J. 2010;714:L64–L67.

[10] Tulej M, Kirkwood DA, Pachkov M, Maier JP. Gas-phase electronic transitions of carbon chain anions coinciding with diffuse interstellar bands. Astrophys J Lett. 1998;506:L69–L73.

[11] Larsson M, Geppert WD, Nyman G. Ion chemistry in space. Rep Prog Phys. 2012;75:066901.

[12] Herbst E. The synthesis of large interstellar molecules. Int Rev Phys Chem. 2017;36:287–331.

[13] Cernicharo J, Guélin M, Agúndez M, Kawaguchi K, McCarthy M, Thaddeus P. Astronomical detection of C_4H^-, the second interstellar anion. Astron Astrophys. 2007;467:L37–L40.

[14] McCarthy M, Gottlieb C, Gupta H, Thaddeus P. Laboratory and astronomical identification of the negative molecular ion C_6H^-. Astrophys J. 2006;652:L141.

[15] Sakai N, Sakai T, Osamura Y, Yamamoto S. Detection of C_6H^- toward the low-mass protostar IRAS 04368 + 2557 in L1527. Astrophys J Lett. 2007;667:L65.

[16] Brünken S, Gupta H, Gottlieb C, McCarthy M, Thaddeus P. Detection of the carbon chain negative ion C_8H^- in TMC-1. Astrophys J Lett. 2007;664:L43.

[17] Remijan AJ, Hollis JM, Lovas FJ, Cordiner MA, Millar TJ, Markwick-Kemper AJ, Jewell PR. Detection of C_8H^- and comparison with C_8H toward IRC+10 216. Astrophys J. 2007;664:L47–L50.

[18] Agúndez M, Cernicharo J, Guélin M, Kahane C, Roueff E, Klos J, Aoiz FJ, Lique F, Marcelino N, Goicoechea JR, Gonzalez Garcia M, Gottlieb CA, McCarthy MC, Thaddeus P. Astronomical identification of CN^-, the smallest observed molecular anion. Astron Astrophys. 2010;517:L2.

[19] Thaddeus P, Gottlieb CA, Gupta H, Bruenken S, McCarthy MC, Agundez M, Guelin M, Cernicharo J. Laboratory and astronomical detection of the negative molecular ion C_3N^-. Astrophys J. 2008;677:1132–9.

[20] Cernicharo J, Guélin M, Agúndez M, McCarthy M, Thaddeus P. Detection of C_5N^- and vibrationally excited C6H in IRC + 10216. Astrophys J Lett. 2008;688:L83.

[21] Millar TJ, Walsh C, Field TA. Negative ions in space. Chem Rev. 2017;117:1765–95.

[22] Petrie S. Novel pathways to CN^- within interstellar clouds and circumstellar envelopes: implications for IS and CS chemistry. Mon Not R Astron Soc. 1996;281:137–44.

[23] Herbst E, Osamura Y. Calculations on the formation rates and mechanisms for C_nH anions in interstellar and circumstellar media. Astrophys J. 2008;679:1670–1679.

[24] Khamesian M, Douguet N, dos Santos SF, Dulieu O, Raoult M, Brigg WJ, Kokoouline V. Formation of CN^-, C_3N^-, and C_5N^- molecules by radiative electron attachment and their destruction by photodetachment. Phys Rev Lett. 2016;117:123001.

[25] Lara-Moreno M, Stoecklin T, Halvick P. Radiative electron attachment and photodetachment rate constants for linear carbon chains. ACS Earth Space Chem. 2019;3:1556–63.

[26] Chandrasekaran V, Prabhakaran A, Kafle B, Rubinstein H, Heber O, Rappaport M, Toker Y, Zajfman D. Formation and stabilization of C_6^- by radiative electron attachment. J Chem Phys. 2017;146:094302.

[27] Kumar SS, Hauser D, Jindra R, Best T, Roučka S, Geppert WD, Millar TJ, Wester R. Photodetachment as a destruction mechanism for CN^- and C_3N^- anions in circumstellar envelopes. Astrophys J. 2013;776:25.

[28] Fuselier SA, Altwegg K, Balsiger H, Berthelier JJ, Bieler A, Briois C, Broiles TW, Burch JL, Calmonte U, Cessateur G, Combi M, De Keyser J, Fiethe B, Galand M, Gasc S, Gombosi TI, Gunell H, Hansen KC, Hässig M, Jäckel A, Korth A, Le Roy L, Mall U, Mandt KE, Petrinec SM, Raghuram S, Rème H, Rinaldi M, Rubin M, Sémon T, Trattner KJ, Tzou C-Y, Vigren E, Waite JH, Wurz P. ROSINA/DFMS and IES observations of 67P: ion-neutral chemistry in the coma of a weakly outgassing comet. Astron Astrophys. 2015;583:A2.

[29] Beth A, Altwegg K, Balsiger H, Berthelier J-J, Combi MR, Keyser JD, Fiethe B, Fuselier SA, Galand M, Gombosi TI, Rubin M, Sémon T. ROSINA ion zoo at comet 67P. Astron Astrophys. 2020; 642:A27.

[30] Cordiner MA, Charnley SB. Negative ion chemistry in the coma of comet 1P/Halley. Meteorit Planet Sci. 2014;49:21–7.

[31] Aplin KL. Composition and measurement of charged atmospheric clusters. Space Sci Rev. 2008;137:213–24.

[32] Dobrijevic M, Loison JC, Hickson KM, Gronoff G. 1D-coupled photochemical model of neutrals, cations and anions in the atmosphere of Titan. Icarus. 2016;268:313–39.

[33] Vuitton V, Yelle RV, Klippenstein SJ, Hörst SM, Lavvas P. Simulating the density of organic species in the atmosphere of Titan with a coupled ion-neutral photochemical model. Icarus. 2019;324: 120–97.

[34] Waite Jr. JH, Young DT, Cravens TE, Coates AJ, Crary FJ, Magee B, Westlake J. The process of tholin formation in Titan's upper atmosphere. Science. 2007;316:870–5.

[35] Coates AJ, Crary FJ, Lewis GR, Young DT, Waite Jr. JH, Sittler Jr. EC. Discovery of heavy negative ions in Titan's ionosphere. Geophys Res Lett. 2007;34:L22103.

[36] Coates AJ. Interaction of Titan's ionosphere with Saturn's magnetosphere. Philos Trans R Soc Math Phys Eng Sci. 2009;367:773–788.

[37] Vuitton V, Lavvas P, Yelle RV, Galand M, Wellbrock A, Lewis GR, Coates AJ, Wahlund JE. Negative ion chemistry in Titan's upper atmosphere. Planet Space Sci. 2009;57:1558–72.

[38] Lavvas P, Yelle RV, Koskinen T, Bazin A, Vuitton V, Vigren E, Galand M, Wellbrock A, Coates AJ, Wahlund J-E, Crary FJ, Snowden D. Aerosol growth in Titan's ionosphere. Proc Natl Acad Sci USA. 2013;110:2729–34.

[39] Williams KL, Martin IT, Fisher ER. On the importance of ions and ion-molecule reactions to plasma-surface interface reactions. J Am Soc Mass Spectrom. 2002;13:518–29.

[40] Armentrout PB. Kinetic energy dependence of ion–molecule reactions related to plasma chemistry. In: Bederson B, Walther H, editors. Advances in atomic, molecular, and optical physics. Academic Press, San Diego, San Francisco, New York Boston, London, Sydney, Tokyo; 2000. p. 187–229. (Fundamentals of plasma chemistry; vol 43). https://reader.elsevier.com/reader/sd/pii/S1049250X08601 172?token=2B54B6E40D1C1915BD54C625B8EFBADDF395E8E7D 3C66AD38D695D9E40238410A469F53DD87726AC40AB8D89151D EF53&originRegion=eu-west-1&originCreation=20210929145455

[41] Imanaka H, Khare BN, Elsila JE, Bakes ELO, McKay CP, Cruikshank DP, Sugita S, Matsui T, Zare RN. Laboratory experiments of Titan tholin formed in cold plasma at various pressures: implications for nitrogen-containing polycyclic aromatic compounds in Titan haze. Icarus. 2004;168:344–66.

[42] Szopa C, Cernogora G, Boufendi L, Correia JJ, Coll P. PAMPRE: a dusty plasma experiment for Titan's tholins production and study. Planet Space Sci. 2006;54:394–404.

[43] Geppert WD, Larsson M. Experimental investigations into astrophysically relevant ionic reactions. Chem Rev. 2013;113:8872–905.

[44] Rowe BR, Dupeyrat G, Marquette JB, Gaucherel P. Study of the reactions $N_2^+ + 2N_2 \to N_4^+ + N_2$ and $O_2^+ + 2O_2 \to O_4^+ + O_2$ from 20–160 K by the CRESU technique. J Chem Phys. 1984;80:4915–21.

[45] Joalland B, Jamal-Eddine N, Papanastasiou D, Lekkas A, Carles S, Biennier L. A mass-selective ion transfer line coupled with a uniform supersonic flow for studying ion–molecule reactions at low temperatures. J Chem Phys. 2019;150:164201.

[46] Rowe BR, Dupeyrat G, Marquette JB, Smith D, Adams NG, Ferguson EE. The reaction $O_2^+ + CH_4 \to CH_3O_2^+ + H$ studied from 20–560 K in a supersonic jet and in a SIFT. J Chem Phys. 1984;80:241–5.

[47] Rowe BR, Marquette JB, Dupeyrat G, Ferguson EE. Reactions of He^+ and N^+ ions with several molecules at 8 K. Chem Phys Lett. 1985;113:403–6.

[48] Bourgalais J, Jamal-Eddine N, Joalland B, Capron M, Balaganesh M, Guillemin J-C, Le Picard SD, Faure A, Carles S, Biennier L. Elusive anion growth in Titan's atmosphere: low temperature kinetics of the $C_3N^- + HC_3N$ reaction. Icarus. 2016;271:194–201.

[49] Adams NG, Smith D. The selected ion flow tube (SIFT); A technique for studying ion-neutral reactions. Int J Mass Spectrom Ion Phys. 1976;21:349–59.

[50] Rowe BR, Marquette J-B, Rebrion C. Mass-selected ion–molecule reactions at very low temperatures. The CRESUS apparatus. J Chem Soc Faraday Trans. 1989;85:1631–41.

[51] Rowe BR, Marquette JB, Dupeyrat G. Measurements of ion-molecule reaction rate coefficients between 8 and 160 K by the CRESU technique. In: Molecular astrophysics–state of the art and future directions. D. Reidel Publishing Company; 1985. pp. 631–8.

[52] Speck T, Mostefaoui TI, Travers D, Rowe BR. Pulsed injection of ions into the CRESU experiment. Int J Mass Spectrom. 2001;208:73–80.

[53] Rowe BR. Private communication; n.d.

[54] Herman Z, Futrell JH. Dynamics of ion–molecule reactions from beam experiments: a historical survey. Int J Mass Spectrom. 2015;377:84–92.

[55] Wester R. Velocity map imaging of ion-molecule reactions. Phys Chem Chem Phys. 2014;16:396–405.

[56] Chandler DW, Houston PL. Two-dimensional imaging of state-selected photodissociation products detected by multiphoton ionization. J Chem Phys. 1987;87:1445–7.

[57] Eppink ATJB, Parker DH. Velocity map imaging of ions and electrons using electrostatic lenses: application in photoelectron and

photofragment ion imaging of molecular oxygen. Rev Sci Instrum. 1997;68:3477–84.

[58] Gerlich D. Experimental investigations of ion-molecule reactions relevant to interstellar chemistry. J Chem Soc Faraday Trans. 1993;89:2199–208.

[59] Luine JA, Dunn GH. Ion-molecule reaction probabilities near 10 K. Astrophys J Lett. 1985;299:L67–L70.

[60] Barlow SE, Dunn GH, Schauer M. Radiative association of CH_3^+ and H_2 at 13 K. Phys Rev Lett. 1984;52:902–5.

[61] Gerlich D. Inhomogeneous RF fields: a versatile tool for the study of processes with slow ions. In: Ng C-Y, Baer M, Prigogine I, Rice SA, editors. Advances in Chemical Physics. John Wiley & Sons, Inc. Hoboken; 1992. vol. 82, p. 1–176. https://onlinelibrary.wiley.com/doi/10.1002/9780470141397.ch1

[62] Gerlich D. Ion-neutral collisions in a 22-pole trap at very low energies. Physica Scripta. 1995;1995:256.

[63] KIDA. KInetic database for astrochemistry. Available from: https://kida.astrochem-tools.org/, 2022.

[64] Anicich VG. An index of the literature for bimolecular gas phase cation-molecule reaction kinetics. JPL Publication 03–19; 2003. pp. 1–1194.

[65] Willitsch S. Chemistry with controlled ions. In: Rice, Stuart A. / Dinner, Aaron R. (Editor). Advances in chemical physics. John Wiley & Sons, Ltd, Hoboken; 2017. pp. 307–40. https://www.wiley-vch.de/en/areas-interest/natural-sciences/advances-in-chemical-physics-978-1-119-32457-7

[66] van de Meerakker SYT, Bethlem HL, Vanhaecke N, Meijer G. Manipulation and control of molecular beams. Chem Rev. 2012;112:4828–78.

[67] Heazlewood BR. Cold ion chemistry within Coulomb crystals. Mol Phys. 2019;117:1934–41.

[68] Kas M, Liévin J, Vaeck N, Loreau J. Cold collisions of C_2^- with Li and Rb atoms in hybrid traps. J Phys Conf Ser. 2020;1412:062003.

[69] Ferguson EE, Fehsenfeld FC, Schmeltekopf AL. Flowing afterglow measurements of ion-neutral reactions. In: Bates DR, Estermann I, editors. Advances in atomic and molecular physics. Academic Press, New York London; 1969. vol 5, pp. 1–56.

[70] Hawley M, Mazely T, Randeniya L, Smith R, Zeng X, Smith M. A free jet flow reactor for ion molecule reaction studies at very low energies. Int J Mass Spectrom. 1990;97:55–86.

[71] Rowe BR, Canosa A, Lepage V. Falp and Cresu studies of ionic reactions. Int J Mass Spectrom Ion Process. 1995;149:573–96.

[72] Marquette JB, Rebrion C, Rowe BR. Proton transfer reactions of H_3^+ with molecular neutrals at 30 K. Astron Astrophys. 1989;213: L29–L32.

[73] Su T, Bowers MT. Ion-polar molecule collisions: the effect of ion size on ion-polar molecule rate constants; The parameterization of the average-dipole-orientation theory. Int J Mass Spectrom Ion Phys. 1973;12:347–56.

[74] Su T, Su ECF, Bowers MT. Ion-polar molecule collisions. Conservation of angular momentum in the average dipole orientation theory. The AADO theory. J Chem Phys. 1978;69:2243–50.

[75] Clary DC. Rates of chemical reactions dominated by long-range intermolecular forces. Mol Phys. 1984;53:3–21.

[76] Rebrion C, Rowe BR, Marquette JB. Reactions of Ar^+ with H_2, N_2, O_2, and CO at 20, 30, and 70 K. J Chem Phys. 1989;91:6142–6147.

[77] Rowe BR, Marquette JB. CRESU studies of ion molecule reactions. Int J Mass Spectom Ion Process. 1987;80:239–54.

[78] Rebrion C, Marquette JB, Rowe BR, Clary DC. Low-temperature reactions of He^+ and C^+ with HCl, SO_2 and H_2S. Chem Phys Lett. 1988;143:130–34.

[79] Dateo CE, Clary DC. Rate constant calculations on the C^+ + HCl reaction. J Chem Phys. 1989;90:7216–28.

[80] Johnsen R, Chen A, Biondi MA. Dissociative charge transfer of He^+ ions with H_2 and D_2 molecules from 78–330 K. J Chem Phys. 1980;72:3085–88.

[81] Langevin P. Une formule fondamentale de théorie cinétique. Ann Chim Phys. 1905;5:245.

[82] Marquette JB, Rebrion C, Rowe BR. Reactions of $N^+(^3P)$ ions with normal, para, and deuterated hydrogens at low temperatures. J Chem Phys. 1988;89:2041–47.

[83] Cheung AC, Rank DM, Townes CH, Thornton DD, Welch WJ. Detection of NH_3 molecules in the interstellar medium by their microwave emission. Phys Rev Lett. 1968;21:1701–5.

[84] Genzel R, Ho PTP, Bieging J, Downes D. NH3 in Orion-KL–A new interpretation. Astrophys J. 1982;259:L103–L107.

[85] Ho PTP, Townes CH. Interstellar ammonia. Ann Rev Astron Astrophys. 1983;21:239–70.

[86] Marquette JB, Rowe BR, Dupeyrat G, Roueff E. CRESU study of the reaction $N^+ + H_2 \rightarrow NH^+ + H$ between 8 and 70 K and interstellar chemistry implications. Astron Astrophys. 1985;147:115–20.

[87] Fukutani K, Sugimoto T. Physisorption and ortho–para conversion of molecular hydrogen on solid surfaces. Prog Surf Sci. 2013;88: 279–348.

[88] Pagani L, Roueff E, Lesaffre P. Ortho-H_2 and the age of interstellar dark clouds. Astrophys J Lett. 2011;739:L35.

[89] Bron E, Le Petit F, Le Bourlot J. Efficient ortho-para conversion of H_2 on interstellar grain surfaces. Astron Astrophys. 2016; 588:A27.

[90] Adams NG, Smith D, Millar TJ. The importance of kinetically excited ions in the synthesis of interstellar molecules. Mon Not R Astron Soc. 1984;211:857–65.

[91] Adams NG, Smith D. A study of the nearly thermoneutral reactions of N^+ with H_2, HD and D_2. Chem Phys Lett. 1985;117:67–70.

[92] Grozdanov TP, Mc Carroll R, Roueff E. Reactions of $N^+(^3P)$ ions with H_2 and HD molecules at low temperatures. Astron Astrophys. 2016;589:A105.

[93] Savin DW, Krsti PS, Haiman Z, Stancil PC. Rate coefficient for $H^+ + H_2(X^1\Sigma_g^+, v = 0, J = 0) \rightarrow H(1s) + H_2^+$ charge transfer and some cosmological implications. Astrophys J. 2004;606:L167–L170.

[94] Haider SA, Abdu MA, Batista IS, Sobral JHA, Sheel V, Molina-Cuberos GJ, Maguire WC, Verigin MI. Zonal wave structures in the nighttime tropospheric density and temperature and in the D region ionosphere over Mars: modeling and observations. J Geophys Res. 2009;114:A12315.

[95] Fox JL, Johnson A, Benna M. The chemistry of hydrogen in the Martian thermosphere/ionosphere. In: AGU fall meeting abstracts. 2018. vol 2018, p. P32B-04. https://ui.adsabs.harvard.edu/abs/2018 AGUFM.P32B..04F/abstract

[96] Luna H, McGrath C, Shah MB, Johnson RE, Liu M, Latimer CJ, Montenegro EC. Dissociative charge exchange and ionization of O_2 by fast H^+ and O^+ ions: energetic ion interactions in Europa's oxygen atmosphere and neutral torus. Astrophys J. 2005;628:1086–1096.

[97] Lindinger W, Howorka F, Lukac P, Kuhn S, Villinger H, Alge E, Ramler H. Charge transfer of $Ar^+ + N_2 \leftrightarrow N_2^+ + Ar$ at near thermal energies. Phys Rev A. 1981;23:2319–26.

[98] Anicich VG. Evaluated bimolecular ion–molecule gas phase kinetics of positive ions for use in modeling planetary atmospheres, cometary comae, and interstellar clouds. J Phys Chem Ref Data. 1993;22:1469–569.

[99] Rakshit AB, Warneck P. Thermal rate coefficients for reactions involving $^2P_{1/2}$ and $^2P_{3/2}$ argon ions and several neutral molecules. J Chem Phys. 1980;73:2673–9.

[100] Liao C-L, Liao C-X, Ng CY. A state-to-state study of the symmetric charge transfer reaction $Ar^+(^2P_{3/2,1/2}) + Ar(^1S_0)$. J Chem Phys. 1985;82:5489–98.

[101] Spalburg MR, Gislason EA. Theoretical state-to-state cross sections for the $Ar^+ + N_2 \leftrightarrow Ar + N_2^+$ system. Chem Phys. 1985;94:339–50.

[102] Gioumousis G, Stevenson DP. Reactions of gaseous molecule ions with gaseous molecules. V. theory. J Chem Phys. 1958;29:294–9.

[103] Theard LP, Hamill WH. The energy dependence of cross sections of some ion–molecule reactions. J Am Chem Soc. 1962;84:1134–9.

[104] Su T, Bowers MT. Theory of ion-polar molecule collisions. Comparison with experimental charge transfer reactions of rare gas ions to geometric isomers of difluorobenzene and dichloroethylene. J Chem Phys. 1973;58:3027–37.

[105] Bass L, Su T, Chesnavich WJ, Bowers MT. Ion-polar molecule collisions. A modification of the average dipole orientation theory: the $\cos(\theta)$ model. Chem Phys Lett. 1975;34:119–22.

[106] Su T, Chesnavich WJ. Parametrization of the ion–polar molecule collision rate constant by trajectory calculations. J Chem Phys. 1982;76:5183–5.

[107] Marquette JB, Rowe BR, Dupeyrat G, Poissant G, Rebrion C. Ion–polar-molecule reactions: a CRESU study of He^+, C^+, $N^+ + H_2O$, NH_3 at 27, 68 and 163 K. Chem Phys Lett. 1985;122:431–5.

[108] Clary DC, Smith D, Adams NG. Temperature dependence of rate coefficients for reactions of ions with dipolar molecules. Chem Phys Lett. 1985;119:320–6.

[109] Quack M, Troe J. Specific rate constants of unimolecular processes II. Adiabatic channel model. Ber Bunsenges Phys Chem. 1974;78:240–52.

[110] Quack M, Troe J. Unimolecular processes IV: product state distributions after dissociation. Ber Bunsenges Phys Chem. 1975;79:469–475.

[111] Loreau J, Lique F, Faure A. An efficient statistical method to compute molecular collisional rate coefficients. Astrophys J Lett. 2018;853:L5.

[112] Smith IWM, Rowe BR. Reaction kinetics at very low temperatures: laboratory studies and interstellar chemistry. Acc Chem Res. 2000;33:261–8.

[113] Clary DC. Rate constants for the reactions of ions with dipolar polyatomic molecules. J Chem Soc Faraday Trans 2. 1987;83:139–48.

[114] Le Page V. Conception et mise au point d'un moyen d'essai CRESU pour l'etude des reactions ion–molecule et applications aux temperatures ultrasbasses (10 K) (phD dissertation), University of Rennes 1 (France); 1995.

[115] Smith D, Adams NG, Miller TM. A laboratory study of the reactions of N^+, N_2^+, N_3^+, N_4^+, O^+, O_2^+, and NO^+ ions with several molecules at 300?K. J Chem Phys. 1978;69:308–18.

[116] Herbst E, van Dishoeck EF. Complex organic interstellar molecules. Ann Rev Astron Astrophys. 2009;47:427–80.

[117] Coustenis A, Jennings DE, Nixon CA, Achterberg RK, Lavvas P, Vinatier S, Teanby NA, Bjoraker GL, Carlson RC, Piani L, Bampasidis G, Flasar FM, Romani PN. Titan trace gaseous composition from CIRS at the end of the Cassini-Huygens prime mission. Icarus. 2010;207:461–76.

[118] Biennier L, Carles S, Cordier D, Guillemin J-C, Le Picard SD, Faure A. Low temperature reaction kinetics of CN^- + HC_3N and implications for the growth of anions in Titan's atmosphere. Icarus. 2014;227:123–31.

[119] Joalland B, Jamal-Eddine N, Kłos J, Lique F, Trolez Y, Guillemin J-C, Carles S, Biennier L. Low-temperature reactivity of $C_{2n+1}N^-$ anions with polar molecules. J Phys Chem Lett. 2016;7:2957–61.

[120] Kołos R, Gronowski M, Botschwina P. Matrix isolation IR spectroscopic and ab initio studies of C_3N^- and related species. J Chem Phys. 2008;128:154305.

[121] Herbst E. The onset of condensation: association reactions. Astrophys Space Sci. 1979;65:13–20.

[122] Barber RJ, Miller S, Dello Russo N, Mumma MJ, Tennyson J, Guio P. Water in the near-infrared spectrum of comet 8P/Tuttle. Mon Not Roy Astron Soc. 2009;398:1593–600.

[123] Goicoechea JR, Cernicharo J. Far-infrared detection of H_3O^+ in Sagittarius B2. Astrophys J. 2001;554:L213–L216.

[124] Faure A, Hily-Blant P, Rist C, Pineau des Forêts G, Matthews A, Flower DR. The ortho-to-para ratio of water in interstellar clouds. Mon Not R Astron Soc. 2019;487:3392–403.

[125] Hamon S, Speck T, Mitchell JBA, Rowe B, Troe J. Experimental and modeling study of the ion–molecule association reaction H_3O^+ + $H_2O(+M) \rightarrow H_5O_2^+(+M)$. J Chem Phys. 2005;123:54303.

[126] Bierbaum VM, Golde MF, Kaufman F. Flowing afterglow studies of hydronium ion clustering including diffusion effects. J Chem Phys. 1976;65:2715–24.

[127] Troe J. Predictive possibilities of unimolecular rate theory. J Phys Chem. 1979;83:114–26.

[128] Troe J. Theory of thermal unimolecular reactions at low pressures. I. Solutions of the master equation. J Chem Phys. 1977;66:4745–4757.

[129] Oref I, Tardy DC. Energy transfer in highly excited large polyatomic molecules. Chem Rev. 1990;90:1407–45.

[130] Dupeyrat G, Marquette JB, Rowe BR, Rebrion C. Reactions of Ar^{2+} (^3P) ions with some neutrals at 30 K. Int J Mass Spectrom Ion Process. 1991;103:149–56.

[131] Marquette JB, Rowe BR, Dupeyrat G, Poissant G. Ion-molecule reaction studies below 80 K by the CRESU technique. Symp–Int Astron Union. 1987;120:19–23.

[132] Rebrion C, Marquette JB, Rowe BR, Adams NG, Smith D. Low-temperature reactions of some atomic ions with molecules of large quadrupole moment: C_6F_6 and c-C_6H_{12}. Chem Phys Lett. 1987;136:495–500.

[133] Le Garrec JL, Sidko O, Queffelec JL, Hamon S, Mitchell JBA, Rowe BR. Experimental studies of cold electron attachment to SF_6, CF_3Br, and CCl_2F_2. J Chem Phys. 1997;107:54–63.

[134] Rowe BR, Marquette JB, Dupeyrat G. Measurements of ion-molecule reaction rate coefficients between 8 and 160 K by the CRESU technique. In: Diercksen GHF, Huebner WF, Langhoff PW, editors. NATO Advanced Science Institutes (ASI) series C. D. Reidel Publishing Company; 1985. vol 157, pp. 631–8.

[135] Rebrion C, Marquette JB, Rowe BR, Chakravarty C, Clary DC, Adams NG, Smith D. Reactions of N^+ and H_3^+ with the structural isomers of $C_2H_2Cl_2$. J Phys Chem. 1988;92:6572–74.

[136] Le Garrec J-L, Lepage V, Rowe BR, Ferguson EE. The temperature dependence of the rate constant for $O^+ + NO \rightarrow NO^+ + O$ from 23–30000 K. Chem Phys Lett. 1997;270:66–70.

[137] Le Garrec J-L, Carles S, Speck T, Mitchell JBA, Rowe BR, Ferguson EE. The ion-molecule reaction of O^+ with N_2 measured down to 23 K. Chem Phys Lett. 2003;372:485–8.

[138] Gaucherel P, Marquette J, Rebrion C, Poissant G, Dupeyrat G, Rowe BR. Temperature dependence of slow charge-exchange reactions: $N_2^+ + O_2$ from 8–163 K. Chem Phys Lett. 1986;132:63–6.

[139] Barlow SE, Van Doren JM, DePuy CH, Bierbaum VM, Dotan I, Ferguson EE, Adams NG, Smith D, Rowe BR, Marquette JB, Dupeyrat G, Durup-Ferguson M. Studies of the reaction of O_2^+ with deuterated methanes. J Chem Phys. 1986;85:3851–9.

[140] Hamon S, Mitchell JBA, Rowe BR. Low-temperature measurements of the atomic association reaction $Ar^+ + 2Ar \rightarrow Ar_2^+ + Ar$. Chem Phys Lett. 1998;288:523–6.

[141] Hamon S, Speck T, Mitchell JBA, Rowe BR, Troe J. Experimental and theoretical study of the ion-molecule association reaction $NH_4^+ + NH_3(+M) \rightarrow N_2H_7^+(+M)$. J Chem Phys. 2002;117:2557–67.

[142] Hays BM, Guillaume T, Hearne TS, Cooke IR, Gupta D, Abdelkader Khedaoui O, Le Picard SD, Sims IR. Design and performance of an E-band chirped pulse spectrometer for kinetics applications: OCS–He pressure broadening. J Quant Spectrosc. Radiat Transf. 2020;250:107001.

[143] Lanucara F, Holman SW, Gray CJ, Eyers CE. The power of ion mobility-mass spectrometry for structural characterization and the study of conformational dynamics. Nat Chem. 2014;6:281–94.

[144] Armentrout PB, Baer T. Gas-phase ion dynamics and chemistry. J Phys Chem. 1996;100:12866–77.

[145] Florescu-Mitchell AI, Mitchell JBA. Dissociative recombination. Phys Rep. 2006;430:277–374.

[146] Larsson M, Orel AE. Dissociative recombination of molecular ions. Cambridge molecular science. Cambridge University Press, Cambridge; 2008.

[147] Strasser D, Lammich L, Krohn S, Lange M, Kreckel H, Levin J, Schwalm D, Vager Z, Wester R, Wolf A, Zajfman D. Two- and three-body kinematical correlation in the dissociative recombination of H_3^+. Phys Rev Lett. 2001;86:779–82.

[148] Strasser D, Lammich L, Kreckel H, Krohn S, Lange M, Naaman A, Schwalm D, Wolf A, Zajfman D. Breakup dynamics and the isotope effect in H_3^+ and D_3^+ dissociative recombination. Phys Rev A. 2002;66:032719.

[149] Strasser D, Lammich L, Kreckel H, Lange M, Krohn S, Schwalm D, Wolf A, Zajfman D. Breakup dynamics and isotope effects in D_2H^+ and H_2D^+ dissociative recombination. Phys Rev A. 2004;69:064702.

[150] Motapon O, Tamo FOW, Urbain X, Schneider IF. Decisive role of rotational couplings in the dissociative recombination and superelastic collisions of H_2^+ with low-energy electrons. Phys Rev A. 2008;77:052711.

[151] Takagi H. In: Dissociative recombination: theory, experiment and applications. Guberman SL, editor. Extension of the quantum defect theory and its application to electron and molecular ion collisions. Kluwer Academic/Plenum Publishers, New York; 2003. pp. 177–86.

[152] Takagi H. Theoretical study of the dissociative recombination of HeH^+. Phys Rev A. 2004;70:022709.

[153] Novotný O, Wilhelm P, Paul D, Kálosi A, Saurabh S, Becker A, Blaum K, George S, Göck J, Grieser M, Grussie F, von Hahn R, Krantz C, Kreckel H, Meyer C, Mishra PM, Muell D, Nuesslein F, Orlov DA, Rimmler M, Schmidt VC, Shornikov A, Terekhov AS, Vogel S, Zajfman D, Wolf A. Quantum-state–selective electron recombination studies suggest enhanced abundance of primordial HeH^+. Science. 2019;365:676–9.

[154] Yousif FB, Mitchell JBA, Rogelstad M, Le Padellec A, Canosa A, Chibisov MI. Dissociative recombination of HeH^+: a reexamination. Phys Rev A. 1994;49:4610–15.

[155] Carles S. Private communication; n.d.

[156] Svensmark H. Influence of cosmic rays on earth's climate. Phys Rev Lett. 1998;81:5027–30.

[157] Svensmark H, Enghoff MB, Shaviv NJ, Svensmark J. Increased ionization supports growth of aerosols into cloud condensation nuclei. Nat Comm. 2017;8:2199.

[158] Kirkby J, Duplissy J, Sengupta K, Frege C, Gordon H, Williamson C, Heinritzi M, Simon M, Yan C, Almeida J, Tröstl J, Nieminen T, Ortega IK, Wagner R, Adamov A, Amorim A, Bernhammer A-K, Bianchi F, Breitenlechner M, Brilke S, Chen X, Craven J, Dias A, Ehrhart S, Flagan RC, Franchin A, Fuchs C, Guida R, Hakala J, Hoyle CR, Jokinen T, Junninen H, Kangasluoma J, Kim J, Krapf M, Kürten A, Laaksonen A, Lehtipalo K, Makhmutov V, Mathot S, Molteni U, Onnela A, Peräkylä O, Piel F, Petäjä T, Praplan AP, Pringle K, Rap A, Richards NAD, Riipinen I, Rissanen MP, Rondo L, Sarnela N, Schobesberger S, Scott CE, Seinfeld JH, Sipilä M, Steiner G, Stozhkov Y, Stratmann F, Tomé A, Virtanen A, Vogel AL, Wagner AC, Wagner PE, Weingartner E, Wimmer D, Winkler PM, Ye P, Zhang X, Hansel A, Dommen J, Donahue NM, Worsnop DR, Baltensperger U, Kulmala M, Carslaw KS, Curtius J. Ion-induced nucleation of pure biogenic particles. Nature. 2016;533:521–6.

[159] Le Garrec J-L, Speck T, Rowe BR, Mitchell JBA. Dissociative electron attachment at low temperature to clusters. In: Guberman SL, editor. Dissociative recombination of molecular ions with electrons. Kluwer Academic/Plenum Publishers, New York; 2003. pp. 451–60.

Chapter 4

Study of Electron Attachment Reactions Using the CRESU Technique

Fabien Goulay[*,‡] and Bertrand R. Rowe[†,§]

*C. Eugene Bennett Department of Chemistry,
West Virginia University,
Prospect Street, 26501 Morgantown, WV, USA
†Rowe Consulting, 22 chemin des moines,
22750 Saint Jacut de la Mer, France
‡fabien.goulay@mail.wvu.edu
§bertrand.rowe@gmail.com

Abstract

Electron attachment reactions using the CRESU technique were investigated from 48–170 K. The high density of electrons in the supersonic flow guarantees a local electronic temperature, although attachment experiments on DI have demonstrated that the temperature of the electrons is larger than that of the gas. The comparison of the experimental branching fractions for attachment to $POCl_3$ to theoretical predictions indicates that the electron temperature ranges from 100 to 450 K depending on the temperature of the gas. After introducing dissociative electron attachment, this chapter describes how the data from the CRESU experiments can all be reconciled with previous data based on the knowledge of the true electron temperature. Attachment on CF_2Cl_2, CF_3Br, and HBr is likely not to be affected as the observed temperature dependence is due to vibrational cooling of the neutral or formation of clusters. The higher electron temperatures are expected to decrease the attachment rate coefficients in the case of SF_6 and more importantly mask the temperature dependence in the case of CH_3I.

Keywords: Electron attachment; Kinetics; Low temperature.

4.1 Introduction

4.1.1 *Generalities on dissociative electron attachment*

Dissociative electron attachment (DEA) is a resonance phenomenon during which an electron is captured by a neutral molecule to form a transient anion intermediate [1, 2]. The so-formed reaction adduct ultimately dissociates to form a neutral radical and an anion. Non-dissociative electron attachment leads to stabilization of the adduct anion by collisional quenching with the buffer gas molecules [3]. Figure 4.1 displays the potential curves of a neutral molecule and its corresponding anion along one of the vibrational coordinate. For a neutral molecule in its ground vibrational level, the vertical electron affinity (VEA) corresponds to the maximum Franck-Condon overlap between the initial and final wave functions.

The difference between the crossing point of the two potential curves and the neutral vibrational ground state is defined as the attachment energy barrier E_b. It is the minimum energy that the system has to overcome in order to form the initial adduct anion. The closer the crossing is from the neutral equilibrium position, the lower the energy barrier will be, to eventually be zero. At infinite separation, the difference ΔE between the asymptote to the anion potential and neutral zero-point energy defines the exothermicity of the attachment process.

There are a wealth of studies aimed at measuring the attachment cross section over a wide range of electron energy [1]. In particular quantum-state-resolved measurements and theoretical calculations of the cross sections have provided very valuable information about the attachment process [4]. Studies have also looked at the attachment cross section as a function of the electron energy for a constant gas temperature, mostly near room temperature [1]. In such measurements, the energy of the attaching electrons is not correlated to the

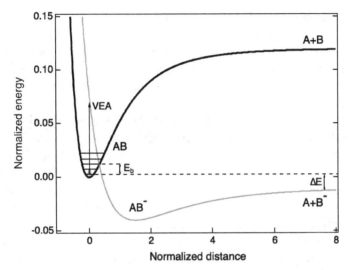

Fig. 4.1.   Schematic representation of the neutral and anion potential energy curve for electron attachment to a molecule AB. The vertical arrow displays the vertical electron affinity (VEA) corresponding to maximum Franck-Condon overlap (blue wave function). The energy E_b is the energy barrier leading to attachment and Δ_E is the exothermicity of the attachment process.

average molecular energy, the latter being defined by a Maxwell–Boltzmann distribution.

For high electron densities, electron–electron collisions are sufficient to guarantee local thermal equilibrium. An electron temperature T_e is defined which may or may not be equal to that of the attaching gas T_g. As discussed in Chapter 1, measurements of attachment cross sections $\sigma(\varepsilon)$ at an electron energy ε can be directly compared to attachment rate coefficients $k_{att}(T_e)$ at a temperature T_e through integration of $\sigma(\varepsilon)$ over the Maxwell–Boltzmann energy distribution $f(\varepsilon)d\varepsilon$

$$f(\varepsilon)d\varepsilon = 2\sqrt{\frac{\varepsilon}{\pi}}\left(\frac{1}{k_B T_e}\right)^{3/2} e^{-\frac{\varepsilon}{k_B T_e}} d\varepsilon \qquad (4.1)$$

where k_B is the Boltzmann constant. When only relative values of the attachment cross section are available, the resulting rate coefficients need to be normalized to a known value at a given temperature.

In the case of an exothermic reaction ($\Delta E < 0$), the presence of an energy barrier leads to an Arrhenius behavior of the measured rate coefficient as a function of the neutral molecule temperature T_g (see Section 4.2.1). The corresponding activation energy inferred from the dependence of the attachment rate coefficient with the temperature often differs from the energy barrier due to quantum effects in the dependence of the attachment cross section with vibrational quantum numbers.

With no barriers to attachment, the rate coefficient will be often close to the so-called capture rate coefficient discussed in Chapter 1. However, due to the small electron mass, its de Broglie wavelength ($\lambda = h/p$), where h is the Planck's constant and p the electron's linear momentum) is large compared to the typical length of inter-molecular forces. Classical calculations are therefore expected not to be accurate, especially at low temperatures.

The kinetic energy of the electron–molecule system is almost that of the electron due to the small electron mass. For neutral molecules in their vibrational ground level ($v = 0$), the attachment cross section therefore depends only on the kinetic energy of the approaching electrons. The electron wave function can be inferred by solving the Schrodinger equation for a charged particle moving in a long-range electron-induced or permanent dipole potential [5–8]. Within the Vogt and Wannier [6] formalism, the attachment cross section is calculated as the flux of electrons reaching the origin of the potential, neglecting any reflective wave. Dashevskaya *et al.* [5] performed an exhaustive modeling of low-energy electron–molecule capture processes. They provide an accurate analytical formula of the electron capture rate coefficient as a function of a reduced temperature in the case of non-polar and polar molecules. Their studies include s-wave (electrons with no angular momentum $l = 0$) and all wave calculations. For non-polar molecules at temperatures below $300\,\mathrm{K}$, the capture coefficient increases rapidly to reach twice the Langevin value. Polar molecules with a large dipole moment display a much larger increase than other molecules.

Techniques used to study DEA are based on the formation of variable-energy electrons in a gas at constant temperature. A detailed description of DEA techniques may be found elsewhere [1, 9]. In laser photoelectron attachment (LPA), the electrons are created by ionizing high-energy metastable argon atoms. The electrons can be formed by tuning the ionizing radiation with an energy resolution as low as $0.4\,\mathrm{meV}$ [4]. Detection of negatively charged products as a function of incident electron energy provides a relative value of the attachment cross section. Flowing Afterglow Langmuir Probe (FALP) techniques have allowed one to study DEA from room temperature up to 1100 K [10]. Under these conditions, the electrons are thermalized to the temperature of the flow. The advantage of the CRESU technique for the study of electron attachment reactions resides in the possibility of generating vibrationally and rotationally cold molecules together with thermalized low energy electrons. This allows the observation of predicted *s*-wave behaviors [5] or resonance features [4] in the attachment process, which are not observable at higher temperatures.

As discussed in Chapters 1 and 10 of this book, besides its fundamental aspect, low and ultralow temperature chemical and physical phenomena are strongly related to cold astrophysical media, such as interstellar clouds, planetary atmospheres, and comets. Ions and electrons often play a key role in these media. The very first model of interstellar chemistry [11, 12] was almost entirely based on positive ion chemistry. In such models, neutral molecules are formed by dissociative recombination of cations with electrons, a process very close to DEA. In addition, the discovery of anions [13, 14] in interstellar clouds and Titan's atmosphere has raised the question of their formation and reactivity at very low temperature, justifying furthermore the interest of electron attachment studies in these conditions.

The most abundant molecule in interstellar clouds is molecular hydrogen. Calculations based on the work of Mozumder [15] showed that the electron relaxation time toward the temperature of the neutrals ranges from a few to tens of years. Compared to the cloud lifetime (tens to hundreds of millions of years, see Chapter 10), it is clear that electrons are in local thermodynamic equilibrium (LTE) in the clouds.

4.1.2   *Using the CRESU technique to measure DEA rate coefficients*

Figure 4.2 displays a schematic of the CRESU experiment dedicated to the measurement of electron attachment rate coefficients. The technique is similar to that described for a reaction of neutral–neutral (Chapter 5) or ion–neutral species (Chapter 3). In the case of electron

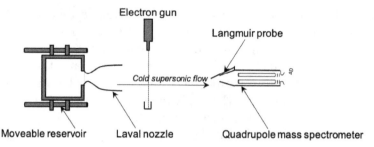

Fig. 4.2. Adaptation of the CRESU technique for the measurement of electron attachment rate coefficients [16].

attachment reactions, the electrons are generated continuously by a 12-keV ($10\,\mu$A) electron gun at a fixed location after the nozzle. The high-energy electrons collide with the buffer gas molecules generating lower energy electrons and buffer gas cations. Argon is usually not used as a buffer gas to avoid formation of metastable species. The secondary electrons then mix with the neutral reactants in the supersonic flow [16]. The electron density is measured as a function of distance between the electron beam and a movable Langmuir probe. Initial secondary electron number densities of up to $10^9\,\mathrm{cm}^{-3}$ can be generated and detected in the supersonic flow. Anions formed during the attachment process are monitored using a movable quadrupole mass spectrometer along the axis of the supersonic jet. For a uniform flow of constant velocity v, assuming electron attachment is the only process leading to electron density decay (n_e), n_e changes exponentially as a function of the distance x between the electron source and the probe. For an initial electron number density n_e^0, neutral species number density n_R, and attachment rate coefficient k_{att}, the electron number density along the axis of the flow x is given by

$$n_e = n_e^0\, exp\left(-\frac{k_{att} n_R x}{v}\right) \qquad (4.2)$$

Since the reactant number density is much greater than that of the electrons at any location x along the flow, it is assumed to be constant. The logarithm of the n_e/n_e^0 ratio as a function of n_R can be fit to a line of slope $-k_{att} n_R/v$. Experimentally, the attachment rate coefficients k_{att} at a given flow temperature are determined by measuring the electron density as a function of the neutral reactant number density for different x values.

In addition to electron attachment, diffusion and electron–ion recombination contribute to the overall decay of the electron density along the supersonic flow. The longitudinal n_e gradient may be expressed as

$$v\frac{dn_e}{dx} = D_{ae}\nabla^2 n_e - \alpha n_+ n_e - k_{att} n_R n_e \qquad (4.3)$$

where D_{ae} is the ambipolar diffusion coefficient, α is the electron–ion recombination rate coefficient, and n_+ is the number density of positively charged ions (mainly buffer gas cations). In the case of electrons at constant pressure, the diffusion coefficient D_{ae} is proportional to the square of the temperature and becomes negligible at low temperature. Under such conditions, the decay rate of electrons due to diffusion is of the order of $10\,\mathrm{s}^{-1}$ compared to attachment rates typically maintained within the 1000–$5000\,\mathrm{s}^{-1}$ range [16].

Regarding electron–ion recombination, reactions with cations formed from ionization of the buffer gas, He^+ or N_2^+, are the most likely ions. Recombination rate coefficients, obtained by integrating measured cross sections, are of the order of 1×10^{-11} and $4.4 \times 10^{-7}\,\mathrm{cm}^3\,\mathrm{s}^{-1}$ for He^+ and N_2^+ (at 45 K) [17], respectively. Under typical experimental conditions $\left(n_e^0 = n_+ = 1 \times 10^9\,\mathrm{cm}^{-3}\right)$, the electron decay rate due to recombination is of the order of $0.01\,\mathrm{s}^{-1}$ in He and $440\,\mathrm{s}^{-1}$ in N_2. Compared to attachment rates, electron–ion recombination processes are negligible in an He buffer gas. In an N_2 buffer, electron decay without attaching reactants is observable along the reaction flow. Numerical analysis of Eq. (4.3) under typical CRESU flow experiments suggests that the error made on the attachment rate coefficient is of the order of 9% when neglecting electron recombination with nitrogen cations. In all subsequent work in CRESU flow, electron–ion and diffusion processes are neglected and the gradient of the electron density along the flows is given by Eq. (4.2).

4.1.3   *Are the electrons thermalized to the temperature of the supersonic flow?*

Because the attachment cross section depends on both the energy of the electrons and on the vibrational and rotational distributions of the attaching molecules, the thermalization condition is paramount to the validity of the CRESU experiments. The attachment cross section $\sigma(\varepsilon, T_g)$ is a function of both ε and T_g which, in most cases, are uncorrelated. The cross section can be integrated to give the attachment rate coefficients $k_{att}(T_e, T_g)$ as followed:

$$k_{att}(T_e, T_g) = \frac{2}{\sqrt{\pi}} \left(\frac{1}{k_B T_e} \right)^{3/2} \int_0^\infty \sigma(\varepsilon, T_g) \sqrt{\varepsilon} e^{-\frac{\varepsilon}{k_B T_e}} d\varepsilon \qquad (4.4)$$

For an electron density n_e, the energy relaxation time due to electron–electron collisions is given by [16]

$$\tau_{ee} = 0.24 \frac{T_e^{3/2}}{n_e ln\Lambda} \qquad (4.5)$$

with

$$\Lambda = 8.26 \times 10^3 \frac{T_e^{3/2}}{n_e^{1/2}} \qquad (4.6)$$

For a typical electron number density of $1 \times 10^8 \, \text{cm}^{-3}$, electron–electron relaxation time remains shorter than $10 \, \mu s$ for T_e up to 1200 K, compared to the typical hydrodynamic time of a CRESU flow ($\sim$100–1000 μs). Such a short relaxation time guarantees that the energy distribution of the electrons follows a Maxwell–Boltzmann distribution very rapidly after their formation.

The thermalization of the electrons to a temperature T_e is not the only requirement. The electron temperature needs to be that of the attaching gas: $T_e = T_g$. The final temperature T_e of the electron's energy distribution is also dependent on energy transfer processes between the electrons and the surrounding molecules. In the presence of energized molecules, the gas–electron system may reach a steady state at an electron temperature, which is higher than that of the supersonic flow.

In a low-pressure supersonic flow, the energy transfer between the electrons and the buffer gas molecules will occur mostly through elastic collisions for monoatomic gases, and via inelastic collisions involving rotational and vibrational transitions for diatomic gases. For these reasons, experiments performed in He as a buffer gas required total number densities 5–10 times greater than that of experiments performed in nitrogen. Le Garrec *et al.* [16] estimated thermalization times of the electrons to the temperature of the CRESU flow ranging from 20–75 μs at $T_g < 170$ K. In nitrogen, these values do not take into account the non-thermalization of the N_2 vibrational levels to the translational energy as well as the possible vibrational excitation of the nitrogen molecules due to interaction with the 12-keV primary electron beams. Highly vibrationally excited nitrogen molecules are likely to heat up the electrons to a temperature higher than that of the flow.

During the initial experiments on electron attachment using the CRESU apparatus [16], the electrons were assumed to be thermalized to the temperature of the supersonic flow. The first reactions that were investigated were attachment to CF_3Br, CF_2Cl_2, and SF_6 [16]. These reactions have been extensively studied and the CRESU technique allowed for an extension of the temperature range down to 48 K. These first CRESU measurements where mostly in agreement with previous experiments at low electron energy and validated the experimental setup [16]. A series of experiments were then dedicated to verifying the low temperature of the electrons by measuring the attachment rate coefficients to HBr, HCl, and CH_3I, as they are expected to change with the electron temperature [18, 19]. Attachment to the halogen halides [18] did not provide clear evidence on the thermalization of the electron due to rapid attachment to dimers at low temperature. The investigation of the attachment to CH_3I in the CRESU experiment was pivotal in determining whether or not the electrons were thermalized to the flow temperature [19]. The CRESU rate coefficients were found to be constant at low temperature [19], while advanced theoretical predictions showed a net increase of the integrated attachment cross section below room temperature [4]. The non-thermalization of the electrons to the temperature of the flow was later confirmed by the comparative study of electron attachment to HI and DI [20]. Subsequent experiments on $POCl_3$ and comparison with theory provided an estimate of the electron temperature in the reaction flow [21]. Although the electrons were found to be hotter than the gas, their energy is still believed to follow a Maxwell–Boltzmann distribution at a temperature below 500 K.

In this chapter, instead of presenting the CRESU DEA experiments in chronological order, the experimental evidence for non-thermalization of electrons to the temperature of the flow (HI/DI) [20] is presented first. The more recent experiments [21] for the attachment to $POCl_3$ are employed to infer a temperature range of the electrons. A simple model [20] is also described in order to determine the contribution of vibrationally excited nitrogen molecules to the electron steady state temperature. This approach allows for the discussion of all the past CRESU attachment experiments with the knowledge of the true electron temperature. Electron attachment to CF_3Br and CF_2Cl_2 for which an Arrhenius behavior is expected [16] with T_g is then presented. In this case, the dependence of the attachment cross section on the rotational or vibrational distribution of the neutral molecule is so pronounced that the higher-energy part of the electron energy distribution plays only a minor role. Attachment to SF_6 [16], HBr/HCl [18], and CH_3I [19] is discussed with an emphasis on the effect of non-thermalization of the electrons. For attachment to HCl and HBr, other phenomena such as the formation of molecular clusters must be taken into account in order to interpret the experimental data.

4.2   Non-Thermalization of the Electrons and Electron Temperature

4.2.1   *Case of an Arrhenius behavior of the attachment process with gas temperature*

Fabrikant and Hotop [22] provided a comprehensive review of electron attachment processes displaying an Arrhenius behavior. Figure 4.3 displays the potential curves of the neutral (solid thick line) and anion along the coordinate A–B in the case of an endothermic reaction without a barrier (thick red dashed line) and exothermic reaction with a barrier (thin blue dashed line). In both cases, the attachment cross section depends on the rotational and vibrational states of the molecules.

In the case of an endothermic reaction with no activation barrier (crossing of the potential curves near the neutral equilibrium position), an electron with a kinetic energy ε will have sufficient energy to overcome the reaction barrier only if

$$\varepsilon > E_{th} - E_n \tag{4.7}$$

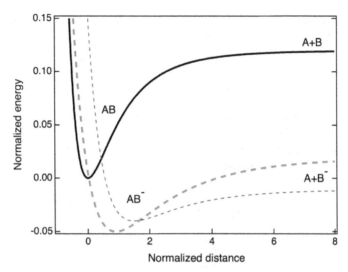

Fig. 4.3. Schematic representation of the neutral and anionic potential curve for exothermic (blue thin dashed line) and endothermic (red thick dashed line) attachment to a molecule AB. In the case of the exothermic reaction, the crossing of the potential curves is above the zero-point energy of the neutral, which is characteristic of a reaction displaying an Arrhenius behavior.

The threshold energy E_{th} is the endothermicity of the reaction and E_n is the energy of a quantum level n of the neutral (rotational and vibrational), corrected for zero-point energy. If the neutral and anion potential curves cross near the minimum equilibrium position of the neutral, the Franck–Condon factors for attachment are favorable and only weakly dependent on the quantum number n and electron energy ε [22]. For a molecule in a quantum state n such that $E_n < E_{th}$, integrating the attachment cross section over a Maxwell–Boltzmann distribution of the electron energy leads to the state-specific attachment rate coefficient:

$$k_{att}(n, T_e) = k_n exp\left(-\frac{E_{th} - E_n}{k_B T_e}\right) \qquad (4.8)$$

where k_n is an attachment rate coefficient which is only weakly dependent on the electron temperature. If the energy of the quantum state is above the threshold energy, corresponding to an exothermic reaction, the state-specific rate coefficient becomes independent of the quantum number n. If n_0 is such that $E_{n_0} < E_{th} < E_{n_0+1}$, the total attachment rate coefficient is given by the sum of all the state-specific rate coefficients weighted by the relative populations of each energy state according to [22]

$$k_{att}(T_e, T_g) = \sum_{n=0}^{n0} \frac{N_n(T_g)}{Q(T_g)} k_n e^{-\frac{E_{th}-E_n}{k_B T_e}} + k_c \sum_{n0+1}^{\infty} \frac{N_n(T_g)}{(T_g)} \qquad (4.9)$$

where $Q(T_g)$ is the partition function, $N_n(T_g)$ the population of the quantum state n at the temperature of the gas T_g, and k_c a rate coefficient weakly dependent on n and T_e. For a quantum level with g_n degeneracy, the attachment rate coefficient becomes

$$k_{att}(T_e, T_g) = \frac{1}{Q(T_g)} \left\{ \sum_0^{n_0} g_n k_n e^{-\frac{E_{th}}{k_B T_e}} e^{-\frac{E_n}{k_B}\left(\frac{1}{T_g}-\frac{1}{T_e}\right)} \right.$$

$$\left. + k_c \sum_{n_0+1}^{\infty} g_n e^{-\frac{E_n}{k_B T_g}} \right\} \qquad (4.10)$$

For $T_e = T_g$, the reaction rate displays an Arrhenius behavior only if $E_n \gg k_B T_g$. For non-thermalized electrons with $T_e \gg T_g$, the overall attachment rate coefficient becomes only weakly dependent on the temperature of the gas due to k_n and k_c being relatively constant on the quantum state n and T_e. Equation (4.10) is used in Section 4.2.2 to model the attachment kinetics to HI and DI [20]. Deviation from the expected Arrhenius behavior is strong evidence of non-thermalization of the electrons to the temperature of the flow.

In the exothermic case with a small barrier E_b (blue thin dashed line in Fig. 4.3), the attachment cross section is non-zero even at very low electron energy. Below the reaction barrier, the corresponding state-specific attachment rate coefficient $k_{att}(n, T_e)$ is weakly dependent on the temperature of the electron but strongly dependent on the quantum number n. For low values of T_e, Fabrikant and Hotop [22] proposed an exponential dependence of $k_{att}(n, T_e)$ with the quantum number n for $E_n < E_b$ and a constant rate coefficient for $E_n > E_b$. As described above, for a quantum number n_0 such that $E_{n0} < E_b < E_{n0+1}$, the overall reaction rate coefficient is given by [22]

$$k_{att}(T_e, T_g) = \frac{1}{Q(T_g)} \left\{ k_1 \sum_0^{n_0} g_n e^{-\frac{an-E_n}{k_B T_g}} + k_2 \sum_{n_0+1}^{\infty} g_n e^{-\frac{E_n}{k_B T_g}} \right\}$$

$$(4.11)$$

where k_1, k_2, and a are constants. Equation (4.11) can be approximated to an Arrhenius behavior within a restricted temperature range which depends on the rotational or vibrational constants of the molecule as well as the constant a [22]. Over this range, the reaction rate becomes independent of T_e and is given by

$$k_{att}(T_g) = k_1 e^{(n_o-1)\alpha} e^{-\frac{E_{n_0-1}}{k_B T_g}} \qquad (4.12)$$

The above equation is valid only over a very restricted temperature range [22]. At the low temperature of the CRESU flow, deviations from the Arrhenius behavior are expected and are discussed in Section 4.2 for attachment to CF_2Cl_2 and CF_3Br.

4.2.2   *Attachment to HI/DI: evidence of non-thermalization*

Electron attachment to the HI molecule is exothermic by 5 meV. The capture rate coefficient [23] is found to be large and to increase slightly as the temperature is lowered. The attachment to the isotope-substituted DI is endothermic by 35 meV corresponding to the rotational level $J = 8$ [24]. As discussed in the previous section, the reaction rate coefficients for attachment to DI with an energy threshold E_{th} are the sum of the rotationally resolved rate coefficients weighted by the relative population N_J of each level. For $T_e = T_g$, the overall rate coefficient will follow an Arrhenius behavior such that [20]

$$k_{att}(T) = \frac{1}{Q(T)} \left\{ \sum_{J=0}^{8} N_J k_{cap}(J, T_g) e^{-\frac{E_{th}-E_J}{k_B T_g}} + \sum_{J\geq 9} N_J k_{cap}(J, T_g) \right\}$$

$$(4.13)$$

where $k_{cap}(J,T)$ is the state-resolved capture rate coefficient.

Figure 4.4 displays the calculated rate coefficients for HI (blue line) and DI (red line) together with CRESU measurements (open blue circles for HI and filled red circles for DI) [20]. The room temperature data are FALP measurements for HI (open blue square) [25] and krypton photoionization measurements for DI (filled red triangle) [24]. The calculated coefficients are based on analytical expression of the capture rate coefficient by Dashevskaya *et al.* [5] for s-wave attachment. The attachment rate coefficient on DI is expected to

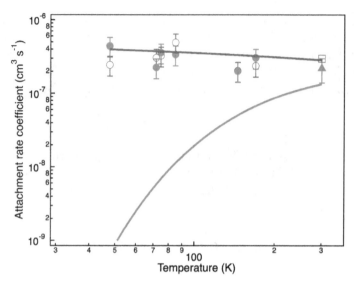

Fig. 4.4.   Measured rate coefficients for attachment to HI (open blue circles) and DI (filled red circles) by the CRESU experiments. The lines are modeled attachment rate coefficients for HI (blue line) and DI (red line) [5, 20]. The room temperature data are FALP measurements for HI (open blue square) [25] and Kr photoionization DI (filled red triangle) [24].

decrease considerably as the temperature is lowered, while the capture rate coefficient for HI increases. The CRESU measurements for DI are in clear disagreement with the predicted temperature dependence. The attachment rate coefficients on DI are however biased due to possible isotopic exchange between DI and adsorbed water molecules in the gas lines and CRESU reservoir. Mass spectra using a quadrupole mass spectrometer have shown that the experiments are performed with a mixture of HI and DI [20]. In this case, the DI may represent only half of the attaching molecules in the reaction flow. Nonetheless, the measured attachment rate coefficients are still expected to decrease as the temperature is decreased, especially at very low temperature at which the DEA process for DI is not efficient.

The comparison of the attachment rates to HI and DI in the CRESU flow provides strong evidence that the temperature of the electrons is higher than that of the neutral. In the next section, the temperature dependence of the attachment to $POCl_3$ is used to obtain a good estimate of the electrons' temperature.

4.2.3   *Attachment to POCl₃: estimate of the electron temperature*

Electron attachment to $POCl_3$ proceeds through the formation of an electron–molecule excited adduct $POCl_3^-$ which can eventually dissociate or be stabilized by collision with the buffer gas [21]. At low temperature, the only accessible (thermoneutral) dissociative channel is the formation of the $POCl_2^-$ anion and a chlorine atom. The formation of the initial adduct has been modeled using the Vogt–Wannier electron capture modified for intramolecular vibrational redistribution [3]. The rate coefficient is large ($\sim 1 \times 10^{-7} \, cm^3 \, s^{-1}$) and depends mostly on the electron temperature [26]. Above room temperature, for $T_e = T_g$, the attachment rate coefficient decreases monotonically from about $2.5 \times 10^{-7} \, cm^3 \, s^{-1}$ at $300 \, K$ to $0.5 \times 10^{-7} \, cm^3 \, s^{-1}$ at $1200 \, K$ [26]. Below room temperature, the attachment process is likely to remain fast, although there are no theoretical calculations or experimental data available.

Measurements in a CRESU flow focused on the temperature dependence of the product branching fractions [21]. Van Doren *et al.* [26] measured and calculated the pressure and temperature dependences of the non-dissociative channel branching fraction. The formation of the stabilized adduct is found to be more favorable as the pressure increases due to energy stabilization by collision with the buffer gas molecules. Temperature is found to have the opposite effect. At room temperature and 1 Torr, non-dissociative attachment accounts for 25% of the product fraction. This value drops to only 2% at $550 \, K$ at which the $POCl_2^- + Cl$ channel becomes dominant [3]. Figure 4.5 displays the dissociative (triangles) and non-dissociative (circles) branching fractions measured at low temperature by the CRESU technique (red markers) [21] and at high temperature using the FALP technique (blue markers) [3]. The data point at $170 \, K$ was recorded at a much lower pressure than for the data at $200 \, K$ leading to a decreased stabilization of the adduct. Overall, the data show an inversion of the branching fraction between the exothermic non-dissociative channel and the thermoneutral dissociative one. Both channels are found to be equally probable between 200 and $300 \, K$.

The trend displayed in Fig. 4.5 is due to the dependence of the branching fraction both on the gas and electron temperatures.

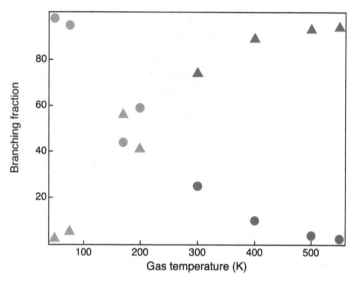

Fig. 4.5. Branching fractions for dissociative (triangles) and non-dissociative (circles) attachment to POCl$_3$. The red markers are measured using the CRESU technique [21], while the blue markers are measured in a high-temperature FALP apparatus [3].

Figure 4.6 displays POCl$_3^-$ branching fractions computed as a function of the electron temperature at $T_g = 50\,\text{K}$ (dotted black line), 75 K (dashed black line), 170 K (solid black line), and 200 K (dashed and dotted black line) [21]. For each gas temperature, the branching ratio is greatly dependent on the electron temperature. The horizontal red lines in Fig. 4.6 correspond to measurements in a CRESU flow at $T_g = 47.7\,\text{K}$ (dotted red line), 74.5 K (dashed red line), 169.7 K (solid red line), and 199.5 K (dashed and dotted red line) [21]. The intersection between the theoretical and experimental lines is indicative of the temperature of the electrons in the supersonic flow. At 170 and 200 K, the experimental data may be reproduced for an electron temperature of about 450 K. At lower temperature, the electron temperature is found to range from 100–150 K.

The temperature dependence of the non-dissociative branching fraction for attachment to POCl$_3$ is consistent with electrons having a higher temperature than that of the neutrals in the supersonic flow. The statistical model indicates that the electron temperature may not exceed 450 K and is lower than room temperature for $T_g < 100\,\text{K}$.

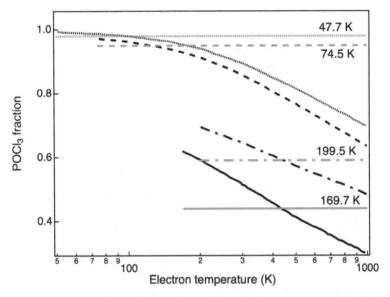

Fig. 4.6.   Predicted $POCl_3$ fraction formed by electron attachment to $POCl_3$ as a function of electron temperature and at gas temperatures of 50 K (dotted black line), 75 K (dashed black line), 170 K (solid black line), and 200 K (dashed and dotted black line). The horizontal lines corresponds to CRESU measurements at $T_g = 47.7$ K (dotted red line), 74.5 K (dashed red line), 169.7 K (solid red line), and 199.5 K (dashed and dotted red line) [21].

4.2.4   *Modeling the electron temperature*

Electron attachment to DI [20] and $POCl_3$ [21] demonstrated the non-thermalization of the electrons to the temperature of the CRESU flow. The electrons are still expected to have a Maxwell–Boltzmann distribution of kinetic energy but at a higher temperature. According to the non-dissociative $POCl_3$ attachment branching fractions, the temperature of the electrons ranges from 100–450 K depending on the temperature of the gas [21]. The high temperature of the electrons is likely to come from vibrational excitation of the nitrogen buffer gas by the high-energy primary electrons from the electron gun. The secondary electrons formed by ionization of the buffer gas then interact with the surrounding molecules through elastic, vibrational, and rotational energy transfer. This problem remains the same when using helium or argon as buffer gases. Metastable atoms are produced by the high-energy electron beam, which ultimately heats up the secondary electrons in the supersonic flow through super-elastic collisions.

In a nitrogen buffer gas, for electrons of energy ε, the rate of energy transfer is given by [20]

$$\frac{d\epsilon}{dt} = \frac{d\epsilon}{dt}\bigg|_{elastic} + \frac{d\epsilon}{dt}\bigg|_{vibration} + \frac{d\epsilon}{dt}\bigg|_{rotation} \qquad (4.14)$$

where the three terms on the right-hand side of the equation account for the elastic, vibrational, and rotational contributions of the buffer gas molecule energy transfer. The elastic contribution to the energy rate may be estimated by integrating the momentum transfer cross section over a Maxwell–Boltzmann distribution of the electron energy. The vibrational and rotational energy rates are determined from the excitation and deexcitation vibrational and rotational cross sections and integrated [20]. The differential equation for the temperature of the electrons in the supersonic flow is solved numerically for different flow conditions [20]. The presence of vibrationally excited buffer gas molecules is modeled by setting the vibrational temperature to 1200 K.

Figure 4.7 displays the modeled electron temperatures for a flow at 169.7 K, with nitrogen density of 0.56×10^{16} cm^{-3} and vibrational temperature of the buffer gas of 300 K (blue line) and 1200 K (red line). For a Maxwell–Boltzmann vibrational distribution of the nitrogen molecules at room temperature or lower, the electrons are found to thermalize to the temperature of the supersonic flow within less than 100 μs. At higher temperature, the modeled electron temperature reaches a steady state at a temperature which is higher than that of the flow. For vibrationally excited nitrogen molecules at 1200 K, the final electron temperature is about 430 K, similar to estimates obtained from the POCl$_3$ attachment branching fractions [21]. For a flow temperature of 75 K, the corresponding stationary electron temperature in the presence of vibrationally excited nitrogen molecules is determined to be 280 K. Although there are large uncertainties on the initial energy distribution of the excited nitrogen molecules, the electron temperature range inferred from the POCl$_3$ study is in agreement with heating of the electrons through vibrational energy transfer from the nitrogen buffer gas. In the case of endothermic attachment (HBr/HCl) or reactions for which the attachment to the ground vibrational level depends on the electron temperature

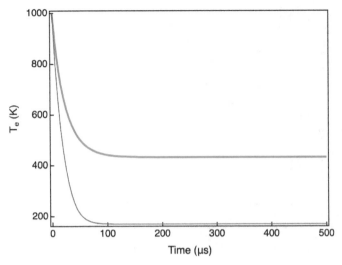

Fig. 4.7. Modeled electron temperature as a function of time in a CRESU flow at $T_g = 169\,\mathrm{K}$ and nitrogen buffer gas density of $0.56 \times 10^{16}\,\mathrm{cm^{-3}}$. The temperatures are models for a nitrogen vibrational distribution at $300\,\mathrm{K}$ (thin blue line) and $1200\,\mathrm{K}$ (thick red line) [20].

(SF_6 and CH_3I), it is important to revisit the CRESU data based on the knowledge of the true electron temperature.

4.3   Electron Attachment Reactions with an Arrhenius Behavior

4.3.1   *Electron attachment to $CF_2 Cl_2$*

The dissociative electron attachment to CF_2Cl_2 is believed to proceed through activation of the ν_3 A_1 symmetric CCl_2 stretch followed by vibrational energy redistribution into the C–Cl stretch [27]. The overall dissociative attachment reaction $CF_2Cl_2 + e^- \rightarrow CF_2Cl + Cl^-$ is exothermic by $237\,\mathrm{meV}$ and the VEA (maximum Franck-Condon overlap transition) to the lowest $CF_2Cl_2^-$ ion state lies $0.9\,\mathrm{eV}$ above the lowest ν_3 vibrational level. The crossing of the neutral and anion potential curves occurs near the neutral equilibrium position leading to significant attachment cross sections close to zero energy. For this reason, the dissociative attachment to CF_2Cl_2 is expected to follow an Arrhenius behavior with the temperature of the neutral [22].

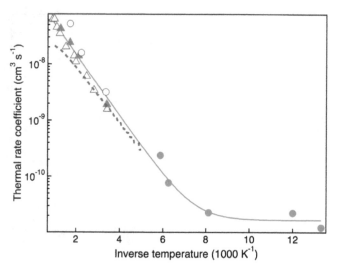

Fig. 4.8. Arrhenius plot of the thermal rate coefficient for electron attachment to CF$_2$Cl$_2$ from electron swarm experiments (purple open triangles) [10], FALP experiments (open blue circles [28] and green triangles [29]), and CRESU experiments down to 45 K (red filled circles) [16]. The overall data are fitted to a modified Arrhenius (solid red line). The blue dashed line is the theoretical prediction by Graupner *et al.* [27].

Figure 4.8 displays an Arrhenius plot of the attachment rate coefficients from electron swarm experiments (purple open triangles) [10], FALP experiments (open blue circles [28] and green triangles [29]), and CRESU experiments down to 45 K [16]. The overall data are fitted to a modified Arrhenius equation (red line) according to

$$k_{att}(T_g) = k_0 + k_1 exp\left(-\frac{E_a}{k_B T_g}\right) \tag{4.15}$$

The blue dashed line is the theoretical prediction by Graupner *et al.* [27] At intermediate temperatures, the attachment rate coefficients measured in a CRESU flow follow the same trend as that observed for the data measured at higher temperatures by Smith *et al.* [28] and more recently by Miller *et al.* [10] Over the 123–1100 K temperature range, the attachment rate coefficients follow an Arrhenius behavior with an activation energy $E_a = 109$ kJ mol^{-1}. This value is found to be slightly smaller than previously published values [27] and close to two quanta of excitation of the ν_3 CCl$_2$ symmetric stretch of the molecule.

The pure Arrhenius behavior is found to be non-valid at temperatures below 170 K. The measured rate coefficient tends toward a constant value $k_0 = 1.7 \times 10^{-11}\,\text{cm}^3\,\text{s}^{-1}$. Graupner *et al.* [27] suggested that the low-temperature limit of the measured rate coefficient reflects the *s*-wave, temperature independent, characteristic of the attachment process at low electron energy. In this case, only the $v = 0$ vibrational levels of the molecule are populated, and the process becomes independent of the temperature of the molecules. The slow attachment process in the CRESU flow below 170 K provides strong evidence that the average temperature of the electrons is well below the barrier for attachment to CF_2Cl_2 (i.e. 1000 K). Although the constant values with decreasing temperature may already reflect the independence of the attachment rate coefficient with temperature, it could also be due to the electron temperature reaching a minimum for a flow temperature at and below 123 K. In such a case, the change in electron temperature with the temperature of the flow is not sufficient to significantly slow down the attachment process. Attachment to impurities could also lead to faster apparent attachment.

The study of the electron attachment process on CF_2Cl_2 in the CRESU apparatus provided a low-temperature validation of the Arrhenius behavior down to 170 K. The fit to the data shows an activation energy of 109 kJ mol^{-1} which is consistent with an exothermic attachment with an energy barrier of the order of 2 quanta of CCl_2 vibrational stretch. For such reactions, the non-thermalization of the electrons discussed in Section 4.3 of this chapter is not expected to have major effects on the Arrhenius behavior with the gas temperature.

4.3.2   *Electron attachment to* CF_3Br

Electron attachment to CF_3Br is a dissociative process leading to the formation of a CF_3 radical and a Br$^-$ anion with an excess energy of 35 meV [30], following the reaction

$$CF_3Br + e^- \rightarrow CF_3 + Br^-$$

Although the VEA is estimated to be close to 1 eV [30], the crossing of the neutral and ionic potentials along the C–Br bond coordinate occurs near the $v = 3$ of the C–Br vibrational level, corresponding to an energy barrier of 129 meV relative to the ground vibrational level. Prior to CRESU experiments, the Arrhenius behavior of the attachment rate coefficient was confirmed over the 173–600 K temperature range [31]. The corresponding activation energy was measured to be within the 75–80 meV range, which is lower than the predicted energy barrier. The difference between the activation energy and the energy barrier is likely to be due to the participation of other vibrational modes of the molecule, especially the CF_3 umbrella mode [30].

Figure 4.9 displays the experimental and calculated rate coefficients for the DEA on CF_3Br measured using a high-temperature discharge flow (open diamond) [32], laser photoelectron attachment (LPA) (filled squares) [30], FALP (open triangles) [31], and the CRESU technique (filled red circles) [16]. The solid line results from calculations using R-matrix theory [30] and the dashed line is the temperature dependence of the thermal rate coefficients calculated by Troe *et al.* [33] using a statistical approach. The theoretical R-Matrix

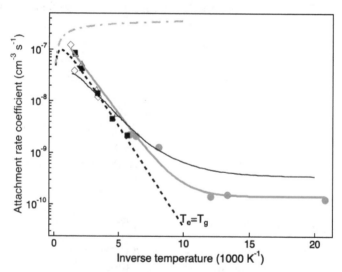

Fig. 4.9. Arrhenius plot of the rate coefficient for electron attachment to CF3Br measured using a high-temperature discharge flow (open diamond) [32], LPA (filled squares) [30], FALP (open triangles) [31], and the CRESU technique (filled red circles) [16]. The red line is a fit to the data using a modified Arrhenius equation. The black lines are modeled rate coefficients using R-matrix theory (solid thin line) [30] and kinetic modeling approach (dashed line) [33]. The thick dashed and dotted line is the electron contribution to the rate coefficient [32].

rate coefficients are found to deviate from the measured values at high and low temperatures. At low temperature, the CRESU data are also found to deviate from a pure Arrhenius behavior. All the experimental data points in Fig. 4.9 are fit to a modified Arrhenius function (Eq. (4.15)), giving an activation energy of 65 meV. The obtained energy barrier is lower than the one reported previously including only experimental points above 173 K.

Troe *et al.* [33] attempted to decouple the dependence of the electron and gas temperatures on the attachment rate coefficients. They empirically fit DEA cross sections as a function of energy and attachment rate coefficients as a function of temperature over the 173–600 K range. They proposed a dependence of the attachment rate coefficient of the form

$$k_{att}(T_e, T_g) = k_{att}(T_e)exp(-920\,K/T_g) \qquad (4.16)$$

where $k_{att}(T_e)$ is the rate coefficient inferred from integrated attachment cross sections derived from an empirical model. The exponential contribution corresponds to the presence of an energy barrier of 79 meV. For $T_e = T_g$ (see Fig. 4.9), there is good agreement between Eq. (4.16) (dashed line) and the experimental data between 173 K and 600 K. At higher temperature, the modeled values are closer to the experimental data than the R-Matrix model. At low temperature, the CRESU values are found to be much higher than those predicted by Troe *et al.* [33] As for CF_2Cl_2, deviation from an Arrhenius behavior may suggest that the molecular contribution becomes constant at low temperature.

The separation of the electron and gas temperature dependences by Troe *et al.* [33] is very useful in order to discuss the effect of the non-thermalization of the electrons on the attachment rate of CF_3Br. Based on the proposed $k_{att}(T_e)$ dependence with temperature [33], the absolute value of the measured rate coefficient at a given gas temperature is expected to decrease with increasing electron temperature. Although Troe *et al.* [33] do not provide values for the electron contribution to the rate coefficient $k_{att}(T_e)$ (thick dashed and dotted line in Fig. 4.9) below 100 K, it is found to decrease by less than 50% between 100 K and 1000 K. The presence of hot electrons in the CRESU flow would therefore be expected to slightly decrease the measured values compared to fully thermalized electrons. Assuming the presence of hot electrons in the CRESU, the low-temperature data in Fig. 4.9 (red circles) would be expected to be closer to the R-Matrix predictions at $T_e = T_g$. Overall, because of the relatively small dependence of the attachment rate coefficient with electron temperature below $T_e = 1000$ K, it is expected that the observed temperature dependence of the CRESU measurements is mostly due to the change in neutral temperature.

4.4 Electron Attachment Reactions with a Non-Arrhenius Behavior: Effect of the Electron Temperature

4.4.1 *Electron attachment to SF$_6$*

Electron attachment to SF$_6$ proceeds through the capture of the electron and formation of electron–SF$_6$ intermediate with excess energy. Vibrational redistribution along the molecular vibrational modes may lead to formation of a highly vibrationally excited SF$_6^-$ anion in its ground electronic state (non-dissociative attachment). The so-formed anion can further be stabilized by collision with the buffer gas or by emission of radiations, or dissociate to SF$_5^-$+F. Other pathways following the initial capture of the electron may form vibrationally excited SF$_6$.

Troe *et al.* [34] parameterized the attachment cross section on SF$_6$ using the transmission coefficient $P(\varepsilon - \varepsilon_0)$ with ε_0 the energy of the system at infinite electron–SF$_6$ distance. The ratio of the rate coefficient for attachment to SF$_6$ over the Langevin rate coefficient (classical capture) is inferred from the transmission coefficient after integration over the Maxwell–Boltzmann distribution of energy. The coefficient $P(\varepsilon - \varepsilon_0)$ has three main contributions such that

$$P = P^{VW} P^{IVR} P^{VEX} \tag{4.17}$$

where P^{VW} is the s-wave contribution, and P^{IVR} and P^{VEX} account for deviation from the s-wave behavior due to incomplete inter-molecular vibrational redistribution (IVR) and vibrational excitation (VEX), respectively. The P^{VW} term is a fit from the accurate Vogt–Wannier calculation, while P^{IVR} and P^{VEX} are parameterized using the empirically determined coefficients obtained by fitting experimental data for $T_e \neq T_g$. Such formalism allows defining s-wave or electronic (VW) and molecular (IVR and VEX) contributions to the overall attachment rate coefficient.

Smirnov and Kosarim [7] employed the Breit–Wigner formula [8] in order to determine the SF_6 attachment cross section as a function of both the electrons (T_e) and gas (T_g) temperatures. For electron energy below 80 meV, and averaging over the Maxwell–Boltzmann distribution function, the temperature dependence of the attachment rate coefficient is given by [7]

$$k_{att}(T_e, T_g) = \frac{2\pi\hbar^2}{m_e}\sqrt{\frac{2\pi}{m_e T_e}}\frac{1}{\sqrt{1 + 4\epsilon_1/\pi T_e}} f_0 \frac{1 - e^{-\frac{E_1}{T_g}}}{1 - e^{-\frac{E_2}{T_g}}} \qquad (4.18)$$

where $\varepsilon_1 = 2$ meV is a threshold energy, f_0 is a constant, E_1 is the typical SF_6 vibrational energy in a given mode, and E_2 is of the order of the attachment energy barrier E_b. The parameters f_0, E_1, and E_2 are inferred from fitting experimental data from 76–700 K, which include CRESU data.

Figure 4.10 displays the electronic (as a function of T_e) and molecular (as a function of T_g) contributions to the SF_6 attachment rate

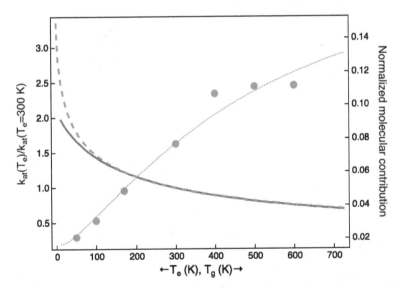

Fig. 4.10. Electronic (as a function of T_e) and molecular (as a function of T_g) contributions to the SF_6 attachment rate coefficient using the formalism from Troe *et al.* [34] (dashed red line and red filled circles, respectively) and Smirnov *et al.* [7] (solid blue line and dotted blue line, respectively).

coefficient using the formalism from Troe *et al.* [34] (dashed red lines and red filled circles) and Smirnov *et al.* [7] (solid blue lines and dotted blue line). The s-wave capture rate coefficient decreases with increasing electron temperature, while the parameterized molecular contribution increases with increasing gas temperature. From Troe's parameterization, at low temperature, any increase of the rate coefficient with temperature is due to incomplete IVR. At room temperature and higher, vibrational excitation increases the absolute value of the rate coefficient.

Figure 4.11 displays the attachment rate coefficient on SF_6 molecules obtained using the CRESU technique (filled red circles) [16], threshold photoelectron (black diamonds) [2, 35], drift tubes

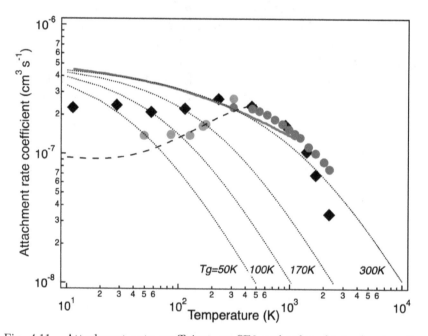

Fig. 4.11. Attachment rate coefficient on SF6 molecules obtained using the CRESU technique (filled red circles) [16], threshold photoelectron (black diamonds) [35], drift tubes (opened blue circles) [36], and laser photoattachment (LPA) (thick purple line) [37, 38]. The dotted lines are the parameterized attachment rate coefficients by Troe *et al.* [34] as a function of electron energy at different gas temperatures. The dashed line is the $T_e = T_g$ model by Smirnov *et al.* [7] at room temperature. The temperature in the horizontal axis may refer to electron or gas temperature depending on the experiments.

(open blue circles) [36], and LPA (thick purple line) [37, 38]. The dotted lines are the parameterized attachment rate coefficients by Troe *et al.* [34] as a function of electron energy at different gas temperatures. The dashed line is the $T_e = T_g$ model by Smirnov and Kosarim [7]. The CRESU values appear to be slightly lower than the threshold photoelectron data measured at $T_g = 300\,\text{K}$, although the room temperature values are in agreement. Based on the electronic contribution to the reaction rate coefficient displayed in Fig. 4.10, a higher electron temperature decreases the attachment rate coefficient. For electrons at a temperature higher than that of the CRESU flow, the rate coefficients are therefore expected to be lower than those measured at $T_e = T_g$. Cooler electrons in the CRESU flow would likely increase the measured value closer to the threshold photoelectron data. The dotted lines in Fig. 4.11, which are parameterized using the CRESU results [16], would increase to be closer to the room temperature values with a weaker electron temperature dependence. The trend of the rate coefficients as a function of temperature observed in the CRESU experiment is therefore mostly due to IVR, and the non-thermalization of the electrons to the temperature of the flow is likely to only slightly decrease the measured values with no major impact on the temperature dependence.

The SF_6 CRESU rate coefficients have also been used as reference values for the measurement of electron attachment to naphthalene [39]. The SF_6-relative naphthalene attachment rate coefficients are found to be of the order of $1 \times 10^{-9}\,\text{cm}^3\,\text{s}^{-1}$ and independent of the gas temperature. As discussed above, for $T_e = T_g$, the SF_6 attachment rate coefficients, and as a consequence those for naphthalene, are expected to be higher than the measured values. A faster attachment rate on polycyclic aromatic hydrocarbon at low temperature would have a major significance for the formation of negatively charged hydrocarbon carriers in the interstellar medium.

4.4.2 *Electron attachment to HBr and HCl — effect of cluster formation*

The reaction rate coefficients for the electron attachment to strong gaseous acids has been studied by Adams *et al.* [40] using the FALP technique. The overall process may be considered as a Brønsted acid/base reaction where the electron acts as the base. In the case of

HBr, the dissociative attachment reaction can be written as

$$HBr + e^- \rightarrow H + Br^-$$

The reaction is found to be endothermic by $\sim 0.4\,eV$ with an energy barrier close to the $v = 2$ H–Br vibrational level. Using FALP experiments, Smith and Adams [25] reported an attachment rate coefficient on HBr of $3.3(\pm 1) \times 10^{-12}\,cm^3\,s^{-1}$ at 300 K and $2.8(\pm 1) \times 10^{-10}\,cm^3\,s^{-1}$ at 510 K. Similarly, the attachment to HCl is endothermic by 0.8 eV and is found to have an attachment rate coefficient below $1 \times 10^{-11}\,cm^3\,s^{-1}$ both at 300 and 510 K. Below room temperature, the rate coefficients for attachment to HBr and HCl are therefore expected to be below $1 \times 10^{-12}\,cm^3\,s^{-1}$, which is close to the CRESU general lower measurement limit.

Figure 4.12 displays the measured attachment rate coefficient on HBr from 170 K down to 48 K [18]. As expected, the attachment rate coefficient at 170 K is below the CRESU measurement limit and an upper value of $1 \times 10^{-12}\,cm^3\,s^{-1}$ is given. The very slow attachment to HBr at 170 K agrees with electrons having an average energy much lower than the endothermicity of the reaction, $\sim 3000\,K$ $(\sim 0.26\,eV)$. Surprisingly, the measured attachment rate coefficients are found to

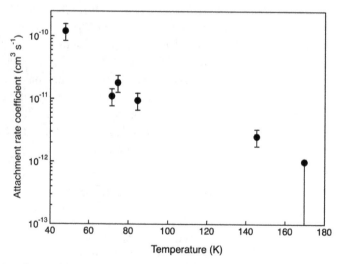

Fig. 4.12.   Apparent electron attachment to HBr measured by the CRESU technique [18]. The data point at 170 K represents an upper value.

increase monotonically by two orders of magnitude as the temperature is lowered down to 48 K. Similar observations were made for HCl [18]. Although HBr and HCl have electric dipoles below the critical value predicted for the formation of dipole bond states [41], Cizek *et al.* [42] calculated a very narrow well in the anion potential near the crossing point. The quasi-bound anionic state is below the asymptotic energy of the dissociative attachment potential and between the $v = 0$ and $v = 1$ HBr vibrational energy levels. For HBr, the crossing of the neutral and quasi-bound potential leads to an energy barrier of about 150 meV, which is lower than that predicted for the dissociative attachment. Even in the case that non-dissociative attachment is the favored pathway at low temperature, an Arrhenius behavior would still lead to negligible attachment at 48 K.

Alternatively, the increase of the attachment rate coefficient observed in Fig. 4.12 for HBr may be explained by the formation of hydrogen halide clusters in the supersonic flow [18]. Under CRESU flow conditions, a cluster $(HBr)_n$ is in a steady state between its formation from the cluster $(HBr)_{n-1}$ and its reaction with HBr to form cluster $(HBr)_{n+1}$. The overall attachment process follows the chemical scheme

$$HBr + e^- \rightarrow H + Br^-$$
$$HBr + HBr \rightarrow (HBr)_2$$
$$(HBr)_n + HBr \rightarrow (HBr)_{n+1}$$
$$(HBr)_n + e^- \rightarrow (Br^- HBr) \cdot (HBr)_{n-2} + H$$

The contribution from attachment to the electron gradient density in the flow (Eq. (4.2)) is given by

$$-k_{att} n_R / v = -\frac{1}{v} \sum_n k_{att}^n n_n = -\frac{1}{v} \sum_n k_{att}^n \frac{k_n}{k_{n+1}} n_{n-1}$$

$$= -\frac{1}{v} \sum_n k_{att}^n \frac{k_n}{k_{n+1}} \frac{k_{n-1}}{k_n} \cdots \frac{k_2}{k_3} n_{HBr}$$

$$= -\frac{1}{v} \sum_n k_{att}^n \frac{k_2}{k_{n+1}} n_{HBr} \qquad (4.19)$$

where k_{att}^n is the attachment rate coefficient on a cluster $(HBr)_n$, n_n is its number density, and k_n is the cluster formation rate coefficient.

Under steady state approximation for each cluster, the overall contribution from attachment to the electron density gradient along the CRESU flow remains directly proportional to the number density of HBr, n_{HBr}. In Fig. 4.12, the increase observed in the measured coefficients is likely to reflect the fact that attachment to HBr clusters is significantly faster than attachment to the monomer and that the fraction of clusters in the reaction flow increases with decreasing temperature. The formation of HBr clusters in the reaction flow was confirmed by measuring the radical decay rate for the OH + HBr reaction [18]. At 48 K and for high HBr concentrations, the pseudo-first-order rate for loss of OH appears to be constant with HBr number density, likely due to the formation of HBr clusters.

Rauk and Armstrong [43] performed ab initio studies on the electron capture to HBr and HCl dimers. In both cases, the attachment process to the ground vibrational level of the neutral is found to be exothermic. In addition, the dissociative attachment threshold decreases from 0.4 eV to 0.0 eV for HBr dimers and 0.81 eV to 0.21 eV for HCl dimers. In the case of trimers of HCl, they calculated that the energy threshold for attachment is less than 0.06 eV depending on the cluster geometry. These calculations confirm that the attachment to dimers and trimers of HCl and HBr is very favorable compared to attachment to the monomers. At 170 K, the temperature at which the concentration of the clusters is expected to be low, the very slow attachment rate coefficient agrees with electrons that have an average energy below the first vibrational energy level of HBr. The non-thermalization of the electrons to the temperature of the flow has therefore likely no observable effects on the measured rate coefficients.

4.4.3   *Electron attachment to* CH_3I

Electron interaction to methyl iodide (CH_3I) occurs through the formation of a negative-ion complex which can dissociate to give the initial molecule in the same or different rovibrational state or form a negatively charged I^- atom and a methyl radical. Figure 4.13 displays the calculated CH_3I (solid black line) and CH_3I^- (dashed red line) potential energy curves along the C–I coordinate [4]. The horizontal lines represent the ν_3 symmetric C–I stretch vibrational levels. The crossing of the anionic and neutral potential curves occurs

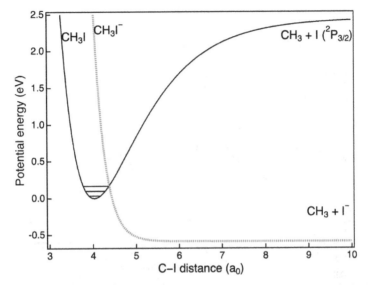

Fig. 4.13.   Neutral (solid black line) and anionic (dashed red line) potential energy curve along the C–I symmetric stretch for electron attachment to CH_3I. The horizontal lines represent the C–I vibrational levels. Reproduced with permission from Ref. [4], Copyright 2011 IOP Publishing Ltd.

above the zero-point energy level, close to the $\nu_3 = 2$ vibrational level. Similar behavior may be observed as for attachment to CF_2Cl_2 (Section 4.3.1) only if the attachment rate coefficient to the vibrational ground state of the neutral is independent of the electron temperature.

Figure 4.14 displays the R-matrix cross section for attachment to CH3I calculated using the potential energy curves displayed in Fig. 4.13. The cross section is found to non-monotonically increase with decreasing energy and displays a broad maximum at about 60 meV and a step at about 130 meV. Schramm *et al.* [4] attributed the peak in the cross section at ∼60 meV to a vibrational Feshbach resonance (VFR) near the first vibrational excitation energy (red dashed vertical line). Such resonance is likely to be due to the temporary capture of the electron by long-range dipolar and polarization potentials of CH_3I leading to the formation of a long-lived quasi-bond anion state. As the energy of the e–CH_3I system becomes resonant with the vibrational levels of the molecules (red dashed vertical lines in Fig. 4.14), there is a sharp decrease of the dissociative attachment

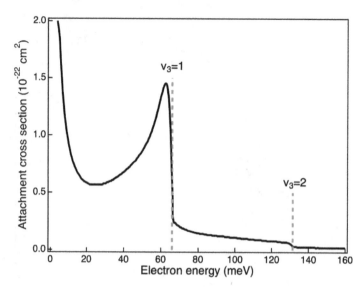

Fig. 4.14. Calculated attachment cross section on CH_3I showing a vibrational Feshbach resonance before the $\nu_3 = 1$ vibrational threshold and sharp decrease around the $\nu_3 = 2$ vibrational threshold [4].

cross section due to a decrease in the lifetime of the negative ion complex and dissociation back to the neutral molecule.

Figure 4.15 displays the CH_3I attachment rate coefficient as a function of temperature measured using in a CRESU flow (filled circles) [19], flowing afterglow (open circles [31] and filled squares [29]), and drift tube (filled triangles) [44] techniques. The attachment cross sections measured using laser photoelectron attachment at $T_g = 300\,K$ are integrated over a Boltzmann distribution and plotted as a function of electron temperature (open diamond) [4]. The solid lines are integrated theoretical attachment cross sections for $T_g = 300\,K$ (thick solid black line) and $T_g = 0\,K$ (thin black line) [45]. For the molecule in its ground vibrational level ($T_g = 0\,K$), the theory predicts a strong decrease of the attachment rate coefficient with increasing temperature up to ~150 K. At higher electron temperature, the rate coefficient increases slightly up to 500 K. For $T_g = 300\,K$, a non-negligible fraction of molecules is in the $\nu_3 = 1$ vibrational level. This leads to a faster attachment process than for the $T_g = 0\,K$ situation and an attachment rate coefficient which

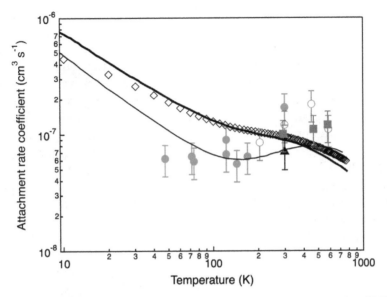

Fig. 4.15. Attachment rate coefficient on CH_3I measured using the CRESU (filled circles) [19], flowing afterglow (open circles [31] and filled squares [29]), and drift tube (filled triangles) [44] techniques. Integrated attachment cross sections measured using laser photoelectron attachment at $T_g = 300$ K are normalized to the data (open diamonds) [4, 46]. The lines are theoretical attachment rates calculated at $T_g = 300$ K (thick black line) and $T_g = T_e$ (thins black line) [45]. The temperature in the horizontal axis may refer to electron or gas temperature depending on the experiments.

decreases monotonically with increasing temperature. At high electron temperatures ($T_e > 300$ K), the absolute value of the attachment rate appears to be only weakly dependent on the gas temperature.

The lack of temperature dependence in the low-temperature CRESU data (Fig. 4.15) can be reconciled with the theoretical prediction and experimental data by Schramm *et al.* [4] based on the higher electron temperature in the CRESU flow. In Fig. 4.15, the attachment rate coefficient to the ground vibrational level of CH_3I depends only weakly on the electron temperature within the expected true electron temperature range: 100–500 K [21] The lack of temperature dependence for attachment to CH_3I below 300 K measured in the CRESU flow reflects the non-thermalization of the electrons to the temperature of the flow with a lower limit of 100 K.

4.5   Moving Forward

For an exothermic DEA process with an energy barrier of the order of few quanta of vibrational excitation, the attachment rate coefficient is independent of the electron temperature as long as the state-resolved attachment rate is constant and only weakly dependent on the quantum state. Because the electrons are not thermalized to the temperature of the reaction flow, the CF_2Cl_2 and CF_3Br rate coefficients at low temperature are likely to be characteristic of attachment with electrons at or below room temperature to the ground vibrational level of the molecules. For an electron temperature lower than $500\,K$, the energy maximum of the Boltzmann distribution remains below the energy barrier of the reaction leading to a low rate coefficient. As the gas temperature increases, additional vibrational levels become populated leading to more efficient state-specific DEA processes and an increasing rate coefficient.

For electron attachment onto HBr and HCl, the higher electronic temperature has no major influence on the results. Attachment to the hydrogen halide monomers is very slow even at high electron temperature and is not observed in the CRESU flow. In this case, the fast and increasing rate coefficient with decreasing gas temperature is due to formation of HCl or HBr clusters [18]. The higher electron temperature may only have an effect on the absolute value of the measured rate coefficient for attachment to clusters. The experimental data on HBr and HCl therefore provide a qualitative understanding of the effect of cluster formation on DEA processes.

In the case of attachment to SF_6 [16], the effect of higher electron temperature may be understood by looking at the parameterized rate coefficients given by Troe *et al.* [34] Even at low gas temperatures, for which the molecules are in their vibrational ground state, the attachment rate is electron temperature dependent due to incomplete IVR in the initial molecular anion. At a given gas temperature, a higher electron temperature leads to a decrease of the attachment rate coefficient. The measured rate coefficients for attachment to SF_6 are therefore likely to increase for electrons thermalized to the flow temperature. In Fig. 4.11, this would bring the CRESU measurements closer to the threshold photoelectron data. The SF_6 CRESU rate coefficients have also been used as reference values for the measurement of electron attachment to naphthalene [39]. Faster attachment rate coefficients on polycyclic aromatic hydrocarbons at low temperature would have a major significance for the formation of negatively charged carrier in the interstellar medium.

The non-thermalization of the electrons to the flow temperature has the most noticeable effects on the attachment to CH_3I. In this case, the DEA process is strongly dependent on the temperature of the electrons below 300 K. The low-temperature measurements reported by Speck *et al.* [19] were in fact performed at higher electron temperatures, significantly decreasing the measured rate coefficient below 100 K. For an electron temperature over the 100–500 K range, the attachment rate coefficient is expected to be only weakly dependent on the temperature of the neutrals.

The CRESU method remains one of the only techniques which potentially allows the study of electron attachment with both the electrons and the neutral molecules being in local thermal equilibrium at low temperature. The measurement on DI and CH_3I, however, demonstrates that the use of a high-energy electron gun generates excited buffer gas molecules, ultimately biasing the electron temperature toward higher values. Although the knowledge of this temperature now allows for reinterpreting previous measurements, it is important to develop a new way of generating low-temperature electrons in the supersonic flow.

Threshold laser photo-ionization is an appealing way of generating low-energy electrons. In this case, a short-wavelength (e.g. 157-nm excimer laser) light pulse is used to ionize a gaseous molecule with an ionization energy close to the radiation energy (<7.9 eV). For the typical number densities of neutral reactive molecules used in the CRESU flow ($1 - 10 \times 10^{12}$ cm^{-3}), and absolute ionization cross section ranging from $0.1 - 1.0 \times 10^{-18}$ cm^2, a laser fluence of 100 mJ/cm^2/pulse (likely from a focused beam) leads to electron number densities ranging from $1 - 100 \times 10^6$ cm^{-3}. These electronic number densities are sufficient to guaranty thermalization.

References

[1] Fabrikant II, Eden S, Mason NJ, Fedor J. Recent progress in dissociative electron attachment: from diatomics to biomolecules. In: Arimondo E, Lin CC, Yelin SF, editors. Advances in atomic, molecular, and optical physics. Elsevier Inc.; 2017. vol. 66, pp. 545–657. https://www.sciencedirect.com/science/article/abs/pii/S1049250X17300034

[2] Chutjian A, Alajajian SH. s-wave threshold in electron-attachment–observations and cross section in CCl_4 and SF_6 at ultralow electron energies. Phys Rev A. 1985;31:2885–92.

[3] Shuman NS, Miller TM, Viggiano AA, Troe J. Electron attachment to $POCl_3$. II. Dependence of the attachment rate coefficients on gas and electron temperature. Int J Mass Spectrom Ion Proc. 2011;306:123–8.

[4] Schramm A, Fabrikant II, Weber JM, Leber E, Ruft MW, Hotop H. Vibrational resonance and threshold effects in inelastic electron collisions with methyl iodide molecules. J Phys B. 1999;32:2153–71.

[5] Dashevskaya EI, Litvin I, Nikitin EE, Troe J. Modelling low-energy electron-molecule capture processes. Phys Chem Chem Phys. 2008;10:1270–6.

[6] Vogt E, Wannier GH. Scattering of ions by polarization forces. Phys Rev. 1954;95:1190–98.

[7] Smirnov BM, Kosarimn AV. Electron attachment to the SF_6 molecule. J Exp Theo Phys. 2015;121:377–84.

[8] Breit G, Wigner E. Capture of slow neutrons. Phys Rev. 1936;49:0519–31.

[9] Hotop H, Ruf MW, Fabrikant II. Resonance and threshold phenomena in low-energy electron collisions with molecules and clusters. Phys Scr. 2004;T110:22–31.

[10] Miller TM, Friedman JF, Schaffer LC, Viggiano AA. Electron attachment to halomethanes at high temperature: CH_2Cl_2, CF_2Cl_2, CH_3Cl, and CF_3Cl attachment rate constants up to 1100 K. J Chem Phys. 2009;131(084302):1–7.

[11] Herbst E, Klemperer W. Formation and depletion of molecules in dense interstellar clouds. Astrophys J. 1973;185:505–33.

[12] Black JH, Hartquist TW, Dalgarno A. Models of inter-stellar clouds .2. Zeta-Persei cloud. Astrophys J. 1978;224:448–52.

[13] McCarthy MC, Gottlieb CA, Gupta H, Thaddeus P. Laboratory and astronomical identification of the negative molecular ion C_6H. Astrophys J. 2006;652:L141–L144.

[14] Millar TJ, Walsh C, Field TA. Negative ions in space. Chem Rev. 2017;117:1765–95.

[15] Mozumder A. Electron thermalization in gases. 4. Relaxation-time in molecular-hydrogen. J Chem Phys. 1982;76:3277–84.

[16] Le Garrec JL, Sidko O, Queffelec JL, Hamon S, Mitchell JBA, Rowe BR. Experimental studies of cold electron attachment to SF_6, CF_3Br, and CCl_2F_2. J Chem Phys. 1997;107:54–63.

[17] Mitchell JBA. The dissociative recombination of molecular-ions. Phys Rep Rev Sec Phys Lett. 1990;186:215–48.

[18] Speck T, Le Garrec JL, Le Picard S, Canosa A, Mitchell JBA, Rowe BR. Electron attachment in HBr and HCl. J Chem Phys. 2001;114:8303–9.

[19] Speck T, Mostefaoui T, Rebrion-Rowe C, Mitchell JBA, Rowe BR. Low-temperature electron attachment to CH_3I. J Phys B. 2000;33: 3575–82.

[20] Goulay F, Rebrion-Rowe C, Carles S, Le Garrec JL, Rowe BR. Electron attachment on HI and DI in a uniform supersonic flow: thermalization of the electrons. J Chem Phys. 2004;121:1303–8.

[21] Carles S, Saidani G, Le Garrec JL, Guen N, Mitchell JBA, Viggiano AA, Shuman NS. Demonstration of the branching ratio inversion for the electron attachment to phosphoryl chloride $POCl_3$ in the gas phase between 300 and 200 K. Chem Phys Lett. 2016;650:144–7.

[22] Fabrikant II, Hotop H. On the validity of the Arrhenius equation for electron attachment rate coefficients. J Chem Phys. 2008;128(124308):1–8.

[23] Clary DC, Henshaw JP. Reaction-rates of electrons with dipolar molecules. Int J Mass Spectrom Ion Proc. 1987;80:31–49.

[24] Alajajian SH, Chutjian A. Measurement of electron-attachment lineshapes, cross-sections, and rate constants in HI and DI at ultralow electron energies. Phys Rev A. 1988;37:3680–84.

[25] Smith D, Adams NG. Studies of the reactions HBr(HI)+e-reversible-$Br^-(I^-)$ H using the FALP and SIFT techniques. J Phys B. 1987;20:4903–13.

[26] Van Doren JM, Friedman JF, Miller TM, Viggiano AA, Denifl S, Scheier P, Mark TD, Troe J. Electron attachment to $POCl_3$: measurement and theoretical analysis of rate constants and branching ratios as a function of gas pressure and temperature, electron temperature, and electron energy. J Chem Phys. 2006;124(124322):1–9.

[27] Graupner K, Haughey SA, Field TA, Mayhew CA, Hoffmann TH, May O, Fedor J, Allan M, Fabrikant II, Illenberger E, Braun M, Ruf MW, Hotop H. Low-energy electron attachment to the dichlorodifluoromethane (CCl_2F_2) molecule. J Phys Chem A. 2010;114: 1474–84.

[28] Smith D, Adams NG, Alge E. Attachment coefficients for the reactions of electrons with CCl_4, CCl_3F, CCl_2F_2, $CHCl_3$, Cl_2 and SF_6 determined between 200 K and 600 K using the FALP technique. J Phys B. 1984;17:461–72.

[29] Burns SJ, Matthews JM, McFadden DL. Rate coefficients for dissociative electron attachment by halomethane compounds between 300 and 800 K. J Phys Chem. 1996;100:19436–40.

[30] Marienfeld S, Sunagawa T, Fabrikant II, Braun M, Ruf MW, Hotop H. The dependence of low-energy electron attachment to CF_3Br on electron and vibrational energy. J Chem Phys. 2006;124(154316): 1–14.

[31] Alge E, Adams NG, Smith D. Rate coefficients for the attachment reactions of electrons with c-C_7F_{14}, CH_3Br, CF_3Br, CH_2Br_2 and CH_3I determined between 200 K and 600 K using the FALP technique. J Phys B. 1984;17:3827–33.

[32] Levy RG, Burns SJ, McFadden DL. Temperature-dependence of rate coefficients for thermal electron-attachment reactions of CH_3Br, CF_3Br and CF_3I. Chem Phys Lett. 1994;231:132–8.

[33] Troe J, Miller TM, Shuman NS, Viggiano AA. Analysis by kinetic modeling of the temperature dependence of thermal electron attachment to CF_3Br. J Chem Phys. 2012;137(024303):1–6.

[34] Troe J, Miller TM, Viggiano AA. Low-energy electron attachment to SF_6. I. Kinetic modeling of nondissociative attachment. J Chem Phys. 2007;127(244303):1–12.

[35] Chutjian A. Experimental SF_6-SF_6 and Cl-$CFCl_3$ electron-attachment cross-section in the energy-range 0–200 meV. Phys Rev Lett. 1981;46:1511–4.

[36] Hunter SR, Carter JG, Christophorou LG. Low-energy electron-attachment to SF_6 in N_2, Ar, and Xe buffer gases. J Chem Phys. 1989;90:4879–91.

[37] Klar D, Ruf MW, Hotop H. Attachment of electrons to molecules at meV resolution. Aust J Phys. 1992;45:263–91.

[38] Klar D, Ruf MW, Hotop H. Attachment of electrons to molecules at submilli electron volt resolution. Chem Phys Lett. 1992;189:448–54.

[39] Moustefaoui T, Rebrion-Rowe C, Le Garrec JL, Rowe BR, Mitchell JBA. Low temperature electron attachment to polycyclic aromatic hydrocarbons. Faraday Discuss. 1998;109:71–82.

[40] Adams NG, Smith D, Viggiano AA, Paulson JF, Henchman MJ. Dissociative attachment reactions of electrons with strong acid molecules. J Chem Phys. 1986;84:6728–31.

[41] Crawford OH, Garrett WR. Electron-affinities of polar-molecules. J Chem Phys. 1977;66:4968–70.

[42] Cizek M, Horacek JT, Allan M, Domcke W. Resonances and threshold phenomena in low-energy electron collisions with hydrogen halides: new experimental and theoretical results. Czech J Phys. 2002;52: 1057–70.

[43] Rauk A, Armstrong DA. Potential energy barriers for dissociative attachment to HF.HF and HCl.HCl: Ab initio study. Int J Quantum Chem. 2003;95:683–96.

[44] McCorkle DL, Christodoulides AA, Christophorou LG, Szamrej I. Electron-attachment to chlorofluormethanes using the electrons-Swarm method. J Chem Phys. 1980;72:4049–57.

[45] Wilde RS, Gallup GA, Fabrikant II. Comparative studies of dissociative electron attachment to methyl halides. J Phys B. 2000;33: 5479–92.

[46] Christophorou LG. Electron-attachment and detachment processes in electronegative gases. Contrib Plasma Phys. 1987;27:237–81.

Chapter 5

Neutral–Neutral Reactions

Kevin M. Hickson[*,‡] and Dwayne E. Heard[†,§]

*Institut des Sciences Moléculaires, CNRS UMR 5255,
Université de Bordeaux, F-33400 Talence, France
†School of Chemistry, University of Leeds,
LS2 9JT Leeds, UK
‡kevin.hickson@u-bordeaux.fr
§d.e.heard@leeds.ac.uk

Abstract

In this chapter, we discuss the progress that has been made in the field of gas phase chemical kinetics since the application of the CRESU method to the investigation of neutral–neutral reactions at low temperatures in the early 1990s. A brief introduction of the advances that motivated the adaptation of this technique to study neutral reactivity at very low temperature is given. The main kinetic analyses used to interpret the results of CRESU-type experiments are described, providing examples for the specific cases of reagent and product detection, allowing temperature-dependent rate coefficients to be derived for use in astrochemical/ photochemical models. The various techniques typically used to generate reactive radical species in these experiments are outlined, in addition to the detection methods employed to follow the reaction kinetics. Finally, we describe the distinct classes of reactions (activated, barrierless, etc.) that have been studied so far with the CRESU method, providing specific examples of potentially important neutral species whose reactions have yet to be examined at low temperature.

Keywords: Atom and radical — molecule reactions; Atom and radical — radical reactions; Bimolecular reactions; Termolecular reactions; Activated reactions; Barrierless reactions.

5.1 Introduction

It is really only since the 1960s that the kinetics of gas phase reactions have been studied accurately in the laboratory, in large part due to the development of pulsed photolysis techniques allowing reactive reagent species to be generated and reliable detection methods to follow the progress of reactions. Since this time, it has been clearly demonstrated that reactions between uncharged species (where uncharged means both closed-shell species and radicals) play critical roles in a wide range of complex environments ranging from combustion systems to the chemistry of planetary atmospheres both at low and high temperature, as well as in the diverse regions of interstellar space.

A major impetus for the measurement of gas phase reactions between uncharged species below room temperature was the growing awareness that such processes were likely to dominate the chemistry of the Earth's stratosphere, in particular where temperatures fall as low as 180 K. From the early 1970s onwards, numerous research groups began to apply cryogenic cooling techniques to study the various processes that would influence the composition of the atmosphere, with the aim of measuring rate coefficients and branching ratios as a function of temperature for the most important atmospheric reactions. Photochemical models [1] were created, with the measured kinetic data as inputs to try to explain the observables, the species concentrations as a function of altitude. For several decades, two independent panels of experts [2, 3] have been reviewing the available kinetic data, providing recommendations concerning the temperature-dependent rate coefficients and product branching ratios (with their associated uncertainties) for the majority of atmospherically important reactions. More recently, the motivation to extend such measurements to even lower temperatures has been provided by the need to reconcile Earth- and satellite-based observations of planetary atmospheres and of the interstellar medium with photochemical models of these regions. As the list of molecules detected in extraterrestrial environments expands with each passing year, an ever increasing number of pathways is required to explain their formation.

Although cryogenic cooling methods are reasonably well suited to study the chemistry of the Earth's temperate atmosphere, the extension of these techniques to even lower temperatures is more problematic. Eventually, critical components of the gas mixture begin to condense on the reactor walls, so that further measurements are impossible with these methods. Indeed, in the absence of an obvious way to circumvent the condensation problem, few reactions were studied below 200 K until the mid-1980s and only a handful of investigations, notably a flow-tube study of H atom recombination [4] and several flash photolysis studies of radical–molecule reactions where the molecules all possess high vapor pressures at low temperature (CO [5], O_2 [6, 7], and NO [8]), allowed kinetic measurements to be performed to temperatures as low as 80 K.

Below 80 K, a vast majority of kinetic studies of neutral–neutral reactions have been performed using the CRESU technique. Although this method had already been applied to investigate the kinetics of ion–molecule reactions in the 1980s [9] (see Chapter 3 for more details), the first studies of neutral–neutral reactions were only conducted in the early 1990s [10] through a joint project between the experimental groups of Bertrand Rowe at the University of Rennes and Ian Smith at the University of Birmingham. During this time, laser-based radical generation and detection methods were coupled with the continuous flow CRESU method for the first time, and this development led to a second CRESU system being built shortly afterward in Birmingham, dedicated to the study of neutral–neutral reactivity and energy transfer processes. Both of these apparatuses are now located at the Institut de Physique de Rennes.

A schematic diagram of the apparatus used for one of the first CRESU studies, measuring rate coefficients for the CN + O_2 reaction down to 26 K, is shown in Fig. 5.1.

Several other groups [11] around the globe have now developed their own CRESU-type apparatuses, dedicated to the study of reactions between neutral species either using pulsed [12–18] or continuous [19] flows. Several representative photographs have been gathered together in the CRESU gallery in Chapter 2.

5.2 Rate Coefficient Determination

A brief introduction to the analysis employed in kinetics experiments is appropriate at this point. As a general rule, reactions between neutral species involve collisions between either two (bimolecular)

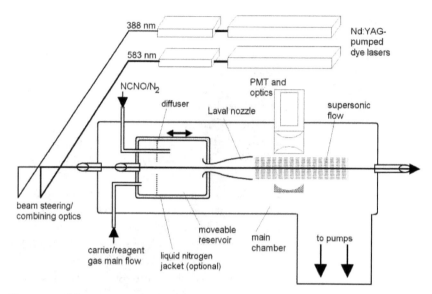

Fig. 5.1. Schematic diagram of the neutral–neutral CRESU apparatus used in Ref. [10]. Reproduced with permission from Ref. [10], Copyright 1992 AIP Publishing.

reagents leading to multiple product species,

$$W + X \rightarrow Y + Z \tag{5.1}$$

or three (termolecular) reagents resulting in association,

$$W + X + M \rightarrow WX + M \tag{5.2}$$

In this case, the probability for all three reagents to collide simultaneously is small (except at very high pressures), so that termolecular reactions are considered to occur through a two-step process involving the initial formation of an energized association complex or intermediate

$$W + X \rightarrow WX^* \tag{5.3}$$

followed by complex stabilization through collisions with the bath gas and/or dissociation to products

$$WX^* + M \rightarrow WX + M \tag{5.4}$$

$$WX^* \rightarrow Y + Z \tag{5.5}$$

The procedures to derive rate coefficients for both bimolecular and termolecular reactions are entirely equivalent. The reactive reagent species X is maintained in excess with respect to the other reactive reagent species W so that the loss of reagent W is reduced to a first-order process (the pseudo-first-order approximation). To simplify the experiment, X is normally chosen to be a stable species (so that its concentration can be easily deduced), whereas W is chosen to be the radical reagent. As a general rule, the excess reagent concentration should be at least an order of magnitude larger than the minor reagent one so that any deviations of the first-order approximation are negligibly small. For termolecular processes, the bath gas concentration is also large and constant with respect to the minor reagent one. Pseudo-first-order rate coefficients can be extracted by two different methods. The first one requires the minor reagent concentration to be followed as a function of time for a given excess reagent concentration, yielding an exponential decay profile

$$[W] = [W]_0 e^{-k't} \tag{5.6}$$

where $[W]$ and $[W]_0$ are the time-dependent and initial concentrations of W, respectively, and k' is the sum of all first-order losses of W. In the absence of secondary losses for W, $k' = k'_1$ where $k'_1 = k_1[X]$. Otherwise,

$$k' = k'_1 + k_A[A] + k_B[B] + \cdots + k_L \tag{5.7}$$

where A and B are other species present in the flow which can react with W, with their associated rate coefficients k_A and k_B. k_L represents any other physical losses of W (such as diffusion). A profile of this type is shown in Fig. 5.2.

The second method for obtaining pseudo-first-order rate coefficients requires one of the product species to be followed as a function of time, yielding an exponential rise profile of the form

$$[Y] = [Y]_0(1 - e^{-k't}) \tag{5.8}$$

in the absence of secondary losses. In reality, secondary losses of both the minor reagent W and product species Y now need to be considered, which results in an additional exponential loss term in the equation describing the temporal evolution of Y

$$[Y] = [Y]_0(e^{-k_{L(Y)}t} - e^{-k't}) \tag{5.9}$$

where $k_{L(Y)}$ represents the combined first-order losses of Y and k' is defined above. This modified profile is shown in Fig. 5.3.

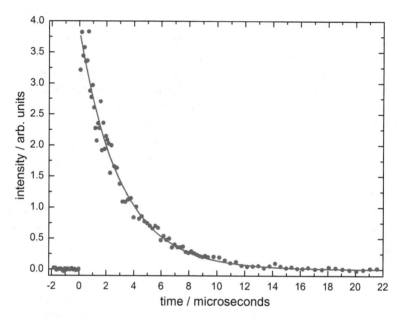

Fig. 5.2.   Pseudo-first-order decay profile of $O(^1D)$ atoms in the presence of D_2 molecules recorded at 50 K.

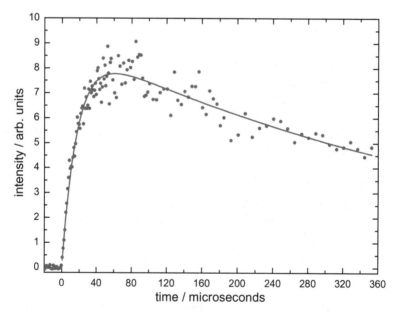

Fig. 5.3.   Monitoring the kinetics of the $C(^1D) + CH_4$ reaction at 50 K by following H-atom production.

As a general rule, as $k_{L(Y)}$ is almost always non-zero, the first method (following the reagent loss rather than product formation as a function of time) is the preferred one for the determination of pseudo-first-order rate coefficients, as fits to this type of profile introduce an extra fitting parameter leading to substantially larger uncertainties in the derived pseudo-first-order rates. Secondary losses (chemical or otherwise) of product Y contribute further uncertainty to the derived pseudo-first-order rate coefficients for the target reaction. Another potential issue in following the kinetics of product formation arises from the fact that the molecular product species may initially be formed over a wide range of vibrational and rotational levels. While vibrational energy transfer is generally rather slow at the densities and timescales of CRESU-type experiments, the same cannot be said of rotational energy transfer which occurs essentially on the collision timescale. In this case, the cascade of high rotational levels back to an equilibrium distribution could interfere with the kinetic analysis of the product rovibrational state used to follow the progress of the reaction. Although there are several clear drawbacks of the biexponential method, it has been successfully employed in several CRESU studies (see for instance [17, 20–22]) where for one reason or another it was difficult or impossible to follow the minor reagent species directly.

Although the second-order rate coefficients for bimolecular reactions can be obtained directly by analyzing single kinetic profiles of the type described above [20], in this instance secondary losses of the minor reagent need to be carefully quantified to determine their contribution to the decay rate. The standard way to determine the second-order rate coefficient is to record pseudo-first-order decays of species W for different values of [X]. Then, a plot of the pseudo-first-order rate coefficient against the corresponding [X] yields the second-order rate coefficient from the slope as shown in Fig. 5.4.

Here, a majority of secondary W losses are represented by the y-axis intercept value (that is, when [X] = 0) as these losses remain constant as a function of [X]. In this way, only reactive impurities introduced with reagent X could lead to significant deviations in the measured second-order rate coefficient.

To determine the termolecular rate coefficient at any given temperature, it is also necessary to repeat this procedure at various bath gas concentrations. While this is a relatively simple procedure

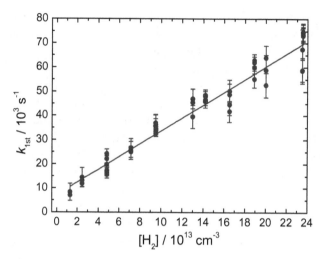

Fig. 5.4.   Pseudo-first-order rate coefficients for the $C(^1D) + H_2$ reaction as a function of $[H_2]$ at 50 K. A weighted linear least-squares fit yields the second-order rate coefficient. The error bars on the ordinate reflect the statistical uncertainties (1σ) obtained by fitting to H atom decay profiles. Reproduced with permission from Ref. [21]. Copyright 2015 American Chemical Society.

in studies where a conventional flow reactor is used (such as those employing the flash photolysis technique) [8], the same cannot be said for those experiments using the CRESU method. Here, each Laval nozzle profile is calculated to yield a supersonic flow with a specified temperature and bath gas concentration. Consequently, several different Laval nozzles must be designed (with each one producing the same specified supersonic flow temperature) to obtain different bath gas concentrations. Although it is possible to tune the bath gas density to some extent with a single Laval nozzle while maintaining the flow temperature at an approximately constant value [23, 24], this usually results in poorer flow quality and shorter hydrodynamic times, leading to larger uncertainties on the measured rate coefficients. This effect is mostly due to the departure from the nominal nozzle operating conditions that consider a specified set of boundary conditions which are now invalid. Chapter 2 develops these aspects and explains the extent to which a single nozzle can be used to obtain supersonic flows at different temperatures and pressures other than the nominal ones.

5.3   Radical Generation Methods

5.3.1   *Pulsed laser photolysis*

An overwhelming majority of CRESU studies of neutral–neutral reactions have employed pulsed laser photolysis as the preferred radical generation method, where the photolysis laser is aligned along the axis of the cold supersonic flow. In these studies, a radical "precursor" molecule is introduced into the Laval nozzle reservoir along with the carrier gas flow and excess co-reagent molecules upstream of the supersonic flow. Assuming that the precursor molecules are well mixed with the carrier gas flow, absorption of the laser beam by the precursor molecules leads to the in situ generation of radicals along the entire length of the supersonic flow, which begin to react with the excess co-reagent molecules also present in the flow. In common with most experimental apparatuses for kinetic measurements, the detector is maintained at a fixed position, while the Laval nozzle is displaced with respect to the detector to a distance which is generally chosen to correspond to the maximum distance for optimal flow conditions to remain valid. This choice provides the maximum time over which kinetic measurements can be exploited. When radicals produced next to the detector are detected, a maximum radical signal is obtained as the reaction time is negligible. Conversely, if we consider those radicals that were produced just downstream of the Laval nozzle, it is easy to visualize that before we can detect these radicals they must move in front of the detector (with a delay inversely proportional to the supersonic flow velocity) during which time a significant fraction will have been lost through reaction with the excess reagent. Consequently, it is important to realize that while the CRESU technique is closely related to earlier conventional kinetic techniques such as the slow-flow flash photolysis method and the flow-tube technique, in its most widely used configuration employing pulsed laser photolysis coupled with laser induced fluorescence detection, it is actually a hybrid of the two. The main criteria to consider for the selection of a good precursor molecule are as follows: (I) The product of the absorption cross section, the flow concentration, and the fluence at a given photolysis laser wavelength (leading to production of the desired radical species) must be sufficiently large to produce an easily detectable quantity of radical reagents without leading to significant levels of absorption along the flow. (II) Does the photodissociation process lead to radical formation with significant internal excitation? (III) Could any potential photodissociation co-products interfere with the investigation of the target reaction? Point (I) is particularly important and specific to CRESU experiments, as the *nascent* radical concentration must be identical at all points along the supersonic flow to avoid significant modification of the kinetic profiles. This fact also means that the photolysis laser beam must be unfocused in such experiments, except when a modification of the usual method is employed as described below. In a series of experiments to investigate the kinetics of silicon and boron atom reactions at low temperature, Le Picard and co-workers [25–30] used a focused photolysis laser at 266 nm to generate ground electronic state $Si(^3P)$ and $B(^2P)$ from precursors tetramethylsilane $Si(CH_3)_4$ and trimethylborate $B(OCH_3)_3$, respectively, due to the lack of a viable alternative source of these atoms. The atoms were produced at a single point along the flow (around the focal point of the laser) and were only detected after they had traveled in front of the detector, which corresponded to delay times as large as hundreds of microseconds depending on the relative positions of the focal point and the detector. In contrast to the method typically used during kinetic measurements, the delay time between photolysis and probe lasers was fixed to the optimal time for these atoms to cover the distance between the focal point and the detector (thereby recording the maximum signal) and instead of varying the reaction time, the co-reagent concentration was modified instead. The atom signal intensity was seen to vary exponentially as a function of the co-reagent concentration ($I = I_0 e^{-k_{1st}[X]}$), where k_{1st} here was expressed in units cm^3. The experiment was repeated at different focal point positions (by moving the Laval nozzle) so that different delay times could be obtained. A plot of k_{1st} against the time delay thus yielded the second-order rate coefficient at a given temperature with conventional units $cm^3 s^{-1}$. In another variant of the usual pulsed laser photolysis method, Nuñez-Reyes and Hickson[31] investigated the reactivity of *chemically* produced metastable $N(^2D)$ atoms, using the $C(^3P) +$ $NO \rightarrow N(^2D) + CO$ reaction as a source. In a similar manner to the previously mentioned studies of Si and B atom reactivity, photolytic sources of $N(^2D)$ atoms are scarce. Here, the initial $C(^3P)$ atoms were produced by the pulsed laser photolysis of CBr_4 precursor molecules at 266 nm, with NO used as the excess reagent species. As the $N(^2D)$ atoms subsequently react with NO molecules themselves, it was possible to extract second-order rate coefficients for the $N(^2D)$ + NO reaction as a function of temperature by fitting to the $N(^2D)$ temporal profiles.

5.3.2 *Microwave discharge*

This specific technique has been successfully used to study the kinetics of gas phase reactions in conventional apparatuses since the 1960s, as an efficient way of producing radical atoms such as N, O, and halogen atoms in particular. Although the range of reactions that can be investigated might seem somewhat limited (being restricted to those gas phase reactions essentially involving atoms that can be generated from gaseous diatomic molecules), this list is considerably extended by the application of *ex situ* chemical transformation techniques so that molecular radicals such as OH and HO_2, for example, are easily produced by the reactions of F with H_2O and H_2O_2, respectively. Despite the widespread use of the microwave discharge technique in gas phase kinetic studies, early attempts to couple it with the CRESU method were unsuccessful. In pulsed apparatuses, the discharge would need to be modulated in a stable, reproducible way. In the original continuous CRESU systems, high carrier gas flows result in large dilution factors, while the large Laval nozzle reservoir volumes (and long residence times) mean that reactive atoms generated upstream of the reservoir are mostly lost through recombination and collisions with the walls before reaching the supersonic flow. The first successful demonstration of a coupled discharge flow CRESU apparatus was realized by the Bordeaux group in 2009, in an investigation of the $N(^4S)$ + NO reaction [32]. The Bordeaux continuous flow CRESU system is characterized by a small reservoir volume of just a few tens of cm^3, while the gas flows are typically five times lower than those of the original CRESU apparatuses, meaning that both short residence times and low dilution factors contribute to a significant fraction of the radical atoms being preserved. A discussion of the mass flow rates employed in CRESU apparatuses can be found in Chapter 1. Here, the discharge cavity was mounted on one of the arms of a quartz tube in the form of a Y. A few percent of the main Ar carrier gas flow containing traces of N_2 was passed into the cavity producing atomic nitrogen in both the ground and excited electronic states. The cavity was situated 75 cm upstream of the Laval nozzle to increase the time for the excited atoms to relax to the ground state as well as allowing the hot gas to cool through collisions with the walls. In order to minimize losses of the minor $N(^4S)$ reagent, the excess co-reagent NO was injected directly into the reservoir, just upstream of the Laval nozzle. As the $N(^4S)$ was generated continuously (as opposed to the usual pulsed photolysis method), it was not possible to change the reaction time in the usual way (by scanning the time between the photolysis and probe pulses). In this instance, the reaction time was altered by modifying the distance between the Laval nozzle and the detection region in a similar manner to conventional discharge-flow experiments. The reactions of $Cl(^2P)$ with saturated alkanes ethane (C_2H_6) and propane (C_3H_8) were also studied using this technique [33]. Although this technique was only employed to generate low concentrations of atomic radicals for these applications, the coupled microwave discharge CRESU apparatus outlined above also allows high radical concentrations to be created within the supersonic flow, a prerequisite for the kinetics of radical–radical reactions to be studied at low temperature. These experiments will be described in more detail in Section 5.5.2.2.

5.4 Detection Methods

5.4.1 *Laser-induced fluorescence*

As a highly sensitive technique for the selective detection of atomic and molecular species, pulsed laser-induced fluorescence (LIF) is one of the most widely used methods in kinetic and dynamic studies of gas phase reactions since its invention in the late 1960s [34]. In common with the pulsed laser photolysis technique for radical generation, LIF is also by far the most commonly used detection technique in CRESU studies of gas phase reactivity since its first application to the investigation of the CN + O_2 reaction in 1992 [10]. The principle of this method is relatively simple. Here, the target radical is excited by a narrowband probe laser whose frequency is tuned to match an absorption line, usually from one electronic state to another and therefore in the ultraviolet or visible wavelength range for the majority of molecular radicals. The desired wavelength is generated from the output of a pulsed narrowband tunable laser (such as a dye laser) whose frequency is further modified if necessary by various sum or difference frequency mixing schemes in nonlinear optical crystalline materials such as beta barium borate (BBO) or potassium diphosphate (KDP). The excited radical species relaxes by emitting photons (fluorescence) which are typically collected by a broadband photosensitive detector such as a photomultiplier tube. In the event that the excited state radical decays exclusively through a non-radiative process, an alternative detection scheme must be employed (see below). The strength of the fluorescence signal for a given excitation wavelength is a sensitive probe of the lower state population; as the radical population decreases due to reaction, the fluorescence signal decreases proportionally. The wavelength dependence of the emitted fluorescence depends on several factors, notably the lifetime of the excited state, which for long lifetimes would allow collisional energy transfer into neighbouring rotational and vibrational levels to the laser-excited level prior to relaxation. In addition, each rovibrational excited state can relax to a range of different ground state rovibrational levels depending on their respective transition probabilities. In such experiments, the detector is usually positioned at right angles to the probe laser axis to minimize the detection of scattered laser light, thereby maximizing the signal-to-noise ratio of the radical fluorescence signal. In cases where the fluorescence emission is shifted to longer wavelength, it is also possible to place a filter in front of the detector or to use a monochromator to eliminate the majority of stray photons from the excitation laser. In this way, in their pioneering CRESU kinetic study of the CN + O_2 reaction, Sims *et al.* [35] excited the $(0,0)$ band of the $CN(B^2\Sigma^+ \leftarrow X^2\Sigma^+)$ transition at 388 nm with fluorescence detection in the $(0, 1)$ band of the $CN(B^2\Sigma^+ \rightarrow X^2\Sigma^+)$ transition at 420 nm using a narrowband interference filter centered at 420 nm. In cases where the transition probabilities for emission to excited vibrational levels in the ground electronic state are too weak, such molecules can also be followed at the same wavelength (on resonance). One such example is the kinetic study of the reactions of $^1C_2$ with various unsaturated hydrocarbons by Daugey and co-workers [36] where these authors excited the $(0,0)$ band of the $C_2(D^1\Sigma_u^+ \leftarrow X^1\Sigma_g^+)$ transition at 231 nm, with on-resonance detection. In this instance, a narrowband interference filter centered on 231 nm was employed, which prevented scattered light from the photolysis laser alone from reaching the detector. Even in this situation, some discrimination is possible between photons from the excitation laser and the fluorescence signal if the lifetime of the excited state radical is longer than the pulse width of the laser. In this situation, the signal integration device can be set to record only photons that arrive after a specified time delay has elapsed.

When the reagent or product species to be detected is an atomic radical, the available electronic transitions sometimes fall in the vacuum ultraviolet range (VUV). Unfortunately, the sum frequency generation methods described above cannot be applied in this spectral region due to significant reabsorption of the VUV radiation within the nonlinear optical material, so that alternative techniques are required to generate light at these wavelengths. In their studies of the kinetics of the reaction of $C(^3P)$ atoms with methanol, Shannon *et al.* [37] used a rare gas frequency tripling method to directly detect $C(^3P)$ atoms by VUV LIF using the $2s^2 2p^2 \; ^3P_2 \rightarrow 2s^2 2p3d \; ^3D_3^0$ transition at 127.755 nm. Here, the frequency-doubled UV output of a dye laser at 383.3 nm was focused into a cell containing 30 Torr of Xe, exciting a three-photon transition of Xe to produce the desired tunable radiation. In later experiments [38], the efficiency of the tripling process was significantly enhanced (an order of magnitude) through phase matching by adding Ar to the cell [39]. Chastaing *et al.* [40] used the related two-photon resonant four-wave mixing method to detect $C(^3P)$ atoms directly in their studies of the reaction of $C(^3P)$ atoms with C_2H_2 through the 3s ^3P–2p ^3P transition around 165.69 nm. The frequency-doubled UV output of a dye laser at 255.94 nm was mixed with the visible output of a second dye laser around 560 nm and focused into a cell containing Xe. The UV frequency (ν_{UV}) was fixed to the Xe $5p^5 6p \; [2^{1/2}, 2]$ two-photon resonance, a procedure which was facilitated by the detection of the Xe^+ multiphoton ionization signal in a separate reference cell. The visible frequency (ν_{VIS}) was tuned such that $\nu_{VUV} = 2\nu_{UV} - \nu_{VIS}$ to access the various spin-orbit levels of $C(^3P)$ over the 165.44–165.88 nm range. As the radiation generated in these studies is divergent, a lens made from a material with a high transmittance in the VUV range (CaF_2, MgF_2, LiF) is used as the output window of the rare gas cell to recollimate the VUV beam. Furthermore, it is essential that the cell is directly attached to the CRESU reactor to prevent absorption losses of the excitation source by atmospheric O_2. On the detection side, a VUV sensitive detector such as a solar blind photomultiplier tube (PMT) is required in this type of experiment. In a similar manner to the rare gas cell placement, losses of the resonant emission are avoided by attaching the PMT directly to the reactor, or as a better solution to prevent degradation of the PMT window by reactive gases within the chamber, through an isolated volume (under vacuum or flushed by a neutral gas) between the reactor and the PMT.

5.4.2 *Resonance fluorescence*

Although a majority of CRESU studies have employed pulsed LIF as the detection method, a handful of experiments have used continuous spontaneous light sources to probe the fate of radical species in the CRESU environment. Bergeat *et al.* [32] applied the resonance fluorescence method to detect ground state atomic nitrogen $N(^4S)$ at 120 nm and oxygen $O(^3P)$ at 130 nm in their investigation of the $N + NO \rightarrow N_2 + O$ reaction. These species have proven difficult to detect due to the inefficiency of the simpler rare gas tripling method described previously at these wavelengths (although $N(^4S)$ detection at 120 nm is known to be efficient in Hg vapor [41] and $O(^3P)$ atoms can be detected directly through four-wave mixing (see, for example, Takahashi *et al.* [42])). Here, a microwave discharge cavity is placed around a quartz tube isolated from the CRESU reactor (at the level of the detector) by an MgF_2 window. Helium gas containing trace amounts of N_2 or O_2 is flowed through the lamp, generating emission from the excited N or O atoms produced by microwave dissociation of the parent diatomic molecules. Light is emitted at discrete wavelengths corresponding to the allowed atomic transitions of the various excited states of N or O produced by the discharge. The emitted light is collimated by a series of baffles before interacting with the reagent N or product O atoms within the cold supersonic flow. The emission from these excited atoms is collected at right angles to minimize the detection of the resonant excitation source. A VUV monochromator or a narrowband interference filter can be used to discriminate against other emission lines, coupled with a solar blind channel photomultiplier operating in photon-counting mode. In later work, Hickson *et al.* [33] studied the kinetics of Cl radical reactions by introducing trace amounts of Cl_2 into the lamp, following ground state $Cl(^2P)$ atoms around 138 nm. Although this has proven to be an extremely sensitive technique for detecting a range of atomic and molecular radicals (radicals such as OH can also be detected [43]), as a continuous method, it must be associated with a continuous radical source such as microwave dissociation, constraining radical production to the region upstream of the Laval nozzle at the present time.

5.4.3 *Chemiluminescence*

In certain circumstances, it is desirable to use a detection method other than LIF in CRESU studies of gas phase reactivity. For instance, it may be that no tunable lasers are available in the laboratory or that a suitable detection scheme could not be established for a particular radical species. In this case, and for certain radical species, the chemiluminescence method can be applied to follow the progress of a reaction. Here, the minor reagent species is generated by pulsed laser photolysis in the usual manner. A second co-reagent species (in excess with respect to the minor reagent) is introduced into the flow, yielding products in an *excited* electronic state, which relax back to the ground state by fluorescence which is followed by a photosensitive detector. A crucial point in the application of this technique is that the radiative lifetime of the product species is relatively short allowing a steady state concentration to be rapidly established that is proportional to both the minor reagent concentration and the fluorescence emission signal. One of the earliest examples of the use of chemiluminescence as a detection method in a CRESU apparatus was applied to an investigation of the reactions of the ethynyl (C_2H) radical with various unsaturated hydrocarbons down to 15 K [44, 45]. As C_2H radicals in the ground $\tilde{X}^2\Sigma^+$ (0, 0, 0) band cannot be probed directly by LIF, Chastaing *et al.* [44] added a fixed excess concentration of O_2 to the supersonic flow so that the reaction between C_2H and O_2 occurred with one of the minor channels leading to $CH(A^2\Delta) + CO_2$ as products. Chemiluminescent emission from the $\Delta\nu = 0$ bands of the $CH(A^2\Delta \rightarrow X^2\Pi)$ transition was recorded using a narrowband interference filter at 428 nm. By varying the hydrocarbon concentration in the flow (but by maintaining a constant O_2 concentration), it was thus possible to extract second-order rate coefficients for the target reactions over a range of temperatures.

In other work, Hickson *et al.* [46] studied the reaction of CH radicals with water by employing a chemiluminescence detection method for CH radicals. As with earlier work on C_2H, a fixed excess concentration of O_2 was added to the flow allowing the CH + O_2 reaction to occur. Here, one of the product channels leads directly to the formation of electronically excited OH radicals, $OH(A^2\Sigma^+)$, which relax back to the ground $X^2\Pi$ state with the emission of photons around 310 nm. To discriminate against scattered light from the photolysis laser and other possible sources of radiation, a narrowband interference filter was employed.

Similar chemiluminescence schemes have also been employed in CRESU studies of atom — molecule reactions such as $O(^3P)$ + alkenes [47] and $C(^1D/^3P)$ + methanol [37].

5.4.4 *Mass spectrometry*

Here, the discussion will be restricted to the use of mass spectrometric detection methods to study the kinetics of neutral–neutral reactions. Other applications, such as those following ion–molecule reactions, nucleation processes, and product formation in the CRESU environment, are dealt with elsewhere (see Chapters 3, 6, and 8).

Although the spectroscopic detection methods described above are among the most sensitive techniques for following the kinetics of radical reactions, they are not universally applicable. In certain cases, the target radical cannot be probed by these schemes which rely on emission from excited states for detection purposes. The CH_3 radical, for example, is characterized by strong predissociation of its electronically excited states preventing its detection by fluorescence. Moreover, the predominantly monochromatic nature of the spectroscopic methods generally restricts the user to the detection of a single species, a situation that is less than ideal for following product formation in multichannel reactions. One alternative is to apply mass spectrometric detection, a technique which allows the simultaneous detection of multiple species when coupled with VUV photoionization. The first application of mass spectrometric detection in the CRESU environment was demonstrated by Lee *et al.* [17]. Here, the kinetics of the $C_2H + C_2H_2$ reaction was studied at 90 K by following the C_4H_2 formation. The pulsed supersonic flow was sampled by a cone-shaped skimmer with a 1 mm diameter. The pulsed photoionization laser at 118 nm crossed the sampled gas in a differentially pumped zone (10^{-5} Torr) where a time-of-flight mass spectrometer was mounted at right angles to both the supersonic flow and photoionization laser. Although this and subsequent measurements [48, 49] clearly demonstrated the promise of the method, signal levels were small and other problems associated with the sampling geometry (notably with respect to the possible generation of shock-waves in the cold supersonic flow) prevented the further application of mass spectrometry in CRESU systems until 2011. In these later experiments, Soorkia *et al.* [50] made several major modifications to the original apparatus by (I) replacing the conventional skimmer by a symmetric airfoil-shaped device with a 450-μm pinhole, to sample the pulsed supersonic flow with minimal perturbation, (II) using a quadrupole mass spectrometer in a linear geometry and (III) coupling the CRESU reactor to tunable VUV synchrotron radiation to improve the identification of the various products and isomers compared to the original pulsed 118-nm laser source. These authors revisited the kinetics of the $C_2H + C_2H_2$ reaction by following the formation of product C_4H_2, with significantly higher signal-to-noise ratios than their earlier work. The modified apparatus has since been used to follow the low-temperature kinetics and product branching ratios for several reactions of the C_2H radical [51, 52]. Supplementary details on this device can be found in Chapter 8.

5.4.5 *Chirped-pulse Fourier transform rotational spectroscopy*

Although significant progress has been made in quantifying the kinetics of gas phase neutral–neutral reactions through photoionization mass spectrometry, detection requires a sample of the reaction mixture to be extracted and ionized, potentially inducing perturbations in the upstream flow in CRESU systems as described above and in Chapter 8. Ideally, we would like to be able to apply a spectroscopic (and therefore non-invasive) method allowing the simultaneous detection of a wide range of species in a sensitive and quantitative manner. One potentially interesting technique that is only just beginning to be coupled with the CRESU method involves detection through rotational emission spectroscopy. Here, a short ($<$1 ms) pulse of radiation is generated in the microwave/millimeter wave frequency range, whose frequency is rapidly scanned such that the pulse has a significant bandwidth (a chirped pulse) [53]. This pulse is injected into the CRESU reactor whereupon molecules with a permanent dipole moment and rotational transitions falling within the frequency range of the chirped pulse are excited (polarized) and emit on resonance. This coherent emission (called the Free Induction Decay) is detected as a function of time and fast Fourier transformed into the frequency domain to recover the sample spectrum. The advantages of this method with respect to conventional spectroscopic techniques such as LIF are (1) the large instrument bandwidth, allowing the simultaneous detection of all species possessing rotational transitions within this frequency range, and (2) the potential to detect any molecular species possessing a permanent dipole moment. Conversely, as the incident photon fluxes are lower than those typically used in LIF-type experiments, much more signal averaging is required to achieve equivalent signal-to-noise levels. Moreover, as collisions lead to rapid rotational relaxation of the polarized sample (rather than emission), such experiments clearly favor the use of low-density supersonic flows. Despite these drawbacks, this method has considerable potential for investigating product formation in gas phase neutral–neutral reactions (this application is described in more detail in Chapter 8). As the chirped pulse is generally short enough to be compatible with the laser-based photolysis techniques currently used to generate radicals in CRESU apparatuses, it is also a potentially interesting method for following the kinetics of neutral–neutral reactions. In this case, the multiplex nature of the detection method is less important as the intensities of individual transitions will be followed as a function of time. Indeed, while a preliminary study of the CN + C_2H_2 reaction at 22 K [54] clearly showed the promise for detecting a wide range of products, the measurement of rate coefficients for a gas phase reaction has yet to be demonstrated.

5.5 Kinetic Studies of Bimolecular Reactions Using the CRESU Technique

A vast majority of measurements performed using the CRESU method have focused on bimolecular reactions between neutral species of the type A + B → P. The various energy transfer processes of the type $A(v, j) + B \rightarrow A(v', j') + B$ that have been studied using the CRESU technique are treated in Chapter 7. Bimolecular reactions can be classified into two distinct types — those which are endothermic requiring an input of energy to proceed and those which are exothermic and occur spontaneously. As the rates of endothermic reactions become negligibly small at low temperatures, few or no endothermic processes have ever been investigated using the CRESU technique, which is mostly limited to the study of reactions with rate coefficients greater than $10^{-12}\,\mathrm{cm^3\,s^{-1}}$. This limit arises from considerations of the hydrodynamic times of the supersonic flows, coupled with the co-reagent concentration levels that can be used in the supersonic flow. Indeed, once the co-reagent concentration exceeds a few percent of the total carrier gas concentration, the potential for reduction of the flow uniformity or modification of the flow characteristics becomes much greater. Considering a carrier gas density of $1 \times 10^{17}\,\mathrm{cm^{-3}}$, it would be appropriate to limit the co-reagent concentration to less than $5\,\%$ of the total flow ($5 \times 10^{15}\,\mathrm{cm^{-3}}$). Assuming that the reaction rate coefficient is $1 \times 10^{-12}\,\mathrm{cm^3\,s^{-1}}$, this would generate decay profiles with a time constant of $5000\,\mathrm{s^{-1}}$. If the supersonic flow has a characteristic hydrodynamic time of $400\,\mu\mathrm{s}$, we have the possibility to follow the progress of this reaction over approximately three half-lives (the half-life $t_{1/2}$ being given by $\ln(2)/k'$) with the initial minor reagent concentration decaying to around $12.5\,\%$ of its initial value for the case where the maximum co-reagent concentration is used. For a co-reagent concentration of $2.5 \times 10^{15}\,\mathrm{cm^{-3}}$, the minor reagent decays by slightly more than $50\,\%$ over the same timescale.

Only reactions which have at least one exothermically accessible (or thermoneutral) channel are relevant at low and very low temperatures, so all CRESU studies of reactivity involve exothermic reactions. Two types of exothermic reactions can be identified as shown in Fig. 5.5.

5.5.1 *Activated reactions*

Type A reactions are characterized by an energy barrier over the potential energy surface (PES) separating reagents from products. These processes exhibit similar behavior to endothermic reactions at room temperature and above, obeying an Arrhenius-type rate law.

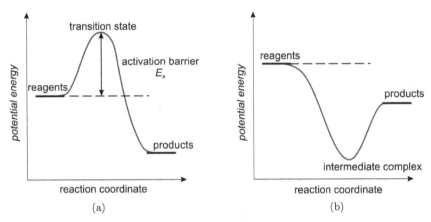

Fig. 5.5. Generic potential energy surfaces for exothermic type reactions.
(a) Activated reactions; (b) barrierless reactions.

There are few CRESU studies of this kind of reaction as such processes are often too slow to be measured at low temperature given the
relatively short duration of the CRESU flows. Nevertheless, Sabbah
et al. [47] studied the reactions of ground state oxygen $O(^3P)$ atoms
with various alkenes over the 23–300 K range using a chemiluminescence method to follow $O(^3P)$ loss (through the $O(^3P) + NO \rightarrow NO_2^*$
reaction). For the $O(^3P)$ + ethene (C_2H_4) reaction, which is thought
to possess an appreciable activation barrier [55], the rate coefficient
was measured to be 7.4×10^{-13} cm^3 s^{-1} at room temperature, while
only an upper limiting value of 1.6×10^{-13} cm^3 s^{-1} could be obtained
at 39 K. In contrast, the reactions of $O(^3P)$ with larger alkenes were
all seen to accelerate at lower temperature, indicative of barrierless
processes. Similarly, Hickson *et al.* [33] measured rate coefficients for
the reactions of ground state chlorine atoms $Cl(^2P)$ with the alkanes
C_2H_6 and C_3H_8. Here, Cl atoms were produced upstream of the Laval
nozzle reservoir by microwave discharge of Cl_2 and these atoms were
followed by resonance fluorescence detection around 138 nm. In this
case, while rate coefficients for the $Cl(^2P) + C_3H_8$ remained large and
constant over the entire 50–800 K temperature range, those obtained
for the $Cl(^2P) + C_2H_6$ reaction decreased at low temperature, as can
be seen in Fig. 5.6, with the most recent theory indicating the likely
presence of an activation barrier for this process [56].

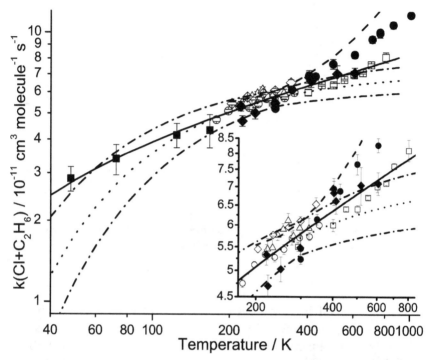

Fig. 5.6. Temperature dependence of the Cl + C_2H_6 reaction rate displayed on a log–log scale. ■ Hickson *et al.* [33], ━━: best fit of all data with the exception of Bryukov *et al.* [57], ● Bryukov *et al.* [57], ◆ Lewis *et al.* [58], □ Pilgrim *et al.* [59], ○ Hickson *et al.* [60], △ Manning and Kurylo [61], ◇ Dobis and Benson [62], ━ ━ Fernandez-Ramos *et al.* [63], · · · Sander *et al.* [3] evaluation with recommended uncertainties (·━·). Inset: expanded view of the region from 180 K to 800 K and k(Cl+C_2H_6) from $(4.5$–$8.5)\times10^{-11}$ cm^3 molecule^{-1} s^{-1}. Reproduced with permission from Ref. [33]. Copyright 2010 American Chemical Society.

It is interesting to note, however, that the measured rate coefficients for the $Cl(^2P) + C_2H_6$ reaction deviate from the predicted Arrhenius behavior at the lowest temperatures. This effect is likely to be due to the influence of quantum mechanical tunneling on the abstraction process due to the light mass of the transferred H atom, leading to rate coefficients somewhat greater than expected.

In recent years, an entire class of reaction has been identified, initially by the Leeds CRESU group, which displays surprising and dramatic increases in the rate coefficient at very low temperatures, despite the presence of an activation barrier on the reaction coordinate [11, 23, 64–66]. These are hydrogen atom transfer (abstraction) reactions of OH with complex organic molecules (COMs) possessing an oxygen-containing functional group, for example, alcohols, ketones, aldehydes, ethers, and esters. These reactions had largely been ignored in astrochemical networks as it was assumed, owing to the Arrhenius behavior that had been observed at higher temperatures, that the rate coefficients would be vanishingly small at temperatures relevant in space. The reactions are all characterized by the presence of a weakly bound hydrogen-bonded complex in the entrance channel of the PES for the reaction. These complexes, whose binding energies vary from 5–30 $kJ\,mol^{-1}$ depending on the reaction, have very short lifetimes at temperatures above $\sim$200 K, owing to their rapid redissociation back to reactants. In addition, under low-pressure conditions of the experiments, which are relevant to conditions in space, the collision rate is insufficient to remove energy from the newly formed complex to lower its internal energy below the dissociation energy back to reactants. However, once the temperatures become low enough within a CRESU expansion, the lifetime of the complex becomes sufficiently long that the probability of quantum mechanical tunneling (QMT) through the barrier to reaction becomes much higher compared with dissociation back to reactants, resulting in a rapid acceleration in the rate of reaction and the formation of products [23]. Kinetics studies of this class of reactions have been studied so far using the pulsed variant of the CRESU method and have been recently reviewed [66], and the general shape of temperature dependence of the rate coefficient for a number of reactions is shown in Fig. 5.7.

Evidence for the occurrence of the long-lived complex–QMT mechanism comes from observation of one of the products of the reaction; for example, the CH_3O radical product was observed for the reaction of OH with CH_3OH [23], which is formed via abstraction of the alcoholic hydrogen atom by OH over the higher barrier to reaction, but associated with a larger imaginary frequency (narrower barrier). The rate coefficient for the reaction measured from the rate of production of the CH_3O product was found to be the same as that determined from the rate of loss of the reagent OH. Further evidence for the mechanism occurring comes from a lack of an observed pressure dependence of the rate coefficient, for example,

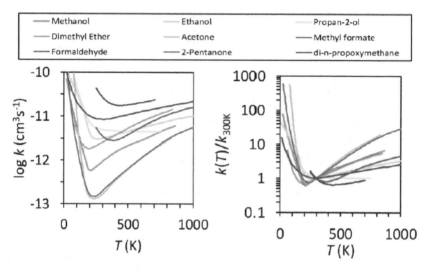

Fig. 5.7. Temperature dependence of the rate coefficient (left panel) and the rate coefficient divided by the room temperature value (right panel) for reactions of OH with a series of oxygenated volatile organic compounds, and which are characterized by a distinct minimum in the rate coefficient. Reproduced with permission from Ref. [66]. Copyright 2018 American Chemical Society.

in OH + methanol [23, 67] with the rate coefficient being significantly smaller than the high-pressure limit determined from a study of the reaction of OH(v = 1) with CH$_3$OH [23]. Hence, the loss of OH is not caused by formation of the complex and its subsequent stabilization via collisions with the bath gas. It must be noted, however, that the range of bath gas density that can be achieved in a CRESU experiment is more limited than in conventional reaction cells used in flash-photolysis kinetic studies. In some cases, for example, the OH + acetone [68] and OH + dimethyl ether [68] reactions, a pressure dependence is observed at ~80 K, demonstrating collisional stabilization of the complex. However, extrapolation of the fall-off curve of the rate coefficients versus bath gas pressure at 80 K shows an intercept at zero pressure, indicative that a bimolecular channel involving QMT could also be operating under conditions relevant to astrochemical environments. In a very recent study of the OH + acetone reaction at lower temperatures down to 11.7 K [69], no pressure dependence was observed at 20 K and 64 K, but a modest dependence was observed around 50 K. The recommended rate coefficient at $10\,K$ for use in models of interstellar molecular clouds for the OH + acetone reaction is $1.8 \times 10^{-10}\,\mathrm{cm}^3\,\mathrm{s}^{-1}$ [69], and hence almost every collision leads to reaction.

Deuteration of the reagent reacting with OH ought to reduce the rate of QMT significantly owing to the heavier mass of the D atoms being transferred, and is commonly used to demonstrate the presence of QMT. However, it proved difficult to cleanly perform such experiments owing to rapid exchange of D atoms from the oxygenated functional group with H atoms on the walls of the vacuum lines used for reagent preparation and delivery. It is also very difficult within a CRESU experiment to measure absolute branching ratios for different product channels. Instead, there is reliance on theory to estimate these as a function of temperature, for example, using transition state theory incorporating QMT, and utilizing accurate ab initio PESs. However, in favorable cases, it is possible to estimate the yield experimentally. For example, the yield of the CH_3OCH_2 from the reaction of OH with dimethyl ether, (DME, $(CH_3)_2O$), was measured indirectly to be 0.53 at $90\,K$ and $[N_2] = 6.5 \times 10^{16}\,\mathrm{cm}^{-3}$ [68]. The yield was measured by the addition of O_2, which is known to react with the product CH_3OCH_2 to form an energized adduct, $CH_3OCH_2O_2^*$, a fraction of which decomposes to reform OH and slow down its decay. By measuring the reduction in the rate coefficient measured via the loss of OH in the presence of O_2 compared with the absence of O_2, the yield of CH_3OCH_2 was measured [68].

The Castilla-La Mancha CRESU group, again using a pulsed Laval apparatus incorporating a rotating slot apparatus [70], has significantly extended the range of temperatures to lower values for this class of reactions. In the case of OH + CH_3OH, the reaction has now been studied down to $11.7\,K$ [67], with the rate coefficient leveling off below $\sim70\,K$, where it is close to the capture limit. The group has also studied reactions of OH with formaldehyde, ethanol, methyl formate, and acetone down to temperatures as low as $11.7\,K$.

The experimental observation of the dramatic increase in the rate coefficient at low temperatures, sometimes increasing by a factor of almost 1000 compared to the value at $\sim200\,K$, has spawned a considerable theoretical interest in this class of reactions. In the case of OH + CH_3OH, 6 theoretical studies have been published in the last couple of years alone [67, 71–75]. The general shape of $k(\mathrm{T})$ with temperature was reproduced using transition state theory incorporating QMT over the range 20–400 K [67], with the theory showing closer agreement for TST calculations performed at the high-pressure limit. TST calculations performed at the low-pressure limit significantly underestimated the rate coefficient at very low temperatures [67, 73, 74]. However, more recently, quantum roaming was suggested as a mechanism for the reactions of OH with methanol and formaldehyde at very low temperatures and in the limit of zero pressure using a ring polymer molecular dynamics (RPMD) approach [72]. In these quantum dynamics calculations, the complex formed between OH and the organic molecule was calculated to have a very long lifetime, longer than 100 ns [72], which is on the same timescale as QMT to form reaction products, enabling acceleration of the reaction rate. Quantum effects are the cause of the formation of the long-lived complexes at zero pressure, facilitating tunneling without the need for collisional stabilization, as required by the TST calculations in the high-pressure limit to provide agreement with experimental measurements [67, 73, 74].

The large enhancements in the rate coefficient at very low temperatures have important implications for interstellar chemistry, as previously this class of reaction had largely been ignored in astrochemical networks. The experimental rate coefficients, and in some cases parameterizations of the rate coefficients in order to make modest extrapolations to temperatures relevant to interstellar environments, have now been used to update gas-grain chemical networks. These networks are incorporated into a number of models appropriate to cold molecular clouds. Inclusion of these reactions has shown a significant impact, and has led to improvements in the level of agreement between observations using telescopes and modeled abundances of key species [76, 77]. For example, inclusion of the OH + CH_3OH reaction doubled the CH_3O radical abundance under conditions typical of a dense molecular cloud [76], and for the cold core Barnard 1b, the CH_3O/CH_3OH abundance ratio was predicted to be similar to astronomical observations [77]. Table 5.1 summarizes all the ground state radical closed-shell molecule reactions to have been investigated using the CRESU technique at the present time (updated from Table 5.2 of Potapov *et al.* [64]). The values in this table represent the lowest temperature for which measurements were performed. Concerning the OH + CH_3OH reaction, Fig. 5.8 presents

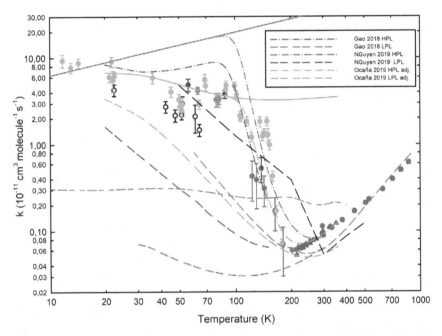

Fig. 5.8. Temperature dependence of the rate coefficients for the OH + CH$_3$OH reaction. Experiments: (solid magenta circles) Shannon *et al.* [23]; (solid orange circles) Ocaña *et al.* [67]; (open black circles) Antiñolo *et al.* [77]; (solid black circles) Gómez Martin *et al.* (CRESU) [117]; (solid purple circles) Gómez Martin *et al.* (flow tube) [117]; (solid green circles) Hess and Tully [133]; (solid olive triangles) Dillon *et al.* [134] Theory: (purple line) Shannon *et al.* [23]; (red lines) Gao *et al.* [74]; (black line) del Mazo-Sevillano *et al.* [72]; (blue lines) Nguyen *et al.* [73]; (light orange lines) Ocaña *et al.* [67]; (grey line) Roncero *et al.* [71] For the Gao *et al.*, Nguyen *et al.*, and Ocaña *et al.* studies, high-pressure limiting (HPL) and low-pressure limiting (LPL) data are represented by dashed dotted and dashed dashed lines, respectively. For the Gao *et al.* and Ocaña *et al.* studies, solid lines represent the collision and capture limits, respectively.

an overview of the current state of experimental and theoretical knowledge regarding this process.

5.5.2 *Barrierless reactions*

Reactions of type B, that is, barrierless exothermic reactions, constitute more than 90% of all CRESU investigations of neutral–neutral reactions. Two basic types can be identified: (I) those occurring between a radical and a closed-shell molecule, and (II) those occurring between two radical species.

5.5.2.1 *Radical + molecule reactions*

As experimental studies of type (I) reactions involve the production of only a single radical species (with the closed-shell molecule generally used as the excess reagent under pseudo-first-order conditions), there are numerous examples of this type of measurement as can be seen from Table 5.1. The case of the stable molecule being a radical itself, for example, O_2, is considered in Section 5.5.2.2.

These include one of the earliest CRESU studies by Sims *et al.* [129] of the reaction of CN with NH_3 over the 25–295 K range. The kinetics of other molecular radical reactions such as those involving the hydrides OH [13, 15, 102, 121], CH [19, 86], and NH [87] with a wide variety of organic molecules with both saturated and unsaturated hydrocarbons have also been extensively studied. In all, the kinetics of ten distinctly different molecular radicals (CH, CH_2, C_2H, C_4H, NH, OH, CN, C_2N, C_3N, and C_2) have been studied to date using the CRESU method. All of these species have been detected in interstellar environments and most if not all are thought to contribute to the chemistry of planetary atmospheres. In the case of the CH_2 radical, almost all CRESU studies concerned the reactivity of the first electronically excited singlet state CH_2 (a^1A_1), whose kinetics have been examined down to 43 K [22, 135] with this species likely to play an important role in the atmosphere of Titan, one of Saturn's moons [136]. For $^1CH_2$, there are two possible fates, either collisional relaxation to the ground triplet state CH_2 (X^3B_1) or a reactive collision to form products. For reactions of $^1CH_2$ with H_2 and CH_4, branching ratios (BRs) for reactive removal of $^1CH_2$ were determined down to 73 K [135], and down to 100 K for C_2H_2, C_2H_4, and C_2H_6 [22]. For these reactions, the H atom products were monitored using VUV LIF at 121 nm, while branching ratios for the electronic quenching of $^1CH_2$ to $^3CH_2$ were also measured. The reactions of excited state radicals with stable molecules that have been measured by the CRESU method to date are listed in Table 5.2. In contrast, $^3CH_2$ is quite unreactive with closed-shell molecules and is difficult to detect [137]. However, for the reaction of $^3CH_2$ with O_2, the rate coefficient has been determined down to 100 K by monitoring the H atom product using VUV LIF at 121 nm [22]. For C_2, the reactions

Table 5.1. Ground state radical + closed-shell molecule reactions to have been measured by the CRESU method. Numbers represent the lowest temperature (K) of the experimental measurement. The second and subsequent entry, in brackets and superscripts, for each reaction, is the reference(s) of the corresponding investigation.

Excess Reagent	Minor Reagent													
	CH	NH	OH	C_2H	$^1C_2$	CN	C_3N	C_4H	$B(^2P)$	$C(^3P)$	$O(^3P)$	$F(^2P)$	$Si(^3P)$	$Cl(^2P)$
H_2	23[78]						24[79]					11[20]		
D_2	13[78]													
HBr			23[80-83]											
H_2O	50[46]									50[84]				
D_2O										50[84]				
H_2O_2			96[85]											
CH_4	23[86]	53[87]		15[44,45,48,50]	24[88]		24[79]	200[89,90]						
C_2H_2	23[86]	53[87]	96[95]	15[44,52,96]	24[36,88]	25[91]	24[79]	39[89,92]	23[30]	15[93,94]		15[29]		
C_2H_4	23[86]	53[87]		96[97]	24[88]	25[91]	24[79]	39[89,92]	23[28]	15[93]	23[47]	15[29]		
C_2H_6	23[86]	53[87]		63[45,98]	77[36]	15[98,99]	24[79]	39[89,90]						48[33]
C_3H_4 (propyne)	77[19]			63[45,98,101]	77[36]	15[98]	24[79]	39[89,92]		15[100]				
C_3H_4 (propadiene)	77[19]							39[92]		15[100]				
C_3H_6	77[19]	53[87]	50[15,95,102,103]	15[44,52,96]	77[36]	23[104,105]	24[79]	39[92]		15[93]	23[47]			
C_3H_8		53[87]		96[97]	24[88]	23[104]	24[79]	39[89,90]						48[33]
C_4H_2				74[106]		23[104]		39[92]						
C_4H_6 (butyne)				104[107]										
1,3-butadiene						23[108]		39[92]						
cis-butene			23[109]	79[51]							23[47]			

(Continued)

Table 5.1. (*Continued*)

Excess Reagent	CH	NH	OH	C$_2$H	^{1}C$_2$	CN	C$_3$N	C$_4$H	B(^{2}P)	C(^{3}P)	O(^{3}P)	F(^{2}P)	Si(^{3}P)	Cl(^{2}P)
						Minor Reagent								
but-1-ene	23[86]		22[14,95,102,103,109]	79[51,96,107]			39[92]				23[47]			
isobutene			23[109]	79[51,107]							23[47]			
trans-but-2-ene				79[51]							23[47]			
isoprene			58[15]											
n-butane				96[97]				39[90]						
isobutane				104[107]										
benzene				105[110]	105[111,112]									
toluene						105[111]								
C$_6$H$_5$C$_2$H		58[114]				123[113]								
anthracene														
H$_2$CO	31[115]		22[116]			17[118]								
CH$_3$OH			22[23,67,77,117]										50[37]	
CH$_3$CH$_2$OH			21[119,120]											
CH$_3$CH(OH)CH$_3$			88[120]											
HC(O)OCH$_3$			22[121]											
(CH$_3$)$_3$COOH			86[13]											
CH$_3$COC$_2$H$_5$			93[65]											

Species					
CH_3CHO	$12^{[122,123]}$				
C_2H_5CHO	$58^{[124]}$				
CH_3COCH_3	$12^{[68,69]}$				
CH_3OCH_3	$63^{[65,68]}$				
L-alanine ester	$58^{[125,126]}$				
L-alanine	$58^{[125,126]}$				
NH_3	$23^{[127]}$	$104^{[128]}$	$25^{[129]}$	$24^{[79]}$	$50^{[38,130]}$
ND_3		$104^{[128]}$			
CH_3NH_2			$23^{[131]}$		
$(CH_3)_2NH$			$23^{[131]}$		
$(CH_3)_3N$			$23^{[131]}$		
HC_3N			$23^{[18]}$		
CH_3CN		$165^{[107]}$	$23^{[132]}$		
C_2H_5CN		$104^{[107]}$			
C_3H_7CN		$104^{[107]}$			

314   *Uniform Supersonic Flows in Chemical Physics*

Table 5.2.   Excited state radical + molecule reactions to have been measured by the CRESU method. Numbers represent the lowest temperature (K) of the experimental measurement. The second and subsequent entry, in brackets and superscripts, for each reaction, is the reference(s) of the corresponding investigation.

Excess Reagent	Minor Reagent					
	$^3C_2$	$^1CH_2$	$C(^1D)$	$N(^2D)$	$O(^1D)$	$S(^1D)$
H_2		43[135]	50[21]		50[168]	6[152]
D_2			50[169]		50[170]	
HD			50[171]		50[172]	
H_2O			75[173]			
CH_4		43[135]	50[159]	127[160]	50[174]	23[153]
C_2H_2	24[138]			50[161]	50[175]	23[153]
C_2H_4	24[138]			50[162]		23[154,155]
C_2H_6	200[138]		50[159]	75[160]	50[175]	
C_3H_8	36[138]			75[160]		
CH_3OH			127[37]			
CO_2			50[176]		50[176]	
Kr					50[177]	
Ar					50[178]	6[156]
He		43[135]				
N_2		43[135]	50[179]		50[178]	

of both the ground singlet ($^1C_2$ $X^1\Sigma_g^+$) and the low-lying first excited triplet ($^3C_2$ $a^3\Pi_u$) electronic states with unsaturated hydrocarbons have been studied over a wide temperature range [36, 138], providing an interesting comparison of the reactivity differences between the triplet and singlet configurations at similar energies. Both $^1C_2$ and $^3C_2$ are potentially important radicals in Titan's upper atmosphere through the photolysis of C_3 molecules, which could be present at high concentrations [136]. Among the many notable absences from this list are certain molecular radicals that have proven difficult to detect in the CRESU environment such as CH_3 and C_3H. As these cannot be followed by LIF, other detection methods will need to be implemented to follow the progress of reactions involving these radicals. Radicals containing elements from the third row of the Periodic Table such as sulphur (radicals such as SH, SO, and C_XS), phosphorous (C_xP and PO), and silicon (SiO and SiS) are entirely absent despite the fact that all of these radicals have been detected in interstellar environments. Moreover, as C_2N radicals have been detected in interstellar space [139] and are likely to be present in planetary atmospheres, the reactions of these radicals with stable molecules should also be investigated. Indeed, this radical can easily be detected by LIF and room temperature measurements have already demonstrated that C_2N reacts rapidly with a range of saturated and unsaturated hydrocarbon molecules [140, 141].

To date, the reactions of eight atomic radicals in their ground electronic states have been studied by the CRESU method ($B(^2P)$, $C(^3P)$, $N(^4S)$, $O(^3P)$, $F(^2P)$, $Si(^3P)$, $Al(^2P)$, and $Cl(^2P)$). Although the reactions of $N(^4S)$ and $Al(^2P)$ are absent from Table 5.1, these are included in Table 5.3 as only radical + radical reactions have been studied by the CRESU method for these species. The specific methods used to generate these atomic reagents in CRESU flows have been described earlier in this chapter (photodissociation for $C(^3P)$,[93] $O(^3P)$ [47], $F(^2P)$ [20], and $Al(^2P)$ [142]; multiphoton photodissociation for $B(^2P)$ [28], $C(^3P)$ [37], and $Si(^3P)$ [29]; and microwave discharge for $N(^4S)$ [32] and $Cl(^2P)$ [33]). During these studies $B(^2P)$, $C(^3P)$, $Si(^3P)$, and $Al(^2P)$ were all detected directly through LIF in UV or VUV wavelength ranges, whereas the kinetics of the $F(^2P) + H_2$ reaction were followed by VUV LIF of the H-atom product at 121.567 nm. $N(^4S)$ and $Cl(^2P)$ atoms were detected by resonance fluorescence, whereas the kinetics of $O(^3P)$ atom reactions were followed by chemiluminescence [47]. In a general sense, it can be seen from Table 5.1 that there are relatively few CRESU investigations of ground state atom reactions, with the exception of $C(^3P)$ and $O(^3P)$, whose reactions with unsaturated hydrocarbons have been extensively investigated. Indeed, only one reaction has been studied experimentally for each of $F(^2P)$ and $Al(^2P)$, although both of these atoms are expected to react rapidly with a wide range of closed-shell molecules. Despite being one of the most abundant species in the ISM and in planetary atmospheres, no $H(^2S)$ atom reactions have ever been studied by the CRESU technique. $H(^2S)$ is expected to react with closed-shell molecules mostly by addition at low temperatures, which partly explains the absence of experimental kinetics studies at low temperature, due to the associated difficulty of varying the pressure in CRESU apparatuses. Nevertheless, bimolecular reactions such as $H(^2S) + CH_3SH$, C_2H_5SH could be non-negligible at low temperatures, and radical–radical reactions such as $H(^2S) + NO_2$ have already been shown to remain rapid above 200 K [3].

While the CRESU technique has been employed in numerous kinetic studies of ground state radical–molecule reactions, excited state radical reactions have received much less attention. Indeed, in the gas phase ISM where densities are low, collisions occur infrequently so that excited state species formed in these regions are expected to radiatively relax to the ground state before reacting. Nevertheless, there is a growing body of evidence pointing toward the potential roles of excited state radical–molecule reactions in the chemistry of interstellar ices produced either through the photodissociation of surface species or by cosmic ray interactions [143–145]. As many of the parameters for the surface reactions used in current reaction networks are derived from the equivalent gas phase ones, the measurement of reaction rates and product branching ratios for these processes is an important step forward for the improvement of current astrochemical models. Although reactions between excited state radicals and molecules have only a limited influence on the gas phase chemistry of the interstellar medium, such processes are particularly relevant to the photochemistry of planetary atmospheres where large fluxes of energetic photons and high densities lead to numerous collisions with the various atmospheric components.

In the Earth's upper atmosphere, excited state atomic nitrogen $N(^2D)$ is produced through the short wavelength photodissociation of N_2 and through N_2 collisions with photoelectrons in the upper atmosphere [146]. The reaction between $N(^2D)$ and O_2 is a major source of thermospheric NO, limiting the overall odd nitrogen concentration at these altitudes. The photolysis of O_3 in the stratosphere and the troposphere produces $O(^1D)$ atoms with high quantum yields[3] which either react with trace molecules in the atmosphere (such as CH_4 or H_2 in the stratosphere or water vapor in the troposphere) or are quenched through non-reactive collisions with O_2 and N_2. In Titan's upper atmosphere, which also has an atmosphere primarily composed of N_2, both $N(^4S)$ and $N(^2D)$ atoms are considered to be important reactive species for the formation of nitrogen-bearing compounds. While $N(^4S)$ atoms react with radicals such as CH_3 [147], to form nitrogen-containing hydrocarbon molecules such as H_2CN (and ultimately HCN) [148], these ground state atoms are largely unreactive with the most abundant closed-shell hydrocarbon molecules (CH_4, C_2H_2, C_2H_4, and C_2H_6) present in Titan's atmosphere. Instead, the more reactive metastable $N(^2D)$ atoms are expected to react with these molecules to form a range of amines, imines, and nitriles [149].

In the Martian atmosphere, $O(^1D)$ atoms are produced with high quantum yields by the photodissociation of CO_2 below 170 nm [150]. These atoms either react with species present in the Martian atmosphere such as H_2 or H_2O or are removed by non-reactive collisions with the major atmospheric constituent CO_2.

The relevance of the reactions of $C(^1D)$ atoms to the chemistry of planetary atmospheres is less obvious. Current models of Titan's atmosphere [136] predict that the radical C_3 could reach high abundances around 900 km. As C_3 is unreactive with small hydrocarbon molecules [151], this species is predominantly lost through photolysis, potentially forming $C(^1D)$ atoms as one of the major products.

The first study of an excited state atom reaction using the CRESU technique was performed by Berteloite *et al.* [152] These authors produced $S(^1D)$ atoms in the cold supersonic flow through the pulsed photolysis of precursor molecule CS_2 at 193 nm. The decay of $S(^1D)$ in the presence of H_2 molecules was followed directly by VUV LIF at 166.67 nm. This investigation is also interesting as it represents the lowest temperature attained so far using the CRESU method. Here, the carrier gas He was pre-cooled to 77 K in the Laval nozzle reservoir allowing a flow temperature of 5.8 K to be reached. Subsequent work has focused on the reactions of $S(^1D)$ with saturated [153] and unsaturated [153–155] hydrocarbons in addition to one investigation of the electronic quenching process $S(^1D) + Ar \rightarrow S(^3P) + Ar$ [156]. These and other energy transfer studies are discussed in more detail in Chapter 7.

Since 2015, the Bordeaux CRESU group has investigated the reactions of both $C(^1D)$ and $O(^1D)$ atoms with a range of molecular reagents over the 50–296 K range. The studies of $O(^1D)$ atom reactions were performed in a similar way to those of $S(^1D)$ atoms described above, with $O(^1D)$ atoms being produced by the pulsed laser photolysis of O_3 at 266 nm and detected by pulsed VUV LIF at 115.215 nm.

For their studies of the reactivity of excited state carbon atoms, $C(^1D)$ was produced by the photodissociation of CBr_4 molecules at 266 nm in a sequential multiphoton process which simultaneously produces $C(^3P)$ atoms as the major product ($C(^1D)$ atoms are produced at the level of 10–15% with respect to $C(^3P)$) [37]. The authors were unable to directly detect $C(^1D)$ atoms in these experiments due to the lack of strong transitions in wavelength ranges accessible with the frequency tripling method used by this group. Instead, two alternative methods were employed to follow the kinetics of $C(^1D)$ reactions. Reactions that led to the formation of ground state hydrogen atoms $H(^2S)$ as products (such as $C(^1D)$ + H_2, D_2, HD, CH_4, C_2H_6, and CH_3OH), were followed through sensitive VUV LIF of the $H(^2S)$ $(D(^2S))$ products at 121.567 nm (121.534 nm) and by fitting the LIF traces with the biexponential function given in Equation (5.9). In a general sense, barrierless excited state atom reactions are considered to display only small reactivity differences as the temperature is varied, given the typically large exothermicities of these processes. Despite this, the reactions of $C(^1D)$ atoms with the closely related saturated hydrocarbon co-reagents CH_4 and C_2H_6 have been shown to be quite different as the temperature is varied below room temperature as shown in Fig. 5.9.

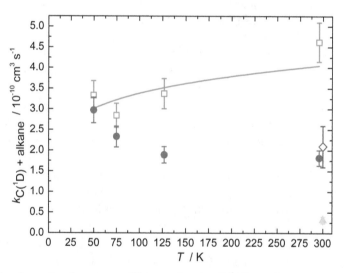

Fig. 5.9. Second-order rate coefficients for the $C(^1D)$ + alkane reactions as a function of temperature. $C(^1D)$ + CH_4 reaction: (cyan solid triangle) Braun *et al.* [157]; (purple open diamond) Husain and Kirsch [158]; (blue solid circles) Nuñez-Reyes *et al.* [153] $C(^1D)$ + C_2H_6 reaction: (red open squares) Nuñez-Reyes *et al.* [159] The red solid line represents the fit to the $C(^1D)$ + C_2H_6 rate data of Nuñez-Reyes *et al.* [159] with a fixed $T^{1/6}$ temperature dependence. Reproduced with permission from Ref. [159]. Copyright 2017 American Chemical Society.

Here, the measured rate coefficients for the $C(^1D) + CH_4$ reaction are clearly seen to increase as the temperature is lowered, while the rate coefficients for the $C(^1D) + C_2H_6$ reaction decrease over the same range. Although the temperature dependence of the $C(^1D) + C_2H_6$ reaction is characteristic of the capture-type behavior dominated by long-range dispersion interactions (with the rate coefficients varying as $T^{1/6}$), the increasing rate coefficients displayed by the $C(^1D) + CH_4$ reaction could be indicative of nonadiabatic effects due to the longer lifetime of the $^1HCCH_3$ intermediate at low temperature.

To study the reactions of $C(^1D)$ with species that do not lead to atomic hydrogen as a product (such as $C(^1D) + N_2$, CO_2, NO, and O_2), a fixed excess concentration of H_2 (or CH_4) was added to the flow in addition to the co-reagent, generating $H(^2S)$ atoms through the $C(^1D) + H_2/CH_4$ reactions. Under these conditions, the $H(^2S)$ temporal profile follows the kinetics of $C(^1D)$ removal by both processes (and not just the kinetics of the $C(^1D) + H_2$ (or CH_4) reaction), so that any *change* in the pseudo-first-order formation rate of H-atoms is entirely due to the change in co-reagent concentration. Consequently, the second-order rate coefficients are obtained in the usual manner, by plotting the pseudo-first-order rate coefficient as a function of the co-reagent concentration. In this case, the $C(^1D) + H_2$ (or CH_4) reaction contributes to the y-axis intercept value of the second-order plot.

Very recently, the Bordeaux CRESU group has also validated a method for the production of $N(^2D)$ atoms, using the reaction of $C(^3P)$ with NO as a source (see Section 5.3.1 for more details) [31, 160–162]. This method, coupled with the direct VUV LIF detection of $N(^2D)$ atoms at 116.745 nm is now being applied to investigate the kinetics of the reactions of $N(^2D)$ with a range of saturated and unsaturated hydrocarbon species of relevance to the atmosphere of Titan. The measured rate coefficients for the $N(^2D) + CH_4$, C_2H_6, and C_3H_8 reactions are displayed as a function of temperature in Fig. 5.10.

Whereas the new kinetic measurements for the $N(^2D) + CH_4$ reaction are seen to be entirely consistent with earlier work between 223 K and 300 K, the previously recommended rate coefficients for the other two reactions were significantly overestimated at low temperature [167]. Such differences clearly illustrate the dangers of extrapolating

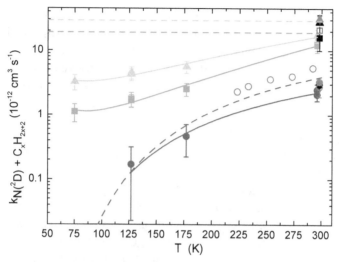

Fig. 5.10.   Second-order rate coefficients for the N(^{2}D) + C$_x$H$_{2x+2}$ reactions as a function of temperature. The N(^{2}D) + CH$_4$ reaction: (blue solid circles) Nuñez-Reyes *et al.* [160]; (black solid circle) Black *et al.* [163]; (red open circle) Fell *et al.* [164]; (purple solid circle) Umemoto *et al.* [165]; (purple open circles) Takayanagi *et al.* [166]. The solid blue line represents best fit to the present experimental data. The N(^{2}D) + C$_2$H$_6$ reaction: (green solid squares) Nuñez-Reyes *et al.* [160]; (black solid square) Fell *et al.* [164]; (blue open square) Umemoto *et al.* [165]. The N(^{2}D) + C$_3$H$_8$ reaction: (cyan solid triangles) Nuñez-Reyes *et al.* [160]; (red solid triangle) Fell *et al.* [164]; (black solid triangle) Umemoto *et al.* [165]. The dashed lines represent the previously recommended values for these reactions by Herron [167]. Reproduced with permission from Ref. [160]. Copyright 2019 PCCP Owner Societies.

data obtained at room temperature to the low-temperature regime for use in astrochemical modeling studies and highlight the continuing need for kinetic measurements over the appropriate temperature range.

5.5.2.2   *Radical + radical reactions*

As the interactions between two open-shell (radical) species give rise to multiple PESs correlating adiabatically with potential product species, barrierless pathways leading exothermically to these products are likely to exist for a wide range of radical–radical reactions, making such processes particularly relevant in low-temperature environments. Indeed, the reactions of ground state atomic carbon,

Table 5.3. Radical + radical reactions to have been measured by the CRESU method. Numbers represent the lowest temperature (K) of the experimental measurement. The second and subsequent entry, in brackets and superscripts, for each reaction, is the reference(s) of the corresponding investigation.

Minor Reagent	Excess Reagent			
	O_2	NO	$N(^4S)$	$O(^3P)$
$B(^2P)$	$24^{[26]}$			
$C(^3P)$	$15^{[93,181,182]}$	$15^{[181,182]}$		
$Al(^2P)$	$23^{[142]}$			
$Si(^3P)$	$15^{[25,27]}$	$15^{[27]}$		
$C(^1D)$	$50^{[183]}$	$50^{[183]}$		
$N(^2D)$		$50^{[31]}$		
$O(^1D)$	$50^{[178]}$			
$N(^4S)$		$48^{[32]}$		
CH	$13^{[127]}$	$13^{[127]}$	$48^{[184]}$	
NH		$53^{[87]}$		
OH			$48^{[185]}$	$39^{[186]}$
$^1CH_2$	$43^{[135]}$			
$^3CH_2$	$100^{[22]}$			
$^1C_2$	$145^{[187]}$	$24^{[187]}$	$48^{[188]}$	
$^3C_2$		$24^{[187]}$		
CN	$13^{[35,129]}$		$48^{[189]}$	
C_2H	$15^{[44,45,49]}$			
C_2N			$48^{[190]}$	
C_3N	$24^{[79]}$			

nitrogen, and oxygen atoms with small molecular radicals such as the hydrides CH, OH, and NH and other carbon bearing radicals such as CN and C_2 are crucial steps for the eventual formation of complex molecules in dense interstellar clouds [180]. Despite their clear relevance to astrochemistry, experimental studies of radical–radical reactions by the CRESU method are quite rare, with the exception of those investigations where one of the open-shell species is stable such as NO or O_2 (see Table 5.3).

Two major problems have to be addressed to study the kinetics of reactions between two unstable radicals with the CRESU method. The first problem, brought about by the relatively short timescales of CRESU flows (several hundreds of microseconds at most), means that large excess radical densities are required to induce a substantial change of the minor reagent concentration. We can illustrate the point by considering the example of a non-specific target radical–radical reaction with a rate coefficient of $\sim 5 \times 10^{-11} \, \text{cm}^3 \, \text{s}^{-1}$. Under pseudo-first-order conditions, the ideal situation would be to follow the decay of the minor reagent back to its baseline value or thereabouts, corresponding to four half-lives or $1/16$ of the initial reagent signal $(t_{1/2} = \ln(2)/k')$. With a supersonic flow that remains exploitable for $250 \, \mu\text{s}$, an excess radical concentration $\sim 2.2 \times 10^{14} \, \text{cm}^{-3}$ would be required to arrive at this value. In conventional flow-tube or flash-photolysis experiments, such high concentrations of radicals are not usually required as the reaction can be followed over longer timescales.

The second major problem to overcome to study the kinetics of radical–radical reactions with the CRESU technique relates to the determination of the excess radical reagent concentration. In particular, titration schemes (where the radical species concentration is determined through its reaction with a titrant in separate calibration experiments conducted under the same conditions), which allow radical concentrations to be determined directly in conventional apparatuses cannot be applied in CRESU experiments. Titration techniques rely on the injection of a known titrant concentration just upstream of the detection region, but in CRESU experiments any attempt to inject gases into the cold jet downstream of the Laval nozzle would result in the immediate destruction of the supersonic flow. Injection of a titrant upstream of the Laval nozzle would not take upstream radical losses (such as those occurring inside the nozzle) into consideration.

The first radical–radical reaction to be studied below 80 K using the CRESU technique was the $O(^3P) + OH$ reaction [186], an important process in dense interstellar clouds as essentially the only gas phase route for O_2 formation. In these experiments, $O(^3P)$ was used as the excess reagent, produced by the pulsed photolysis of molecular oxygen at 157.6 nm. OH radicals were generated by the reaction of precursor H_2O with excited state $O(^1D)$ atoms, which were also produced during O_2 photodissociation. Large O_2 densities and a high laser fluence generated large $O(^3P)$ concentrations ($[O(^3P)]$ $\leq 3.7 \times 10^{14} \, \text{cm}^{-3}$), allowing OH kinetic decays to be measured under pseudo-first-order conditions. However, as significant photodissociation occurred (10% or more of the O_2 precursor molecules were photolyzed in some experiments), the gas temperature was calculated to increase by several Kelvin as a result of the exothermic energy release of the photolysis step. Although such an increase in temperature was not a major obstacle to the measurement of low-temperature rate coefficients, estimations had to be made regarding the $O(^3P)$ density as a function of distance along the gas jet due to attenuation of the photolysis radiation. Despite the clear drawbacks of the photolysis method, it could be applied to the study of other low-temperature radical–radical reactions involving oxygen atoms. Nevertheless, this technique does not allow other atomic radicals such as nitrogen, hydrogen, and carbon to be generated from their molecular counterparts given the negligible absorption cross sections for these species in an accessible wavelength range (although such experiments might be feasible if coupled to an intense tunable synchrotron radiation source). Different methods are clearly required to extend the potential range of radical–radical reactions that can be investigated by the CRESU technique.

One alternative solution has been implemented by the Bordeaux CRESU group, whereupon *atomic* radicals such as H, N, and O are generated from their molecular counterparts upstream of the Laval nozzle by microwave discharge. The motivation to study the reactions of atoms with radicals at low temperatures arose from the knowledge that atom–radical reactions are likely to be among the most important processes in dense interstellar clouds. Although this method permits the creation of large concentrations ($< 10^{15} \, \mathrm{cm}^{-3}$) of a range of reactive atoms and also molecular radicals (through subsequent in situ chemical transformation), it had never been successfully coupled with the CRESU technique. Indeed, the large reservoir volumes and high gas flows used by the original continuous flow CRESU systems meant that the concentration of reactive atoms reaching the supersonic flow was too low to apply the pseudo-first-order approximation (using this species as the *excess* reagent). In pulsed apparatuses, the discharge would have to be pulsed in a stable manner (although this might be possible in such systems by using a mechanical chopper downstream of the microwave source). The Bordeaux continuous CRESU apparatus overcomes such issues through reduced gas flows and a small reservoir volume of a few tens of cm^3 to limit the upstream radical residence time. The quartz inlet tubes and all surfaces upstream of the reservoir are coated with halocarbon wax to inhibit heterogeneous atom recombination so that atoms created

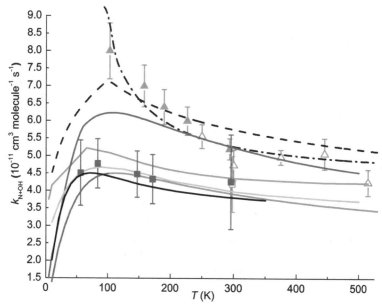

Fig. 5.12. Rate coefficients for the $N(^4S) + OH(^2\Pi)$ reaction as a function of temperature. Experimental values: (▲) Smith and Stewart [193]; (△) Howard and Smith [197]; ☐ Brune *et al.* [198]; ■ Daranlot *et al.* [185]. Theoretical values: —— Edvardsson *et al.* [199]; —— Jorfi *et al.* [200]; —— Li *et al.* [201]; —— Ge *et al.* [202]; Daranlot *et al.* [185]; —— Time-independent quantum mechanical calculations with the J–shifting approximation; —— Time-dependent quantum mechanical calculations based on all J. Recommended values: — • — Woodall *et al.* [203]; — — Wakelam *et al.* [204] Reproduced with permission from Ref. [185]. Copyright 2011 American Association for the Advancement of Science.

results as both minor reagents follow single exponential decay profiles. However, careful test experiments were required to ensure that neither the precursor molecules for the minor reagent species nor their photolysis products were involved in secondary reactions that could interfere with the target reaction investigation.

Low-temperature rate coefficients have now been measured for all of the reactions implicated in the molecular nitrogen formation cycles presented above. Although these measurements were not made at temperatures directly applicable to dense interstellar clouds, the fact that rate coefficients have been obtained down to 56 K allows us to make better estimates of the values at 10 K. When these values are tested in astrochemical models [190], the dark cloud abundances of N_2 are seen to be as much as an order of magnitude lower than previously thought.

Interestingly, given the general nature of the microwave discharge method for the production of large densities of atomic radicals from their gaseous diatomic precursors, it could also be applied to study the reactions of both hydrogen $H(^2S)$ and oxygen $O(^3P)$ atoms (which are both highly abundant in dense interstellar clouds) with molecular radicals. Indeed, current astrochemical models use large rate coefficients for the reactions of $O(^3P)$ atoms with carbon-containing radicals such as C_2 and CH, thereby inhibiting complex molecule formation by tying up a large fraction of the available carbon as CO. As neither of these processes has ever been studied experimentally below room temperature (nor has any H-atom reactions!), there is a clear need for further kinetic measurements of atom–radical reactions using the CRESU technique to clarify the importance of such processes in astrochemistry.

5.6 Kinetic Studies of Termolecular Reactions Using the CRESU Technique

Given the added complications arising from the measurement of termolecular reactions in the CRESU environment, it is unsurprising that there are significantly fewer CRESU studies of termolecular reactions than bimolecular ones. It is worth noting, however, that the seminal paper on the CRESU technique presented by Rowe *et al.* [9] described a series of measurements of ion–molecule association reactions such as $N_2^+ + 2N_2 \rightarrow N_4^+ + N_2$. There are essentially two types of termolecular processes to have been studied with the CRESU method. The first involves cluster formation or nucleation phenomena promoted by the low temperatures and high densities typically found within the supersonic flow. These processes are treated in detail in Chapter 6. The second type of process involves the formation of a chemically bound adduct. Measurements of termolecular reactions performed using the CRESU technique are summarized in Table 3 of Potapov *et al.* [64]. A majority of these studies have been radical associations involving OH and CH. The studies by Brownsword *et al.* [78] of the CH + H_2 reaction and those of Le Picard *et al.* [205, 206] of the CH + N_2 and CH + CO reactions are for example) have since been revealed as potentially important mechanisms for the formation of complex organic molecules.

Despite this, our knowledge of several areas continues to be lacking. If we examine the studies listed in Tables 5.1–5.3 more closely, it becomes immediately apparent that there are only a limited number of reactions that have been investigated at temperatures lower than 22 K, with very few reactions to have been studied below 15 K. As the dense regions of many interstellar clouds have characteristic temperatures of less than 10 K, it is clearly desirable to extend kinetic measurements below this value.

Below 15 K, the Mach number increases so much that pumping capacities larger than the most powerful systems available worldwide are required to generate a uniform flow with a sufficiently low density (to avoid clustering). A solution to tackle this issue is to cool the reservoir down to liquid nitrogen temperature (77 K) for a Laval nozzle having a moderate Mach number. Employing this method, gas phase reactions have already been studied at 13 K and 6 K in the Rennes group. However, as the reservoir temperature is low, this limits the potential co-reagent and precursor molecules that can be employed due to condensation issues. Consequently, a majority of reactions have not been investigated at temperatures below 15 K. This frontier has been recently lowered down to 12 K at the University of Castilla-La Mancha in Spain thanks to the new pulsed CRESU apparatus developed there [14] coupled to a moderate pumping group (see Chapter 2). This opens the possibility to routinely extend the present available kinetic data to temperatures pertinent to dense interstellar regions. Combining a pulsed uniform flow with an even larger pumping system will make the generation of uniform flows at temperatures below 10 K possible without the necessity to pre-cool the reservoir. Alternatively, increasing the gain of a pulsed CRESU apparatus (the ratio of the buffer gas mass flow consumed in the continuous regime compared with the pulsed one) for a given pumping capacity should lead to the same result.

A quick look at Table 5.3 also confirms that our understanding of the reactivity between unstable radicals at low temperature is rather poor, with only six measurements of this type among the hundreds of other neutral–neutral reactions to have been investigated by the CRESU method. Although theoretical studies have made a significant contribution in this area, it is important to at least validate the calculated rate coefficients with selected CRESU measurements of important radical–radical reactions, such as those of atomic hydrogen and oxygen with other small carbonaceous radicals.

A vast majority of reactions to have been studied by the CRESU method so far have involved radicals which can be easily detected by fluorescence. Indeed, reactions involving non-fluorescing or weakly fluorescing radical species such as CH_3, $^3CH_2$, or larger hydrocarbon radicals that are expected to be critical to the chemistry of certain planetary atmospheres (Titan in particular) have largely been ignored. Consequently, the use of CRESU systems employing mass spectrometric detection, rotational spectroscopy, or cavity ring-down spectroscopy (CRDS) (as recently demonstrated by Suas-David *et al.* [105] in the infrared range) would be of great benefit for kinetic studies of such reactions at low temperature.

Finally, these three types of apparatuses also fulfill many of the requirements for performing quantitative measurements of product formation in neutral–neutral reactions (particularly for those reactions with multiple product channels), with an accurate knowledge of the low-temperature branching ratios for multichannel processes sorely lacking in current photochemical and astrochemical models. Hopefully, future measurements of neutral–neutral reactions with the CRESU technique can address some or all of these issues, allowing this method to continue its important contribution to our understanding of low-temperature environments.

References

[1] Crutzen PJ. Ozone production rates in an oxygen-hydrogen-nitrogen oxide atmosphere. J Geophys Res. 1971;76:7311–7327.
[2] Atkinson R, Baulch DL, Cox RA, Hampson JRF, Kerr JA, Rossi MJ, Troe J. Evaluated kinetic and photochemical data for atmospheric chemistry II, organic species. Atmos Chem Phys. 2006;6:3625–4055.
[3] Sander SP, Abbatt J, Barker JR, Burkholder JB, Friedl RR, Golden DM, Huie RE, Kolb CE, Kurylo MJ, Moortgat GK, Orkin VL, Wine PH. Chemical kinetics and photochemical data for use in atmospheric studies, evaluation no. 17. Pasadena: JPL Publication 10-6, Jet Propulsion Laboratory; 2011. Available from: http://jpldataeval.jpl.nasa.gov

[4] Trainor DW, Ham DO, Kaufman F. Gas-phase recombination of hydrogen and deuterium atoms. J Chem Phys. 1973;58:4599–4609.

[5] Frost MJ, Sharkey P, Smith IWM. Reaction between OH (OD) radicals and CO at temperatures down to 80 K: experiment and theory. J Phys Chem. 1993;97:12254–12259.

[6] Rawlins WT, Caledonia GE, Armstrong RA. Dynamics of vibrationally excited ozone formed by 3-body recombination. 2. Kinetics and mechanism. J Chem Phys. 1987;87:5209–5221.

[7] Sims IR, Smith IWM. Rate constants for the radical-radical reaction between CN and O_2 at temperatures down to 99 K. Chem Phys Lett. 1988;151:481–484.

[8] Sims IR, Smith IWM. Pressure and temperature-dependence of the rate of reaction between CN radicals and NO over the range $99 \leq T/K \leq 450$. J Chem Soc Faraday Trans. 1993;89:1–5.

[9] Rowe BR, Dupeyrat G, Marquette JB, Gaucherel P. Study of the reactions $N_2^+ + 2N_2 \rightarrow N_4^+ + N_2$ and $O_2^+ + 2O_2 \rightarrow O_4^+ + O_2$ from 20–160 K by the CRESU technique. J Chem Phys. 1984;80:4915–4921.

[10] Sims IR, Queffelec JL, Defrance A, Rebrion-Rowe C, Travers D, Rowe BR, Smith IWM. Ultra-low temperature kinetics of neutral-neutral reactions — the reaction $CN + O_2$ down to 26 K. J Chem Phys. 1992;97:8798–8800.

[11] Cooke IR, Sims IR. Experimental studies of gas-phase reactivity in relation to complex organic molecules in star-forming regions. ACS Earth Space Chem. 2019;3:1109–1134.

[12] Atkinson DB, Smith MA. Design and characterization of pulsed uniform supersonic expansions for chemical applications. Rev Sci Instrum. 1995;66:4434–4446.

[13] Taylor SE, Goddard A, Blitz MA, Cleary PA, Heard DE. Pulsed Laval nozzle study of the kinetics of OH with unsaturated hydrocarbons at very low temperatures. Phys Chem Chem Phys. 2008;10:422–437.

[14] Jiménez E, Ballesteros B, Canosa A, Townsend TM, Maigler FJ, Napal V, Rowe BR, Albaladejo J. Development of a pulsed uniform supersonic gas expansion system based on an aerodynamic chopper for gas phase reaction kinetic studies at ultra-low temperatures. Rev Sci Instrum. 2015;86:045108.

[15] Spangenberg T, Kohler S, Hansmann B, Wachsmuth U, Abel B, Smith MA. Low-temperature reactions of OH radicals with propene and isoprene in pulsed Laval nozzle expansions. J Phys Chem A. 2004;108:7527–7534.

[16] Oldham JM, Abeysekera C, Joalland B, Zack LN, Prozument K, Sims IR, Park GB, Field RW, Suits AG. A chirped-pulse Fourier-transform microwave/pulsed uniform flow spectrometer. I. The low-temperature flow system. J Chem Phys. 2014;141:154202.

[17] Lee S, Hoobler RJ, Leone SR. A pulsed Laval nozzle apparatus with laser ionization mass spectroscopy for direct measurements of rate coefficients at low temperatures with condensable gases. Rev Sci Instrum. 2000;71:1816–1823.

[18] Cheikh Sid Ely S, Morales SB, Guillemin J-C, Klippenstein SJ, Sims IR. Low temperature rate coefficients for the reaction $CN + HC_3N$. J Phys Chem A. 2013;117:12155–12164.

[19] Daugey N, Caubet P, Retail B, Costes M, Bergeat A, Dorthe G. Kinetic measurements on methylidyne radical reactions with several hydrocarbons at low temperatures. Phys Chem Chem Phys. 2005;7:2921–2927.

[20] Tizniti M, Le Picard SD, Lique F, Berteloite C, Canosa A, Alexander MH, Sims IR. The rate of the $F + H_2$ reaction at very low temperatures. Nat Chem. 2014;6:141–145.

[21] Hickson KM, Loison J-C, Guo H, Suleimanov YV. Ring-polymer molecular dynamics for the prediction of low-temperature rates: an investigation of the $C(^1D) + H_2$ reaction. J Phys Chem Lett. 2015;6:4194–4199.

[22] Douglas KM, Blitz MA, Feng W, Heard DE, Plane JMC, Rashid H, Seakins PW. Low temperature studies of the rate coefficients and branching ratios of reactive loss vs quenching for the reactions of $^1CH_2$ with C_2H_6, C_2H_4, C_2H_2. Icarus. 2019;321:752–766.

[23] Shannon RJ, Blitz MA, Goddard A, Heard DE. Accelerated chemistry in the reaction between the hydroxyl radical and methanol at interstellar temperatures facilitated by tunnelling. Nat Chem. 2013;5:745–749.

[24] Le Picard SD, Tizniti M, Canosa A, Sims IR, Smith IWM. The thermodynamics of the elusive HO_3 radical. Science. 2010;328:1258–1262.

[25] Le Picard SD, Canosa A, Pineau des Forêts G, Rebrion-Rowe C, Rowe BR. The $Si(^3P_J)+O_2$ reaction: a fast source of SiO at very low temperature; CRESU measurements and interstellar consequences. Astron Astrophys. 2001;372:1064–1070.

[26] Le Picard SD, Canosa A, Geppert W, Stoecklin T. Experimental and theoretical temperature dependence of the rate coefficient of the $B(^2P_{1/2,3/2})+O_2(X^3\Sigma_g^-)$ reaction in the 24–295 K temperature range. Chem Phys Lett. 2004;385:502–506.

[27] Le Picard SD, Canosa A, Reignier D, Stoecklin T. A comparative study of the reactivity of the silicon atom $Si(^3P_J)$ towards O_2 and NO molecules at very low temperature. Phys Chem Chem Phys. 2002;4:3659–3664.

[28] Canosa A, Le Picard SD, Geppert WD. Experimental kinetics study of the reaction of boron atoms, $B(^2P_J)$, with ethylene at very low temperatures (23–295 K). J Phys Chem A. 2004;108:6183–6185.

[29] Canosa A, Le Picard SD, Gougeon S, Rebrion-Rowe C, Travers D, Rowe BR. Rate coefficients for the reactions of $Si(^3P_J)$ with C_2H_2 and C_2H_4: experimental results down to 15 K. J Chem Phys. 2001;115:6495–6503.

[30] Geppert WD, Goulay F, Naulin C, Costes M, Canosa A, Le Picard SD, Rowe BR. Rate coefficients and integral cross-sections for the reaction of $B(^2P_J)$ atoms with acetylene. Phys Chem Chem Phys. 2004;6:566.

[31] Nuñez-Reyes D, Hickson KM. A low temperature investigation of the gas-phase $N(^2D)$ + NO reaction. Towards a viable source of $N(^2D)$ atoms for kinetic studies in astrochemistry. Phys Chem Chem Phys. 2018;20:17442–17447.

[32] Bergeat A, Hickson KM, Daugey N, Caubet P, Costes M. A low temperature investigation of the $N(^4S)$ + NO reaction. Phys Chem Chem Phys. 2009;11:8149–8155.

[33] Hickson KM, Bergeat A, Costes M. A low temperature study of the reactions of atomic chlorine with simple alkanes. J Phys Chem A. 2010;114:3038–3044.

[34] Tango WJ, Link JK, Zare RN. Spectroscopy of K_2 using laser-induced fluorescence. J Chem Phys. 1968;49:4264–4268.

[35] Sims IR, Queffelec JL, Defrance A, Rebrion-Rowe C, Travers D, Rowe BR, Smith IWM. Ultra-low temperature kinetics of neutral–neutral reactions: the reaction $CN+O_2$ down to 26 K. J Chem Phys. 1992;97:8798–8800.

[36] Daugey N, Caubet P, Bergeat A, Costes M, Hickson KM. Reaction kinetics to low temperatures. Dicarbon + acetylene, methylacetylene, allene and propene from $77 \leq T \leq 296$ K. Phys Chem Chem Phys. 2008;10:729–737.

[37] Shannon RJ, Cossou C, Loison J-C, Caubet P, Balucani N, Seakins PW, Wakelam V, Hickson KM. The fast $C(^3P)$ + CH_3OH reaction as an efficient loss process for gas-phase interstellar methanol. RSC Adv. 2014;4:26342.

[38] Hickson KM, Loison J-C, Bourgalais J, Capron M, Le Picard SD, Goulay F, Wakelam V. The $C(^3P)$ + NH_3 reaction in interstellar chemistry. II. Low temperature rate constants and modeling of NH, NH_2 and NH_3 abundances in dense interstellar clouds. Astrophys J. 2015;812:107.

[39] Hilbig R, Wallenstein R. Enhanced production of tunable vuv radiation by phase-matched frequency tripling in krypton and xenon. IEEE J Quantum Electron. 1981;17:1566–1573.

[40] Chastaing D, Le Picard SD, Sims IR, Smith IWM. Rate coefficients for the reactions of $C(^3P_J)$ atoms with C_2H_2, C_2H_4, $CH_3C{\equiv}CH$ and $H_2C{=}C{=}CH_2$ at temperatures down to 15 K. Astron Astrophys. 2001;365:241–247.

[41] Nakayama T, Takahashi K, Matsumi Y, Shibuya K. $N(^4S)$ formation following the 193.3-nm ArF laser irradiation of NO and NO_2 and its application to kinetic studies of $N(^4S)$ reactions with NO and NO_2. J Phys Chem A. 2005;109:10897–10902.

[42] Takahashi K, Hayashi S, Matsumi Y, Taniguchi N, Hayashida S. Quantum yields of $O(^1D)$ formation in the photolysis of ozone between 230 and 308 nm. J Geophys Res Atmos. 2002;107:8.

[43] Hickson KM, Keyser LF, Sander SP. Temperature dependence of the $HO_2{+}ClO$ reaction. 2. Reaction kinetics using the discharge-flow resonance-fluorescence technique. J Phys Chem A. 2007;111:8126–8138.

[44] Chastaing D, James PL, Sims IR, Smith IWM. Neutral–neutral reactions at the temperatures of interstellar clouds rate coefficients for reactions of C_2H radicals with O_2, C_2H_2, C_2H_4 and C_3H_6 down to 15 K. Faraday Discuss. 1998;109:165–181.

[45] Vakhtin AB, Heard DE, Smith IWM, Leone SR. Kinetics of reactions of C_2H radical with acetylene, O_2, methylacetylene, and allene in a pulsed Laval nozzle apparatus at T = 103 K. Chem Phys Lett. 2001;344:317–324.

[46] Hickson KM, Caubet P, Loison J-C. Unusual low-temperature reactivity of water: the $CH + H_2O$ reaction as a source of interstellar formaldehyde? J Phys Chem Lett. 2013;4:2843–2846.

[47] Sabbah H, Biennier L, Sims IR, Georgievskii Y, Klippenstein SJ, Smith IWM. Understanding reactivity at very low temperatures: the reactions of oxygen atoms with alkenes. Science. 2007;317:102–105.

[48] Lee S, Samuels DA, Hoobler RJ, Leone SR. Direct measurements of rate coefficients for the reaction of ethynyl radical (C_2H) with C_2H_2 at 90 and 120 K using a pulsed Laval nozzle apparatus. J Geophys Res Planets. 2000;105:15085–15090.

[49] Lee S, Leone SR. Rate coefficients for the reaction of C_2H with O_2 at 90 K and 120 K using a pulsed Laval nozzle apparatus. Chem Phys Lett. 2000;329:443–449.

[50] Soorkia S, Liu CL, Savee JD, Ferrell SJ, Leone SR, Wilson KR. Airfoil sampling of a pulsed Laval beam with tunable vacuum ultraviolet synchrotron ionization quadrupole mass spectrometry: application to low-temperature kinetics and product detection. Rev Sci Instrum. 2011;82:124102.

[51] Bouwman J, Fournier M, Sims IR, Leone SR, Wilson KR. Reaction rate and isomer-specific product branching ratios of $C_2H + C_4H_8$: 1-butene, cis-2-butene, trans-2-butene, and isobutene at 79 K. J Phys Chem A. 2013;117:5093–5105.

[52] Bouwman J, Goulay F, Leone SR, Wilson KR. Bimolecular rate constant and product branching ratio measurements for the reaction of C_2H with ethene and propene at 79 K. J Phys Chem A. 2012;116:3907–3917.

[53] Dian BC, Brown GG, Douglass KO, Pate BH. Measuring picosecond isomerization kinetics via broadband microwave spectroscopy. Science. 2008;320:924–928.

[54] Abeysekera C, Zack LN, Park GB, Joalland B, Oldham JM, Prozument K, Ariyasingha nm, Sims IR, Field RW, Suits AG. A chirped-pulse Fourier-transform microwave/pulsed uniform flow spectrometer. II. Performance and applications for reaction dynamics. J Chem Phys. 2014;141:214203.

[55] Nguyen TL, Vereecken L, Hou XJ, Nguyen MT, Peeters J. Potential energy surfaces, product distributions and thermal rate coefficients of the reaction of $O(^3P)$ with $C_2H_4(X^1A_g)$: a comprehensive theoretical study. J Phys Chem A. 2005;109:7489–7499.

[56] Rangel C, Espinosa-Garcia J. Full-dimensional analytical potential energy surface describing the gas-phase $Cl + C_2H_6$ reaction and kinetics study of rate constants and kinetic isotope effects. Phys Chem Chem Phys. 2018;20:3925–3938.

[57] Bryukov MG, Slagle IR, Knyazev VD. Kinetics of reactions of Cl atoms with ethane, chloroethane, and 1,1-dichloroethane. J Phys Chem A. 2003;107:6565–6573.

[58] Lewis RS, Sander SP, Wagner S, Watson RT. Temperature-dependent rate constants for the reaction of ground-state chlorine with simple alkanes. J Phys Chem. 1980;84:2009–2015.

[59] Pilgrim JS, McIlroy A, Taatjes CA. Kinetics of Cl atom reactions with methane, ethane, and propane from 292 to 800 K. J Phys Chem A. 1997;101:1873–1880.

[60] Hickson KM, Keyser LF. Kinetics of the $Cl(^2P_j) + C_2H_6$ reaction between 177 and 353 K. J Phys Chem A. 2004;108:1150–1159.

[61] Manning RG, Kurylo MJ. Flash photolysis resonance fluorescence investigation of the temperature dependencies of the reactions of chlorine (^2P) atoms with methane, chloromethane, fluoromethane, excited fluoromethane, and ethane. J Phys Chem. 1977;81:291–296.

[62] Dobis O, Benson SW. Temperature coefficients of the rates of chlorine atom reactions with C_2H_6, C_2H_5, and C_2H_4. The rates of disproportionation and recombination of ethyl radicals. J Am Chem Soc. 1991;113:6377–6386.

[63] Fernández-Ramos A, Martínez-Núñez E, Marques JMC, Vázquez SA. Dynamics calculations for the Cl+C$_2$H$_6$ abstraction reaction: thermal rate constants and kinetic isotope effects. J Chem Phys. 2003;118:6280–6288.

[64] Potapov A, Canosa A, Jiménez E, Rowe B. Uniform supersonic chemical reactors: 30 years of astrochemical history and future challenges. Angew Chem-Int Edit. 2017;56:8618–8640.

[65] Shannon RJ, Taylor S, Goddard A, Blitz MA, Heard DE. Observation of a large negative temperature dependence for rate coefficients of reactions of OH with oxygenated volatile organic compounds studied at 86–112 K. Phys Chem Chem Phys. 2010;12:13511–13514.

[66] Heard DE. Rapid acceleration of hydrogen atom abstraction reactions of OH at very low temperatures through weakly bound complexes and tunneling. Acc Chem Res. 2018;51:2620–2627.

[67] Ocaña AJ, Blázquez S, Potapov A, Ballesteros B, Canosa A, Antiñolo M, Vereecken L, Albaladejo J, Jiménez E. Gas-phase reactivity of CH$_3$OH toward OH at interstellar temperatures (11.7–177.5 K): experimental and theoretical study. Phys Chem Chem Phys. 2019;21:6942–6957.

[68] Shannon RJ, Caravan RL, Blitz MA, Heard DE. A combined experimental and theoretical study of reactions between the hydroxyl radical and oxygenated hydrocarbons relevant to astrochemical environments. Phys Chem Chem Phys. 2014;16:3466–3478.

[69] Blázquez S, González D, García-Sáez A, Antiñolo M, Bergeat A, Caralp F, Mereau R, Canosa A, Ballesteros B, Albaladejo J, Jiménez E. Experimental and theoretical investigation on the OH + CH$_3$C(O)CH$_3$ reaction at interstellar temperatures (T = 11.7–64.4 K). ACS Earth Space Chem. 2019;3:1873–1883.

[70] Canosa A, Ocaña AJ, Antiñolo M, Ballesteros B, Jiménez E, Albaladejo J. Design and testing of temperature tunable de Laval nozzles for applications in gas-phase reaction kinetics. Exp Fluids. 2016;57:152.

[71] Roncero O, Zanchet A, Aguado A. Low temperature reaction dynamics for CH$_3$OH + OH collisions on a new full dimensional potential energy surface. Phys Chem Chem Phys. 2018;20:25951–25958.

[72] del Mazo-Sevillano P, Aguado A, Jiménez E, Suleimanov YV, Roncero O. Quantum roaming in the complex-forming mechanism of the reactions of OH with formaldehyde and methanol at low temperature and zero pressure: a ring polymer molecular dynamics approach. J Phys Chem Lett. 2019;10:1900–1907.

[73] Nguyen TL, Ruscic B, Stanton JF. A master equation simulation for the •OH + CH$_3$OH reaction. J Chem Phys. 2019;150:084105.

[74] Gao LG, Zheng J, Fernández-Ramos A, Truhlar DG, Xu X. Kinetics of the methanol reaction with OH at interstellar, atmospheric, and combustion temperatures. J Am Chem Soc. 2018;140:2906–2918.

[75] Naumkin F, del Mazo-Sevillano P, Aguado A, Suleimanov YV, Roncero O. Zero- and high-pressure mechanisms in the complex forming reactions of OH with methanol and formaldehyde at low temperatures. ACS Earth Space Chem. 2019;3:1158–1169.

[76] Acharyya K, Herbst E, Caravan RL, Shannon RJ, Blitz MA, Heard DE. The importance of OH radical-neutral low temperature tunnelling reactions in interstellar clouds using a new model. Mol Phys. 2015;113:1–12.

[77] Antiñolo M, Agúndez M, Jiménez E, Ballesteros B, Canosa A, El Dib G, Albaladejo J, Cernicharo J. Reactivity of OH and CH$_3$OH between 22 and 64 K: modeling the gas phase production of CH$_3$O in Barnard 1b. Astrophys J. 2016;823:25.

[78] Brownsword RA, Canosa A, Rowe BR, Sims IR, Smith IWM, Stewart DWA, Symonds AC, Travers D. Kinetics over a wide range of temperature (13–744 K): rate constants for the reactions of CH($\nu = 0$) with H$_2$ and D$_2$ and for the removal of CH($\nu = 1$) by H$_2$ and D$_2$. J Chem Phys. 1997;106:7662–7677.

[79] Fournier M. Reactivity of C$_3$N and C$_2$H at low temperature: applications for the interstellar medium and Titan [Ph.D Thesis]. Université de Rennes 1; 2014.

[80] Jaramillo VI, Smith MA. Temperature-dependent kinetic isotope effects in the gas-phase reaction: OH+HBr. J Phys Chem A. 2001;105:5854–5859.

[81] Jaramillo VI, Gougeon S, Le Picard SD, Canosa A, Smith MA, Rowe BR. A consensus view of the temperature dependence of the gas phase reaction: OH+HBr → H$_2$O+Br. Int J Chem Kinet. 2002;34:339–344.

[82] Sims IR, Smith IWM, Clary DC, Bocherel P, Rowe BR. Ultra-low temperature kinetics of neutral-neutral reactions — new experimental and theoretical results for OH+HBr between 295 K and 23 K. J Chem Phys. 1994;101:1748–1751.

[83] Atkinson DB, Jaramillo VI, Smith MA. Low-temperature kinetic behavior of the bimolecular reaction OH + HBr (76–242 K). J Phys Chem A. 1997;101:3356–3359.

[84] Hickson KM, Loison J-C, Nuñez-Reyes D, Méreau R. Quantum tunneling enhancement of the C + H$_2$O and C + D$_2$O reactions at low temperature. J Phys Chem Lett. 2016;7:3641–3646.

[85] Vakhtin AB, McCabe DC, Ravishankara AR, Leone SR. Low-temperature kinetics of the reaction of the OH radical with hydrogen peroxide. J Phys Chem A. 2003;107:10642–10647.

[86] Canosa A, Sims IR, Travers D, Smith IWM, Rowe BR. Reactions of the methylidine radical with CH_4, C_2H_2, C_2H_4, C_2H_6, and but-1-ene studied between 23 and 295 K with a CRESU apparatus. Astron Astrophys. 1997;323:644.

[87] Mullen C, Smith MA. Low temperature $NH(X^3\Sigma^-)$ radical reactions with NO, saturated, and unsaturated hydrocarbons studied in a pulsed supersonic Laval nozzle flow reactor between 53 and 188 K. J Phys Chem A. 2005;109:1391–1399.

[88] Canosa A, Páramo A, Le Picard SD, Sims IR. An experimental study of the reaction kinetics of $C_2(X^1\Sigma_g^+)$ with hydrocarbons (CH_4, C_2H_2, C_2H_4, C_2H_6 and C_3H_8) over the temperature range 24–300 K: implications for the atmospheres of Titan and the giant planets. Icarus. 2007;187:558–568.

[89] Berteloite C, Le Picard SD, Birza P, Gazeau M-C, Canosa A, Benilan Y, Sims IR. Low temperature (39–298 K) kinetics study of the reactions of the C_4H radical with various hydrocarbons observed in Titan's atmosphere. Icarus. 2008;194:746–755.

[90] Berteloite C, Le Picard SD, Balucani N, Canosa A, Sims IR. Low temperature rate coefficients for reactions of the butadiynyl radical, C_4H, with various hydrocarbons. Part I: reactions with alkanes (CH_4, C_2H_6, C_3H_8, C_4H_{10}). Phys Chem Chem Phys. 2010;12:3666–3676.

[91] Sims IR, Queffelec J-L, Travers D, Rowe BR, Herbert LB, Karthäuser J, Smith IWM. Rate constants for the reactions of CN with hydrocarbons at low and ultra-low temperatures. Chem Phys Lett. 1993;211:461–468.

[92] Berteloite C, Le Picard SD, Balucani N, Canosa A, Sims IR. Low temperature rate coefficients for reactions of the butadiynyl radical, C_4H, with various hydrocarbons. Part II: reactions with alkenes (ethylene, propene, 1-butene), dienes (allene, 1,3-butadiene) and alkynes (acetylene, propyne and 1-butyne). Phys Chem Chem Phys. 2010;12:3677–3689.

[93] Chastaing D, James PL, Sims IR, Smith IWM. Neutral-neutral reactions at the temperatures of interstellar clouds: rate coefficients for reactions of atomic carbon, $C(^3P)$, with O_2, C_2H_2, C_2H_4 and C_3H_6 down to 15 K. Phys Chem Chem Phys. 1999;1:2247–2256.

[94] Hickson KM, Loison J-C, Wakelam V. Temperature dependent product yields for the spin forbidden singlet channel of the $C(^3P)$ + C_2H_2 reaction. Chem Phys Lett. 2016;659:70–75.

[95] Vakhtin AB, Murphy JE, Leone SR. Low-temperature kinetics of reactions of OH radical with ethene, propene, and 1-butene. J Phys Chem A. 2003;107:10055–10062.

[96] Vakhtin AB, Heard DE, Smith IWM, Leone SR. Kinetics of C_2H radical reactions with ethene, propene and 1-butene measured in a pulsed Laval nozzle apparatus at T = 103 and 296 K. Chem Phys Lett. 2001;348:21–26.

[97] Murphy JE, Vakhtin AB, Leone SR. Laboratory kinetics of C_2H radical reactions with ethane, propane, and n-butane at T = 96 − 296 K: implications for Titan. Icarus. 2003;163:175–181.

[98] Carty D, Le Page V, Sims IR, Smith IWM. Low temperature rate coefficients for the reactions of CN and C_2H radicals with allene (CH_2 =C=CH_2) and methyl acetylene ($CH_3C\equiv CH$). Chem Phys Lett. 2001;344:310–316.

[99] Abeysekera C, Joalland B, Ariyasingha N, Zack LN, Sims IR, Field RW, Suits AG. Product branching in the low temperature reaction of CN with propyne by chirped-pulse microwave spectroscopy in a uniform supersonic flow. J Phys Chem Lett. 2015;6:1599–1604.

[100] Chastaing D, Le Picard SD, Sims IR, Smith IWM, Geppert WD, Naulin C, Costes M. Rate coefficients and cross-sections for the reactions of $C(^3P_j)$ atoms with methylacetylene and allene. Chem Phys Lett. 2000;331:170–176.

[101] Soorkia S, Leone SR, Wilson KR. Radical-neutral chemical reactions studied at low temperature with VUV synchrotron photoionization mass spectrometry. In: 28th International Symposium on Rarefied Gas Dynamics 2012; 2012. vols 1 and 2. p. 1365–1372.

[102] Daranlot J, Bergeat A, Caralp F, Caubet P, Costes M, Forst W, Loison J-C, Hickson KM. Gas-phase kinetics of hydroxyl radical reactions with alkenes: experiment and theory. Chemphyschem. 2010;11:4002–4010.

[103] Vakhtin AB, Lee S, Heard DE, Smith IWM, Leone SR. Low-temperature kinetics of reactions of the OH radical with propene and 1-butene studied by a pulsed Laval nozzle apparatus combined with laser-induced fluorescence. J Phys Chem A. 2001;105:7889–7895.

[104] Morales SB, Le Picard SD, Canosa A, Sims IR. Experimental measurements of low temperature rate coefficients for neutral-neutral reactions of interest for atmospheric chemistry of Titan, Pluto and Triton: reactions of the CN radical. Faraday Discuss. 2010;147: 155–171.

[105] Suas-David N, Thawoos S, Suits AG. A uniform flow-cavity ring-down spectrometer (UF-CRDS): a new setup for spectroscopy and kinetics at low temperature. J Chem Phys. 2019;151:244202.

[106] Soorkia S, Trevitt AJ, Selby TM, Osborn DL, Taatjes CA, Wilson KR, Leone SR. Reaction of the C_2H radical with 1-butyne (C_4H_6): low-temperature kinetics and isomer-specific product detection. J Phys Chem A. 2010;114:3340–3354.

[107] Nizamov B, Leone SR. Kinetics of C_2H reactions with hydrocarbons and nitriles in the 104–296 K temperature range. J Phys Chem A. 2004;108:1746–1752.

[108] Morales SB, Bennett CJ, Le Picard SD, Canosa A, Sims IR, Sun BJ, Chen PH, Chang AHH, Kislov VV, Mebel AM, Gu X, Zhang F, Maksyutenko P, Kaiser RI. A crossed molecular beam, low-temperature kinetics, and theoretical investigation of the reaction of the cyano radical (CN) with 1,3-butadiene (C_4H_6). A route to complex nitrogen-bearing molecules in low-temperature extraterrestrial environments. Astrophys J. 2011;742:26–35.

[109] Sims IR, Smith IWM, Bocherel P, Defrance A, Travers D, Rowe BR. Ultra-low temperature kinetics of neutral-neutral reactions — rate constants for the reactions of OH radicals with butenes between 295 K and 23 K. J Chem Soc Faraday Trans. 1994;90:1473–1478.

[110] Goulay F, Leone SR. Low-temperature rate coefficients for the reaction of ethynyl radical (C_2H) with benzene. J Phys Chem A. 2006;110:1875–1880.

[111] Trevitt AJ, Goulay F, Taatjes CA, Osborn DL, Leone SR. Reactions of the CN radical with benzene and toluene: Product detection and low-temperature kinetics. J Phys Chem A. 2009;114:1749–1755.

[112] Cooke IR, Gupta D, Messinger JP, Sims IR. Benzonitrile as a proxy for benzene in the cold ISM: low-temperature rate coefficients for $CN + C_6H_6$. Astrophys J Lett. 2020;891:L41.

[113] Bennett CJ, Morales SB, Le Picard SD, Canosa A, Sims IR, Shih YH, Chang AHH, Gu X, Zhang F, Kaiser RI. A chemical dynamics, kinetics, and theoretical study on the reaction of the cyano radical CN ($X^2\Sigma^+$) with phenylacetylene (C_6H_5CCH; X^1A_1). Phys Chem Chem Phys. 2010;12:8737–8749.

[114] Goulay F, Rebrion-Rowe C, Biennier L, Le Picard SD, Canosa A, Rowe BR. Reaction of anthracene with CH radicals: an experimental study of the kinetics between 58 and 470 K. J Phys Chem A. 2006;110:3132–3137.

[115] West NA, Millar TJ, Van de Sande M, Rutter E, Blitz MA, Decin L, Heard DE. Measurements of low temperature rate coefficients for the reaction of CH with CH_2O and application to dark cloud and AGB stellar wind models. Astrophys J. 2019;885:134.

[116] Ocaña AJ, Jiménez E, Ballesteros B, Canosa A, Antiñolo M, Albaladejo J, Agúndez M, Cernicharo J, Zanchet A, del Mazo P, Roncero O, Aguado A. Is the gas-phase $OH+H_2CO$ reaction a source of HCO in interstellar cold dark clouds? A kinetic, dynamic, and modeling study. Astrophys J. 2017;850:12.

[117] Gómez-Martin JC, Caravan RL, Blitz MA, Heard DE, Plane JMC. Low temperature kinetics of the CH_3OH + OH reaction. J Phys Chem A. 2014;118:2693–2701.

[118] Gupta D, Cheikh Sid Ely S, Cooke IR, Guillaume T, Abdelkader Khedaoui O, Hearne TS, Hays BM, Sims IR. Low temperature kinetics of the reaction between methanol and the CN radical. J Phys Chem A. 2019;123:9995–10003.

[119] Ocaña AJ, Blázquez S, Ballesteros B, Canosa A, Antiñolo M, Albaladejo J, Jiménez E. Gas phase kinetics of the OH + CH_3CH_2OH reaction at temperatures of the interstellar medium (T = 21 − 107 K). Phys Chem Chem Phys. 2018;20:5865–5873.

[120] Caravan RL, Shannon RJ, Lewis T, Blitz MA, Heard DE. Measurements of rate coefficients for reactions of OH with ethanol and propan-2-ol at very low temperatures. J Phys Chem A. 2015;119:7130–7137.

[121] Jiménez E, Antiñolo M, Ballesteros B, Canosa A, Albaladejo J. First evidence of the dramatic enhancement of the reactivity of methyl formate ($HC(O)OCH_3$) with OH at temperatures of the interstellar medium: a gas-phase kinetic study between 22 K and 64 K. Phys Chem Chem Phys. 2016;18:2183–2191.

[122] Vöhringer-Martinez E, Hansmann B, Hernandez-Soto H, Francisco JS, Troe J, Abel B. Water catalysis of a radical-molecule gas-phase reaction. Science. 2007;315:497–501.

[123] Blázquez S, González D, Neeman EM, Ballesteros B, Agúndez M, Canosa A, Albaladejo J, Cernicharo J, Jiménez E. Gas-phase kinetics of CH_3CHO with OH radicals between 11.7 and 177.5 K. Phys Chem Chem Phys. 2020;22:20562–20572.

[124] Vöhringer-Martinez E, Tellbach E, Liessmann M, Abel B. Role of water complexes in the reaction of propionaldehyde with OH radicals. J Phys Chem A. 2010;114:9720–9724.

[125] Hansmann B, Abel B. Kinetics in cold Laval nozzle expansions: from atmospheric chemistry to oxidation of biomolecules in the gas phase. Chemphyschem. 2007;8:343–356.

[126] Liessmann M, Hansmann B, Blachly PG, Francisco JS, Abel B. Primary steps in the reaction of OH radicals with amino acids at low temperatures in Laval nozzle expansions: perspectives from experiment and theory. J Phys Chem A. 2009;113:7570–7575.

[127] Bocherel P, Herbert LB, Rowe BR, Sims IR, Smith IWM, Travers D. Ultralow-temperature kinetics of $CH(X^2\Pi)$ reactions: rate coefficients for reactions with O_2 and NO (T = 13 − 708 K), and with NH_3 (T = 23 − 295 K). J Phys Chem. 1996;100:3063–3069.

[128] Nizamov B, Leone SR. Rate coefficients and kinetic isotope effect for the C_2H reactions with NH_3 and ND_3 in the 104–294 K temperature range. J Phys Chem A. 2004;108:3766–3771.

[129] Sims IR, Queffelec JL, Defrance A, Rebrion-Rowe C, Travers D, Bocherel P, Rowe BR, Smith IWM. Ultralow temperature kinetics of neutral-neutral reactions. The technique and results for the reactions $CN + O_2$ down to 13 K and $CN + NH_3$ down to 25 K. J Chem Phys. 1994;100:4229–4241.

[130] Bourgalais J, Capron M, Kailasanathan RKA, Osborn DL, Hickson KM, Loison J-C, Wakelam V, Goulay F, Le Picard SD. The $C(^3P) + NH_3$ reaction in interstellar chemistry. I. Investigation of the product formation channels. Astrophys J. 2015;812:106.

[131] Sleiman C, El Dib G, Talbi D, Canosa A. Gas phase reactivity of the CN radical with methyl amines at low temperatures (23–297 K): a combined experimental and theoretical investigation. ACS Earth Space Chem. 2018;2:1047–1057.

[132] Sleiman C, González S, Klippenstein SJ, Talbi D, El Dib G, Canosa A. Pressure dependent low temperature kinetics for $CN + CH_3CN$: competition between chemical reaction and van der Waals complex formation. Phys Chem Chem Phys. 2016;18:15118–15132.

[133] Hess WP, Tully FP. Hydrogen-atom abstraction from methanol by hydroxyl radical. J Phys Chem. 1989;93:1944–1947.

[134] Dillon TJ, Hölscher D, Sivakumaran V, Horowitz A, Crowley JN. Kinetics of the reactions of HO with methanol (210–351 K) and with ethanol (216–368 K). Phys Chem Chem Phys. 2005;7:349–355.

[135] Douglas K, Blitz MA, Feng WH, Heard DE, Plane JMC, Slater E, Willacy K, Seakins PW. Low temperature studies of the removal reactions of $^1CH_2$ with particular relevance to the atmosphere of Titan. Icarus. 2018;303:10–21.

[136] Hébrard E, Dobrijevic M, Loison J-C, Bergeat A, Hickson KM, Caralp F. Photochemistry of C_3H_p hydrocarbons in Titan's stratosphere revisited. Astron Astrophys. 2013;552:A132.

[137] Kraus H, Oehlers C, Temps F, Wagner HG. Kinetics of the reactions of $CH_2(X^3B_1)$ with selected polycyclic aromatic-hydrocarbons at temperatures between 296 K and 690 K. J Phys Chem. 1993;97:10989–10995.

[138] Páramo A, Canosa A, Le Picard SD, Sims IR. Rate coefficients for the reactions of $C_2(a^3\Pi_u)$ and $C_2(X^1\Sigma_g^+)$ with various hydrocarbons (CH_4, C_2H_2, C_2H_4, C_2H_6 and C_3H_8): a gas-phase experimental study over the temperature range 24–300 K. J Phys Chem A. 2008;112:9591–9600.

[139] Anderson JK, Ziurys LM. Detection of CCN($X^2\Pi_r$) in IRC+10216: constraining carbon-chain chemistry. Astrophys J Lett. 2014;795:6.

[140] Zhu Z, Zhang Z, Huang C, Pei L, Chen C, Chen Y. Kinetics of CCN radical reactions with a series of normal alkanes. J Phys Chem A. 2003;107:10288–10291.

[141] Zhu Z, Ji S, Zhang SW, Chen YT. Temperature-dependent kinetics of the reactions of the CCN ($X^2\Pi$) radical with C_2H_2 and C_2H_4. The 235th ACS National Meeting. New Orleans, Louisiana; 2008 Apr. Available from: http://oasys2.confex.com/acs/235nm/techprogram/P1157562.HTM

[142] Le Picard SD, Canosa A, Travers D, Chastaing D, Rowe BR, Stoecklin T. Experimental and theoretical kinetics for the reaction of Al with O_2 at temperatures between 23 and 295 K. J Phys Chem A. 1997;101:9988–9992.

[143] Bergner JB, Öberg KI, Rajappan M. Methanol formation via oxygen insertion chemistry in ices. Astrophys J. 2017;845:29.

[144] Shingledecker CN, Herbst E. A general method for the inclusion of radiation chemistry in astrochemical models. Phys Chem Chem Phys. 2018;20:5359–5367.

[145] Shingledecker CN, Tennis J, Le Gal R, Herbst E. On cosmic-ray-driven grain chemistry in cold core models. Astrophys J. 2018;861:15.

[146] Gérard J-C. Thermospheric odd nitrogen. Planet Space Sci. 1992;40:337–353.

[147] Marston G, Nesbitt FL, Nava DF, Payne WA, Stief LJ. Temperature dependence of the reaction of nitrogen atoms with methyl radicals. J Phys Chem. 1989;93:5769–5774.

[148] Hébrard E, Dobrijevic M, Loison J-C, Bergeat A, Hickson KM. Neutral production of hydrogen isocyanide (HNC) and hydrogen cyanide (HCN) in Titan's upper atmosphere. Astron Astrophys. 2012;541:A21.

[149] Loison J-C, Hébrard E, Dobrijevic M, Hickson KM, Caralp F, Hue V, Gronoff G, Venot O, Bénilan Y. The neutral photochemistry of nitriles, amines and imines in the atmosphere of Titan. Icarus. 2015;247:218–247.

[150] Schmidt JA, Johnson MS, Schinke R. Carbon dioxide photolysis from 150 to 210 nm: singlet and triplet channel dynamics, UV-spectrum, and isotope effects. Proc Natl Acad Sci USA. 2013;110:17691–17696.

[151] Nelson HH, Helvajian H, Pasternack L, McDonald JR. Temperature dependence of $C_3(X^1\Sigma_g^+)$ reactions with alkenes and alkynes, 295–610 K. Chem Phys. 1982;73:431–438.

[152] Berteloite C, Lara M, Bergeat A, Le Picard SD, Dayou F, Hickson KM, Canosa A, Naulin C, Launay J-M, Sims IR, Costes M. Kinetics and dynamics of the $S(^1D_2) + H_2 \rightarrow SH + H$ reaction at very low temperatures and collision energies. Phys Rev Lett. 2010;105:203201.

[153] Berteloite C, Le Picard SD, Sims IR, Rosi M, Leonori F, Petrucci R, Balucani N, Wang X, Casavecchia P. Low temperature kinetics, crossed beam dynamics and theoretical studies of the reaction $S(^1D) + CH_4$ and low temperature kinetics of $S(^1D) + C_2H_2$. Phys Chem Chem Phys. 2011;13:8485–8501.

[154] Leonori F, Petrucci R, Balucani N, Casavecchia P, Rosi M, Berteloite C, Le Picard SD, Canosa A, Sims IR. Observation of organosulfur products (thiovinoxy, thioketene and thioformyl) in crossed-beam experiments and low temperature rate coefficients for the reaction $S(^1D) + C_2H_4$. Phys Chem Chem Phys. 2009;11:4701–4706.

[155] Leonori F, Petrucci R, Balucani N, Casavecchia P, Rosi M, Skouteris D, Berteloite C, Le Picard SD, Canosa A, Sims IR. Crossed-beam dynamics, low-temperature kinetics, and theoretical studies of the reaction $S(^1D) + C_2H_4$. J Phys Chem A. 2009;113:15328–15345.

[156] Lara M, Berteloite C, Paniagua M, Dayou F, Le Picard SD, Launay JM. Experimental and theoretical study of the collisional quenching of $S(^1D)$ by Ar. Phys Chem Chem Phys. 2017;19:28555–28571.

[157] Braun W, Bass AM, Davis DD, Simmons JD. Flash photolysis of carbon suboxide: absolute rate constants for reactions of $C(^3P)$ and $C(^1D)$ with H_2, N_2, CO, NO, O_2 and CH_4. Proc R Soc London Ser A. 1969;312:417–434.

[158] Husain D, Kirsch LJ. Study of electronically excited carbon atoms, $C(2^1D_2)$, by time-resolved atomic absorption at 193.1 nm. Part 2. Reactions of $C(2^1D_2)$ with molecules. Trans Faraday Soc. 1971;67:3166–3175.

[159] Nuñez-Reyes D, Hickson KM. Kinetic and product study of the reactions of $C(^1D)$ with CH_4 and C_2H_6 at low temperature. J Phys Chem A. 2017;121:3851–3857.

[160] Nuñez-Reyes D, Loison J-C, Hickson KM, Dobrijevic M. A low temperature investigation of the $N(^2D) + CH_4$, C_2H_6 and C_3H_8 reactions. Phys Chem Chem Phys. 2019;21:6574–6581.

[161] Nuñez-Reyes D, Loison J-C, Hickson KM, Dobrijevic M. Rate constants for the $N(^2D) + C_2H_2$ reaction over the 50–296 K temperature range. Phys Chem Chem Phys. 2019;21:22230–22237.

[162] Hickson KM, Bray C, Loison J-C, Dobrijevic M. A kinetic study of the $N(^2D) + C_2H_4$ reaction at low temperature. Phys Chem Chem Phys. 2020;22:14026–14035.

[163] Black G, Slanger TG, St John GA, Young RA. Vacuum-ultraviolet photolysis of N_2O. IV. Deactivation of $N(^2D)$. J Chem Phys. 1969;51:116–121.

[164] Fell B, Rivas IV, McFadden DL. Kinetic study of electronically metastable nitrogen atoms, $N(2^2D_J)$, by electron spin resonance absorption. J Phys Chem. 1981;85:224–228.

[165] Umemoto H, Hachiya N, Matsunaga E, Suda A, Kawasaki M. Rate constants for the deactivation of $N(^2D)$ by simple hydride and deuteride molecules. Chem Phys Lett. 1998;296:203–207.

[166] Takayanagi T, Kurosaki Y, Sato K, Misawa K, Kobayashi Y, Tsunashima S. Kinetic studies on the $N(^2D, {}^2P) + CH_4$ and CD_4 reactions: the role of nonadiabatic transitions on thermal rate constants. J Phys Chem A. 1999;103:250–255.

[167] Herron JT. Evaluated chemical kinetics data for reactions of $N(^2D)$, $N(^2P)$, and $N_2(A^3\Sigma_u^+)$ in the gas phase. J Phys Chem Ref Data. 1999;28:1453–1483.

[168] Hickson KM, Suleimanov YV. Low-temperature experimental and theoretical rate constants for the $O(^1D) + H_2$ reaction. J Phys Chem A. 2017;121:1916–1923.

[169] Hickson KM, Suleimanov YV. An experimental and theoretical investigation of the $C(^1D) + D_2$ reaction. Phys Chem Chem Phys. 2017;19:480–486.

[170] Nuñez-Reyes D, Hickson KM, Larrégaray P, Bonnet L, González-Lezana T, Suleimanov YV. A combined theoretical and experimental investigation of the kinetics and dynamics of the $O(^1D) + D_2$ reaction at low temperature. Phys Chem Chem Phys. 2018;20:4404–4414.

[171] Wu Y, Cao J, Ma H, Zhang C, Bian W, Nuñez-Reyes D, Hickson KM. Conical intersection–regulated intermediates in bimolecular reactions: insights from $C(^1D) + HD$ dynamics. Sci Adv. 2019; 5:eaaw0446.

[172] Nuñez-Reyes D, Hickson KM, Larrégaray P, Bonnet L, González-Lezana T, Bhowmick S, Suleimanov YV. Experimental and theoretical study of the $O(^1D) + HD$ reaction. J Phys Chem A. 2019;123:8089–8098.

[173] Hickson KM. Low-temperature rate constants and product-branching ratios for the $C(^1D) + H_2O$ reaction. J Phys Chem A. 2019;123:5206–5213.

[174] Meng QY, Hickson KM, Shao KJ, Loison J-C, Zhang DH. Theoretical and experimental investigations of rate coefficients of $O(^1D) + CH_4$ at low temperature. Phys Chem Chem Phys. 2016;18:29286–29292.

[175] Nuñez-Reyes D, Hickson KM. Rate constants and H-atom product yields for the reactions of $O(^1D)$ atoms with ethane and acetylene from 50 to 296 K. J Phys Chem A. 2018;122:4696–4703.

[176] Nuñez-Reyes D, Hickson KM. Kinetics of the gas-phase $O(^1D)$ + CO_2 and $C(^1D)$ + CO_2 reactions over the 50–296 K range. J Phys Chem A. 2018;122:4002–4008.

[177] Nuñez-Reyes D, Kłos J, Alexander MH, Dagdigian PJ, Hickson KM. Experimental and theoretical investigation of the temperature dependent electronic quenching of $O(^1D)$ atoms in collisions with Kr. J Chem Phys. 2018;148:124311.

[178] Grondin R, Loison J-C, Hickson KM. Low temperature rate constants for the reactions of $O(^1D)$ with N_2, O_2, and Ar. J Phys Chem A. 2016;120:4838–4844.

[179] Hickson KM, Loison J-C, Lique F, Kłos J. An experimental and theoretical investigation of the $C(^1D)$ + N_2 → $C(^3P)$ + N_2 quenching reaction at low temperature. J Phys Chem A. 2016;120:2504–2513.

[180] Agúndez M, Wakelam V. Chemistry of dark clouds: databases, networks, and models. Chem Rev. 2013;113:8710–8737.

[181] Chastaing D, Le Picard SD, Sims IR. Direct kinetic measurements on reactions of atomic carbon, $C(^3P)$, with O_2 and NO at temperatures down to 15 K. J Chem Phys. 2000;112:8466–8469.

[182] Geppert WD, Reignier D, Stoecklin T, Naulin C, Costes M, Chastaing D, Le Picard SD, Sims IR, Smith IWM. Comparison of the cross-sections and thermal rate constants for the reactions of $C(^3P_j)$ atoms with O_2 and NO. Phys Chem Chem Phys. 2000;2:2873–2881.

[183] Nuñez-Reyes D, Hickson KM. The reactivity of $C(^1D)$ with oxygen bearing molecules NO and O_2 at low temperature. Chem Phys Lett. 2017;687:330–335.

[184] Daranlot J, Hu X, Xie C, Loison J-C, Caubet P, Costes M, Wakelam V, Xie D, Guo H, Hickson KM. Low temperature rate constants for the $N(^4S)$ + $CH(X^2\Pi_r)$ reaction. Implications for N_2 formation cycles in dense interstellar clouds. Phys Chem Chem Phys. 2013;15:13888–13896.

[185] Daranlot J, Jorfi M, Xie C, Bergeat A, Costes M, Caubet P, Xie D, Guo H, Honvault P, Hickson KM. Revealing atom-radical reactivity at low temperature through the N + OH reaction. Science. 2011;334:1538–1541.

[186] Carty D, Goddard A, Kohler SPK, Sims IR, Smith IWM. Kinetics of the radical-radical reaction, $O(^3P_j)$ + $OH(X^2\Pi_\Omega)$ → O_2 + H, at temperatures down to 39 K. J Phys Chem A. 2006;110:3101–3109.

[187] Páramo A, Canosa A, Le Picard SD, Sims IR. An experimental study of the intersystem crossing and reactions of $C_2(X^1\Sigma_g^+)$ and $C_2(a^3\Pi_u)$ with O_2 and NO at very low temperature (24–300 K). J Phys Chem A. 2006;110:3121–3127.

[188] Loison J-C, Hu X, Han S, Hickson KM, Guo H, Xie D. An experimental and theoretical investigation of the $N(^4S) + C_2(^1\Sigma_g^+)$ reaction at low temperature. Phys Chem Chem Phys. 2014;16:14212–14219.

[189] Daranlot J, Hincelin U, Bergeat A, Costes M, Loison J-C, Wakelam V, Hickson KM. Elemental nitrogen partitioning in dense interstellar clouds. Proc Natl Acad Sci USA. 2012;109:10233–10238.

[190] Stubbing JW, Vanuzzo G, Moudens A, Loison J-C, Hickson KM. Gas-phase kinetics of the $N + C_2N$ reaction at low temperature. J Phys Chem A. 2015;119:3194–3199.

[191] Viala YP. Chemical-equilibrium from diffuse to dense interstellar clouds. 1. Galactic molecular clouds. Astron Astrophys Suppl Ser. 1986;64:391–437.

[192] Womack M, Ziurys LM, Wyckoff S. Estimates of N_2 abundances in dense molecular clouds. Astrophys J. 1992;393:188–192.

[193] Smith IWM, Stewart DWA. Low-temperature kinetics of reactions between neutral free radicals. Rate constants for the reactions of OH radicals with N atoms (103–294 K) and with O atoms (158–294 K). J Chem Soc Faraday Trans. 1994;90:3221–3227.

[194] Brownsword RA, Gatenby SD, Herbert LB, Smith IWM, Stewart DWA, Symonds AC. Kinetics of reactions between neutral free radicals — rate constants for the reaction of CH radicals with N atoms between 216 and 584 K. J Chem Soc Faraday Trans. 1996;92:723–727.

[195] Lee JH, Michael JV, Payne WA, Stief LJ. Absolute rate of reaction of $N(^4S)$ with NO from 196–400 K with DF-RF and FP-RF techniques. J Chem Phys. 1978;69:3069–3076.

[196] Whyte AR, Phillips LF. Rate of reaction of N with CN (v = 0, 1). Chem Phys Lett. 1983;98:590–593.

[197] Howard MJ, Smith IWM. Direct rate measurements on the reactions $N + OH \rightarrow NO + H$ and $O + OH \rightarrow O_2 + H$ from 250 to 515 K. J Chem Soc Faraday Trans. 1981;77:997–1008.

[198] Brune WH, Schwab JJ, Anderson JG. Laser magnetic resonance, resonance fluorescence, and resonance absorption studies of the reaction kinetics of $O + OH \rightarrow H + O_2$, $O + HO_2 \rightarrow OH + O_2$, $N + OH \rightarrow H + NO$ and $N + HO_2 \rightarrow$ Products at 300 K between 1 and 5 Torr. J Phys Chem. 1983;87:4503–4514.

[199] Edvardsson D, Williams CF, Clary DC. Rate constant calculations on the $N(^4S)+OH(^2\Pi)$ reaction. Chem Phys Lett. 2006;431:261–266.

[200] Jorfi M, Honvault P, Halvick P. Quasi-classical determination of integral cross-sections and rate constants for the $N + OH \rightarrow NO + H$ reaction. Chem Phys Lett. 2009;471:65–70.

[201] Li A, Xie C, Xie D, Guo H. A global ab initio potential energy surface for HNO (a^3A'') and quantum mechanical studies of vibrational states and reaction dynamics. J Chem Phys. 2011;134:194309.

[202] Ge M-H, Chu T-S, Han K-L. An accurate quantum dynamics study of the N + OH reaction. J Theor Comput Chem. 2008;7:607–613.

[203] Woodall J, Agúndez M, Markwick-Kemper AJ, Millar TJ. The UMIST database for astrochemistry 2006. Astron Astrophys. 2007;466:1197–1204.

[204] Wakelam V, Smith IWM, Herbst E, Troe J, Geppert W, Linnartz H, Öberg K, Roueff E, Agúndez M, Pernot P, Cuppen HM, Loison J-C, Talbi D. Reaction networks for interstellar chemical modelling: Improvements and challenges. Space Sci Rev. 2010;156:13–72.

[205] Le Picard SD, Canosa A, Rowe BR, Brownsword RA, Smith IWM. Determination of the limiting low pressure rate constants of the reactions of CH with N_2 and CO: a CRESU measurement at 53 K. J Chem Soc Faraday Trans. 1998;94:2889–2893.

[206] Le Picard SD, Canosa A. Measurement of the rate constant for the association reaction CH + N_2 at 53 K and its relevance to Triton's atmosphere. Geophys Res Lett. 1998;25:485–488.

[207] Brownsword RA, Herbert LB, Smith IWM, Stewart DWA. Pressure and temperature dependence of the rate constants for the association reactions of CH radicals with CO and N_2 between 202 and 584 K. J Chem Soc Faraday Trans. 1996;92:1087–1094.

[208] Troe J. Theory of thermal unimolecular reactions at low pressures. II strong collision rate constants. Applications. J Chem Phys. 1977;66:4758–4775.

[209] Liessmann M, Miller Y, Gerber RB, Abel B. Reaction of OH and NO at low temperatures in the presence of water: the role of clusters. Z Phys Chem. 2011;225:1129–1144.

https://doi.org/10.1142/9781800610996_0006

Chapter 6

Nucleation Studies in Supersonic Flow

Barbara Wyslouzil[*,‡] and Ruth Signorell[†,§]

*Department of Chemical and Biochemical Engineering
Department of Chemistry and Biochemistry, Ohio State University
151 W. Woodruff Ave., 43210 Columbus, OH, USA
†ETH Zurich, Department of Chemistry and Applied Biosciences
Vladimir-Prelog-Weg 2, 8093 Zurich, Switzerland
‡wyslouzil.1@osu.edu
§rsignorell@ethz.ch

Abstract

Supersonic flow offers a unique opportunity to follow the evolution of an aerosol on the microsecond timescale. The initial appearance of the droplets/particles and their further development can involve a number of nucleation steps: homogeneous nucleation from the supersaturated vapor to the (supercooled) liquid, nucleation of a solid phase from the supercooled liquid, and even heterogeneous nucleation of other species onto pre-existing particles. In this chapter, we present basic concepts from nucleation theory, the experimental techniques used to characterize the thermodynamic state of the flow as well as the size, state and composition of the aerosol, and the approaches used to determine nucleation rates. The nucleation experiments that have been conducted both within and at the exit of Laval nozzles are summarized, and the importance of experiments conducted in supersonic flows to advancing the field of nucleation is discussed.

Keywords: Nucleation; Supersonic flow; Phase transitions.

6.1 Introduction

Nucleation plays a crucial role initiating phase transitions in both ambient and industrial settings. In the atmosphere, multicomponent, nanometer-sized particles (droplets) can form by vapor phase nucleation and condensation [1, 2]. Some of these can act as cloud condensation nuclei [3] that "activate" at low water supersaturations to form cloud droplets, and cloud droplets, in turn, must freeze in order to grow large enough to fall as rain [4]. In industry, understanding aerosol formation and evolution in high speed flows is important to controlling water vapor condensation in steam turbines [5], avoiding condensation in hypersonic wind tunnels [6], and developing supersonic separators [7]. Thus, measuring the conditions required to initiate phase transitions, the rates at which they occur, the composition of the first fragments of the new phase, and characterizing intermediate phases that may form, are all crucial for developing accurate models of climate, atmospheric chemistry, and industrial processes.

Nucleation studies in supersonic nozzles are also important from the viewpoint of advancing fundamental physical science. In particular, when aerosols form under the extreme supersaturations that develop in high speed flows, the nucleation rates from the vapor are on the order of 10^{17} cm^{-3}s^{-1}, the critical cluster sizes are small and experimental conditions approach or in some cases overlap with those typically required in molecular dynamics (MD) simulations[a] [8–10]. If conditions are extreme enough, experiments may provide a way to approach the vapor–liquid spinodal or pseudo-spinodal region [11–14], where the spinodal corresponds to the thermodynamic limit of stability of a fluid. The high homogeneous nucleation rates also ensure that heterogeneous nucleation is unlikely to compete, and, thus, most experiments produce a high number density of extremely

---

[a]Although MD simulations of nucleation from a supersaturated vapor can match experimental temperatures, most run at much higher partial pressures of the condensable. This is to ensure nucleation occurs within ~10 nanoseconds in a simulation box with a characteristic length of ~10 nanometers. Thus, MD nucleation rates are typically on the order of 10^{22}–10^{26} cm^{-3}s^{-1}, i.e. roughly 5–9 orders of magnitude higher than those measured in experiments. Recent large-scale simulations have begun to bridge this gap [8], simulating nucleation rates for Ar as low as 10^{15} cm^{-3}s^{-1}.

pure particles in the nanometer size range. Since, under a very broad range of experimental conditions, the first phase to condense is the supercooled liquid [15–23], one can also probe this metastable state and furthermore follow the liquid to solid nucleation kinetics on a microsecond time scale [15–22]. In nanodroplets, typical crystallization rates are on the order of 10^{23} cm^{-3}s^{-1} [15–22] and again approach those characteristic of MD simulations [24–26]. If cooling rates are high enough, crystallization may proceed via intermediate metastable states [27], be hindered [19], or, potentially, be avoided altogether [24].

6.2 Basic Nucleation Theory

Phase transitions, between the vapor, liquid, and solid states, are often initiated by nucleation, and, as illustrated in Fig. 6.1, nucleation can proceed via homogeneous or heterogeneous routes. In either case, the basic picture is that monomers of the more stable phase form clusters within the metastable "mother" phase, that need to reach a size large enough in order to grow spontaneously.

Nucleation is driven by the change in chemical potential $\Delta\mu$ of the metastable phase with respect to the more stable phase, but

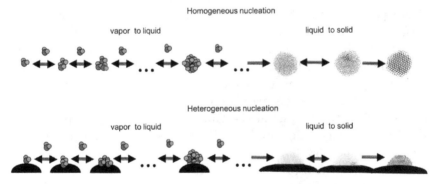

Fig. 6.1. Nucleation processes are classified as heterogeneous if they involve an existing surface, and homogeneous if they do not. Clusters of the new phase change size by the addition or loss of monomers. Once clusters are large enough, they rapidly grow into the new phase. MD simulations are reproduced with permission from (top) Ref. [24], Copyright 2012 ACS and (bottom) Ref. [28], Copyright 2014 ACS.

hindered by the cost of forming the interface between them [29]. Within the framework of Classical Nucleation Theory (CNT), the Gibbs free energy change (ΔG_{CNT}) on forming a cluster of the new phase containing n molecules is

$$\Delta G_{\text{CNT}}(n) = n\Delta\mu + \sigma A_c \qquad (6.1)$$

where σ is the interfacial free energy (surface tension) and A_c is the surface area of the cluster. For the vapor–liquid phase transition, one generally assumes the gas phase is ideal and the liquid incompressible, and thus

$$\Delta\mu = -k_B T \ln S \qquad (6.2)$$

where the supersaturation $S = p_v/p_{\text{ve,l}}$ is the ratio of the actual vapor pressures p_v to the equilibrium vapor pressure above the liquid $p_{\text{ve,l}}$, k_B is the Boltzmann constant, and T is the temperature. For the liquid-crystal phase transition, $S = p_{\text{ve,l}}/p_{\text{ve,s}}$ where $p_{\text{ve,s}}$ is the equilibrium vapor pressure above the solid [30]. Alternatively, $\Delta\mu$ can be written in terms of the enthalpy of melting ΔH_m and the degree of supercooling $\Delta T = T_m - T$ as $\Delta\mu = -\Delta H_m \Delta T/T_m$ where T_m is the melt temperature [31].

Figure 6.2 illustrates the free energy required to form clusters of the new phase as a function of size and supersaturation. When $S \leq 1$, $\Delta G_{\text{CNT}}(n)$ is a strictly increasing function of n, but when $S > 1$, ΔG_{CNT} exhibits a maximum at n^*. The cluster of size n^* is called the critical cluster, and this is the cluster of the emerging phase that is in unstable equilibrium with the supersaturated "mother" phase. As the supersaturation increases, the critical cluster size as well as the value of ΔG_{CNT}^* decrease rapidly. In the limit of the pseudo-spinodal [32], the change in the Gibbs free energy is comparable to the thermal fluctuation $k_B T$, and at the spinodal the free energy barrier vanishes [33, 34].

The nucleation rate J can be written as

$$J = K \exp\left(-\frac{\Delta G_{\text{CNT}}^*}{k_B T}\right) \qquad (6.3)$$

where J corresponds to the rate of production of critical clusters and K is a kinetic prefactor that depends on the nucleation process under

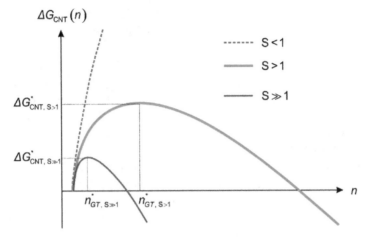

Fig. 6.2. $\Delta G(n)$ only exhibits a maximum if $S > 1$. Both the barrier height and number of molecules in the critical cluster decrease as S increases.

consideration. For details on the derivation of Eq. (6.3) and the exact expressions used for K and ΔG^* under different nucleation scenarios, the reader is referred to Kashchiev [29], Kelton and Greer [31] as well as Pruppacher and Klett [35].

A major assumption in CNT is that the critical cluster is a compact spherical object whose physical properties reflect those of the bulk material. Writing the cluster surface area in terms of n and the molecular volume of the new phase v, and finding the maximum of $\Delta G(n)$, yields the change in Gibbs free energy for the critical cluster

$$\Delta G^*_{\text{CNT}} = \frac{16\pi\sigma^3 v^2}{3(k_B T \ln S)^2} \tag{6.4}$$

as well as the Gibbs–Thomson equation for the number of molecules in the critical cluster n^*_{GT}

$$n^*_{GT} = \frac{32\pi\sigma^3 v^2}{3(k_B T \ln S)^3} \tag{6.5}$$

or the critical cluster radius r^*_{GT}

$$r^*_{GT} = \frac{2\sigma v}{k_B T \ln S} \tag{6.6}$$

Although CNT has many limitations, it remains attractive because the assumptions involved are easy to understand, the nucleation rates can be predicted using well defined, bulk physical properties, and the trends agree with those observed in experiments.

One way to test the predictions of CNT is to directly compare the measured rates to the predictions of classical theory. Further molecular level insight is possible by comparing theory predictions for the critical cluster properties to those derived from measurements. For vapor to liquid phase transitions, two approaches can be used to determine the experimental critical cluster size n^*_{exp}. These are direct interrogation of the nucleation event [11, 14] via mass spectroscopic sampling or further analysis of accurate nucleation rate measurements. The latter approach relies on the first nucleation theorem [36–38], a general result that relates the size of the critical cluster to the isothermal nucleation rates. For nucleation of a single species,

$$n^*_{\text{exp}} \cong \left(\frac{\partial \ln J}{\partial \ln S}\right)_T - \left(\frac{\partial \ln K}{\partial \ln S}\right)_T \qquad (6.7)$$

Since the kinetic prefactor K in Eq. (6.3) depends more weakly on supersaturation than the exponential term, the second term is often neglected. Generalizations to multicomponent systems are also available [39, 40].

The second nucleation theorem [51, 52] provides an estimate for the excess internal energy of the cluster relative to the bulk liquid and is given by

$$E_x(n^*_{\text{exp}}) \cong kT^2 \left[\left(\frac{\partial \ln J}{\partial T}\right)_{\ln S} - \left(\frac{\partial \ln K}{\partial T}\right)_{\ln S}\right]$$

$$\cong k_B T^2 \left[\left(\frac{\partial \ln J}{\partial T}\right)_{\ln S} - L + k_B T\right] \qquad (6.8)$$

where L is the latent heat of vaporization for the bulk liquid. For the second nucleation theorem, the derivative of the kinetic prefactor is not negligible. Within the framework of CNT, Ford [51, 52] also showed that the excess internal energy of the cluster relative to the bulk liquid is given by

$$E_x(n^*_{\text{exp}}) = \left(\sigma - T\frac{d\sigma}{dT}\right) A_c \qquad (6.9)$$

Other forms of both nucleation theorems are available in Kashchiev [29].

In Fig. 6.3, the values for n^*_{exp} and $E_x(n^*_{\mathrm{exp}})$ derived from water nucleation rate measurements are compared to the predictions of CNT. For water, n^*_{exp} values (Eq. (6.7)) deviate significantly from n^*_{GT} values predicted by Eq. (6.5) for $n^*_{\mathrm{exp}} > \sim 40$. For both isotopes of water, experimental $E_x(n^*_{\mathrm{exp}})$ values agree reasonably well with the predictions of CNT (Eq. (6.9)) when $n^*_{\mathrm{exp}} < \sim 40$.

6.3 Supersonic Nozzle Experiments for Nucleation Studies

To date, most supersonic nozzle experiments designed to investigate nucleation use nozzles that expand continuously [15–20, 27, 50, 53–59]. Furthermore, the nozzle is often a sandwich design with a rectangular cross section, flat sidewalls, and contoured upper and lower plates. The operating conditions are chosen so that condensation and, potentially, freezing occur within the nozzle.

A typical setup is illustrated in Fig. 6.4 (top) and shows the key processes that influence aerosol evolution. In particular, particle formation is driven by vapor to liquid nucleation, as (illustrated in Fig. 6.1), droplet growth occurs via monomer condensation, coagulation involves the merger of colliding droplets, and freezing is initiated by liquid to solid nucleation events within individual droplets (see Fig. 6.1).

More recently, nozzles have been developed to explore nucleation under more uniform conditions [11, 55, 60]. As illustrated in Fig. 6.4 (bottom), these axisymmetric (conical) nozzles are designed to ensure a uniform flow at the nozzle exit, and the pressure at the nozzle exit is adjusted to match that of the flow leaving the nozzle. This results in a region of stable flow, up to a few tens of cm long, that ensures nucleation takes place under essentially isothermal conditions. The design also allows for alternative characterization techniques, in particular direct sampling of the evolving cluster distribution via mass spectroscopy [14, 57–60].

Nozzles can be run under continuous [20, 50, 53, 55, 57, 58, 61] or pulsed flow conditions [11, 14, 60]. The advantage of the former includes stable operating conditions that can be probed using

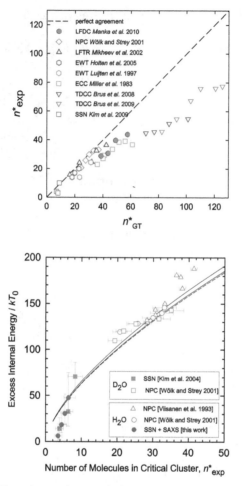

Fig. 6.3. (Top) Experimental critical cluster sizes n^*_{exp} are often smaller than those predicted by CNT, n^*_{GT}. Data are from the following references: LFDC (Laminar Flow Diffusion Chamber) [41]; NPC (Nucleation Pulse Chamber) [42]; LFTR (Laminar Flow Tube Reactor) [43]; EWT (Expansion Wave Tube) [44, 45]; ECC (Expansion Cloud Chamber) [46]; TDCC (Thermal Diffusion Cloud Chamber) [47–49]; and SSN (Supersonic Nozzle) [50]. Reproduced with permission from Ref. [41], Copyright 2010 AIP Publishing. (Bottom) The experimental $E_x(n^*_{exp})$ values as a function of the number of molecules in the critical cluster, (n^*_{exp}). The upper (lower) solid line corresponds to Eq. (6.9) evaluated for D_2O at 260 K (210 K). The dashed line is for H_2O at 210 K. Reprinted from Overview: Homogeneous nucleation from the vapor phase — the experimental science. Reproduced with permission from Ref. [40], Copyright 2016 AIP Publishing.

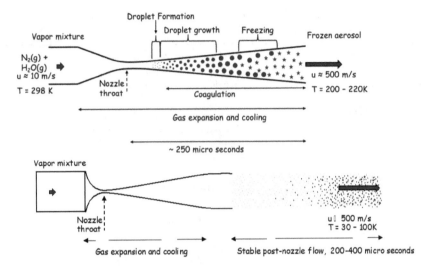

Fig. 6.4. (Top) Supersonic nozzle experiments that investigate nucleation have typically chosen experimental conditions to ensure particle formation and growth are completed inside the nozzle. (Bottom) Pressure matching at low pressure can, however, yield a stable and uniform post-nozzle flow regime where nucleation occurs at essentially constant temperature.

methods that require longer integration times. The disadvantages involve the high pumping capacities required and the high consumption rate of materials. Pulsed nozzles are less demanding in terms of pumping capacity and material flows (see Chapters 1 and 2) but require careful synchronization of the flows and the characterization methods, as well as instruments with rapid response times, typically less than a few milliseconds.

6.4   Characterizing the Flow and the Aerosol/Clusters

Many different techniques have been coupled with supersonic nozzle experiments in order to determine the thermodynamic state of the flow and characterize the clusters or particles that are produced. One challenging aspect of supersonic flow experiments is that one cannot easily extract a sample of the particles or clusters in order to count them using conventional aerosol methods or examine them using standard *ex situ* techniques. The particles only exist because the flow is cold, and the flow is only cold because the velocity is high. If the velocity decreases significantly, the particles simply re-evaporate.

Table 6.1. A summary of the experimental methods that have been used to characterize nucleation in supersonic flows and the questions they address.

Experimental Question	Technique/Representative References
What is the thermodynamic state of the flow and where is heat released?	• Static or impact pressure using probes [11, 12, 61] • Flow density via interferometry [62, 63] • Molecular density and rotational temperature via Rayleigh/Raman scattering [64, 65] • Temperature via Tunable Diode Laser Absorption Spectroscopy (TDLAS) [53, 66]
What are the particle size distribution parameters, i.e. average particle size $\langle r \rangle$, distribution width δ and particle number density N or mass fraction condensate g?	• Visible light scattering [55, 56, 67, 68] • Rayleigh/Raman scattering [64, 65] • Multicolor visible light (7 color) absorption [63] • Small angle neutron scattering (SANS) [69] • Small angle X-ray scattering (SAXS) [70, 71] • Mass spectrometry [11–14, 57, 58, 60]
What is the composition of the aerosol or, equivalently, the partitioning between the vapor and condensed states?	• TDLAS [66, 72, 73] • Fourier Transform Infrared Spectroscopy (FTIR) [15–19, 53, 55, 56, 74–76] • Mass spectrometry [14, 76]
What is the crystal structure of the particle?	• FTIR [18, 19, 55, 56, 76] • Wide angle X-ray scattering (WAXS) [17, 19, 27]
What is the structure of the particle, i.e. core shell? Well mixed? Shape?	• SANS, SAXS [77, 78] • FTIR [75, 79]

Thus, all characterization techniques must be *in situ*, or must sample the clusters in a way that does not significantly perturb them.

Table 6.1 summarizes characterization techniques that have proven particularly useful for nucleation studies, and Fig. 6.5 illustrates how they have been implemented into the experimental setup. For experiments inside the nozzle, a well-defined path length has favored the use of parallel sidewalls and a rectangular nozzle cross

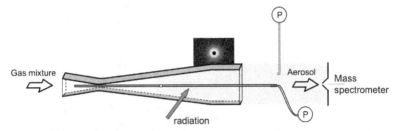

Fig. 6.5. Experimental techniques used to characterize the flow and the resultant aerosol. For experiments conducted within the nozzle, static pressure is typically measured along the centerline using a movable static pressure probe (shown here) or via pressure taps in a side wall. For experiments in the post-nozzle flow, the impact pressure is measured as a function of position relative to the nozzle exit and the center line of the flow by moving the nozzle relative to the fixed probe. The resultant aerosol can be characterized using mass spectrometry as well as different sources of radiation (X-rays, neutrons, light, IR radiation) with wavelengths ranging from 0.1 nm to 10 μm.

section. For external sampling, the symmetry inherent in a conical nozzle design is advantageous.

6.5 Characterizing the Thermodynamic State of the Flow

Characterizing the flow is critical to determining the conditions associated with the phase transition and, in many cases, to pinpoint where the phase transitions occur. In the absence of heat addition, the key variables in the supersonic flow (mass density ρ, velocity u, pressure p, temperature T, effective flow area ratio A/A_{th}) are governed by the conservation of mass, momentum and energy together with an equation of state (see also Chapters 1 and 2). In the presence of condensation, the equations are modified to account for changes in the condensate mass fraction g and the heat released to the flow due to the latent heat of condensation and/or freezing $L(T)$.

For the case of a single species condensing without freezing, the modified equations in differential form are [66]:

$$d\left(\frac{\rho}{\rho_r}\right) = \left[\frac{1}{\gamma_{th}}\left(\frac{u_{th}}{u}\right)^2 \frac{T_r}{T_{th}}\right] d\left(\frac{p}{p_r}\right) - \left(\frac{\rho}{\rho_r}\right) d\ln\left(\frac{A}{A_{th}}\right) \quad (6.10)$$

$$d\left(\frac{T}{T_r}\right) = \frac{\rho_r}{\rho}\left[w_r\left(g\right) - \frac{1}{\gamma_{th}}\left(\frac{u_{th}}{u}\right)^2\frac{T}{T_{th}}\right]d\left(\frac{p}{p_r}\right)$$

$$+ \left(\frac{T}{T_r}\right) \times \left[d\ln\left(\frac{A}{A_{th}}\right) + w(g)dg\right] \quad (6.11)$$

$$\left[\frac{L\left(T\right)}{C_pT_r} - \frac{T}{T_r}w(g)\right]dg = \frac{\rho_r}{\rho}\left[h - \frac{1}{\gamma_{th}}\left(\frac{u_{th}}{u}\right)^2\frac{T}{T_{th}}\right]d\left(\frac{p}{p_r}\right)$$

$$+ \left(\frac{T}{T_r}\right)d\ln\left(\frac{A}{A_{th}}\right) \quad (6.12)$$

and

$$u = \frac{u_{th}\rho_{th}A_{th}}{\rho A} \quad (6.13)$$

where the subscript denotes the values of the variable at the nozzle throat, the subscript r indicates the value at the nozzle inlet (stagnation conditions), γ_{th} is the specific heat ratio at the throat, and C_p is the heat capacity. The quantities $w_r(g)$, $w(g)$, and h are defined by

$$w_r(g) \equiv \bar{\mu}/[\bar{\mu}_r(1-g)] \quad (6.14)$$

$$w(g) \equiv \bar{\mu}/[\bar{\mu}_v(1-g)] \quad (6.15)$$

$$h \equiv w_r(g) - \frac{C_{pr}}{C_p}\left(\frac{\gamma_r - 1}{\gamma_r}\right) \quad (6.16)$$

where $\bar{\mu}$ is the mean molecular weight of the gas mixture, $\bar{\mu}_r$ is the value of $\bar{\mu}$ at the stagnation condition, and $\bar{\mu}_v$ is the mean molecular weight of the condensable vapor.

Equations (6.10)–(6.13), involve six variables, and, thus, in addition to determining the stagnation conditions, p_r, T_r, and initial composition of the gas mixture y_r, two variables are measured as a function of position. The remaining variables are then estimated by integrating the flow equations. Typically, the change in pressure (density or temperature) in the absence of condensation is first used to determine $(A/A_{th})_{\mathrm{dry}}$. A second measurement is made in the presence of condensation and the equations are solved using $(A/A_{th})_{\mathrm{dry}}$ and pressure (density or temperature) as the known variables. The implicit assumption here is that the expansion is unaffected by condensation. For small nozzles ($A_{th} \sim 30$–60 mm^2) working at relatively

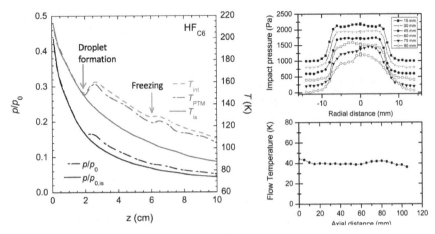

Fig. 6.6. (Left) The static pressure (p/p_r) and temperature profiles (T_{int} or T_{PTM}) of the condensing flow are compared to those expected for an isentropic expansion ($p/p_{r,is}$ and T_{is}). The difference between the temperature based on the more accurate integrated data analysis, T_{int}, and the temperature based on pressure measurements alone, T_{PTM}, reflects the change in the boundary layer brought on by condensation. Reproduced with permission from Ref. [75], Copyright 2019 PCCP Owner Societies. The location $z = 0$ corresponds to the throat. (Right) Impact pressures measured in the post nozzle flow as a function of radial position and distance from the nozzle exit (top), and the corresponding flow temperature at a radial distance of 0 mm (bottom). The origin for the axial distance is the nozzle exit. Reproduced with permission from Ref. [60], Copyright 2015 PCCP Owner Societies.

low densities ($\rho_r <\sim 1\ \mathrm{kg\,m^{-3}}$), this is not necessarily true, and more sophisticated approaches, measuring g in addition to p, for example, are required to accurately determine the other variables of the flow — in particular the flow temperature [66].

Figure 6.6 (left) illustrates the results of an experiment in which hexane condenses from a hexane–Ar gas mixture and the droplets subsequently freeze (inside the nozzle). Heat addition corresponding to the two distinct phase transitions are clearly observed. In this case, the flow equations were modified to include both phase transitions, and, in addition to p, g_l and g_s were measured via FTIR. The importance of using the integrated data analysis approach to accurately determine the temperature is clearly illustrated in Fig. 6.6 (left).

For experiments conducted in the post-nozzle flow, the flow is characterized by measuring the stagnation conditions, p_r, T_r, the initial composition of the gas phase y_0, and the impact pressure p_i in the post-nozzle flow. The p_i measurements are made as a function of the distance from the nozzle exit and the radial distance from the nozzle centerline by translating the impact pressure probe vertically and horizontally. The Mach number M is determined from p_i and from the stagnation pressure p_r using the Rayleigh-Pitot formula [80] (see also Chapter 2)

$$\frac{p_i}{p_r} = \left\{ \frac{(\gamma+1)M^2}{(\gamma-1)M^2+2} \right\}^{\frac{\gamma}{\gamma-1}} \left\{ \frac{\gamma+1}{2\gamma M^2 - \gamma + 1} \right\}^{\frac{1}{\gamma-1}} \qquad (6.17)$$

where $\gamma = C_p/C_v$ is the ratio of the heat capacities at constant pressure and constant volume, respectively, of the sample gas mixture. The flow temperature T_F and the flow pressure p_F are calculated from

$$\frac{T_r}{T_F} = 1 + \frac{\gamma - 1}{2} M^2 \qquad (6.18)$$

and

$$\frac{p_r}{p_F} = \left(\frac{T_r}{T_F} \right)^{\frac{\gamma}{\gamma-1}} \qquad (6.19)$$

assuming isentropic flow conditions. The number concentrations of the carrier gas and condensable vapor molecules in the uniform post-nozzle flow are obtained from the ideal gas law. As illustrated in Fig. 6.6 (right), axially and radially resolved p_i measurements verify that the post-nozzle flow is highly uniform.

6.6 Particle Size Parameters via Light, X-Ray, and Neutron Scattering

Particles scatter light and, therefore, visible light scattering has been used to detect the appearance of the condensed phase from the vapor [55, 56, 67, 68, 81]. Under most experimental operating conditions (see references in Tables 6.1–6.3), particles formed in a supersonic nozzle have diameters d less than $\sim$50 nm. Thus, for visible light the wavelength λ is such that the size parameter d/λ is small and scattering does not contain enough information to uniquely retrieve the particle size. Nevertheless, Stein and co-workers [68] were able to estimate particle sizes and number densities by combining the light scattering intensity $I \propto N\langle r^6 \rangle$ with the estimate for mass fraction of condensate $g \propto N\langle r^3 \rangle$ based on static pressure measurements. For an aerosol that is relatively monodisperse, the ratio of these quantities yields an estimate for $\langle r^6 \rangle / \langle r^3 \rangle \approx \langle r \rangle^3$. For slightly larger particles with diameters of ~150 nm, Lamanna *et al.* [63] used multi-wavelength extinction to determine the maximum particle size observed in oscillating supersonic flows.

Accurate particle sizing is possible if the wavelength of the radiation more closely matches the particle size. For nanoparticles, this criterion is met by both X-rays and neutrons. Lab based X-ray sources are not yet intense enough to make measurements on the dilute aerosol samples (volume fractions $\cong 10^{-6}$), but small angle X-ray scattering (SAXS) experiments conducted at modern synchrotrons can generate very good spectra in a matter of seconds [70, 71]. For routine particle size measurement, SAXS is ideal. Although neutrons cover about the same wavelength range, even at a world class user facility the flux of neutrons is many orders of magnitude lower than the flux of X-rays at a modern synchrotron. Thus, small angle neutron scattering (SANS) spectra require many hours of integration time to collect an equivalent spectrum [50]. An important advantage of SANS, however, is that this technique can be extremely sensitive to particle structure if selective deuteration enhances the contrast from different regions of the particle. For example, SANS experiments were able to detect scattering from the alcohol rich shell of deuterated butanol — H_2O droplets [77].

Figure 6.7(a) shows that SAXS and SANS measurements of aerosol produced under the same conditions agree very well [70, 71]. Furthermore, the ~1 mm wide SAXS beam makes it possible to follow changes in droplet size, polydispersity, number density, as well as the volume or mass fraction of the condensate on the microsecond timescale (Figs. 6.7(b) and (c)) [82]. These data can then be used to test droplet growth laws in the free molecular regime, estimate droplet temperatures, calculate nucleation and coagulation rates and examine the internal structure of the droplets.

Another critical use of the position resolved condensate mass fraction measurements is to improve the flow analysis. By solving the flow equations using p and g as input, the assumption that the area

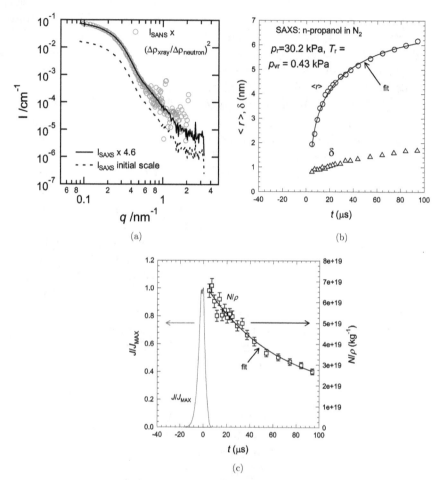

Fig. 6.7. (a) SAXS and SANS spectra agree quantitatively after being adjusted for differences in scattering cross section and absolute calibration. Here q is the scattering vector that is related to the scattering angle θ and the wavelength λ as $q = (4\pi/\lambda)\sin(\theta/2)$. Reproduced with the permission from Ref. [70], Copyright 2007 PCCP Owner Societies. (b, c) SAXS can follow aerosol evolution on the microsecond time scale. In the absence of coagulation N/ρ is constant, thus, under these conditions both condensation and coagulation contribute to particle growth. Here the time origin, $t = 0$, corresponds to the peak in the nucleation rates. Enhanced growth rates of nanodroplets in the free molecular regime: The role of long-range interactions. Reproduced with the permission from Ref. [82], Copyright 2016 Taylor & Francis Ltd.

ratio of expansion during condensation $(A/A_{th})_{\text{wet}}$ is the same as the area ratio of expansion in the absence of condensation $(A/A_{th})_{\text{dry}}$ is no longer necessary. Depending on the level of heat release to the flow, this integrated data analysis approach [66] yields flow (and particle) temperatures that can be significantly higher than those calculated assuming the boundary layers are not affected by condensation (see Fig. 6.6 (left)). The radii and g values can also be combined to estimate the droplet temperature during the period of rapid growth [82, 83]. Finally, the values of g derived from SAXS are required to calibrate the IR absorption measurements of the condensed phases [18].

6.7 Particle Size Parameters via Mass Spectroscopy

For molecular clusters with sizes below about 1 nm, i.e. in the size range that is important for nucleation and early growth processes, SAXS measurements are very challenging and SANS measurements are not currently possible. Furthermore, the scattering signal cannot give detailed information regarding the relative abundances of clusters that differ by a monomer or two. Fortunately, mass spectrometry [11–14, 57, 58, 76, 84, 85] can characterize clusters in the desired size range with the required level of resolution. This is done by sampling the post-nozzle flow into the mass spectrometer (see Fig. 6.5) using a skimmer, ionizing the neutral clusters, and quantitatively detecting the cluster ions.

Although the details are not yet fully understood, recent work has shown that the ionization step is critical [86–88], and a soft ionization method is required to (approximately) preserve the original cluster size distribution. Single-photon ionization using vacuum ultraviolet (VUV) light, with a photon energy just above the lowest ionization threshold of the cluster forming species and with a low photon flux, has proven to be the best compromise at least for the range of compounds investigated to date [11–14, 76, 84, 85, 89]. In contrast, even at low electron kinetic energy, electron ionization does not appear to be suitable [87, 88]. Unfortunately, a perfect and general ionization method, suitable for all compounds of interest, does not appear to exist. For larger clusters, quantitative detection can also be an issue since the multichannel plate (MCP) detectors typically used in mass spectrometry have mass dependent detection efficiencies.

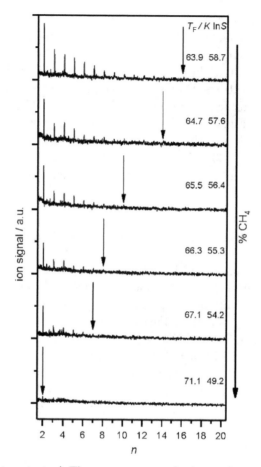

Fig. 6.8. (Bottom to top) The mass spectra of toluene clusters shift to larger sizes, n, as the temperature is decreased and saturation increased during a nucleation event. n is the number of molecules in the cluster. The arrows indicate the largest clusters size n_{max} that was detected in each mass spectrum. Reproduced with permission from Ref. [13], Copyright 2017 ACS.

These challenges can be circumvented in part by applying high extraction voltages or by applying correction functions [11, 60, 90].

Figure 6.8 shows a series of mass spectra that were recorded during a nucleation event induced by systematically decreasing the flow temperature T_F and, thus, systematically increasing the supersaturation S [13]. The small temperature increases were achieved by adding small amounts of methane gas to the flow to decrease the heat capacity ratio of the flow. As explained below in Section 6.10.1, a series of such spectra allows one to retrieve critical cluster sizes directly from molecular-level experimental data.

6.8 Vibrational Spectroscopy

Molecules, and the state they are in, can be identified by their vibrational spectra using techniques such as IR absorption or Raman scattering. Raman scattering has been used to investigate both the physics of free jet expansions, and to investigate nucleation phenomena in them [64, 65, 91]. FTIR spectroscopy [15–17, 19, 53, 55, 56, 74–76, 79] and Tunable Diode Laser Absorption Spectroscopy (TDLAS) [66, 72, 73] are complementary techniques that have both been used to make spatially-resolved composition and temperature measurements focused on nucleation within supersonic nozzles.

TDLAS provides very high resolution spectra over a very restricted wavenumber (~ 1 cm^{-1}) range and can measure the vapor phase concentration of small molecules like water or methane. More importantly, by measuring changes in the absorption intensity of lines with different ground state energies, the temperature of a tracer molecule (methane) and, thus, the temperature of the gas stream can be determined [66, 72]. Tanimura *et al.* used this technique to explore the effect condensation has on boundary layer development in small nozzles [66]. In contrast, the wide wavenumber range covered by FTIR spectroscopy is useful to determine the physical state (vapor/liquid/solid) of molecules [15–19, 53, 55, 56, 75, 76] and, with some calibration, can then be used to quantify the phase (vapor/liquid/solid) distribution of the condensable [15–19].

Two examples of IR spectroscopy used to study nucleation are shown in Fig. 6.9. Figure 6.9 (top) demonstrates that during binary condensation of D_2O and ethanol [73] the alcohol monomer concentration starts to decrease somewhat before the D_2O monomer concentration does, perhaps due to the formation of small alcohol clusters. Furthermore, the difference in the condensation rates of the two molecules demonstrates that the composition of the droplets can change dramatically as the aerosol evolves. In this case, the droplets becoming progressively richer in D_2O as condensation proceeds. In Fig. 6.9 (bottom), FTIR measurements in the hydrogen bonded OH

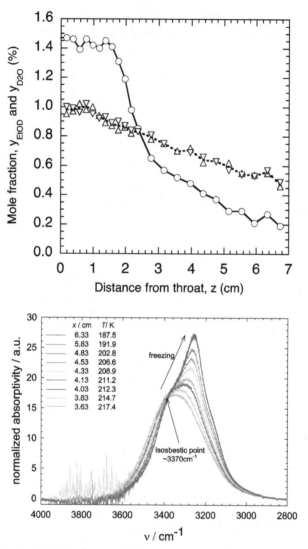

Fig. 6.9. (Top) TDLAS was used to follow the vapor composition during binary condensation of D_2O (circles) and ethanol (triangles). Reproduced with permission from Ref. [73], Copyright 2007 AIP Publishing. (Bottom) FTIR spectra in the hydrogen bonded OH stretch region change rapidly as supercooled water freezes. Reproduced with permission from Ref. [15], Copyright 2012 PCCP Owner Societies.

stretch region show a distinct transition as the supercooled liquid water freezes to form ice [15].

The presence of an isosbestic point — i.e. a point of constant absorption intensity at a fixed wavenumber — is consistent with a 1:1 transformation, in this case water transforming to ice. Fitting the intermediate spectra, as a linear combination of the coldest liquid spectrum and the warmest solid spectrum yields the fraction of the aerosol droplets that are frozen. This information then yields the crystal nucleation rate as described in more detail as follows.

6.9 Particle Crystal Structure via Wide Angle X-Ray Scattering or Electron Scattering

When phase transitions occur under highly non-equilibrium conditions they often proceed via states whose stability lies between that of the initial and most stable states, in accordance with Ostwald's rule of stages. Thus, condensation from the supersaturated vapor at temperatures below the triple point, almost always leads to the liquid state — at least briefly — before the stable crystalline solid appears. Likewise, the crystal that first forms from a supercooled liquid may be a less stable structure than expected. For example, at 1 atmosphere the ice that first forms from highly supercooled water is stacking disordered ice I_{sd} with a high fraction of cubic sequences rather than the more stable ice I_h that consists of purely hexagonal sequences [24, 27, 92].

Vibrational spectroscopy may not always be able to distinguish between different crystalline states, and in this case diffraction experiments using X-rays or electrons are required to determine the crystal structure. Such experiments are extremely challenging for particles inside the supersonic nozzle both because volume fraction of sample is $\sim 10^{-6}$, and because the carrier gas and windows produce significant parasitic scattering.

To decrease parasitic scattering, Bartell and co-workers [21, 22, 93, 94] skimmed the flow exiting a small supersonic nozzle into a vacuum chamber where evaporative cooling reduced the droplet temperature enough that freezing ensued. They used electron diffraction to follow the phase transition from liquid to solid and estimate the volume based nucleation rates. In Bartell's work, droplet size

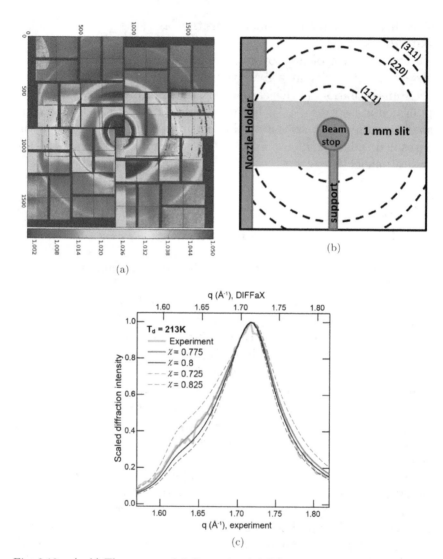

Fig. 6.10.   (a, b) The measured 2-dimensional diffraction pattern of frozen water droplets only shows rings corresponding to those expected for cubic ice. (c) Fits to the first ring using the DIFFaX [92] program suggests 78% of the sequences are cubic in nature. Here q is the scattering vector as defined in the caption of Fig. 6.7(a). Reproduced with permission from Ref. [27], Copyright 2017 ACS.

was estimated from the width of the diffraction peaks, and droplet temperature was estimated using cooling laws. To both confirm and improve upon these measurements, Amaya *et al.* [27] used the free electron laser at the Stanford Linear Accelerator Laboratory (SLAC) to characterize the crystallinity of both water ice, and frozen *n*-decane nanoparticles, inside the nozzle and in the presence of a carrier gas. Particle size was independently determined using SAXS. Using this information, Modak *et al.* [19] were able to confirm that *n*-decane nanodroplets froze to the expected triclinic crystal structure characteristic of bulk solid *n*-decane, whereas, as illustrated in Fig. 6.10, Amaya *et al.* [27] found the water ice was stacking disordered ice I_{sd} with ∼80% cubic sequences.

6.10　Characterizing Nucleation

6.10.1　*Vapor to condensed phase*

As detailed in the other chapters of this book, supersonic expansions have long been used to rapidly cool molecules in order to study the kinetics of gas phase reactions. In these experiments, condensation is not desired. The use of supersonic nozzles to study condensation began in 1934 with the steam condensation experiments of Yellot [95]. Because phase transitions release heat to the flow, and the increase in temperature is accompanied by an increase in pressure relative to the expected isentropic expansion, static pressure measurements have been used from the beginning to detect the presence of a phase transition.

Prior to the ∼1960s, vapor–liquid nucleation experiments did not report quantitative nucleation rates. Rather, the experimental data comprised the conditions — the critical vapor pressures, temperatures, and supersaturations — corresponding to the onset of condensation. The latter, in turn, reflected the characteristic nucleation rates for the particular experiment, and ranged from ∼1 cm^{-3}s^{-1} for thermal diffusion cloud chambers to ∼10^{17} cm^{-3}s^{-1} for a supersonic expansion. For nozzles, the nucleation rate estimates were inferred from 1-D models that coupled the supersonic flow equations to models for nucleation and nanodroplet growth [96, 97].

Table 6.2 summarizes the vapor–liquid condensation experiments conducted in supersonic nozzle that report results in terms of the

Table 6.2. Homogeneous and heterogeneous vapor phase nucleation studies that characterized nucleation via the onset of condensation.

Unary Experiments: Substance (References)

Ammonia [99, 100]; Ar [101–104]; Benzene [105]; n-Butanol [81, 106];
  Carbon Dioxide [107]; Chloroform ($CHCl_3$) [105];
  Ethanol [62, 105, 108, 109]; Freon (CCl_3F) [105]; Nitrogen [110, 111];
  n-Pentanol [106]; n-Propane [12]; n-Propanol [106]; SF_6 [112–115];
  Toluene [13]; Water [62, 95, 98, 99, 116–119]; Zinc [120]

Binary Experiments: Substances (References)

Water–ethanol [121]; Water–ethanol–propanol [61]; H_2O–D_2O [122]

Heterogeneous Experiments: Substances (References)

Water on smoke [123, 124]; CO_2 on water ice [125]; CO_2 on alkanes [75]

"onset" conditions. Although the definition of "onset" can vary, in nozzles it has been defined in terms of the degree of condensation [98], or the deviation of the flow's static pressure [68] or temperature [61] from the expected isentropic value.

As early as 1965, nucleation rate measurements [126] using other experimental apparatuses became quantitative as experimentalists developed ways to accurately determine the characteristic time (Δt) associated with the nucleation event and the number density N of the resulting aerosol. These values yield the nucleation rate directly via

$$J(S,T) = \frac{N}{\Delta t} \times f \qquad (6.20)$$

where f accounts for any changes in the aerosol number density between the time of particle formation and the time of measurement. In an expanding flow this correction is primarily due to continued expansion of the flow, and to a lesser extent, coagulation [127].

In continuously expanding supersonic nozzles, particle formation is a spontaneous, self-quenching process and the characteristic time can be found by evaluating [128]

$$\Delta t = \int \frac{J_{\text{theory}}(S,T)}{J_{\text{theory, max}}(S,T)}\, dt \qquad (6.21)$$

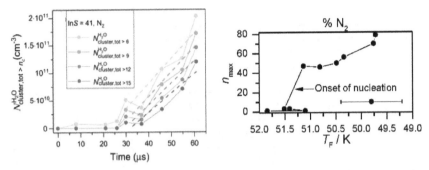

Fig. 6.11. (Left) Determination of the nucleation/growth rate J from linear fits to total cluster number concentrations at early times. Reproduced with permission from Ref. [14], Copyright 2018 AIP Publishing. (Right) The maximum cluster size n_{max} as a function of the flow temperature T_F for propane nucleation. The critical cluster size can be determined from the mass spectra around the onset of nucleation (see text). Reprinted with permission from Ref. [12], Copyright 2016 AIP Publishing.

using the experimentally determined saturation and temperature profiles, $S(t)$ and $T(t)$. Although different theoretical nucleation rate expressions can yield rates that vary by many orders of magnitude, the normalization ensures that the values of Δt are remarkably insensitive to the particular expression used.

When mass spectrometry is used, nucleation is not quenched by increases in T and an approach based on MD simulations has been adapted to determine the nucleation/growth rate J from the measured mass spectra [14, 129]. In particular, J is set equal to the slope of the total cluster number concentration $N_{cluster,tot>n^*}$ as a function of time, where $N_{cluster,tot>n^*}$ is the concentration of clusters that are larger than the critical cluster n^*. In the experiment, t is varied by varying the axial distance in the post-nozzle flow in fine steps (see Figs. 6.4 and 6.5). Figure 6.11 (left) shows a typical example for water condensation at extreme supersaturations; as in MD simulations, the nucleation rates are determined directly from the molecular-level experimental data, i.e. the rate of change of the number density of clusters larger than a given size with time.

As mentioned in Section 6.7 mass spectra can also be used to extract critical cluster sizes directly from the molecular-level data. As explained in more detail in Ref. [11], and illustrated in Fig. 6.11

Table 6.3. Homogeneous nucleation rates measured in supersonic nozzles.

Unary Experiments: Substance (References)

H_2O and D_2O [14, 50, 130, 131]; C3–C5 Alcohols [127, 128]; C5–C9 Alkanes [20, 132, 133]; CO_2 [76, 79, 84]

Binary Experiments: Substances (References)

H_2O–D_2O [131]; water–ethanol [134–136]; water–butanol [137]

(right) at the onset of nucleation both the average and maximum cluster sizes should increase abruptly. This behavior is clearly visible in Fig. 6.11 (right), as a sudden increase of n_{max}, which clearly defines the "onset of nucleation". The range of critical cluster sizes can then be determined from the n_{max} values determined in the mass spectra recorded just before and just after nucleation starts [11, 12]. A recent study [85], however, on water nucleation at extreme supersaturations but low vapor concentrations, hints that similar onset behavior might also arise from kinetic bottlenecks in the cluster growth even in the absence of a true nucleation barrier.

Table 6.3 summarizes the systems for which nucleation rates have been measured in supersonic flows — both within nozzles and in post-nozzle flows.

6.10.2 *Supercooled liquid to solid*

Nucleation from the vapor phase is generally thought to proceed via the liquid state, at least initially, even when the condensate forms below the triple point. If the clusters grow large enough to support crystallization, but remain liquid, they provide an interesting environment in which to characterize liquid to solid nucleation for the following reasons. First, reducing the sample size greatly reduces the probability of contamination by a foreign seed particle. In the supersonic nozzle experiments, droplet purity is further enhanced by the condensation process itself. For example, even if the entering gas stream contained $\sim 10^6$ particles/cm^3 of a condensed impurity, the final aerosol typically contains $\sim 10^{11}$ droplets/cm^3 and, therefore, only 1 particle in 10^5 can be contaminated. Second, liquids dispersed as nanodroplets can reach a high degree of supersaturation.

Thus, when the phase transition occurs the nucleation rate is high enough to put the experimental results in a physical region accessible to MD simulations. Third, although suspended droplets are often touted as providing a "containerless" environment, they still have an interface where surface organization can either enhance or suppress nucleation. If surface phenomena play an important role, freezing can be investigated as a function of the mean droplet size and hence the surface to volume ratio. A related challenge is that small enough droplets also explore higher pressures because the pressure across the highly curved interface is governed by the Laplace equation, $\Delta p = 2\sigma/r$. Thus, disentangling these two effects requires careful analysis [17, 25, 26, 138].

The basic assumption in droplet based $l \to s$ nucleation experiments is that only 1 nucleation event occurs per droplet. The analysis is also simplified if the droplets are monodisperse enough that they are all characterized by the same droplet volume $\langle V \rangle$. Both assumptions are quite good for the droplets produced in supersonic nozzles, and solid nucleation rates can be measured by determining the fraction of droplets that have frozen, F_s, as a function of position using, for example, FTIR spectroscopy [15–19] or electron diffraction [21, 22]. These measurements are then placed on a time basis using the local velocity and the relationship $dt = dz/u$. When $J_{l \to s}$ is not a strong function of temperature, or the measurements are isothermal, the volume based nucleation rate can be derived by fitting all of the data to the functional form expected for a stochastic nucleation process in which the critical nucleus appears randomly throughout the droplet's volume,

$$F_s = 1 - \exp(-J_{l \to s} \langle V \rangle t) \tag{6.22}$$

where t is the time from the onset of ice nucleation. Alternatively, $J_{l \to s}$ can be determined by evaluating

$$J_{l \to s} \langle V \rangle = \frac{1}{t_2 - t_1} \ln \left(\frac{1 - F_{\text{ice}}(t_1)}{1 - F_{\text{ice}}(t_2)} \right) \tag{6.23}$$

Figure 6.12 illustrates the current water ice nucleation rates measured in supersonic nozzles and compares them to those measured in larger droplets. The elevated pressures P in nanodroplets arises from their high curvature and is estimated via the Laplace equation as discussed above. Although the role of pressure on the $l \to s$ nucleation

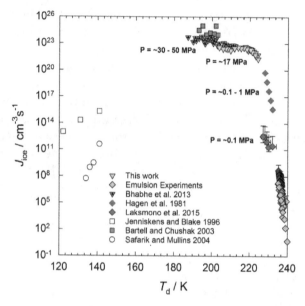

Fig. 6.12.   Freezing rates of supercooled liquid water as a function of the droplet temperature T_d. Solid symbols correspond to freezing from the supercooled liquid droplets. Open symbols correspond to freezing from amorphous films. Data gathered in the present figure are available in the following Refs. [16, 17, 22, 138–141]. The pressures noted in the figure refer to the estimated internal pressures of the droplets that increases with decreasing droplet size. The emulsion data references are available in Ref. [15]. Reprinted with permission from Ref. [17], Copyright 2018 AIP Publishing.

process cannot be neglected, the ability of experiments in supersonic nozzles to reach extreme degrees of supercooling is impressive.

6.10.3   *Heterogeneous nucleation*

Heterogeneous nucleation in supersonic or high speed flows is a largely unexplored area of research. In part, this is because heterogeneous nucleation in supersonic nozzle flows only competes successfully with homogeneous nucleation if the seed particles are small ($\sim$10 nm) and number densities are greater than about 10^7 cm^{-3} [142]. Nevertheless, understanding heterogeneous nucleation under these highly non-equilibrium conditions is important for a wide range of technological applications including energy extraction from low-pressure steam turbines [5], dehydration and heavy hydrocarbon removal from natural gas [7], and the development of hypersonic wind tunnels [6].

There are two ways to explore heterogeneous nucleation. The first is to produce the particles externally and introduce them into the nozzle [123, 124]. The second is to produce the seed aerosol within the nozzle by homogeneous nucleation of one species, and then condensing a second species onto these seed particles at lower temperatures. To date, the only systematic studies of heterogeneous condensation in supersonic flow using the second approach are those of Tanimura *et al.* [125] and Park [75]. Both studied heterogeneous condensation of CO_2 using seed particles comprised of water ice or n-alkanes, respectively. In retrospect, the D_2O-nonane co-condensation studies of Pathak *et al.* [74] could also be considered a case of heterogeneous nucleation. In this case the onset of nucleation was almost always controlled by the appearance of the D_2O droplets, onto which the highly supersaturated nonane rapidly condensed.

Figure 6.13 (left) illustrates representative temperature traces for CO_2 condensation onto water ice [125]. The two distinct heat release events correspond to the formation of the seed particles and the heterogeneous condensation of CO_2 onto them. In the absence of water CO_2 did not condense under the conditions present in this nozzle. Similar behavior was observed by Park for CO_2 condensation onto n-hexane particles [75].

Characterizing heterogeneous nucleation in supersonic nozzle experiments is currently limited to defining the onset of the heterogeneous condensation event. Tanimura *et al.* defined onset as the conditions corresponding to the first detection of heat release [125], whereas Park [75] used the maximum CO_2 saturation reached in the nozzle. When all of the available CO_2 heterogeneous nucleation results for nanoparticles, including those of Nachbar *et al.* [143] on silica and iron oxide, are summarized in Fig. 6.13 (right) two important features are clear. The first is that the saturation levels required to initiate heterogeneous nucleation are extremely high. The second is that the characteristic saturation with respect to solid CO_2, $S_{CO_2,ice}$, always lies relatively close to the extrapolated vapor–liquid equilibrium line, regardless of the seed particles used. In the heterogeneous nucleation literature for H_2O ice, this behavior is taken as an indication that liquid water first nucleates before the water

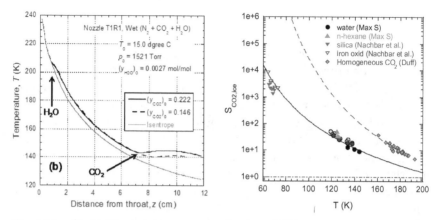

Fig. 6.13. (Left) The heterogeneous nucleation of CO_2 onto water ice particles occurs well downstream of the formation of the water particles. Onset occurs at higher temperatures as the concentration of CO_2 in the gas phase is increased. Reproduced with permission from Ref. [125], Copyright 2015 The Royal Society of Chemistry. (Right) The onset of heterogeneous condensation of CO_2 appears to follow the extrapolated vapor-liquid line (solid black line) over a wide temperature range and irrespective of the seed particle size or composition. Reproduced with permission from Ref. [75], Copyright 2019 PCCP Owner Societies. The dashed grey line and grey diamonds correspond to the homogenous nucleation experiments of Duff. Reproduced with permission from Ref. [107], Copyright 1966 Massachusetts Institute of Technology.

freezes and growth continues via deposition. Whether these data imply the same mechanism for CO_2, is still an open question that would benefit from molecular simulations.

6.11 Summary and Outlook

Over the past 25 years, supersonic nozzles have developed into reliable, quantitative tools for nanoparticle research, including integrated studies of particle nucleation, growth, coagulation and structure. Most work to date has focused on characterizing both the flow and the particles within the nozzle using a broad range of *in situ* techniques including pressure, temperature, and density measurements, light, X-ray and neutron scattering, and vibrational spectroscopy. Complementary experiments that produce the aerosol in a well characterized flow at the nozzle exit, now make it possible to explore nucleation more directly via mass spectroscopy in the uniform post-nozzle flow.

A major goal of nucleation studies in supersonic nozzles and other apparatuses, is to better understand the "nucleation rate problem". Although the "onset of condensation" can be roughly estimated, there are generally large deviations between measured homogeneous nucleation rates and theory even for the simplest systems [40]. Since CNT, and many of its variants, have a stronger temperature dependence than most experiments, the deviations can be many orders of magnitude in either direction.

Whether the issues arise from the experiments, the theory, or from both remains unclear, but the general consensus is that much of the problem stems from the behavior of the smallest, least bulk-like, clusters. Thus, experiments using supersonic expansions — where critical clusters generally contain fewer than ∼10–20 molecules — are particularly important. Furthermore, it would be most useful to compare nucleation rates measured in supersonic flows under comparable conditions but using different characterization methods. For nucleation from the vapor phase, for example, this could mean directly comparing rates determined via mass spectrometry in the post-nozzle flow[14] to those determined from pressure trace and SAXS measurements inside the Laval nozzle [20, 50]. A major challenge to such complementary experiments arises from the difficulty of matching the experimental conditions exactly, and reducing the gap between the two experimental approaches is the first step.

Another big advantage of nucleation studies in supersonic expansions are the high nucleation rates that characterize these methods. This makes it easier to bridge the gap between experiments and large scale molecular simulations — where nucleation rates are generally separated by ∼5–9 orders of magnitude [8–10] — a gap that should close in the foreseeable future with larger scale simulations. The comparison of experiment with molecular modelling is particularly attractive because it also allows for direct comparison of cluster-size resolved data [14] (cluster size distributions, critical cluster sizes). The caveat remains that comparisons must be made in appropriate scaled units to account for differences between the physical properties of the real material and the model material.

Acknowledgments

Support was provided by the National Science Foundation under grant numbers CHE-1464924 and CBET-1511498 and by the Swiss National Science Foundation (SNSF) under grant number 200020_172472. We thank R. Schüpbach for her help with figures and manuscript preparation.

References

[1] Zhang R, Khalizov A, Wang L, Hu M, Xu W. Nucleation and growth of nanoparticles in the atmosphere. Chem Rev. 2012;112(3):1957–2011.

[2] Kerminen VM, Chen XM, Vakkari V, Petaja T, Kulmala M, Bianchi F. Atmospheric new particle formation and growth: review of field observations. Environ Res Lett. 2018;13(10):103003.

[3] Rose C, Sellegri K, Moreno I, Velarde F, Ramonet M, Weinhold K, Krejci R, Andrade M, Wiedensohler A, Ginot P, Laj P. CCN production by new particle formation in the free troposphere. Atmos Chem Phys. 2017;17(2):1529–41.

[4] Storelvmo T, Tan I. The Wegener-Bergeron-Findeisen process — its discovery and vital importance for weather and climate. Meteorol Z (Contrib Atm Sci). 2015;24(4):455–61.

[5] Hughes FR, Starzmann J, White AJ, Young JB. A comparison of modeling techniques for polydispersed droplet spectra in steam turbines. J Eng Gas Turbines Power. 2016;138(4):042603.

[6] Daum FL. Air condensation in a hypersonic wind tunnel. AIAA J. 1963;1(5):1043–46.

[7] Haghighi M, Hawboldt KA, Abdi MA. Supersonic gas separators: review of latest developments. J Nat Gas Sci Eng. 2015;27:109–21.

[8] Diemand J, Angelil R, Tanaka KK, Tanaka H. Large scale molecular dynamics simulations of homogeneous nucleation. J Chem Phys. 2013;139(7):074309.

[9] Angelil R, Diemand J, Tanaka KK, Tanaka H. Homogeneous SPC/E water nucleation in large molecular dynamics simulations. J Chem Phys. 2015;143(6):064507.

[10] Halonen R, Zapadinsky E, Vehkamaki H. Deviation from equilibrium conditions in molecular dynamic simulations of homogeneous nucleation. J Chem Phys. 2018;148(16):164508.

[11] Ferreiro JJ, Gartmann TE, Schläppi B, Signorell R. Can we observe gas phase nucleation at the molecular level? Z Phys Chem. 2015;229(10–12):1765–80.

[12] Ferreiro JJ, Chakrabarty S, Schläppi B, Signorell R. Observation of propane cluster size distributions during nucleation and growth in a Laval expansion. J Chem Phys. 2016;145(21):211907.

[13] Chakrabarty S, Ferreiro JJ, Lippe M, Signorell R. Toluene cluster formation in Laval expansions: nucleation and growth. J Phys Chem A. 2017;121(20):3991–4001.

[14] Lippe M, Chakrabarty S, Ferreiro JJ, Tanaka KK, Signorell R. Water nucleation at extreme supersaturation. J Chem Phys. 2018;149(24):244303.

[15] Manka A, Pathak H, Tanimura S, Wölk J, Strey R, Wyslouzil BE. Freezing water in no-man's land. Phys Chem Chem Phys. 2012;14(13):4505–16.

[16] Bhabhe A, Pathak H, Wyslouzil BE. Freezing of heavy water (D_2O) nanodroplets. J Phys Chem A. 2013;117(26):5472–82.

[17] Amaya AJ, Wyslouzil BE. Ice nucleation rates near similar to 225 K. J Chem Phys. 2018;148(8):084501.

[18] Modak VP, Pathak H, Thayer M, Singer SJ, Wyslouzil BE. Experimental evidence for surface freezing in supercooled n-alkane nanodroplets. Phys Chem Chem Phys. 2013;15(18):6783–95.

[19] Modak VP, Amaya AJ, Wyslouzil BE. Freezing of supercooled n-decane nanodroplets: from surface driven to frustrated crystallization. Phys Chem Chem Phys. 2017;19(44):30181–94.

[20] Ogunronbi KE, Sepehri A, Chen B, Wyslouzil BE. Vapor phase nucleation of the short-chain n-alkanes (n-pentane, n-hexane and n-heptane): experiments and Monte Carlo simulations. J Chem Phys. 2018;148(14):144312.

[21] Huang JF, Bartell LS. Kinetics of homogeneous nucleation in the freezing of large water clusters. J Phys Chem. 1995;99(12):3924–31.

[22] Bartell LS, Chushack YG. Nucleation of ice in large water clusters: experiment and simulation. In: Buch V, Devlin JP, Editors. Water in confining geometries. Berlin: Springer Verlag; 2003. pp. 399–424.

[23] Ishizuka S, Kimura Y, Yamazaki T, Hama T, Watanabe N, Kouchi A. Two-step process in homogeneous nucleation of alumina in supersaturated vapor. Chem Mater. 2016;28(23):8732–41.

[24] Johnston JC, Molinero V. Crystallization, melting, and structure of water nanoparticles at atmospherically relevant temperatures. J Am Chem Soc. 2012;134(15):6650–59.

[25] Espinosa JR, Zaragoza A, Rosales-Pelaez P, Navarro C, Valeriani C, Vega C, Sanz E. Interfacial free energy as the key to the

pressure-induced deceleration of ice nucleation. Phys Rev Lett. 2016;117(13):135702.

[26] Espinosa JR, Vega C, Sanz E. Homogeneous ice nucleation rate in water droplets. J Phys Chem C. 2018;122(40):22892-6.

[27] Amaya AJ, Pathak H, Modak VP, Laksmono H, Loh ND, Sellberg JA, Sierra RG, McQueen TA, Hayes MJ, Williams GJ, Messerschmidt M, Boutet S, Bogan MJ, Nilsson A, Stan CA, Wyslouzil BE. How cubic can ice be? J Phys Chem Lett. 2017;8(14):3216-22.

[28] Lupi L, Hudait A, Molinero V. Heterogeneous nucleation of ice on carbon surfaces. J Am Chem Soc. 2014;136(8):3156-64.

[29] Kashchiev D. Nucleation: basic theory with applications. Oxford: Butterworth-Heinemann; 2000.

[30] Murray BJ, Broadley SL, Wilson TW, Bull SJ, Wills RH, Christenson HK, Murray EJ. Kinetics of the homogeneous freezing of water. Phys Chem Chem Phys. 2010;12(35):10380-7.

[31] Kelton KL, Greer AL. Nucleation in condensed matter: applications in materials and biology. Pergamon, Oxford, UK: Elsevier. 2010.

[32] Kalikmanov VI. Generalized Kelvin equation and pseudospinodal in nucleation theory. J Chem Phys. 2008;129(4):044510.

[33] Wilemski G, Li JS. Nucleation near the spinodal: limitations of mean field density functional theory. J Chem Phys. 2004;121(16):7821-8.

[34] Wedekind J, Chkonia G, Wölk J, Strey R, Reguera D. Crossover from nucleation to spinodal decomposition in a condensing vapor. J Chem Phys. 2009;131(11):114506.

[35] Pruppacher HR, Klett JD. Microphysics of clouds and precipitation. Netherlands: Springer; 2010.

[36] Kashchiev D. On the relation between nucleation work, nucleus size, and nucleation rate. J Chem Phys. 1982;76(10):5098-102.

[37] Strey R, Wagner PE, Viisanen Y. The problem of measuring homogeneous nucleation rates and the molecular contents of nuclei: progress in the form of nucleation pulse measurements. J Phys Chem. 1994;98(32):7748-58.

[38] Kashchiev D. Forms and applications of the nucleation theorem. J Chem Phys. 2006;125(1):014502.

[39] Strey R, Viisanen Y. Measurement of the molecular content of binary nuclei. Use of the nucleation rate surface for ethanol–hexanol. J Chem Phys. 1993;99(6):4693-04.

[40] Wyslouzil BE, Wölk J. Overview: homogeneous nucleation from the vapor phase-the experimental science. J Chem Phys. 2016;145(21):211702.

[41] Manka AA, Brus D, Hyvarinen AP, Lihavainen H, Wölk J, Strey R. Homogeneous water nucleation in a laminar flow diffusion chamber. J Chem Phys. 2010;132(24):244505.

[42] Wölk J, Strey R. Homogeneous nucleation of H_2O and D_2O in comparison: the isotope effect. J Phys Chem B. 2001;105(47):11683–701.

[43] Mikheev VB, Irving PM, Laulainen NS, Barlow SE, Pervukhin VV. Laboratory measurement of water nucleation using a laminar flow tube reactor. J Chem Phys. 2002;116(24):10772–86.

[44] Holten V, Labetski DG, van Dongen MEH. Homogeneous nucleation of water between 200 and 240 K: new wave tube data and estimation of the Tolman length. J Chem Phys. 2005;123(10):104505.

[45] Luijten CCM, Bosschaart KJ, van Dongen MEH. High pressure nucleation in water/nitrogen systems. J Chem Phys. 1997;106(19): 8116–23.

[46] Miller RC, Anderson RJ, Kassner JL, Hagen DE. Homogeneous nucleation rate measurements for water over a wide range of temperature and nucleation rate. J Chem Phys. 1983;78(6):3204–11.

[47] Brus D, Zdimal V, Smolik J. Homogeneous nucleation rate measurements in supersaturated water vapor. J Chem Phys. 2008;129(17):174501.

[48] Brus D, Zdimal V, Smolik J. Homogeneous nucleation rate measurements in supersaturated water vapor (vol 129, 174501, 2008). J Chem Phys. 2009;130(21):219902.

[49] Brus D, Zdimal V, Uchtmann H. Homogeneous nucleation rate measurements in supersaturated water vapor II. J Chem Phys. 2009;131(7):074507.

[50] Kim YJ, Wyslouzil BE, Wilemski G, Wölk J, Strey R. Isothermal nucleation rates in supersonic nozzles and the properties of small water clusters. J Phys Chem A. 2004;108(20):4365–77.

[51] Ford IJ. Thermodynamic properties of critical clusters from measurements of vapour-liquid homogeneous nucleation rates. J Chem Phys. 1996;105(18):8324–32.

[52] Ford IJ. Nucleation theorems, the statistical mechanics of molecular clusters, and a revision of classical nucleation theory. Phys Rev E. 1997;56(5):5615–29.

[53] Laksmono H, Tanimura S, Allen HC, Wilemski G, Zahniser MS, Shorter JH, Nelson D, McManus JB, Wyslouzil BE. Monomer, clusters, liquid: an integrated spectroscopic study of methanol condensation. Phys Chem Chem Phys. 2011;13(13):5855–71.

[54] Hamon S, Le Picard SD, Canosa A, Rowe BR, Smith IWM. Low temperature measurements of the rate of association to benzene dimers in helium. J Chem Phys. 2000;112(10):4506–16.

[55] Bonnamy A, Georges R, Benidar A, Boissoles J, Canosa A, Rowe BR. Infrared spectroscopy of $(CO_2)_N$ nanoparticles ($30 < N < 14500$) flowing in a uniform supersonic expansion. J Chem Phys. 2003;118(8):3612–21.

[56] Bonnamy A, Georges R, Hugo E, Signorell R. IR signature of $(CO_2)_N$ clusters: size, shape and structural effects. Phys Chem Chem Phys. 2005;7(5):963–9.

[57] Sabbah H, Biennier L, Klippenstein SJ, Sims IR, Rowe BR. Exploring the role of PAHs in the formation of soot: pyrene dimerization. J Phys Chem Lett. 2010;1(19):2962–7.

[58] Bourgalais J, Roussel V, Capron M, Benidar A, Jasper AW, Klippenstein SJ, Biennier L, Le Picard SD. Low temperature kinetics of the first steps of water cluster formation. Phys Rev Lett. 2016;116(11):113401.

[59] Bourgalais J, Durif O, Le Picard SD, Lavvas P, Calvo F, Klippenstein SJ, Biennier L. Propane clusters in Titan's lower atmosphere: insights from a combined theory/laboratory study. MNRAS. 2019;488(1): 676–84.

[60] Schläppi B, Litman JH, Ferreiro JJ, Stapfer D, Signorell R. A pulsed uniform Laval expansion coupled with single photon ionization and mass spectrometric detection for the study of large molecular aggregates. Phys Chem Chem Phys. 2015;17(39):25761–71.

[61] Wyslouzil BE, Heath CH, Cheung JL, Wilemski G. Binary condensation in a supersonic nozzle. J Chem Phys. 2000;113(17):7317–29.

[62] Wyslouzil BE, Wilemski G, Beals MG, Frish MB. Effect of carrier gas pressure on condensation in a supersonic nozzle. Phys Fluids. 1994;6(8):2845–54.

[63] Lamanna G, van Poppel J, van Dongen MEH. Experimental determination of droplet size and density field in condensing flows. Exp Fluids. 2002;32(3):381–95.

[64] Ramos A, Fernandez JM, Tejeda G, Montero S. Quantitative study of cluster growth in free-jet expansions of CO_2 by Rayleigh and Raman scattering. Phys Rev A. 2005;72(5):053204.

[65] Ramos A, Tejeda G, Fernández JM, Montero S. Cluster growth in supersonic jets of CO_2 through a slit nozzle. AIP Conf Proc. 2012;1501(1):1383–9.

[66] Tanimura S, Zvinevich Y, Wyslouzil BE, Zahniser M, Shorter J, Nelson D, McManus B. Temperature and gas-phase composition measurements in supersonic flows using tunable diode laser absorption spectroscopy: the effect of condensation on the boundary-layer thickness. J Chem Phys. 2005;122(19):194304.

[67] Stein GD, Moses CA. Rayleigh scattering experiments on the formation and growth of water clusters nucleated from the vapor phase. J Colloid Interface Sci. 1972;39(3):504–12.

[68] Moses CA, Stein GD. On the growth of steam droplets formed in a Laval nozzle using both static pressure and light scattering measurements. J Fluids Eng Trans. ASME. 1978;100(3):311–22.

[69] Wyslouzil BE, Cheung JL, Wilemski G, Strey R. Small angle neutron scattering from nanodroplet aerosols. Phys Rev Lett. 1997;79(3): 431–4.

[70] Wyslouzil BE, Wilemski G, Strey R, Seifert S, Winans RE. Small angle X-ray scattering measurements probe water nanodroplet evolution under highly non-equilibrium conditions. Phys Chem Chem Phys. 2007;9(39):5353–8.

[71] Wyslouzil BE, Wilemski G, Strey R, Seifert S, Winans RE. Small angle X-ray scattering measurements probe water nanodroplet evolution under highly non-equilibrium conditions (vol 9, p. 5353, 2007). Phys Chem Chem Phys. 2008;10(48):7327–8.

[72] Paci P, Zvinevich Y, Tanimura S, Wyslouzil B, Zahniser M, Shorter J, Nelson D, McManus B. Spatially resolved gas phase composition measurements in supersonic flows using tunable diode laser absorption spectroscopy. J Chem Phys. 2004;121(20):9964–70.

[73] Tanimura S, Wyslouzil BE, Zahniser MS, Shorter JH, Nelson DD, McManus B. Tunable diode laser absorption spectroscopy study of CH_3CH_2OD/D_2O binary condensation in a supersonic Laval nozzle. J Chem Phys. 2007;127(3):034305.

[74] Pathak H, Wölk J, Strey R, Wyslouzil BE. Co-condensation of nonane and D_2O in a supersonic nozzle. J Chem Phys. 2014;140(3):034304.

[75] Park Y, Wyslouzil BE. CO_2 condensation onto alkanes: unconventional cases of heterogeneous nucleation. Phys Chem Chem Phys. 2019;21(16):8295–313.

[76] Lippe M, Szczepaniak U, Hou G-L, Chakrabarty S, Ferreiro JJ, Chasovskikh E, Signorell R. Infrared spectroscopy and mass spectrometry of CO_2 clusters during nucleation and growth. J Phys Chem A. 2019;123(12):2426–37.

[77] Wyslouzil BE, Wilemski G, Strey R, Heath CH, Dieregsweiler U. Experimental evidence for internal structure in aqueous — organic nanodroplets. Phys Chem Chem Phys. 2006;8(1):54–57.

[78] Pathak H, Obeidat A, Wilemski G, Wyslouzil B. The structure of D_2O-nonane nanodroplets. J Chem Phys. 2014;140(22):224318.

[79] Dingilian KK, Halonen R, Tikkanen V, Reischl B, Vehkamäki H, Wyslouzil BE. Homogeneous nucleation of carbon dioxide in supersonic nozzles I: experiments and classical theories. Phys Chem Chem Phys. 2020;22:19282–98.

[80] Atkinson DB, Smith MA. Design and characterization of pulsed uniform supersonic expansions for chemical applications. Rev Sci Instrum. 1995;66(9):4434–46.

[81] Rütten F, Alxneit I, Tschudi HR. Kinetics of homogeneous nucleation and condensation of n-butanol at high supersaturation. Atmos Res. 2009;92(1):124–30.

[82] Park Y, Tanimura S, Wyslouzil BE. Enhanced growth rates of nanodroplets in the free molecular regime: the role of long-range interactions. Aerosol Sci Technol. 2016;50(8):773–80.

[83] Pathak H, Mullick K, Tanimura S, Wyslouzil BE. Nonisothermal droplet growth in the free molecular regime. Aerosol Sci Technol. 2013;47(12):1310–24.

[84] Krohn J, Lippe M, Li C, Signorell R. Carbon dioxide and propane nucleation: the emergence of a nucleation barrier. Phys Chem Chem Phys. 2020;22(28):15986–98.

[85] Li C, Lippe M, Krohn J, Signorell R. Extraction of monomer-cluster association rate constants from water nucleation data measured at extreme supersaturations. J Chem Phys. 2019;151(9):094305.

[86] Yoder BL, Litman JH, Forysinski PW, Corbett JL, Signorell R. Sizer for neutral weakly bound ultrafine aerosol particles based on sodium doping and mass spectrometric detection. J Phys Chem Lett. 2011;2(20):2623–28.

[87] Bobbert C, Schütte S, Steinbach C, Buck U. Fragmentation and reliable size distributions of large ammonia and water clusters. Eur Phys J D. 2002;19(2):183–92.

[88] Lengyel J, Pysanenko A, Poterya V, Kocisek J, Farnik M. Extensive water cluster fragmentation after low energy electron ionization. Chem Phys Lett. 2014;612:256–61.

[89] Signorell R. Verification of the vibrational exciton approach for CO_2 and N_2O nanoparticles. J Chem Phys. 2003;118(6):2707–15.

[90] Schläppi B, Ferreiro JJ, Litman JH, Signorell R. Sodium-sizer for neutral nanosized molecular aggregates: quantitative correction of size-dependence. Int J Mass Spectrom. 2014;372:13–21.

[91] Beck RD, Hineman RF, Nibler JW. Stimulated Raman probing of supercooling and phase transitions in large N_2 clusters formed in free jet expansions. J Chem Phys. 1990;92(12):7068–78.

[92] Malkin TL, Murray BJ, Brukhno AV, Anwar J, Salzmann CG. Structure of ice crystallized from supercooled water. Proc Natl Acad Sci USA. 2012;109(4):1041–5.

[93] Bartell LS, Harsanyi L, Valente EJ. Phases and phase changes of molecular clusters generated in supersonic flow. J Phys Chem. 1989;93(16):6201–5.

[94] Bartell LS. Nucleation rates in freezing and solid-state transitions. Molecular clusters as model systems. J Phys Chem. 1995;99(4): 1080–9.

[95] Yellott J. Supersaturated steam. Trans ASME. 1934;56:411–30.

[96] Oswatitsch K. Kondensationserscheinungen in Überschalldüsen. ZAMM — J Appl Math Mech/Z Angew Math Mech. 1942;22(1):1–14.

[97] Wegener PP, Parlange JY. Condensation by homogeneous nucleation in the vapor phase. Naturwissenschaften. 1970;57(11):525–33.

[98] Wegener PP, Pouring AA. Experiments on condensation of water vapor by homogeneous nucleation in nozzles. Phys Fluids. 1964;7(3):352–61.

[99] Jaeger HL, Willson EJ, Hill PG, Russell KC. Nucleation of super-saturated vapors in nozzles. I. H_2O and NH_3. J Chem Phys. 1969;51(12):5380–8.

[100] Kremmer M, Okouronmu O. Condensation of ammonia vapor during rapid expansion. MIT Gas Turbine Lab; 1963.

[101] Pierce T, Sherman PM, McBride DD. Condensation of Argon in a supersonic stream. Astronaut Acta. 1971;16(1):1–4.

[102] Wu BJC, Wegener PP, Stein GD. Homogeneous nucleation of argon carried in helium in supersonic nozzle flow. J Chem Phys. 1978;69(4):1776–7.

[103] Sinha S, Bhabhe A, Laksmono H, Wölk J, Strey R, Wyslouzil B. Argon nucleation in a cryogenic supersonic nozzle. J Chem Phys. 2010;132(6):064304.

[104] Sinha S, Laksmono H, Wyslouzil BE. A cryogenic supersonic nozzle apparatus to study homogeneous nucleation of Ar and other simple molecules. Rev Sci Instrum. 2008;79(11):114101.

[105] Dawson DB, Willson EJ, Hill PG, Russell KC. Nucleation of super-saturated vapors in nozzles. II. C_6H_6, $CHCl_3$, CCl_3F, and C_2H_5OH. J Chem Phys. 1969;51(12):5389–97.

[106] Gharibeh M, Kim Y, Dieregsweiler U, Wyslouzil B, Ghosh ED, Strey R. Homogeneous nucleation of *n*-propanol, *n*-butanol, and *n*-pentanol in a supersonic nozzle. J Chem Phys. 2005;122(9):094512.

[107] Duff K. Non-equilibrium condensation of carbon dioxide in supersonic nozzles. ScD., Massachusetts Institute of Technology; 1966.

[108] Wu BJC, Wegener PP, Clumpner JA. Nucleation and growth of ethanol drops in condensation. Bull Am Phys Soc. 1972;17(4):544–5.

[109] Wegener PP, Wu BJC, Clumpner JA. Homogeneous nucleation and growth of ethanol drops in supersonic flow. Phys Fluids. 1972;15(11):1869–76.

[110] Koppenwallner G, Dankert C. Homogeneous condensation in N_2, Ar, and H_2O vapor free jets. J Phys Chem. 1987;91(10):2482–6.

[111] Bhabhe A, Wyslouzil B. Nitrogen nucleation in a cryogenic supersonic nozzle. J Chem Phys. 2011;135(24):244311.

[112] Wu BJC, Wegener PP, Stein GD. Condensation of sulfur hexafluoride in steady supersonic nozzle flow. J Chem Phys. 1978;68(1):308–18.

[113] Wu BJC, Wegener PP, Stein GD. Determination of onset of SF_6 condensation by homogeneous nucleation. Bull Am Phys Soc. 1977;22(10):1274–74.

[114] Kim SS, Stein GD. Evidence for mixed clusters formed during sulfur hexafluoride expansions in an argon carrier gas. J Appl Phys. 1980;51(12):6419–21.

[115] Abraham O, Kim SS, Stein GD. Homogeneous nucleation of sulfur hexafluoride clusters in Laval nozzle molecular beams. J Chem Phys. 1981;75(1):402–11.

[116] Rettaliata JT. Undercooling in steam nozzles. Trans ASME. 1936;58:599–605.

[117] Binnie M, Woods MW. The pressure distribution in a convergent-divergent steam nozzle. Proc Inst Mech Eng. 1938;138(1):229–66.

[118] Binnie AM, Green JR. An electrical detector of condensation in high-velocity steam. Proc R Soc London Ser. 1942;181(A985):0134–54.

[119] Gyarmathy G, Meyer H. Spontane Kondensation. VDI-Verlag; 1965.

[120] McBride DD, Sherman PM, Pierce TH. Condensation of zinc vapor in a supersonic carrier gas. Appl Sci Res. 1971;25(1–2):83–96.

[121] Wegener PP, Wu BJC. Homogeneous and binary nucleation: new experimental results and comparison with theory. Faraday Discuss. 1976;61:77–82.

[122] Heath CH, Streletzky K, Wyslouzil BE, Wölk J, Strey R. H_2O-D_2O condensation in a supersonic nozzle. J Chem Phys. 2002;117(13):6176–85.

[123] Pouring AA. Effects of heterogeneous nucleation of water vapor in nozzles. J Basic Eng. 1970;92(4):689–94.

[124] Buckle ER, Pouring AA. Effects of seeding on the condensation of atmospheric moisture in nozzles. Nature. 1965;208(5008):367–9.

[125] Tanimura S, Park Y, Amaya A, Modak V, Wyslouzil BE. Following heterogeneous nucleation of CO_2 on H_2O ice nanoparticles with microsecond resolution. RSC Adv. 2015;5(128):105537–50.

[126] Allard EF, Kassner JL. New cloud-chamber method for the determination of homogeneous nucleation rates. J Chem Phys. 1965;42(4):1401–5.

[127] Mullick K, Bhabhe A, Manka A, Wölk J, Strey R, Wyslouzil BE. Isothermal nucleation rates of n-Propanol, n-Butanol, and n-Pentanol in supersonic nozzles: critical cluster sizes and the role of coagulation. J Phys Chem B. 2015;119(29):9009–19.

[128] Ghosh D, Manka A, Strey R, Seifert S, Winans RE, Wyslouzil B. Using small angle x-ray scattering to measure the homogeneous nucleation rates of n-propanol, n-butanol, and n-pentanol in supersonic nozzle expansions. J Chem Phys. 2008;129(12):124302.

[129] Yasuoka K, Matsumoto M. Molecular dynamics of homogeneous nucleation in the vapor phase. II. Water. J Chem Phys. 1998;109(19):8463–70.

[130] Khan A, Heath CH, Dieregsweiler UM, Wyslouzil BE, Strey R. Homogeneous nucleation rates for D_2O in a supersonic Laval nozzle. J Chem Phys. 2003;119(6):3138–47.

[131] Heath CH, Streletzky KA, Wyslouzil BE, Wölk J, Strey R. Small angle neutron scattering from D_2O-H_2O nanodroplets and binary nucleation rates in a supersonic nozzle. J Chem Phys. 2003;118(12):5465–73.

[132] Ghosh D, Bergmann D, Schwering R, Wölk J, Strey R, Tanimura S, Wyslouzil BE. Homogeneous nucleation of a homologous series of n-alkanes (C_iH_{2i+2}, $i = 7$–10) in a supersonic nozzle. J Chem Phys. 2010;132(2):024307.

[133] Ogunronbi KE, Wyslouzil BE. Vapor-phase nucleation of n-pentane, n-hexane, and n-heptane: critical cluster properties. J Chem Phys. 2019;151(15):154307.

[134] Tanimura S, Dieregsweiler UM, Wyslouzil BE. Binary nucleation rates for ethanol/water mixtures in supersonic Laval nozzles. J Chem Phys. 2010;133(17):174305.

[135] Tanimura S, Dieregsweiler UM, Wyslouzil BE. Binary nucleation rates for ethanol/water mixtures in supersonic Laval nozzles (vol 133, 174305, 2010). J Chem Phys. 2010;133(19):199901.

[136] Tanimura S, Pathak H, Wyslouzil BE. Binary nucleation rates for ethanol/water mixtures in supersonic Laval nozzles: analyses by the first and second nucleation theorems. J Chem Phys. 2013;139(17):174311.

[137] Mullick KA. Binary nucleation of n-Butanol and deuterium oxide conducted in supersonic nozzles [PhD]. Ohio State University; 2012.

[138] Laksmono H, McQueen TA, Sellberg JA, Loh ND, Huang C, Schlesinger D, Sierra RG, Hampton CY, Nordlund D, Beye M, Martin AV, Barty A, Seibert MM, Messerschmidt M, Williams GJ, Boutet S, Arnann-Winkel K, Loerting T, Pettersson LGM, Bogan MJ, Nilsson A. Anomalous behavior of the homogeneous ice nucleation rate in "No-Man's Land". J Phys Chem Lett. 2015;6(14):2826–32.

[139] Hagen DE, Anderson RJ, Kassner JL. Homogeneous condensation — freezing nucleation rate measurements for small water droplets in an expansion cloud chamber. J Atmos Sci. 1981;38(6):1236–43.

[140] Jenniskens P, Blake DF. Crystallization of amorphous water ice in the solar system. Astrophys J. 1996;473(2):1104–13.

[141] Safarik DJ, Mullins CB. The nucleation rate of crystalline ice in amorphous solid water. J Chem Phys. 2004;121(12):6003–10.

[142] Heiler M. Instationäre Phänomene in homogen/heterogen kondensierenden Düsen- und Turbinenströmungen [Doctoral Thesis]. Universität Karlsruhe; 1999.

[143] Nachbar M, Duft D, Mangan TP, Gomez Martin JC, Plane JMC, Leisner T. Laboratory measurements of heterogeneous CO_2 ice nucleation on nanoparticles under conditions relevant to the Martian mesosphere. J Geophys Res Planets. 2016;121(5):753–69.

Chapter 7

Collisional Energy Transfer in Uniform Supersonic Flows

Ilsa R. Cooke*,†,‡ and Ian R. Sims*,§

*CNRS, IPR (Institut de Physique de Rennes)-UMR 6251,
Université de Rennes, F-35000 Rennes, France
†Department of Chemistry, University of British Columbia, Canada
‡icooke@chem.ubc.ca
§ian.sims@univ-rennes1.fr

Abstract

An overview is given of collisional energy transfer measurements made in uniform supersonic flows. The fundamentals of molecular collisions and electronic, vibrational, and rotational energy transfer are briefly described, followed by a summary of the collisional systems that have been studied to date. The measurement of total and state-to-state collisional rate constants and cross sections is described and illustrated by reference to case studies for rotational, vibrational, and electronic energy transfer. These experimental data are critical for radiative transfer modeling of astrophysical environments as well as for fundamental tests of theoretical calculations.

Keywords: CRESU; Energy transfer; Collisions.

7.1 Introduction

Collisions lie at the very heart of chemical physics and kinetics. Under Earth's atmospheric conditions at sea level, molecules undergo

393

$\sim 10^{10}$ collisions every second, which could be reactive — producing new molecules; elastic — with no exchange of energy; or inelastic — involving energy exchange. The type of collision depends on the interaction potential of the collision partners and the collision conditions, such as temperature and pressure. This chapter focuses on inelastic collisions in which energy is transferred between the collisional partners and how the kinetics of such collisions has been studied over the past ca. 25 years using uniform supersonic flows supplied by the CRESU technique. Information on collisional energy transfer is required in order to predict the rates of both unimolecular dissociation (or isomerization) and termolecular association reactions. Unimolecular rate constants are pressure dependent as collisional energy transfer is necessary to activate and/or deactivate the vibrationally excited reactants. At the low-pressure limit, energy transfer is rate limiting, introducing a level of complexity that makes predictions much more challenging than for simple bimolecular reactions that do not have a pressure dependence.

Collisional energy transfer is fundamental to the establishment of local thermodynamic equilibrium (LTE). Under LTE conditions, molecular and atomic energy levels will be populated following the Boltzmann distribution, where the excitation temperature (i.e. the temperature at which we would expect to find a system with a ratio of level populations, n_u/n_l, given by the Boltzmann equation) is assumed to equal the kinetic temperature. In general, the LTE assumption only holds when collisions are frequent, i.e. at high pressures or densities. An area in which the establishment (or lack thereof) of LTE conditions is particularly important is the analysis of spectral lines observed toward astronomical sources. The determination of molecular abundances from telescope data is a key outcome of astronomical observations and is coupled to the estimation of the physical conditions of astrophysical sources, in particular, their kinetic temperature and the volume density. The molecular abundances can be used to determine, with the aid of astrochemical kinetics models, fundamental parameters such as the ages of the molecular clouds or the extent of ionization rate by cosmic rays.

At low densities, radiative decay competes with collisions and the excitation temperature is typically less than the kinetic temperature. At these low densities, often found in the interstellar medium, LTE Boltzmann distributions for the level populations are no longer

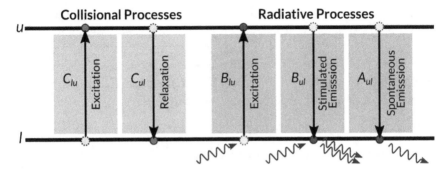

Fig. 7.1. Schematic representations of collisional and radiative processes leading to a change in energy level in a two-level system. Collisional excitation (C_{lu}) and relaxation or deexcitation (C_{ul}) are shown on the left in green and radiative excitation/absorption (B_{lu}), stimulated emission (B_{ul}), and spontaneous emission (A_{ul}) are shown on the right in purple.

good approximations. More sophisticated non-LTE methods have been developed which solve explicitly for the balance between excitation and deexcitation of molecular energy levels by collisional and radiative processes [1–3]. A two-level system is shown in Fig. 7.1 to illustrate the collisional and radiative processes that can occur, leading to a change in energy level ($u \rightarrow l$ or $l \rightarrow u$): collisional excitation (C_{lu}) or relaxation (C_{ul}), radiative excitation (B_{lu}), stimulated emission (B_{ul}), and spontaneous emission (A_{ul}). These methods require molecular collision data as input, as well as the more widely available spectroscopic data. Collisional data, especially laboratory measurements, exist for only a very small subset of astrophysically relevant species.

From a more fundamental standpoint, energy transfer studies are motivated by the desire to understand how the rates of energy transfer are connected to the intermolecular potential, in particular, the roles of the long-range attractive interaction and short-range repulsion. Investigations of inelastic collisions at low temperature (and therefore low collision energy) are important given that these collisions will be governed by weak long-range forces. Energy transfer in such low-temperature collisions involving closed shell species, when the well depth becomes comparable to the collision energy, is controlled by the intermolecular attraction. Experiments involving low-energy collisions are necessary to benchmark theoretical calculations, as outlined in Chapter 11. Uniform supersonic flows produced by the

CRESU technique provide an ideal environment for studying collisional energy transfer processes under thermal conditions and without wall collisions. In the following sections, we review experiments on energy transfer that have been conducted using the CRESU technique and discuss how this technique may be applied in the future to address remaining problems in the field.

7.2   Energy Transfer Fundamentals

When an excited molecule undergoes an inelastic collision, it will lose some of its energy; in order to conserve the total energy of the system, the energy lost will be transferred to the collision partner. Energy transfer processes can involve exchange of electronic, vibrational, rotational, and translational energy. Figure 7.2 shows schematically the relative energies absorbed or emitted during each of these transitions: electronic transitions involving the largest energy difference, typically in the ultraviolet, followed by vibrational transitions in the infrared, and rotational transitions in the microwave region. Electronic fine-structure (spin–orbit) transitions are an exception to this ordering, with transition energies typically corresponding to the microwave region. Theoretical models used to describe these inelastic collisional energy transfer processes are discussed in detail in Chapter 11.

Inelastic **rotational energy transfer** (RET) concerns collisions that involve changes in rotational energy as a result of transitions from one rotational state to another, ΔJ, where J is the quantum number for total rotational angular momentum. RET differs from vibrational and electronic energy transfer due to the fact that rotational energy levels are generally spaced much closer than kT, thus many transitions are energetically accessible. These transitions typically occur in the microwave region of the electromagnetic spectrum.

Resonant rotational energy transfer occurs when $J_i + J_j = J_i' + J_j'$ in the collision between two species A and B, $A(J_i) + B(J_j) \rightarrow A(J_i') + B(J_j')$. Transitions induced by collisions between polar molecules or between a polar and a non-polar molecule generally obey the dipole selection rules commonly seen in spectroscopy; for a linear molecule, these rules give $\Delta J = 0, \pm 1$ and $\Delta M = 0, \pm 1$. The selection rules

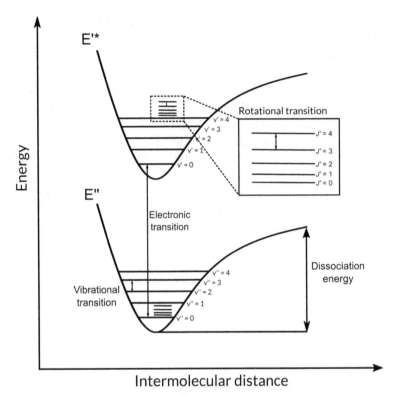

Fig. 7.2. Schematic showing the different types of transitions that can be involved in collisional energy transfer and their relative energy spacings. Transitions can also occur where the change in quantum number is greater or less than 1.

were first derived for RET using microwave–microwave double resonance (DR) methods, pioneered by Oka and co-workers [4]. These collisional selection rules are often called *propensity rules* since forbidden transitions do occur. Subsequent work has identified other propensity rules, for example, $\Delta J =$ even for NO–rare gas collisions [5] and $\Delta J =$ odd for CO–He collisions [6].

Vibrational energy transfer, resulting in a change in vibrational level(s), is thought to occur by two main mechanisms. If the collision partner can accept a substantial portion of the energy that is transferred into one of its own vibrations, the mechanism is known as V-V energy transfer or exchange. If, instead, the collision partner is, for example, a noble gas atom with zero rovibrational degrees of freedom into which the energy can be dissipated, then the energy transfer manifests as an increase in translational energy (V-T transfer). If the collision partner is a molecule that does not have any vibrations of similar frequency to the initially excited vibration, the vibrational energy may still be transferred to the internal degrees of freedom of the collision partner. In this case, some of the energy is transferred into rotational energy within the collision partner and the remainder is converted to translational energy. This mechanism is known as vibration-to-rotation, translation (V-R,T) energy transfer. In general, V-R,T energy transfer is slow, making it difficult to measure using the CRESU technique.

In the process of V-R,T transfer, a diatomic molecule B in vibrational state v_i collides with a molecule or atom A, and the relative collision velocity is given by $\vec{v}(A|B)_i$. After collision, B is in vibrational state v_j, and the relative velocity of separation is $\vec{v}(A|B)_f$,

$$A(\vec{v}(A|B)_i) + B(v_i) \longrightarrow A(\vec{v}(A|B)_f) + B(v_j) \qquad (7.1)$$

In the case where rotational transitions are neglected, the vibrational energy of B is exchanged for relative translational energy of A and B. In *resonant* V-V energy transfer, no energy is exchanged between vibrational and translational modes, and like in the case of resonant RET, $v_i + v_j = v_i' + v_j'$. Often, there is a small V-T energy exchange if the molecular vibrations are not perfectly matched, but this is usually very minor.

At low temperatures, the long-range attractive potential is critical to the efficiency of vibrational energy transfer. In some cases, these attractive forces may be strong enough to form a transient collision complex, which can greatly increase the rate of energy transfer.

Electronic energy transfer (EET) involves relaxation or excitation occurring between electronic energy levels. EET comprises a number of different mechanisms involving electronic transitions between states E'* and E'', as shown in Fig. 7.2. Electronic energy transfer in the form of collisional deactivation of an electronically excited state is often referred to as quenching. Electronic energy transfer, somewhat confusingly, has also been referred to as resonance energy transfer (RET), especially in connection to fluorescence quenching in the condensed phase. The energy of an electronically excited molecule (D*) is transferred to an acceptor molecule, A, facilitated by dipole–dipole couplings between the two species:

$$D^* + A \longrightarrow D + A^* \qquad (7.2)$$

$$D^* + A \longrightarrow D + A \qquad (7.3)$$

Electronic energy transfer can also take place by the formation of long-lived collision complexes which can facilitate non-radiative energy redistribution. In the latter case, the energy transfer does not involve electronic excitation of A, but rather energy is transferred into nuclear motion or by emission of a virtual photon. Electronic energy transfer can then be further categorized into triplet–triplet, i.e. $^1A + ^3D^* \rightarrow ^3A^* + ^1D$, or singlet–singlet, i.e. $^1A + ^1D^* \rightarrow ^1A^* + ^1D$. Energy transfer between singlet and triplet states (or states of different multiplicity more generally) must involve spin–orbit coupling.

Experiments on collisional energy transfer can be placed in two distinct categories: beam experiments and bulb experiments. Crossed molecular beam experiments rely on the preparation of one of the collision partners (typically a molecular species) in a very-low-energy quantum state by supersonic expansion and cooling. Collisions with another supersonic molecular beam of the collision partner (typically monatomic or diatomic gases) result in population transfer that can be observed by, for example, Resonance-Enhanced Multi-Photon Ionization (REMPI) or Laser-induced Fluorescence (LIF) probing. These experiments usually give access to relative integral and differential cross sections for energy transfer as a function of collisional energy, sometimes (if merged beams can be employed) to very-low-collision energy [7, 8].

While such energy-resolved experiments can reveal phenomena such as scattering resonances in exquisite detail, they cannot, in general, provide absolute measurements of energy transfer cross sections. In contrast, bulb measurements, are usually carried out by exciting a small concentration of one of the collision partners (typically molecular) in a known excess density of a bath gas under LTE conditions. Such conditions enable the determination of absolute rate constants for energy transfer. The CRESU experiments that are the subject of this chapter, despite being made in supersonic flows, take place under LTE / bulb conditions.

There are two essential ingredients required for experimentally measuring energy transfer in bulb experiments: preparation of the initial state and detection of the final states that are populated after collisional energy transfer. These experiments typically involve a *pump* to initiate a specific out-of-equilibrium state. In the absence of the pump, the molecules undergo constant collisions and exchange energy, but the system ensemble is in a steady state with an equilibrium distribution among various energy levels and thus these transitions cannot be monitored directly. The pump is introduced to bring a specific energy level out of equilibrium and its effect on the neighboring levels can then be monitored using a *probe* and related to collisional energy transfer.

7.3 Summary of Energy Transfer Measurements Made in Uniform Supersonic Flows

The past ca. 25 years have seen huge advances in the measurement of low-temperature collision energy transfer, facilitated by the CRESU technique. The measurements that have been made in uniform supersonic (and hypersonic) flows produced by Laval nozzles are summarized in Table 7.1. Soon after Sims and Smith [9] suggested that measurements of energy transfer should be possible within CRESU flows, the first measurements of vibrational and rotational energy transfer were reported using the Rennes and Birmingham CRESU experiments. The collisional relaxation of $CH(X^2\Pi, v = 1)$ by CO and N_2 was published in 1996 [10], and represented the first experiments involving vibrationally excited species carried out in a CRESU apparatus. In 1997, the Birmingham CRESU came online with the first application of a CRESU to the study of rotational energy transfer, involving collisions between NO $(X^2\Pi, v = 3)$ and He down to 15 K [11]. These measurements were soon extended down to 7 K for He, and to collisions with N_2 and Ar buffer gases [5] Vibrational energy transfer was subsequently measured for $CH(v = 1)$ in collisions with H_2 and D_2 [12], for the vibrational self-relaxation of $NO(v = 3)$ [13] and for the relaxation of highly excited toluene in collisions with He, Ar, and N_2 [14] Rotational energy transfer measurements have been extended to CO $(v = 2)$ in collisions with He [6] and with Ar [15, 16]. Measurement of electronic energy transfer constitute a large proportion of the measurements made in CRESU flows [17–26] by various mechanisms such as charge transfer, intersystem

Table 7.1. Summary of collisional energy transfer processes studied using CRESU experiments.

Molecule	Collision Partner	Type of ET	Method	Temperature Range (K)	Ref.
CH ($v = 1$)	N_2	VET	PLP-LIF	23–295	Herbert, 1996 [10]
	CO	VET	PLP-LIF	23–295	Herbert, 1996 [10]
	H_2/D_2	VET	PLP-LIF	23–295	Brownsword, 1997 [12]
NO ($v = 3$)	NO	VET	IRUVDR	7–85	James, 1997 [13]
NO ($v = 3, J$)	He	RET	IRUVDR	7–295	James, 1997 [11], 1998 [5]
	N_2	RET	IRUVDR	47–86	James, 1998 [5]
	Ar	RET	IRUVDR	27–139	James, 1998 [5]
Al($^2P_{3/2}$)	Ar	EET:S-O	PLP-LIF	44–137	Le Picard, 1998 [17]
C_7H_8	He	VET	TR-IRF	38–300	Wright, 2000 [14]
	N_2	VET	TR-IRF	74–300	Wright, 2000 [14]
	Ar	VET	TR-IRF	112–300	Wright, 2000 [14]
Si(^3P)	He	EET	PLP-LIF	15–49	Le Picard, 2002 [18]
C(^3P)	He	EET	PLP-LIF	15	Le Picard, 2002 [18]
CO ($v = 2, J$)	He	RET	IRUVDR	15–294	Carty, 2004 [6]
	H_2	RET	IRUVDR	5.5–293	Labiad 2020 [16]
$^3C_2$	NO	EET:ISC	PLP-LIF	24–300	Páramo, 2006 [19]
	O_2	EET:ISC	PLP-LIF	24–300	Páramo, 2006 [19]
NO($A^2\Sigma^+$)	NO	EET	LIF	34–109	Sánchez-González, 2014 [20]
	O_2	EET	LIF	34–109	Sánchez-González, 2014 [20]
	C_6H_6	EET	LIF	130–300	Winner, 2018 [21]
	C_6F_6	EET	LIF	130–300	Winner, 2018 [21]
O(^1D)	N_2	EET	LIF	50–296	Grondin, 2016 [22]
	O_2	EET	LIF	50–296	Grondin, 2016 [22]
	Ar	EET	LIF	50–296	Grondin, 2016 [22]
	Kr	EET	LIF	50–296	Nuñez Reyes, 2018 [20]
C(^1D)	N_2	EET	LIF	50–296	Hickson, 2016 [24]
S(^1D)	Ar	EET	PLP-LIF	5.8–298	Lara, 2017 [28]
CO($v = 2, J$)	Ar	RET	IRUVDR	30.5–293	Mertens, 2018 [15]
$^1CH_2$	He	EET	PLP-LIF	43–298	Douglas, 2018 [25]
	N_2	EET	PLP-LIF	43–298	Douglas, 2018 [25]
	O_2	EET	PLP-LIF	43–298	Douglas, 2018 [25]

(*Continued*)

Table 7.1. (*Continued*)

Molecule	Collision Partner	Type of ET	Method	Temperature Range (K)	Ref.
	H_2	EET	PLP-LIF*	73–298	Douglas, 2018 [25], 2019 [26]
	CH_4	EET	PLP-LIF*	73–298	Douglas, 2018 [25], 2019 [26]
	C_2H_6	EET	PLP-LIF*	100–298	Douglas, 2019 [26]
	C_2H_4	EET	PLP-LIF*	100–298	Douglas, 2019 [26]
	C_2H_2	EET	PLP-LIF*	100–298	Douglas, 2019 [26]
$SO(X^3\Sigma', v = 0, 1, 2)$	SO_2, He	VET	CPUF	22	Abeysekera, 2014 [27]

Note: *Rate constants for relaxation were determined indirectly via titration to OH or H.

Key: VET = vibrational energy transfer, RET = rotational energy transfer, S-O = spin orbit, EET = electronic energy transfer (quenching), ISC = intersystem crossing, (PLP)-LIF = (pulsed-laser photolysis) laser-induced fluorescence, IRUVDR = infrared ultraviolet double resonance, TR-LIF = time-resolved infrared fluorescence, CPUF = chirped-pulse in uniform flow.

crossing, and spin–orbit relaxation. Recently, vibrational relaxation of $SO(X^3\Sigma^-, v = 0, 1, 2)$ has been investigated in a pulsed CRESU using chirped-pulse microwave spectroscopy [27].

7.4   Rotational Energy Transfer

Rotational energy transfer has been a topic of considerable interest in chemical physics due to its application to wide range of fields. Quantitative rotational energy transfer rates are required to understand a range of gas phase phenomena such as the evolution of rotational population in combustion processes, spectral line broadening, transport of sound waves, gas viscosity, and the efficiency of molecular lasers [29, 30]. At low temperature, RET is of particular importance to processes occurring in the interstellar medium. It is considered to be the principal mechanism to cool down interstellar clouds via collisional excitation followed by emission of radiation, and plays a major role in the gravitational collapse of these clouds leading to star formation [31]. Knowledge of RET rates is also of great importance in deriving column densities in the interstellar medium. In addition, low-temperature experiments on RET provide a rigorous test of theory at collision energies less than the well depth of the intermolecular potential between the collision partners and comparable to the spacing between neighboring rotational levels.

Rotational energy transfer has been measured in CRESU flows for collisions involving NO and CO; specifically, NO ($X^2\Pi$; $\Omega = 1/2$; $v = 3$; $J = 0.5, 3.5$ or 6.5) and CO ($X^1\Sigma^+$, $v = 2$, $J = 0, 1, 4, 6$). CRESU experiments have been conducted at temperatures as low as $5.5\,\mathrm{K}$ and $7\,\mathrm{K}$ for collisions involving CO [16] and NO [5], respectively.

7.4.1 *Rotational energy transfer studies involving NO ($X^2\Pi$)*

Rotational energy transfer in collisions between NO ($X^2\Pi$) and He, published in 1997 by the group of Ian Sims and Ian Smith, was the first study carried out using the CRESU apparatus built in Birmingham and was the first application of the CRESU technique to the study of RET [11]. Shortly after, the measurements were extended down to $7\,\mathrm{K}$, by cryogenic cooling of the reservoir and Laval nozzle, as well as to the study of RET in collisions with N_2 and Ar [5].

Measurements of collisions between NO ($X^2\Pi$) and noble gases (He and Ar) represent a major testing ground for theoretical studies of rotational energy transfer involving molecules in Π electronic states. NO is the only commonly known stable diatomic molecule with an orbital angular momentum, resulting in two close-lying ground electronic states $^2\Pi_{3/2}$ and $^2\Pi_{1/2}$ which are separated by $\sim 120\,\mathrm{cm}^{-1}$. Because of its open-shell electronic structure, coupling of the translational motion of the collision partners with the rotation motion and the internal spin and electronic angular momenta makes theoretical treatment more complicated than that of closed-shell molecules. Prior to the CRESU studies in Birmingham, several groups investigated collisional energy transfer of NO with rare gases using infrared ultraviolet double resonance (IRUVDR) in flow cells [32–34] or using crossed-molecular beam experiments [35, 36]. The measurements made in Birmingham, facilitated by the newly built CRESU apparatus, were the first direct measurements of RET carried out at temperatures below $80\,\mathrm{K}$.

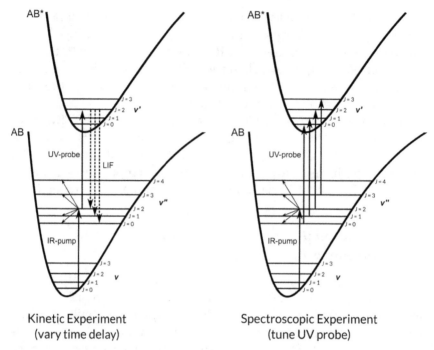

Fig. 7.3.   The two types of double resonance experiments conducted to measure total removal and state-to-state rate constants for rotational energy transfer.

Double resonance techniques are well developed in molecular spectroscopy and are widely used to study the interconnectivity of states. In IRUVDR kinetics experiments, *double* resonance refers to the use of two lasers to excite two interconnected transitions. An infrared pump pulse is used to excite a vibrational transition within the electronic ground state and populate specific J rotational levels. The UV probe laser is then used to excite an electronic transition, from which the fluorescence can be measured, in order to monitor the fate of a state-specific subset of the excited molecule (see Fig. 7.3). Variations in the experiment permit different information to be gained. By changing the time delay between the pump and probe lasers, the total inelastic decay from an initial rotational state J_i to *all* of the possible final rotational states J_f and the rate constant for the *total removal* of J_i can be determined in what has been called the *kinetic experiment*, represented in the left-hand panel of Fig. 7.3. In order to measure the state-to-state rate constants, that is, the rate of energy transfer from a specific J_i to a specific J_f, the UV probe laser is scanned to produce a spectrum containing all of the possible final rotational states. In this so-called *spectroscopic experiment* (see the right-hand panel of Fig. 7.3), a spectrum is taken at very short time delay between the pump and probe pulses (tens of nanoseconds) and another at a long time delay (microseconds) after rotational equilibrium has been reached, but before the much slower process of vibrational relaxation takes place to a significant degree.

Total and state-to-state rate constants for the collisional energy transfer rate constants were determined for NO ($X^2\Pi$; $\Omega = 1/2$; $v = 3$; $J = 0.5, 3.5$ or 6.5) collisions with He, initially down to $15\,\mathrm{K}$ and extended to $7\,\mathrm{K}$, with Ar down to $27\,\mathrm{K}$, and for N_2 down to $47\,\mathrm{K}$.

The $J = 0.5, 3.5$, and 6.5 rotational levels within the $v = 3$ vibrational level of the $X^2\Pi_{1/2}$ ground state were selectively populated by tuning infrared radiation generated by difference frequency mixing at ca. $1.8\,\mu\mathrm{m}$. Population in these levels was then probed by exciting LIF in the (0, 3) band of the ($A^2\Sigma^+$–$X^2\Pi$) system, using 258–259 nm radiation. The total relaxation from a particular J_i level within the NO ($X^2\Pi\Omega = 1/2$, $v = 3$) state was measured in kinetic experiments by populating that level with the IR pump pulse and fixing the UV laser to a transition originating from that level. The LIF intensity, which is directly correlated to the population in ($\Omega = 1/2$, $v = 3$, J_i), was then recorded while varying the time delay between the two laser pulses.

The state-to-state relaxation was measured in spectroscopic experiments, in which the time delay between the lasers was fixed to a few 10s of nanoseconds, corresponding to a small fraction of the relaxation time from J_i, and the UV probe laser was scanned to produce a spectrum. A second spectrum was then recorded at a delay of $5\,\mu\mathrm{s}$ to allow sufficient time for rotational and spin-orbit relaxation while avoiding significant vibrational relaxation.

The thermally averaged cross sections for total removal were derived from the observed first-order rate constants, k_{1st}, and did not display significant variation with temperature or with the initial rotational state. The state-to-state rate constants varied with ΔJ and displayed a decrease with increasing ΔJ, generally favoring even ΔJ transitions over odd ΔJ. At lower temperatures, decreases in J were found to be favorable over increases in J and a narrower distribution of rate constants with ΔJ was observed. The experimental rate constants for collisions with He and Ar were compared with those theoretically calculated using close-coupled and coupled states calculations. The agreement of both the total and the state-to-state rate constants was found to be excellent over the temperature range covered in the experiment. In contrast with comparisons of relative energy-dependent cross sections obtained in crossed molecular beams experiments, it should be emphasized that there is no scaling between the absolute values obtained by theory and by experiment, underlining the quality of the agreement.

7.4.2 *Rotational energy transfer studies involving* CO ($X^1\Sigma^+$)

Experiments on inelastic collisions between CO ($X^1\Sigma^+, v = 2, J_i$) and He were conducted by Carty *et al.* [6] covering the temperature range from 15–294 K and the initial rotational levels $J_i = 0, 1, 4$, and 6. Similar to the work on NO, rotational states of a vibrationally excited state (here $v = 2$) were selectively excited using a tunable IR radiation (tunable between 2.274 and 2.408 μm) obtained by difference frequency mixing of a narrowband dye laser and an injection-seeded 1064-nm Nd:YAG laser in a lithium niobate crystal (see Fig. 7.4). Compared to NO ($X^2\Pi$), the lowest electronic transition of CO lies at much higher energy, and it is necessary to generate VUV laser radiation in order to use LIF to observe CO in specific rotational levels. VUV laser radiation was generated using the two-photon resonant four-wave mixing method in xenon gas, which produces radiation between 165.29 and 165.50 nm. This was used to excite LIF in the (0, 2) band of the CO ($A^1\Pi$–$X^1\Sigma^+$) system.

The thermally averaged cross sections for the total removal of J_i were found to be essentially independent of temperature and of J_i down to 27 K and showed a slight variation at 15 K for the $J_i = 4$ and 6 levels. The state-to-state rate constants, shown in Fig. 7.5, were found to agree remarkably well with the close-coupling calculations of Cecchi-Pestellini *et al.* [37], confirming the accuracy of the CO-He potential energy surface used in their calculations [38].

Rotational energy transfer between CO and Ar was studied by Mertens *et al.* [15] in the temperature range of 30–293 K. In comparison to He, the van der Waals forces are stronger for Ar leading to increased pressure broadening of spectra. Theoretical calculations

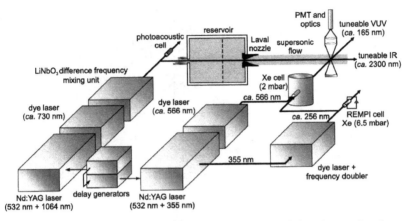

Fig. 7.4. A schematic of a CRESU apparatus adapted for the study of rotational energy transfer in collisions between CO ($X^1\Sigma^+$) and He using the IRU-VDR technique. The CO rotational level population was probed using VUV laser radiation generated by two-photon resonant four-wave mixing processes in xenon gas, which produce radiation at ca. 165 nm. Reproduced with permission from Ref. [6], Copyright 2014 AIP Publishing.

involving the CO-Ar complex are challenging as the potential, and particularly its anisotropic part, is largely determined by the balance between dispersion and repulsion, requiring accurate calculation of the dispersion interaction, including correlation effects. In the same manner as the CO-He studies, an infrared pulse (generated in these later experiments by a mid-infrared optical parametric oscillator) was used to selectively populate a J-level of the $v = 2$ vibrational level within the $X^1\Sigma^+$ ground state and a VUV pulse at ca. 165 nm to excite LIF in the (0,2) band of the ($A^1\Pi$–$X^1\Sigma^+$) system. The cross sections for total removal were found to be 2–3 times larger than those for CO-He collisions. This increase is well explained by the relative magnitudes of the CO-He and CO-Ar van der Waals wells, which are calculated to be 22.34 cm^{-1} and 107.1 cm^{-1}, respectively. The strength of the interaction between CO and the noble gas, and therefore the likelihood of inelastic collisions, is proportional to the well depth.

Trends in the state-to-state rate constants were similar to those found for the CO-He system. The rate constants were larger when ΔJ was small, and decreased as ΔJ was increased, as shown in Fig. 7.5 for $J_i = 0, 1, 4$, and 6 at 111 K. The rovibrationally inelastic rate

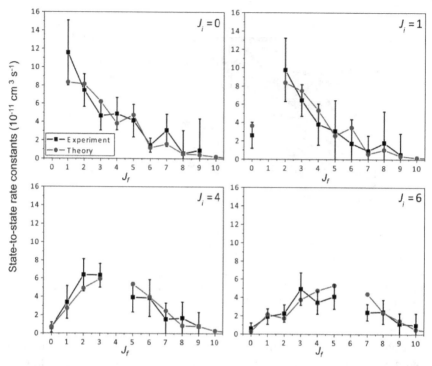

Fig. 7.5. Experimental state-to-state rate constants for collisional quenching of CO ($X^1\Sigma^+$, $v = 2$) for $J_i = 0, 1, 4$, and 6 by Ar at 111 K (squares) compared to theoretical state-to-state rate constants (circles). Reproduced with permission from Ref. [15], Copyright 2017 Elsevier B.V.

constants and cross sections were calculated at the close-coupling level, and good agreement was found between the theoretical predictions and the experimentally measured rate constants.

Total removal cross sections and state-to-state rate constants for rotational energy transfer for CO ($X^1\Sigma^+$, $v = 2, J_i$) in collisions with H_2 have been recently measured using the CRESU apparatus in Rennes [16]. These are the only measurements of rotational energy transfer involving H_2 that have been made in a CRESU to date and thus are an important step toward addressing missing input data that are critical for astrophysical models. Total and state-to-state rate constants were measured for $J_i = 0, 1, 4$, and 6 at 293 K, $J_i = 0, 1$, and 4 at 20 K, and for $J_i = 0$ and 1 at 10 and 5.5 K. Similar to the previous studies, the state-to-state rate constants increased as ΔJ decreased. A slight propensity for $\Delta J = 2$ transitions was observed at room temperature (for $J_i = 0, 1$, and 4) and at $20\,\mathrm{K}$ (for $J_i = 0$ and 4), which was not seen at lower temperatures.

7.5   Vibrational Energy Transfer

Vibrational energy transfer (VET) plays a major role in chemical reaction dynamics. The products of exothermic reactions are typically born in vibrationally excited states and VET is the process by which that stored energy is transferred to the bath gas, sometimes in competition with further unimolecular dissociation and, in some cases, infrared emission. VET plays a key role in association reactions, where collisional relaxation of vibrationally excited association complex competes with redissociation. In astrophysics, just as RET collisional rate constants are required inputs for radiative transfer models, VET can also play a role especially for lower-frequency vibrations. VET is typically two or three orders of magnitude slower than RET in small molecules, precluding its study by the CRESU technique where relaxation times are limited by the hydrodynamic flow time of a few hundred microseconds (see Chapter 2). However, VET has been studied in certain favorable cases, involving V-V transfer or V-R,T transfer via strongly bound association complexes, as well as relaxation of larger, highly excited molecules. These studies are summarized below.

7.5.1   *Vibrational relaxation of CH* $(v = 1)$

Herbert *et al.* [10] reported the first measurements of the kinetics of a vibrationally excited species within a CRESU flow. In this study, the kinetics of $CH(v = 1)$ vibrational relaxation in collisions with N_2 and with CO were measured using PLP-LIF. The PLP-LIF technique is described in detail in Chapter 5. Briefly, radicals are created by photodissociation of a suitable precursor molecule using a pulsed laser, usually at an ultraviolet wavelength. The time evolution of the radical can then be monitored by exciting an electronic transition using a tunable laser and detecting its fluorescence emission to a vibrational level in the electronic ground state. The fluorescence is detected using a photomultiplier tube (PMT), which converts the photon intensity to an electrical signal that can be amplified and read by a computer.

The rate constants were recorded in both heated and cryogenically cooled cells (between 584 and 86 K, respectively), and in a CRESU apparatus (from 295 K down to 23 K). The authors found the vibrational relaxation to be rapid, with a rate constant on the order of $1 \times 10^{-10}\,\mathrm{cm^3\,s^{-1}}$ at 23 K, and showing a slight negative temperature dependence (Fig. 7.6). The rapid relaxation of CH($v = 1$) is

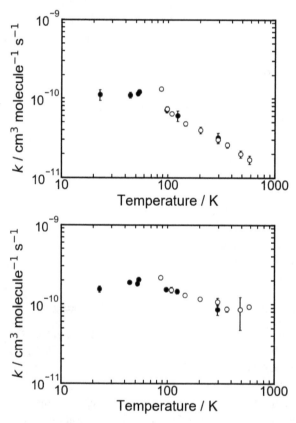

Fig. 7.6.   Rate constants for the relaxation of CH($\nu = 1$) by CO (top) and N$_2$ (bottom) as a function of temperature. The results obtained in the CRESU apparatus are shown in the filled circles. The open circles show those obtained in the heated or cryogenically cooled cell. Reproduced with permission from Ref. [10], Copyright 1996 ACS.

strongly related to the reactions of $CH(v = 0)$ with N_2 and CO, which were studied extensively by the group of Ian Smith [39]. These reactions were found to be termolecular and occur by association, involving collisional stabilization of the excited adducts. The removal of $CH(v = 1)$ was found to be much faster than that of $CH(v = 0)$, and the measured rate constants, in contrast to those for reaction of $CH(v = 0)$ with N_2 and CO, were found to be independent of total pressure. It was proposed that the vibrational relaxation of $CH(v = 1)$ by N_2 and CO is unusually fast due to the formation of strongly bound complexes, which facilitate intramolecular energy transfer within the lifetime of the complex.

The magnitudes of the rate constants were found to approach the collisional rate, consistent with the rate being determined by the capture ability of the lowest intermolecular potentials between CH and CO or N_2. CO was found to relax $CH(v = 1)$ more efficiently compared to N_2, which was explained by the relative strengths of their chemical interaction. CH is known to form datively bonded HCN_2 and HCCO in collisions with N_2 and CO, respectively. The HC-CO bond strength is much stronger than that of HC-N_2 and therefore the chemical interaction at large separations should be greater, leading to higher rates of adiabatic capture.

The relaxation of $CH(v = 1)$ by N_2 could be compared to the limiting high-pressure rate constants of Fulle and Hippler [40], who measured the association between $CH(v = 0)$ and N_2 at total pressures of up to 150 bar. At room temperature, the two measurements were found to agree, supporting the proposed relaxation mechanism and demonstrating that such relaxation measurements can be used to estimate rate constants in the high-pressure limit. Studies of the relaxation of $CH(v = 1)$ were later extended to collisions with H_2 and D_2 [12] In a similar manner to the CO and N_2 experiments, the rate constants were measured over a wide temperature (23–584 K) using heated and cryogenically cooled cells in Birmingham and a CRESU apparatus in Rennes. The rates of removal of $CH(v = 1)$ in collisions with both H_2 and D_2 were found to be rapid and independent of total pressure, with a mild negative temperature dependence. It was concluded that the relaxation rates are essentially those for the formation of strongly bound CH_3^* or CHD_2^* energized collision complexes in which intramolecular vibrational relaxation is facile.

7.5.2 *Vibrational self-relaxation of NO ($v = 3$)*

The vibrational self-relaxation of NO ($X^2\Pi$, $v = 3$) was measured between 7 and 85 K using IRUVDR in the CRESU apparatus in Birmingham [13]. The self-relaxation of vibrationally excited NO can occur by either V-V or V-R,T energy transfer. Vibrational levels above $v = 1$ have been shown to relax via V-V energy exchange, while relaxation of $v = 1$ must occur by V-R,T since there is no effective V-V channel [33, 34].

The vibration relaxation of NO ($X^2\Pi$, $v = 3$) was probed in a similar manner to the RET experiments, using IRUVDR. The IR-pump laser at ca. 1.8 μm was used to excite a single rotational level in the $v = 3$ vibrational level of the lower spin-orbit component ($\Omega = 1/2$) of the $^2\Pi$ electronic ground state of NO. The population in the $v = 3$ level of the $X^2\Pi$ state was observed by exciting LIF in the (0,3) band of the ($A^2\Sigma^+$–$X^2\Pi$) system using 258–259 nm radiation. The vibrational relaxation could be separated from rotational relaxation as rotational and spin-orbit relaxations were expected to be rapid ($\sim$2 μs), while vibrational relaxation with the buffer gases (He, Ar, and N_2) would be slow.

The authors found a rapid increase in the rate constants as the temperature was lowered, increasing from $(3.9 \pm 0.5) \times 10^{-12}$ cm^3 s^{-1} at 85 K to $(25.8 \pm 1.8) \times 10^{-12}$ cm^3 s^{-1} at 7 K, as shown in Fig. 7.7. The relatively large rate constants for relaxation of NO($v > 1$) previously measured above 77 K had been ascribed to near-resonant V-V energy exchange [33, 34]. However, in the case of NO($v = 3$), V-V energy exchange is endothermic by $\sim$56 cm^{-1}, corresponding to $\Delta E/k_B = 80$ K, and therefore V-V energy exchange cannot be responsible for the rapidly accelerating rate of self-relaxation of NO($v = 3$) at low temperature. It was proposed instead that the increased rate constants could be explained by V-R,T energy transfer facilitated by the formation of (NO)$_2$ collision complexes.

At the lowest temperature (7 K), $D_0/k_B T \approx 150$, and transient collision complexes formed between NO($v = 3$) and NO($v = 0$) are expected to have long enough lifetimes with respect to redissociation for intramolecular vibrational relaxation to occur. However, unlike in the case of CH($v = 1$) relaxation, the pressure-dependent rates of association of two NO molecules to form (NO)$_2$ had not been measured and therefore the relaxation rates could not be assessed by comparison to the high-pressure limit.

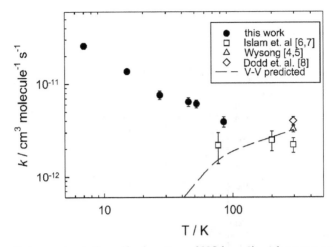

Fig. 7.7.   Rate constants for self-relaxation of $NO(v = 3)$ with temperature measured by James *et al.* [13] as well as those obtained in previous studies. The dashed line shows the predicted temperature dependence for the V-V exchange mechanism. Reproduced with permission from Ref. [13], Copyright 1997 Elsevier Science BV Amsterdam.

7.5.3   *Vibrational relaxation of highly excited toluene*

The collisional relaxation of highly excited toluene molecules was studied at low temperature for the first time, using the CRESU apparatus in Birmingham [14]. This study represents one of the very few measurements made of energy transfer involving collisions with polyatomic molecules in a uniform flow. The vibrational relaxation in collisions with He, Ar, and N_2 was monitored by observing time-resolved infrared fluorescence (TR-IRF) in transition of the C-H stretching modes. Similar to PLP-LIF, time-resolved infrared fluorescence experiments are initiated by a pump laser which excites a specific transition, and the decay is followed by monitoring fluorescence from the upper state. In TR-IRF experiments, the fluorescence is in the infrared and is monitored using an infrared detector, usually a semiconductor photovoltaic device.

Highly excited toluene molecules were produced by the photoisomerization of cycloheptatriene at 248 nm. The TR-IRF signals were used to determine the relationship between the average energy removed per collision, $\langle\langle\Delta E\rangle\rangle$, and the average energy contained in the toluene molecule, $\langle\langle E\rangle\rangle$. They found that $\langle\langle\Delta E\rangle\rangle$ did not vary strongly with temperature for any of the collision pairs (Fig. 7.8).

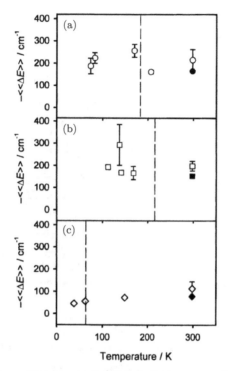

Fig. 7.8.   Plots from Wright *et al.* 2000 [14] showing $-\langle\langle\Delta E\rangle\rangle$ at $\langle\langle E\rangle\rangle$ = $30000\,\mathrm{cm}^{-1}$ versus temperature for highly excited toluene collisions with (a) N_2, (b) Ar, and (c) He in a CRESU apparatus (open symbols) and a flow cell (filled symbols). Reproduced with permission from Ref. [14], Copyright 1997 ACS.

For the toluene-He system, $\langle\langle\Delta E\rangle\rangle$ decreased slightly with temperature, whereas for toluene-Ar and toluene-N_2, $\langle\langle\Delta E\rangle\rangle$ was essentially temperature independent. The toluene-He well depth ($\epsilon/k_B = 64\,\mathrm{K}$) in the intermolecular potential is significantly lower than that of toluene-N_2 ($\epsilon/k_B = 183\,\mathrm{K}$) and toluene-Ar ($\epsilon/k_B = 216\,\mathrm{K}$). The authors suggested that the trends $\langle\langle\Delta E\rangle\rangle$ with temperature may reflect the relative influence of the van der Waals forces in facilitating energy transfer (see Chapter 11).

7.5.4   *Vibrational relaxation of photoproduced SO ($X^3\Sigma^-$) using chirped-pulse spectroscopy*

The CRESU technique has recently been coupled to chirped-pulse microwave spectroscopy [27, 41], which offers unique capabilities to study energy transfer at low temperature. The chirped-pulse detection method has advantages over LIF, particularly in terms of its ability to detect multiple species at the same time. Likewise, the vibrational population of multiple levels can be followed during a single experiment. The group of Arthur Suits (University of Missouri) has demonstrated this capability in proof-of-principle experiments probing the nascent vibration distribution following 193-nm photodissociation of SO_2 in a helium buffer. Spectra were obtained at 5-μs intervals following an initial 10-μs delay between the laser trigger and the first chirped-pulse excitation. Ten different spectra were obtained in order to probe the time evolution of the $v = 2, 1$, and 0 vibrational levels via the $N_J = 1_0 - 0_1$ pure rotational transition of SO near 30 GHz. The time evolution of the population of these vibrational levels is shown in Fig. 7.9. The $v = 2$ level was found to dominate when SO first appears from the photolysis followed by rapid vibrational relaxation to $v = 0$. The rapid vibrational relaxation was proposed to be caused by a near resonance of the SO fundamental vibration with the SO_2 symmetric stretch, resulting in efficient V-V energy transfer. Interestingly, it was observed that $v = 2$ was always more populated than $v = 1$, even after $v = 0$ became dominant, possibly suggesting additional resonant relaxation pathways.

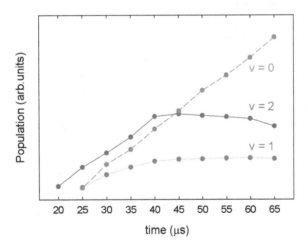

Fig. 7.9. The populations of the $v = 2, 1$, and 0 vibrational levels of $SO(X^3\Sigma^-)$, monitored by their rotational transitions in chirped-pulse microwave spectra, are shown over time. Reproduced with permission from Ref. [27]. Copyright 2017 AIP Publishing.

7.6   Electronic Energy Transfer

7.6.1   *Quenching of electronically excited states*

Quenching of electronically excited species has received a lot of interest due to its application to many scientific fields such as the atmospheric chemistry of Earth and other planetary atmospheres, combustion, photochemistry, and interstellar chemistry. Collisional quenching of a number of electronically excited atoms and molecules in collisions with noble gases (He, Ar, Kr), diatomics (N_2, O_2, H_2) as well as larger species (C_6H_6, C_6F_6) has been studied using the CRESU technique and is summarized below.

7.6.1.1   *Quenching of $O(^1D)$, $C(^1D)$, and $S(^1D)$ atoms*

Collisional and reactive processes involving $O(^1D)$ [22, 23] and $C(^1D)$ [24] atoms have been studied extensively from room temperature down to $50\,K$ by the group of Kevin Hickson in Bordeaux using a miniature continuous flow CRESU apparatus. $O(^1D)$ atoms are produced by photolysis of ozone in the Earth's atmosphere and are known to initiate a rich chemistry. The concentration of $O(^1D)$ atoms in the Earth's atmosphere is controlled by the balance between its photolytic production and its loss by electronic quenching to $O(^3P)$ due to collisions and, dependent on the region of the atmosphere, its reaction with water vapor, methane, and hydrogen to form OH.

The kinetics of the gas phase quenching of $O(^1D)$ was measured in Bordeaux for N_2, O_2, Ar, and Kr and for $C(^1D)$ with N_2. $O(^1D)$ atoms were generated by 266-nm photolysis of O_3 and detected by on-resonance VUV LIF using the $3s^1D-2p^1D$ transition at $115.215\,nm$. Further details of these experiments can be found in Chapter 5. The rate constants for deactivation of $O(^1D)$ in collisions with Ar were found to be relatively constant between 50 and $296\,K$, ranging from $(5.6 \pm 0.6) \times 10^{-13}$ to $(6.8 \pm 0.7) \times 10^{-13}\,cm^3\,s^{-1}$. The lack of temperature dependence is consistent with the lack of an energy barrier through a bound ArO intermediate; however, the low quenching rate constants indicate that spin−orbit coupling is relatively inefficient for this system. A negative temperature dependence was observed in collisions with N_2, with a rate constant at $50\,K$ of $(7.0 \pm 0.7) \times 10^{-11}\,cm^3\,s^{-1}$, about double the room-temperature rate constant. The authors related the increased quenching rate to the longer lifetime of the bound $^1\Sigma^+$ state of the N_2O intermediate. Similarly, the rate constants for the $O(^1D)$ + O_2 reaction were seen to increase as the temperature fell, though less pronounced than for N_2. The mechanism for quenching is more complex than in the case of N_2 as it can give rise to $O(^3P)$ as well as electronically excited O_2 in either $b^1\Sigma_g^+$ or $a^1\Delta_g$ states. While less clear than for $O(^1D)$ + N_2, it is likely that an energized O_3 intermediate is formed, but there are several potential pathways that could be responsible for the negative temperature dependence. The quenching kinetics of $O(^1D)$ in collisions with Kr were reported in a subsequent publication [23]. Quenching of $O(^1D)$ was found to be more efficient than for Ar due to more efficient nonadiabatic transitions between the singlet and triplet states through increased spin—orbit coupling. This trend was seen to continue for collisions with xenon using a quantum close-coupling treatment.

Experimental measurements of $C(^1D)$ reactions are less numerous than for $O(^1D)$, in part due to the lack of convenient clean sources for its production (see Chapter 5). Prior to the study of Hickson *et al.* [24], kinetic measurements of non-reactive processes involving $C(^1D)$ had only been reported at room temperature. For both $O(^1D)$ and $C(^1D)$ collisions with N_2, no exothermic product channels are available, thus quenching to the triplet ground state occurs through intersystem crossing. $C(^1D)$ atoms were produced by the pulsed multiphoton dissociation of CBr_4 precursor molecules. Unlike the $O(^1D)$ measurements, rate constants for $C(^1D)$ quenching by N_2 were measured using H atom production as a chemical tracer of $C(^1D)$ population, whereby the $C(^1D)$ + $H_2 \rightarrow CH$ + H reaction was monitored by VUV LIF of the atomic hydrogen product at 121.567 nm. The quenching rate constants were seen to increase significantly with decreasing temperature and were explained by the effect of the CNN intermediate complex lifetime on the efficiency of intersystem crossing.

Kinetic measurements of the collisional quenching of $S(^1D)$ by Ar were performed at temperatures from 5.8 K to 298 K using the continuous flow CRESU apparatus in Rennes. The total removal of $S(^1D)$ atoms was monitored by VUV laser-induced fluorescence at 166.67 nm. Ab initio electronic structure calculations using internally contracted MRCI methodology were performed to describe the interaction. A $\sim T^{1/6}$ temperature dependence was observed suggesting that the quenching process is dominated by an attractive potential $\sim R^{-6}$ at long range. The rate constant for quenching of $S(^1D)$ by Ar was found to be approximately an order of magnitude larger than the rate constant for $O(^1D)$ quenching by Ar measured in Bordeaux. This is consistent with the trend seen in Ar, Kr, and Xe collisions, reflecting the larger size of the spin-orbit coupling in the heavier S atom.

7.6.1.2 *Collisional quenching of NO ($A^2\Sigma^+$)*

Electronic collisional quenching of NO ($A^2\Sigma^+$) has been studied extensively by the group of Simon North at Texas A&M using a pulsed CRESU apparatus. The quenching rate constants in collisions with NO and O_2 were measured down to $34\,\text{K}$ [20] and with C_6H_6 and C_6F_6 down to $130\,\text{K}$ [21]. The rate constants for quenching by NO and O_2 were found to slightly increase with temperature; however, the thermally averaged quenching cross sections displayed a clear increase with decreasing temperature. Including previous data at higher temperatures, they found that the temperature dependence of the quenching cross sections for NO and O_2 can be fit using an inverse power law that could be described by the long-range interactions.

The quenching of NO ($A^2\Sigma^+$) has been explained to occur by a charge transfer model [42]. An electron from NO is transferred to the collision partner at a rate that is defined by their relative velocity, the probability of NO ($A^2\Sigma^+$) and the collider undergoing a curve crossing to an NO^+-M^- surface, as well as a second curve crossing to produce NO ($X^2\Pi_{1/2}$) from the NO^+-M^- ionic surface. For this charge transfer mechanism to be efficient, the quenching partner must have a sufficiently high electron affinity. More details on the theory of charge transfer can be found in Chapter 11.

Measurements of NO ($A^2\Sigma^+$) quenching in collisions with C_6H_6 and C_6F_6 offered a test of the charge transfer mechanism because the electron affinity of C_6F_6 is significantly higher than C_6H_6 ($1.2\,\text{eV}$ and $-1.5\,\text{eV}$, respectively). In this mechanism, the quenching cross section by C_6F_6 was expected to be much higher than by C_6H_6. In contrary, the authors found that both benzene and hexafluorobenzene possess large collision cross sections for quenching of NO ($A^2\Sigma^+$), despite the large differences in their electron affinity (Fig. 7.10). Instead, a resonant electronic energy transfer mechanism

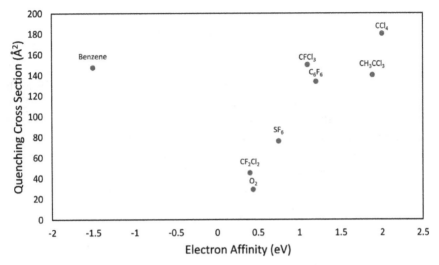

Fig. 7.10.   Total collisional quenching cross sections at 300 K for quenching of the NO A-state fluorescence by C_6H_6 and C_6F_6 plotted against their electron affinity, as well as other molecules expected to undergo the charge transfer mechanism. Reproduced with permission from Ref. [21], Copyright 2018 Elsevier B.V.

was proposed, in which the energy transfer is proportional to the electronic coupling between the NO A-state and the acceptor state. This mechanism requires the collisional quencher to have an electronic excited state lower in energy than the NO A-state, which is satisfied by both molecules. The predicted temperature dependence of this mechanism, $T^{-1/3}$ for C_6H_6 and C_6F_6, was found to be consistent with the experimental data.

7.6.1.3   *Electronic relaxation of $^1CH_2$*

The rate constants for electronic relaxation of $^1CH_2$ with He, N_2, O_2, H_2, CH_4, C_2H_6, C_2H_4, and C_2H_2 have been measured between 40 and 134 K by PLP-LIF in a pulsed CRESU apparatus in Leeds as well as at 160 and/or 198 K in a jacketed flow cell [25, 26]. In addition, because $^1CH_2$ may react with CH_4 and H_2 at a comparable rate to its relaxation, the branching ratios between reaction and relaxation were determined at 73 K and 160 K for H_2 and CH_4 and at 100 K, 160 K, 195 K, and 298 K. The temperature-dependent branching ratios were determined by an indirect method requiring pairs of experiments to be carried out. $^3CH_2$ produced by the relaxation of $^1CH_2$ was titrated to OH or H atoms by reaction with O_2:

$$^3CH_2 + O_2 \rightarrow OH + HCO \tag{7.4}$$

$$^3CH_2 + O_2 \rightarrow H + Products \tag{7.5}$$

The OH or H produced was monitored by its LIF both with and without the reactive co-reagent (H_2, CH_4, C_2H_6, C_2H_4, or C_2H_2). For collisions with He, N_2, and O_2, relaxation is the only loss process on the timescale of the experiments, whereas in collisions with the other species, some of the $^1CH_2$ will be consumed in reactions, lowering the $^3CH_2$ available to be titrated to OH/H. Figure 7.11 shows how this approach was used to determine the ratio of $^1CH_2$ electronic relaxation versus reaction in collisions with H_2.

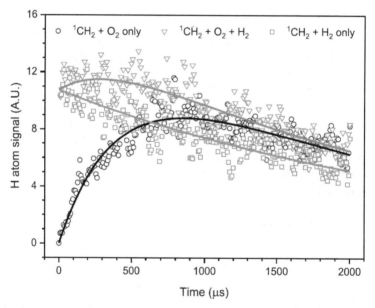

Fig. 7.11.   H atom LIF signal as a function of time due to $^1CH_2$ relaxation to $^3CH_2$ in the presence of O_2 (black circles), in the presence of H_2 (green squares), and in the presence of both O_2 and H_2 (red triangles). In the case of $^1CH_2 + O_2$, all $^1CH_2$ loss can be attributed to relaxation to $^3CH_2$, which then reacts with O_2 to form H atoms. The H atom signal is monitored both with and without the co-reagents present (here H_2) to determine the branching ratio between reaction and relaxation. Reproduced with permission from Ref. [26], Copyright 2019 Elsevier Inc.

All of the rate constants, with the exception of He, increased as the temperature was lowered. The negative temperature dependence was explained to be correlated with the well depth of the $^1CH_2$-collider complex. While the magnitude of $^1CH_2$ removal increased as the temperature was reduced, the fraction of reactive removal was found to decrease, with electronic removal to $^3CH_2$ becoming the dominant channel for H_2, C_2H_2, and C_2H_4 at 100 K. For CH_4 and C_2H_6, removal of $^1CH_2$ by electronic relaxation reaches around 45% of the total removal at 100 K, increased from around 15% at room temperature. The electronic relaxation was found to account to around 65% of $^1CH_2$ removal in collisions with CH_4 at 73 K.

7.6.1.4 *Relaxation of $^3C_2$ by intersystem crossing*

The C_2 molecule has been observed in a wide range of environments — from cold regions such as interstellar clouds [43, 44], circumstellar envelopes [45], and comets [46] to hot media such as hydrocarbon flames [47], plasmas [48], and stellar atmospheres [49]. Low-temperature (300–24 K) rate constants for the collisional relaxation of $C_2(a^3\Pi_u)$ were reported by Páramo *et al.* [19], as well as for reactions of $C_2(X^1\Sigma_g^+)$, in collision with O_2 and NO. C_2 is unusual as its first electronically excited state, $C_2(a^3\Pi_u)$, is very low lying, separated in energy from the electronic ground state by only 610 cm^{-1}.

Both $^1C_2$ and $^3C_2$ were formed by the 193-nm laser photolysis of C_2Cl_4. $^1C_2$ and $^3C_2$ were detected by LIF in the Mulliken $(D^1\Sigma_u^+ \leftarrow X^1\Sigma_g^+)$ and Swan $(d^3\Pi_g \leftarrow a^3\Pi_u)$ bands, respectively. In the presence of increasing concentrations of O_2, the LIF signal from $^1C_2$ was seen to increase at short time delays, while the level of the signal at long time delays was virtually unaffected (Fig. 7.12). The change in the LIF signal was interpreted as relaxation of the $^3C_2$ state by collisions with O_2, providing a source of $^1C_2$. This rapid quenching of $^3C_2$ by intersystem crossing, induced by collisions with O_2, was found to remain very efficient down to 50 K. In contrast, reaction of $^1C_2$ with O_2 became very inefficient below room temperature, which was exploited to measure reactions between $^1C_2$ and NO (and later, hydrocarbons [50]) where O_2 was used to quench the $^3C_2$ produced in the photolysis (see also Chapter 5).

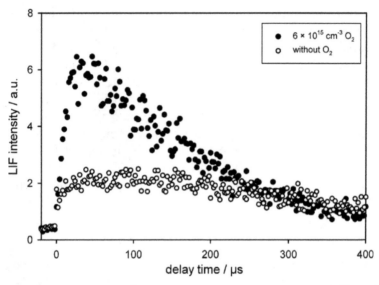

Fig. 7.12. LIF signal from $^1C_2$ at $145\,\mathrm{K}$ in the presence of $6 \times 10^{15}\,\mathrm{cm}^{-3}$ O_2 (filled circles) and without addition of oxygen (open circles). Reproduced with permission from Ref. [19], Copyright 2019 ACS.

7.6.2 *Spin-orbit energy transfer*

Collisional quenching of fine structure levels in neutral or ionized atoms has not been well explored. Despite this, transitions between these levels play an important role in the interstellar medium. Fine structure transitions are known to be coolants in HII (ionized hydrogen) regions and in diffuse HI (neutral hydrogen) clouds, where the atoms collide with electrons and rotationally (or translationally) excited hydrogen [51]. In these collisions, energy is transferred to the atom followed by radiative decay. Rates for energy transfer within fine structure levels are therefore necessary as inputs into radiative transfer models of photodissociation regions or in shock wave models in circumstellar regions to determine how quickly the gas is relaxed. In addition, theoretical studies predict that the reactivity of open-shell atoms will be affected by the fine structure population. Collisions involving intramultiplet relaxation (or excitation) of atomic species are a simple case of inelastic nonadiabatic transitions, where more than one adiabatic potential is involved, and thus offer a unique opportunity to test the accuracy of theoretical calculations.

However, to our knowledge, collisions involving only three different atoms have been studied at low temperature in uniform supersonic flows. These studies concern the spin–orbit relaxation of $Al(^2P_j)$ atoms in collision with argon [17], and of $Si(^3P_j)$ and $C(^3P_j)$ by He [18].

7.6.2.1 *Fine structure relaxation of $Al(^2P)$ by Ar*

Le Picard *et al.* [17] reported, for the first time below room temperature, rate constants for collision-induced spin-orbit transitions — in a study of $Al(^2P_j)$ collisions with Ar between 44 and 137 K. The Al-Ar system was chosen for experimental reasons and because of the availability of codes to compute atom–atom potential energy curves. Al atoms in the $^2P_{1/2}$ and $^2P_{3/2}$ doublet ground state were produced by multiphoton photolysis of trimethylaluminium at 266 nm. On-resonance LIF was observed from the $^2S_{1/2}$ to $^2P_{1/2}$ (394.40 nm) and $^2P_{3/2}$ (396.15 nm). Figure 7.13 shows the LIF signal from the two states at short (0.5 μs) and long (50 μs) time delays, demonstrating the spin-orbit relaxation of the $^2P_{3/2}$ to the $^2P_{1/2}$ state. The spin-orbit relaxation rate constants were found to increase with temperature.

A quantum collisional treatment was developed to calculate cross section and rate constants using the close-coupling formalism of Mies [52] The interatomic potential between $Al(^2P)$ and $Ar(^1S)$ in their ground electronic state leads to two molecular states with $^2\Pi$ and $^2\Sigma^+$ symmetry. It was found that the rate constants are very sensitive to the potential depth of the $^2\Sigma^+$ state — a variation of 15 cm^{-1} (16% of the well depth) changed the rate constant by 350%. The temperature dependence of the rate constants was fairly well reproduced, while their absolute values were overpredicted by ~30% (see Chapter 11).

7.6.2.2 *Intramultiplet transitions in $Si(^3P)$ and $C(^3P)$ collisions with He*

$C(^3P_j)$ is abundant in the ISM, where its intramultiplet transitions contribute to cooling. Rate constants for energy transfer between these fine structure states are required, particularly in collisions with abundant collision partners such as H_2 and He. Le Picard *et al.* [18] studied the collisional relaxation of the spin–orbit distribution of $C(^3P_j)$ and $Si(^3P_j)$ in collisions with He.

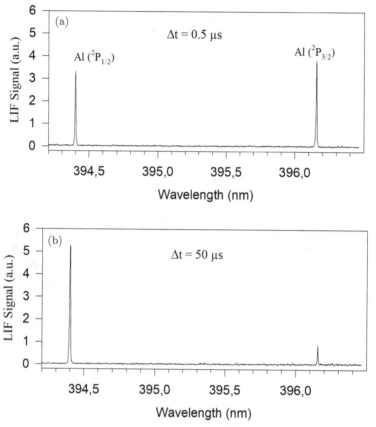

Fig. 7.13. LIF spectra of the $(^2S_{1/2}-{}^2P_{1/2})$ and $(^2S_{1/2}-{}^2P_{3/2})$ lines of Al recorded by Le Picard *et al.* at 52.8 K at 0.5 μs (a) and 50 μs (b) time delays. Reproduced with permission from Ref. [17], Copyright 1998 AIP Publishing.

$C(^3P_j)$ and $Si(^3P_j)$ were generated by the 193-nm photolysis of C_3O_2 and the 266-nm photolysis of $Si(CH_3)_4$, respectively. $C(^3P_j)$ was detected by on-resonance VUV-LIF via both excitation (at ~166 nm) and observation of fluorescence in the $(2s^22p3s\ ^3P_j-2s^22p^2\ ^3P_j)$ transitions at ~160 nm. $Si(^3P_j)$ was detected by exciting at around 251 nm $(3s^23p4s\ ^3P_j-3s^23p^2\ ^3P_j)$ and detecting fluorescence at ~250 nm. By tuning the excitation pulse, three different j transitions were probed for both carbon and silicon. The electronic quenching of $Si(^1D)$, also produced in the photolysis of $Si(CH_3)_4$, was monitored via LIF at ~288 nm.

Rate constants were reported for collisions of $Si(^3P_j)$ with He at 15, 23, 24, and 49 K, and for $C(^3P_j)$ at 15 K. The measurements were limited to these temperatures because the experiments performed at higher temperatures (>15 K for $C(^3P_j)$ and >49 K for $Si(^3P_j)$) displayed nascent spin-orbit distributions following the photolysis, which were too close to a Boltzmann distribution to determine rate constants from their population evolutions.

Similar to the Al-Ar system discussed above, interaction potentials were generated for the $^3\Pi$ and $^3\Sigma^+$ electronic states of $C(^3P_j)+$ He and $Si(^3P_j)+$ He, and a quantum collisional treatment was developed to calculate the relaxation rate constants. Three sets of interaction potentials were calculated: using a fully *ab initio* complete active-space second-order perturbation theory approach (CASPT2) with corrected basis set superposition errors (BSSE), CASPT2-BSSE with additional midbond functions (named CASPT2-MBF), and a hybrid complete active space self-consistent field (CASSCF) + "dispersion energy" approach, in which the dispersion energy was included via a term in C_6/R^6. For $Si(^3P_j)+$ He, the temperature dependences of the rate constants for $1 \to 0$, $2 \to 0$, and $2 \to 1$ transitions were well reproduced by the dynamical calculations for all three sets of potentials (see Fig. 7.14). Good quantitative agreement between the theoretical and experiment rate constant was only achieved with the highest-level *ab initio* potential. However, in the case of the $C(^3P_j)+$ He system (Fig. 7.14), the agreement between the rate constants was not good, especially for the $2 \to 0$ and $2 \to 1$ transitions, where the theoretical rate constants were a factor of ~ 3 larger than the experimental ones. It has since been suggested that this disagreement could be explained by the formation of translationally excited $C(^3P)$ in the 193-nm photolysis of carbon suboxide. Additional details can be found in Chapter 11.

7.7   Future Perspectives

The CRESU technique has opened up new avenues for the study of energy transfer processes at low temperature. Nevertheless, VET and RET rate constants have only been measured for a handful of collision partners. VET at very low temperatures is required for theoretical prediction of reactive rate constants; however, the timescales of such

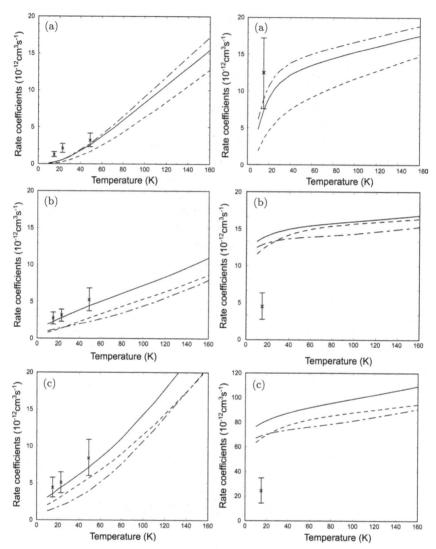

Fig. 7.14.   Total relaxation rate constants of Si(^{3}P) (left) and C(^{3}P) (right) fine structure transitions k_{10} (a), k_{20} (b), and k_{21} (c) by collisions with He using CASPT2-BSSE (dashed line), CASPT2-BSSE+MBF (solid line), or hybrid (dot-dashed line) potentials. The experimental values are shown (crosses) for comparison. Reproduced with permission from Ref. [18], Copyright 2002 AIP Publishing.

processes make their measurement in CRESU flows challenging. RET rate constants at very low temperatures have only been measured to date for non-reactive potential energy surfaces, but measurements of RET on reactive surfaces, e.g. H_2 collisions with radicals, are particularly important for radiative transfer models of the ISM, as well as for fundamental tests of theory.

7.7.1 *Measurements involving collisions with H_2*

For implementation in the spectral analysis of molecules in cold cores, rate constants for rotational energy transfer should ideally be obtained with molecular hydrogen as the collision partner and they should be measured or calculated at low temperatures. However, using pure H_2 flows in the CRESU apparatus poses considerable challenges. The large spacing of rotational energies in H_2 ($B = 86.5\,\mathrm{K}$), combined with the selection rule restriction $\Delta J = \pm 2$, arising from nuclear spin statistics, results in very slow H_2–H_2 rotational energy transfer. Within CRESU expansions, RET is a requirement in order to maintain thermal equilibration during gas transport through the Laval nozzle. Because of its slow RET, this would not occur in the case of H_2. If the flow starts at room temperature, the ratio of specific heat capacities ($\gamma = c_p/c_v$, see Chapter 2) would vary during the expansion, making the calculation of suitable Laval nozzles extremely challenging.

To circumvent this problem, a temperature range in which γ is constant can be used, which can be achieved by pre-cooling the reservoir by continuous flow of liquid N_2. A caveat in these measurements is that H_2 is produced as normal hydrogen (with an ortho-to-para ratio of 3), whereas molecular hydrogen in the ISM is usually in local thermodynamic equilibrium at the very low temperatures of those environments (i.e. pure para). The use of such cryogenically cooled Laval nozzles was pioneered by Tizniti *et al.* [53] in their study of the F + H_2 reaction, and their use has been extended by Labiad *et al.* [16] to the study of CO–H_2 RET.

7.7.2 *Energy transfer in polyatomic molecules*

As yet, there have been very few measurements of rate constants for energy transfer involving polyatomic molecules at low temperatures.

These measurements are required for radiative transfer models of the interstellar medium, with particular relevance to collisions involving complex organic molecules (COMs). Rotational energy transfer rate constants have only been measured using the CRESU technique for collisions with diatomic molecules. RET in triatomics and small poly-atomics will be measurable by IRUVDR for those species possessing appropriate fluorescence spectra. These systems present a challenge in terms of dimension; a large number of rotational levels are thermally populated in polyatomics, leading to a multitude of accessible collision channels.

Polyatomics typically do not possess suitable LIF detection schemes, but HCN is an exception and RET has been studied by the IRUVDR technique for this prototypical linear polyatomic molecule in cell experiments at room temperature by Warr and Smith [54]. It would be interesting to extend these measurements to low temperatures with He, Ar, N_2, and H_2 collision partners in CRESU flows.

VET is typically too slow to be studied by the CRESU technique, as explained above, but some low-frequency bending modes of long-chain molecules such as cyanopolyynes can be relaxed more efficiently. Rate constants for these processes are needed for radiative transfer models of these important astrophysical molecules, and it would be interesting to study VET for the first member of the cyanopolyyne series, HC_3N, using infrared laser-induced fluorescence detection.

7.7.3   *Chirped-pulse microwave spectroscopy for measurement of energy transfer and reaction product detection*

Chirped-pulse Fourier transform microwave spectroscopy (CP-FTMW), developed by Brooks Pate and co-workers [55], has enabled substantial progress in a number of fields due to the large reduction in time required to collect broadband microwave spectra. Chirped-pulse spectroscopy has been recently coupled to pulsed [27, 41] and continuous CRESU flows [56, 57], permitting energy transfer processes to be observed by analysis of micro/millimeter wave spectra. Chirped-pulse spectroscopy has recently been applied to pump-probe

DR experiments to derive state-to-state rate constants for rotational energy transfer in NH_3 self-collisions [58]. This method has the potential to be applied to a large number of molecules due to the broad frequency coverage of current chirped-pulse instruments.

Product channel-specific reaction kinetics are beginning to be explored at low temperature in uniform supersonic flows (see Chapter 8). These measurements present a significant experimental challenge in order to monitor multiple species simultaneously as well as sensitively. Chirped-pulse microwave spectroscopy has been proposed as a potential solution as each reaction product possesses a unique microwave signal. However, collisions with high-density buffer gas, required for thermalization, reduce the sensitivity of the technique through pressure broadening. Pressure broadening manifests in both the time domain, where the length of the free induction decays (FIDs) is reduced and in the frequency domain where the linewidths are increased. The magnitude of the pressure broadening depends on the density of the colliders and their intermolecular potentials and includes contributions from inelastic collisions (R-T energy transfer) and elastic collisions. It is therefore critical that collisional processes within CRESU flows are well characterized to improve the sensitivity for reaction product detection.

Furthermore, both CP-FTMW and infrared absorption techniques, which have also been proposed for studying reaction product branching ratios, detect specific vibrational states of the targeted molecules. Many reactions of interest for astrochemistry are highly exothermic, and the nascent reaction products are highly vibrationally excited, often with non-Boltzmann population distributions depending on the specific and detailed reaction dynamics. It will be important to characterize the rates of vibrational relaxation in collisions with CRESU buffer gases at low temperature in order to interpret correctly the measurement of product branching, especially in the case of infrared absorption between levels that may not be relaxed. Even photoionization mass spectrometric detection can display some sensitivity to the vibrational state of the targeted species, and so VET of small to medium-sized molecules that are the typical subjects of such product branching measurements will be an important area of study in the future.

References

[1] van der Tak FFS, Black JH, Schöier FL, Jansen DJ, van Dishoeck EF. A computer program for fast non-LTE analysis of interstellar line spectra — with diagnostic plots to interpret observed line intensity ratios. Astron Astrophys. 2007;468(2):627–35.

[2] Hogerheijde MR, van der Tak FFS. An accelerated Monte Carlo method to solve two-dimensional radiative transfer and molecular excitation. With applications to axisymmetric models of star formation. Astron Astrophys. 2000;362:697–710.

[3] Brinch C, Hogerheijde MR. LIME — a flexible, non-LTE line excitation and radiation transfer method for millimeter and far-infrared wavelengths. Astron Astrophys. 2010;523:A25.

[4] Daly PW, Oka T. Microwave studies of collision-induced transitions between rotational levels. VII collisions between NH_3 and nonpolar molecules. J Chem Phys. 1970;53(8):3272–8.

[5] James PL, Sims IR, Smith IWM, Alexander MH, Yang M. A combined experimental and theoretical study of rotational energy transfer in collisions between NO ($X^2\Pi$, v = 3) and He, Ar and N_2 at temperatures down to 7 K. J Chem Phys. 1998;109:3882.

[6] Carty D, Goddard A, Sims IR, Smith IWM. Rotational energy transfer in collisions between CO ($X^1\Sigma^+$, v = 2, J = 0, 1, 4, and 6) and He at temperatures from 294 to 15 K. J Chem Phys. 2004;121(10):4671–4683.

[7] Bergeat A, Faure A, Morales SB, Moudens A, Naulin C. Low-energy water–hydrogen inelastic collisions. J Phys Chem A. 2020;124(2): 259–64.

[8] de Jongh T, Besemer M, Shuai Q, Karman T, van der Avoird A, Groenenboom GC, *et al.* Imaging the onset of the resonance regime in low-energy NO-He collisions. Science. 2020;368(6491):626–30.

[9] Sims IR, Smith IWM. Gas-phase reactions and energy transfer at very low temperatures. Ann Rev Phys Chem. 1995;46(1):109–137.

[10] Herbert LB, Sims IR, Smith IWM, Stewart DWA, Symonds AC, Canosa A, *et al.* Rate constants for the relaxation of CH($X^2\Pi$, $\nu = 1$) by CO and N_2 at temperatures from 23 to 584 K. J Phys Chem. 1996;100(36):14928–35.

[11] James PL, Sims IR, Smith IWM. Total and state-to-state rate coefficients for rotational energy transfer in collisions between NO($X^2\Pi$) and He at temperatures down to 15 K. Chem Phys Lett. 1997;272:412–8.

[12] Brownsword RA, Canosa A, Rowe BR, Sims IR, Smith IWM, Stewart DWA, *et al.* Kinetics over a wide range of temperature (13–744 K): rate constants for the reactions of CH($\nu = 0$) with H_2 and D_2 and for the

removal of CH($\nu = 1$) by H_2 and D_2. J Chem Phys. 1997;106(18):7662–7677.

[13] James PL, Sims IR, Smith IWM. Rate coefficients for the vibrational self-relaxation of NO($X^2\Pi$, v = 3) at temperatures down to 7 K. Chem Phys Lett. 1997;276:423–9.

[14] Wright SMA, Sims IR, Smith LWM. Vibrational relaxation of highly excited toluene in collisions with He, Ar, and N_2 at temperatures down to 38 K. J Phys Chem A. 2000;104(45):10347–55.

[15] Merten LA, Labiad H, Denis-Alpizar O, Fournier M, Carty D, Le Picard SD, *et al.* Rotational energy transfer in collisions between CO and Ar at temperatures from 293 to 30 K. Chem Phys Lett. 2017;683:521–8.

[16] Labiad H, Fournier M, Mertens LA, Faure A, Carty D, Stoecklin T, *et al.* Absolute measurements of state-to-state rotational energy transfer between CO and H_2 at interstellar temperatures. Phys. Rev. A. 2022;105, L020802.

[17] Le Picard SD, Bussery-Honvault B, Rebrion-Rowe C, Honvault P, Canosa A, Launay JM, *et al.* Fine structure relaxation of aluminum by atomic argon between 30 and 300 K: an experimental and theoretical study. J Chem Phys. 1998;108(24):10319–26.

[18] Le Picard SD, Honvault P, Bussery-Honvault B, Canosa A, Laubé S, Launay JM, *et al.* Experimental and theoretical study of intramultiplet transitions in collisions of C(^{3}P) and Si(^{3}P) with He. J Chem Phys. 2002;117(22):10109–20.

[19] Páramo A, Canosa A, Le Picard SD, Sims IR. An experimental study of the intersystem crossing and reactions of $C_2(X^1\Sigma_g^+)$ and $C_2(a^3\Pi_u)$ with O_2 and NO at very low temperature (24–300 K). J Phys Chem A. 2006;110(9):3121–7.

[20] Sánchez-González R, Eveland WD, West NA, Mai CLN, Bowersox RDW, North SW. Low-temperature collisional quenching of NO $A^2\Sigma^+$(v' = 0) by NO($X^2\Pi$) and O_2 between 34 and 109 K. J Chem Phys. 2014;141(7):204302.

[21] Winner JD, West NA, McIlvoy MH, Buen ZD, Bowersox RDW, North SW. The role of near resonance electronic energy transfer on the collisional quenching of NO ($A^2\Sigma^+$) by C_6H_6 and C_6F_6 at low temperature. Chem Phys. 2018;501:86–92.

[22] Grondin R, Loison JC, Hickson KM. Low temperature rate constants for the reactions of O(^{1}D) with N_2, O_2, and Ar. J Phys Chem A. 2016;120(27):4838–4844.

[23] Nuñez-Reyes D, Kłos J, Alexander MH, Dagdigian PJ, Hickson KM. Experimental and theoretical investigation of the temperature dependent electronic quenching of $O(^1D)$ atoms in collisions with Kr. J Chem Phys. 2018;148:124311.

[24] Hickson KM, Loison J-C, Lique F, Kłos J. An experimental and theoretical investigation of the $C(^1D) + N_2 \rightarrow C(^3P) + N_2$ quenching reaction at low temperature. J Phys Chem A. 2016;120(16):2504–2513.

[25] Douglas K, Blitz MA, Feng W, Heard DE, Plane JMC, Slater E, *et al.* Low temperature studies of the removal reactions of $^1CH_2$ with particular relevance to the atmosphere of Titan. Icarus. 2018;303:10–21.

[26] Douglas KM, Blitz MA, Feng W, Heard DE, Plane JMC, Rashid H, *et al.* Low temperature studies of the rate coefficients and branching ratios of reactive loss vs quenching for the reactions of $^1CH_2$ with C_2H_6, C_2H_4, C_2H_2. Icarus. 2019;321:752–766.

[27] Abeysekera C, Zack LN, Park GB, Joalland B, Oldham JM, Prozument K, *et al.* A chirped-pulse Fourier-transform microwave/pulsed uniform flow spectrometer. II. Performance and applications for reaction dynamics. J Chem Phys. 2014;141(21):214203.

[28] Lara M, Berteloite C, Paniagua M, Dayou F, Le Picard SD, Launay J-M. Experimental and theoretical study of the collisional quenching of $S(^1D)$ by Ar. Phys Chem Chem Phys. 2017;19:28555.

[29] Heck EL, Dickinson AS. Traditional transport properties of CO. Physica A: Stat Mech Appl. 1995;217(1):107–123.

[30] Kerber RL, Brown RC, Emery KA. Rotational nonequilibrium mechanisms in pulsed $H_2 + F_2$ chain reaction lasers. 2: Effect of VR energy exchange. Appl Opt. 1980;19(2):293–300.

[31] Goldsmith PF, Langer WD. Molecular cooling and thermal balance of dense interstellar clouds. Astrophys J. 1978;222:881–95.

[32] Esherick P, Anderson RJM. Multiphoton ionization of molecules in selectively excited rovibrational states. Chem Phys Lett. 1980;70(3):621–3.

[33] Islam M, Smith IWM, Wiebrecht JW. Infrared-ultraviolet double-resonance measurements on the temperature dependence of relaxation from specific rovibronic levels in $NO(X^2\Pi$, v = 2, J) and $(X^2\Pi$, v = 3, J). J Phys Chem. 1994;98(37):9285–90.

[34] Frost MJ, Islam M, Smith IWM. Infrared–ultraviolet double resonance measurements on the temperature dependence of rotational and vibrational self-relaxation of $NO(X^2\Pi$, v = 2,j). Can J Chem. 1994;72(3):606–11.

[35] Andresen P, Joswig H, Pauly H, Schinke R. Resolution of interference effects in the rotational excitation of NO (N = O) by Ar. J Chem Phys. 1982;77(4):2204–205.

[36] Meyer H. Electronic fine structure transitions and rotational energy transfer of $NO(X^2\Pi)$ in collisions with He: a counterpropagating beam study. J Chem Phys. 1995;102(8):3151–68.

[37] Cecchi-Pestellini C, Bodo E, Balakrishnan N, Dalgarno A. Rotational and vibrational excitation of CO molecules by collisions with ^{4}He atoms. Astrophys J. 2002;571(2):1015–20.

[38] Heijmen TGA, Moszynski R, Wormer PES, van der Avoird A. A new He–CO interaction energy surface with vibrational coordinate dependence. I. Ab initio potential and infrared spectrum. J Chem Phys. 1997;107(23):9921–28.

[39] Brownsword RA, Herbert LB, Smith IWM, Stewart DWA. Pressure and temperature dependence of the rate constants for the association reactions of CH radicals with CO and N_2 between 202 and 584 K. J Chem Soc Faraday Trans. 1996;92(7):1087–94.

[40] Fulle D, Hippler H. The high-pressure range of the reaction of $CH(^2\Pi)$ with N_2. J Chem Phys. 1996;105(13):5423–30.

[41] Oldham JM, Abeysekera C, Joalland B, Zack LN, Prozument K, Sims IR, *et al.* A chirped-pulse Fourier-transform microwave/pulsed uniform flow spectrometer. I. The low-temperature flow system. J Chem Phys. 2014;141(15):154202.

[42] Paul PH, Gray JA, Durant Jr JL, Thoman Jr JW. Collisional quenching corrections for laser-induced fluorescence measurements of NO $A^2\Sigma^+$. AIAA J. 1994;32(8):1670–75.

[43] Cecchi-Pestellini C, Dalgarno A. C_2 absorption-line diagnostics of diffuse interstellar clouds. MNRAS. 2002;331:L31–L34.

[44] van Dishoeck EF, Black JH. Interstellar C_2, CH, and CN in translucent molecular clouds. Astrophys J. 1989;340:273–97.

[45] Souza SP, Lutz BL. Detection of C_2 in the interstellar spectrum of Cygnus OB2 number 12 (IV Cygni number 12). Astrophys J. 1977;216:L49–L51.

[46] Donati GB. Schreiben des Herrn Prof. Donati an den Herausgeber. Astron Nachr. 1864;62:375.

[47] Baronavski AP, McDonald JR. Measurement of C_2 concentrations in an oxygen–acetylene flame: an application of saturation spectroscopy. J Chem Phys. 1977;66(7):3300–311.

[48] Harilal SS, Issac RC, Bindhu CV, Nampoori VPN, Vallabhan CPG. Emission characteristics and dynamics of C_2 from laser produced graphite plasma. J App Phys. 1997;81(8):3637–43.

[49] Lambert DL. The abundances of the elements in the solar photosphere — VIII. Revised abundances of carbon, nitrogen and oxygen. MNRAS. 1978;182(2):249–72.

[50] Páramo A, Canosa A, Le Picard SD, Sims IR. Rate coefficients for the reactions of $C_2(a^3\Pi_u)$ and $C_2(X^1\Sigma_g^+)$ with various hydrocarbons (CH_4, C_2H_2, C_2H_4, C_2H_6, and C_3H_8): a gas-phase experimental study over the temperature range 24–300 K. J Phys Chem A. 2008;112(39):9591–600.

[51] Dalgarno A, Rudge MRH. Cooling of interstellar gas. Astrophys J. 1964;140:800.

[52] Mies FH. Molecular theory of atomic collisions: fine-structure transitions. Phys Rev A. 1973;7(3):942–57.

[53] Tizniti M, Le Picard SD, Lique F, Berteloite C, Canosa A, Alexander MH, *et al.* The rate of the $F + H_2$ reaction at very low temperatures. Nat Chem. 2014;6(2):141–5.

[54] Smith IWM, Warr JF. Tunable infrared–tunable ultraviolet double-resonance experiments on HCN. J Chem Soc Faraday Trans. 1991;87(1):205–206.

[55] Brown GG, Dian BC, Douglass KO, Geyer SM, Shipman ST, Pate BH. A broadband Fourier transform microwave spectrometer based on chirped pulse excitation. Rev Sci Instrum. 2008;79(5):053103.

[56] Hays BM, Guillaume T, Hearne TS, Cooke IR, Gupta D, Abdelkader Khedaoui O, Le Picard SD, Sims IR. Design and performance of an E-band chirped pulse spectrometer for kinetics applications: OCS — He pressure broadening. J Quant Spectrosc Radiat Transfer. 2020;250:107001.

[57] Hearne TS, Abdelkader Khedaoui O, Hays BM, Guillaume T, Sims IR. A novel Ka-band chirped-pulse spectrometer used in the determination of pressure broadening coefficients of astrochemical molecules. J Chem Phys. 2020;153(8):084201.

[58] Endres CP, Caselli P, Schlemmer S. State-to-state rate coefficients for NH_3–NH_3 collisions from pump–probe chirped pulse experiments. J Phys Chem Lett. 2019;10(17):4836–41.

https://doi.org/10.1142/9781800610996_0008

Chapter 8

Low-Temperature Product Detection and Branching Ratios

Sébastien D. Le Picard[*,‡], Abdessamad Benidar[*,§]
and Fabien Goulay[†,¶]

*CNRS, IPR (Institut de Physique de Rennes)-UMR 6251,
Université de Rennes, F-35000 Rennes, France
†C. Eugene Bennett Department of Chemistry,
West Virginia University, Prospect Street,
26501 Morgantown, WV, USA
‡sebastien.le-picard@univ-rennes1.fr
§abdessamad.benidar@univ-rennes1.fr
¶fabien.goulay@mail.wvu.edu

Abstract

The quantitative detection of reaction products at low temperature is crucial for the understanding of low-temperature chemistry, in part due to the increased importance of quantum effects as the temperature is lowered. Product information and branching ratios at very low temperatures are also vital input for astrochemical models. Early mass spectrometry experiments have allowed the detection of reaction products in a pulsed CRESU flow down to 90 K, while spectroscopic techniques have enabled the measurement of H atom branching ratios down to 50 K. Coupling CRESU flows to multiplex mass spectrometry at synchrotron facilities has already provided branching ratios between isomers formed by the same reactions below 100 K. The newly designed CRESUSOL apparatus using photoion–photoelectron coincidence detection scheme has the ability of detecting reaction products down to 10 K. Chirped-pulsed microwave spectroscopy has also appeared as a sensitive and universal technique for the detection of isomer-resolved products.

The cold rotational distributions in the CRESU flow provide ideal conditions for detection of products and also give the advantage of being able to distinguish isomers of products through their rotational spectra. These universal, isomer-resolved detection techniques and others, like frequency comb spectroscopy, have only just started to be implemented to CRESU apparatuses and promise to provide a wealth of information about gas phase chemical and physical processes occurring close to the absolute zero.

Keywords: Free radicals; Kinetics; Low temperature, Branching ratios.

8.1 Introduction

Most chemical reactions can form different sets of products, including isomers. Identification of their chemical structure, as well as knowledge of the branching ratio of each individual channel, gives valuable insight about the mechanisms of the elementary reactions leading to their formation. Such information is crucial for modeling the chemistry of gaseous environments.

Let us consider the simple case of a bimolecular reaction where three exit channels are accessible:

$$A + B \rightarrow C + D \tag{8.1}$$

$$\rightarrow E + F \tag{8.2}$$

$$\rightarrow G + H \tag{8.3}$$

The total rate coefficient of the reaction is $k(T) = k_{R_1}(T) + k_{R_2}(T) + k_{R_3}(T)$ and the branching ratios[a] for the product channels leading to C + D, E + F, and G + H are defined, respectively, by the following equations:

$$BR(C + D) = \frac{k_{R_1}(T)}{k(T)} \tag{8.4}$$

$$BR(E+F) = \frac{k_{R_2}(T)}{k(T)} \tag{8.5}$$

$$BR(G + H) = \frac{k_{R_3}(T)}{k(T)} \tag{8.6}$$

---

[a]May also be referred to as branching fraction. In this case, a branching ratio may refer to the ratio between two branching fractions [1].

where $k_{R_1}(T)$, $k_{R_2}(T)$, and $k_{R_3}(T)$ are the rate coefficients for reaction 8.1, 8.2, and 8.3, respectively. Both theoretical and experimental determinations of the exit channel branching ratios are challenging, and the amount of accurate data about elementary reactions is still limited, especially at low temperature ($T < 100$ K). Exploring the low-temperature regime, however, is of particular interest for understanding the role of quantum effects such as tunneling, which is of major importance in modeling astrochemical environments. A current key question paramount to the understanding of the chemistry in space is how complex organic molecules (COMs) are formed and consumed in cold ($T < 100$ K) or very cold ($T < 10$ K) interstellar clouds. The temperature dependence of the rate coefficients for these reactions can take a variety of forms depending on the type of potential describing the interaction during the collision [2]. In some cases, a reaction rate coefficient can be up to 1000 times faster at very low temperature than at room temperature, even in the presence of an energy barrier to the formation of the products [3–6].

Reactions of astrochemical interest frequently involve either ions or free radicals as reactants. They often form reaction intermediates before decomposing to products. For such reactions, it remains quite challenging to accurately include quantum effects, especially as they are amplified at low temperature, ultimately leading to large uncertainties on the estimation of the corresponding low-temperature rate coefficients [7]. The knowledge of reaction products and their branching ratios is an important piece of information for understanding the type of mechanism at play during such reactions.

Experimental determination of the exit channel branching is very challenging since at any temperature it often requires absolute concentration measurements of unstable species as well as a high sensitivity, as some chemically important species can be present at trace concentrations. Furthermore, molecular and radical species may be formed in various electronic, vibrational, and rotational excitation states, which increases the difficulty of determining branching ratios. In the last few years, experimental devices have been developed to quantitatively address the issue of product branching ratios under a large variety of physical conditions. Quantitative information can be obtained by several approaches, including photoionization or electron spectroscopy combined with mass spectrometry, optical spectroscopy, and microwave spectroscopy. These techniques, which are complementary to laser-induced fluorescence (LIF), are also able to detect simultaneously a wide range of different products with sufficient time resolution to enable elementary reaction kinetics to be followed in real time. Because products and/or reactants may have potentially very different abundances, the dynamic range of the detection also needs to be high. For low-temperature experiments, an additional difficulty lies in the generation of a gas at very low temperature, and the CRESU reactor appears to be the method of choice.

In this chapter, we present an overview of experimental methods using the CRESU technique to estimate branching ratios of elementary chemical reactions at low temperatures. This chapter is organized as follows. Section 8.2 presents the specific approach using the highly sensitive pulsed laser-induced fluorescence (LIF) technique for detecting H atoms formed by a reaction. In Section 8.3, techniques coupling a CRESU experiment with photoionization and mass spectrometry are described, with a particular emphasis on the CRESUSOL experimental set up which is being developed at the SOLEIL (Source optimisée de lumière d'énergie intermédiaire du LURE) synchrotron. Section 8.4 focuses on the chirped-pulse Fourier transformed microwave spectroscopy, describing past and future experimental setups. Finally, the perspectives section discusses upcoming innovative developments for the multiplex detection of reaction products at low temperature using frequency comb spectroscopy.

8.2 Low-Temperature Hydrogen Atom Branching Ratios

8.2.1 *VUV laser-induced fluorescence of H atoms*

Because of its high sensitivity, the pulsed laser-induced fluorescence (LIF) technique is widely used in kinetic and dynamic studies of gas phase processes (see Chapters 5 and 7). In some favorable cases, such as reactions of free radicals with small molecules, the branching ratios of reactions can be determined by monitoring the H atom production compared to a reference reaction for which the H atom yield is known [9]. Although it does not directly provide a full identification of the co-product isomeric structure, it can be combined with reactant decay rates and computational studies in order to infer the most likely reaction mechanisms. At low temperature, the technique has recently been employed to investigate the products of carbon atoms, both in their ground (^{3}P) and first excited (^{1}D) states, and excited oxygen atoms (^{1}D) reacting with small hydrocarbons [9–12], H_2/D_2 [13, 14], water [15, 16], methanol [17], and ammonia [18], as well as singlet CH_2 with a series of hydocarbons [19, 20]. In the case of carbon atoms reacting with ammonia, the Laval nozzle experiments were complemented by tunable photoionization mass spectrometry (PIMS) investigations at room temperature to confirm the isomeric structure of the co-product [18].

In the supersonic flow, H atoms are detected in their ground electronic state (^{2}S) by laser-induced fluorescence through the $(2p)^2P \leftarrow (2s)^2S$ Lyman-α transition line [17]. The 121.657-nm laser radiation is obtained by focusing a frequency-doubled dye laser radiation at 364.7 nm into a krypton/xenon gas cell perpendicular to the supersonic flow. The resonant LIF emission is detected by a solar-blind photomultiplier tube as a function of the delay time between the photolysis and LIF lasers. Figure 8.1 displays the Lyman-α LIF signal detected in a supersonic flow at 106 K upon 266-nm irradiation of a CBr_4 mixture in nitrogen with NH_3 (blue open circles) and a

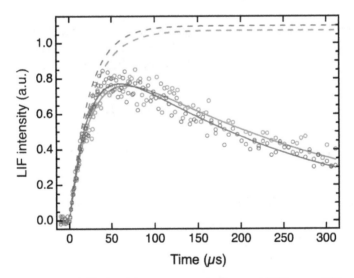

Fig. 8.1. Lyman-α LIF signal in a supersonic flow at 106 K upon 266-nm irradiation of a CBr_4 mixture in nitrogen with NH_3 (blue open circles) and C_2H_4 (red open circles). The solid lines are fits to the experimental data using double exponential functions and the dashed lines are modeled profiles assuming no H atom loss mechanisms. Adapted from Ref. [18].

reference compound, C_2H_4 (red open circles) [18]. The reactant concentrations are adjusted in order to obtain similar pseudo-first-order formation rates for both reactions. The solid lines are fits to the LIF signal S_H using the double exponential function from a sequential reaction scheme [21] as shown in Eq. (8.7):

$$S_H = \frac{k_{1st}S_H^\infty}{k_{1st} - k_{loss}}(e^{-k_{loss}t} - e^{-k_{1st}t}) \tag{8.7}$$

The first-order rate k_{loss} represents the loss of H atoms due to diffusion and reactions, while k_{1st} represents the H atom formation through the $C(^3P) + NH_3$ reaction. The coefficient S_H^∞ is the theoretical LIF signal in the absence of any loss phenomena. It is directly proportional to the amount of H atoms generated by the source reaction. Assuming that the loss mechanisms are the same for all studied reactions and for equal pseudo-first-order rates k_{1st}, the ratio of the pre-exponential factors from the fits to both signals is the ratio of H atom branching ratios for both reactions. Knowledge of the H atom branching ratio of the reference reaction provides an absolute value for that of the studied reaction.

For the $C(^3P) + NH_3$ reaction, Bourgalais *et al.* [18] reported a branching ratio close to unity which remains constant over the 50–296 K temperature range. The H atom co-product was confirmed to be H_2CN using tunable PIMS at room temperature. The overall reaction mechanism is as follows:

$$C(^3P) + NH_3 \rightarrow H_2CN + H(^2S) \tag{8.8}$$

Because the measured H branching ratio does not change with temperature, it is likely that the isomer structure of the co-products remains unchanged at lower temperature. The detection of both co-products in the $C(^3P) + NH_3$ reaction provides valuable information for the modeling of cold interstellar medium. Including reaction 8.8 in astrophysical chemical schemes leads to a drastic decrease of the NH_3 relative abundance at the time when the carbon atom abundance is high in the cloud [22].

8.2.2 *Temperature dependence of the H atom branching ratios*

Low-temperature H atom branching ratios for reactions of $C(^3P/^1D)$ and $O(^1D)$ studied in supersonic flows often do not display a significant temperature dependence [9, 11, 15, 17, 23]. In the case of the C(^{1}D) + H$_2$O reaction [15], the branching ratio is close to the theoretical maximum of two over the 75–206 K temperature range. The reaction is believed to form internally excited HCO through H loss. The molecular product then rapidly dissociates to give CO + H + H as final products. At 50 K, the measured value increases significantly, likely due to an additional contribution from the C(^{3}P) + H$_2$O reaction [16], which was shown to accelerate at very low temperature. Similarly, the branching ratios for C(^{1}D) reactions with alkanes [11] are slightly higher than unity due to side reaction contributions, but do not show any statistically significant changes when the temperature is lowered down to 50 K.

The reaction of ground state carbon C(^{3}P) with acetylene was previously studied using crossed molecular beam experiments [24–27]. The reaction is believed to have strong implications for astrochemistry as a source of C$_3$ cluster through the spin-forbidden channel:

$$C(^3P) + C_2H_2(X^1\Sigma_g^+) \rightarrow C_3(X^1\Sigma_g^+) + H_2(X^1\Sigma_g^+) \qquad (8.9)$$

The other two exothermic channels form the linear and cyclic doublet C$_3$H isomers and the H atom in its ground state [25]. The branching ratio from reaction 8.9 can be directly inferred by subtracting the H atom branching ratios from unity.

Figure 8.2 displays branching ratios for the C$_3$ + H$_2$ channel measured in CRESU experiments using H atom LIF [9], and crossed molecular beam experiments using Doppler–Fizeau spectral broadening (open black circles) [25], and electron impact mass spectrometry detection (blue triangles and filled black squares) [24, 27]. In the case of the dynamics experiments, the temperature is obtained by assuming that the collision energy is equivalent to the mean energy of a Maxwell-Boltzmann distribution. The high-energy measurements display a C$_3$+H$_2$ branching ratio that monotonically increases as the energy is lowered. At intermediate energy, the data of Gu et al. [27] are consistent with the room-temperature values of both Bergeat and Loison [28] and Hickson et al. [9], both measured in a subsonic flow. At lower energy, the CRESU data are found to remain close to the room-temperature value with a slight decrease with decreasing temperature. These values are, however, much lower than crossed-beam data obtained by deconvolution of H atom Doppler–Fizeau

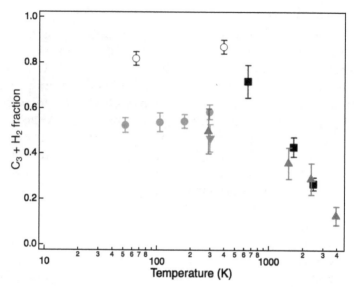

Fig. 8.2. Temperature dependence of the $C_3 + H_2$ channel from the $C(^3P) + C_2H_2$ reaction measured in a supersonic flow (filled red circles) [9], fast flow reactor (green inverted triangle) [28], and crossed molecular beams using Doppler–Fizeau spectrum broadening (open black circles) [25] or electron impact mass spectrometry detection (blue triangles and filled black squares) [24, 27].

profiles [25]. The large difference in absolute value may be due to erroneous data analysis in the crossed-beam data [9].

The temperature independence of the spin-forbidden channel measured in a CRESU flow differs from the negative temperature dependence observed in crossed molecular beam experiments at high collision energy (blue triangles and filled black squares in Fig. 8.2) [24, 27]. The change in temperature behavior may suggest a maximum in the efficiency of the intersystem crossing near room temperature, although there are no available theoretical studies to support this hypothesis. Alternatively, Hickson *et al.* [9] discussed a possible effect of the initial spin-orbit distribution of the ground state carbon atom. In the CRESU flow, the high number of collisions leads to a thermal distribution of the ^3P_J state population, while the low density of the crossed molecular beam is more likely to maintain the nascent distribution from the CBr_4 laser photolysis. In this case, the high-energy data may be characteristic of carbon atom higher spin-orbit state reactions. The difference between the supersonic flow data and the crossed-beam data could also reflect a different C_2H_2 rotational distribution. Due to the initial supersonic expansion, the crossed-beam molecules may be rotationally cooler than ones in the CRESU flow. Nonetheless, the H atom product branching ratios provide valuable data in order to model the density of small C2 and C3 hydrocarbons in the interstellar medium.

8.3 Product Detection Using Mass Spectrometry

8.3.1 *Sampling of the supersonic flow*

8.3.1.1 *Airfoil and skimmer sampling*

A gas flow reactor whose density is usually relatively high $(10^{16}-10^{18}\,\text{cm}^{-3})$ cannot be probed directly by mass spectrometry as long as the low-pressure conditions required for good operation (high voltage, detector saturation) of the detectors are not fulfilled. To reduce the pressure by orders of magnitude, the flow is sampled using an aperture of typically few hundred microns. Furthermore, such a sampling is expected to rapidly "freeze" the chemistry in order to detect only nascent products. After sampling, the products and reactants are ionized and detected by mass spectrometry. For a quasi-static cell or subsonic flow, the sampling may be performed simply with a pinhole, either perpendicularly or collinear to the flow. The pressure differential after the pinhole (from 1–10 torr to $< 1 \times 10^{-6}$ torr) generates an effusive beam that contains all the flow components with unchanged molar fractions. In the case of a supersonic flow, the sampling must be collinear to the gas velocity with minimum disturbance to the flow.

Any interaction of an object with a supersonic flow will generate a shock wave and lead to an abrupt increase in pressure and temperature [29]. Skimmer insertion can induce some perturbations and affect the density and temperature of the gas in the probed region. Another important criterion that will determine the final design of a sampling device is the required instrument time response [30]. Although in supersonic expansions the Maxwell-Boltzmann lateral distribution of velocities is greatly reduced, it may still lead to a decrease of the temporal resolution, especially if the ionization source is far from the aperture.

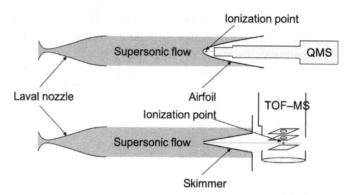

Fig. 8.3.   Schematic representations of the airfoil [31] and skimmer design [32] for sampling of a pulsed CRESU flow. In the design of Lee *et al.* [32], the skimmer is 6 cm long and has a 30° entrance angle.

Most CRESU experiments coupled to mass spectrometry (quadrupole or time-of-flight) have used a conical skimmer placed at variable distances from the nozzle exit [32–35]. One experiment used an airfoil design for sampling the supersonic flow [31]. Figure 8.3 displays a schematic representation of the airfoil (top) [31] and skimmer (bottom) [32] used in supersonic flow sampling. In both schematics, the red dot represents the ionizing radiation. The advantage of an airfoil sampling relative to a skimmer is to increase the spacing available near the pinhole aperture. In the case of a quadrupole mass spectrometer (QMS), this allows bringing the instrument inlet and the ionization point very close to the sampling aperture where the molecular beam is less divergent, which is likely to increase the ion number density and improve the measurement temporal resolution (see Section 8.3.1.2).

Figure 8.4 displays the profiles of the skimmer (black lines) and airfoil (red lines) used in the sampling of pulsed supersonic flows. When a supersonic flow encounters a conical object, a conical shock wave will form at its surface. The oblique waves converge at the apex of the cone (attached wave) leading to the direct sampling of the unperturbed isentropic core of the flow. Maccoll [36] has calculated that an attached shock wave will form for a cone angle up to 115.2° (blue line in Fig. 8.4).

In the case of a greater angle or in the case of airfoil sampling, the shock wave is found to be detached from the sampling object [29].

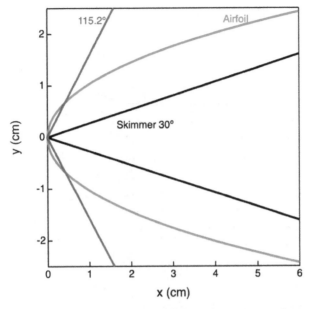

Fig. 8.4. Skimmer (black line) and airfoil (red line) profiles used in pulsed Laval experiments coupled to mass spectrometry [36]. The blue line shows the maximum angle leading to conical attached shock wave by interaction of a supersonic flow with a conical object.

There is a region in space between the shock wave and the pinhole where the gas flow becomes subsonic ($0.4 < M < 1$) and heated to a temperature that is close to that of the reservoir [29]. The location and strength of the detached shock wave depend on several factors, for example, the Mach number, the profile of the sampling body, and the properties of the flowing gas. Figure 8.5 displays a schematic representation of a detached shock wave on an airfoil object [37]. When sampling the supersonic flow using an airfoil, the products are found to be at higher temperatures than during their formation in the supersonic flow. The heating of the products may have an effect on their identification when it involves unstable molecules or molecular clusters that isomerize or decompose at higher temperatures. In the case of the reactions studied using airfoil sampling [38, 39], the fast temperature rise is unlikely to affect the nascent product detection as they are stable closed-shell molecules. In addition, the distance between the detached shock wave and the pinhole is likely to be small

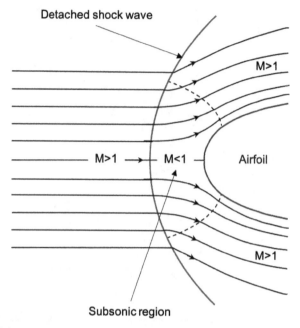

Fig. 8.5. Schematic representation of a detached shock wave by interactionof a supersonic flow with an airfoil. Adapted from Ref. [37].

relative to the dimensions of the system minimizing any bimolecular processes at higher temperature.

More recently, in the framework of the design of the CRESU-SOL apparatus (see Section 8.3.3) [40], the Rennes group performed Computational Fluid Dynamics (CFD) calculations showing that the critical feature in the sampling of the supersonic uniform expansion was not only the shape of the skimmer but also the distance from the apex of the skimmer to the flat wall orthogonal to the flow direction on which the skimmer is mounted. The shape of the skimmer (Beam Dynamics) selected for the CRESUSOL setup is characterized by an orifice diameter of 0.1 mm, an orifice edge thickness of 10 μm, and a total included angle at orifice of 25° internal and 30° external (see Fig. 8.6).

Calculations were carried out for two skimmer lengths (15 and 25 mm) using different carrier gases (Ar, He, and N_2) and different flow temperatures and density conditions. It was shown that when the skimmer is mounted onto a flat flange, a shock wave detached

Fig. 8.6. Picture of the skimmer (beam dynamics) used for the CRESUSOL setup (orifice diameter: 0.1 mm, orifice edge thickness: $10\,\mu$m, total included angle at orifice of 25° internal and 30° external).

from the surface of the flange is generated. Under certain operating conditions of the gas flow reactor, it can be localized well upstream of the skimmer and implies a critical sampling condition. The calculations show that the position of the detached shock wave is actually governed by inertial forces, which depend on the number density and the Mach number of the supersonic flow, as well as the angle of attack (90° for a flat flange). For a given gas, the faster the flow (i.e. the larger is the Mach number), the smaller the detachment distance and the closer the shock wave from the obstacle (flat flange). Figure 8.7 displays the results of CFD simulations for two different uniform supersonic flow conditions. In case (a), uniform supersonic flow of N_2 at $T = 75$ K (Mach number 3.8, total number density 1.7×10^{16} cm^{-3}), the detachment distance is less than the skimmer height and the shock wave remains attached to the tip of the skimmer, allowing for an optimized sampling of the flow's isentropic core. In case (b), uniform supersonic flow of He at $T = 36$ K (Mach number 4.6, total number density 5.3×10^{16} cm^{-3}), the detachment distance is more than the skimmer height and a detached normal shock wave is localized upstream of the skimmer, implying critical sampling as discussed above.

In the framework of the CRESUSOL apparatus design, the obstacle profile formed by the skimmer support flange was optimized with the help of CFD calculations. The resulting flange profile geometry facilitates the evacuation of residual gases at the obstacle surface

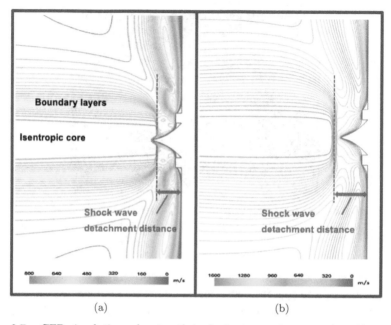

(a)   (b)

Fig. 8.7. CFD simulations showing the velocity maps of supersonic uniform gas flows sampled by a 15-mm-long skimmer mounted on a flat flange. (a) For a uniform supersonic flow of N_2 at $T = 75\,K$ (Mach number 3.8, total number density $1.7 \times 10^{16}\,cm^{-3}$): the shock wave remains attached to the tip of the skimmer. (b) For a uniform supersonic flow of He at $T = 36\,K$ (Mach number 4.6, total number density $5.3 \times 10^{16}\,cm^{-3}$): a detached normal shock wave is generated upstream of the skimmer by the expansion of helium at very high velocity [40].

and leads to the reduction of the shock wave detachment distance. This contributes to obtaining a shock wave attached to the top of the skimmer over a wide range of supersonic flows. Figure 8.8 displays the results of CFD simulations when sampling the uniform supersonic flow of He at 36 K; the streamlined geometry of the skimmer-supporting flange induces the disappearance of the detached layer by facilitating the residual gas evacuation.

8.3.1.2 *Flow sampling and temporal resolution*

The sampling of neutral products in a supersonic flow aims at measuring product branching ratios as well as kinetic rate coefficients. In the latter case, it is important to know the instrument response of the sampling system. In an effusive beam, the instrument response

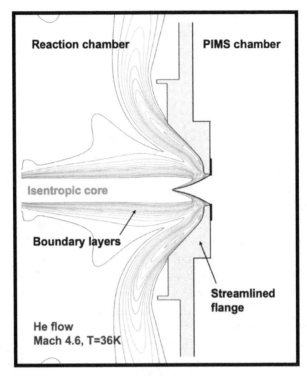

Fig. 8.8. CFD simulations showing the velocity map of a uniform supersonic flow of He at T = 36 K (Mach number 4.6, total number density $5.3 \times 10^{16}\,\mathrm{cm^{-3}}$) sampled by a 25-mm-long skimmer mounted onto a streamlined separating wall. The shock remains fully attached to the skimmer ensuring the sampling of the isentropic core of the cold uniform supersonic flow [40].

is greatly affected by the velocity spread of the gas. This effect is minimized in a supersonic flow, therefore allowing measurements of fast transient phenomena. This is crucial for CRESU experiments as the instrument response must be negligible compared to the aerodynamic time for the flow. Taatjes [30] defined the full-width spread in arrival time, δt, of molecules sampled by a conical skimmer of length x from a supersonic flow of velocity u:

$$\delta t = \frac{\sqrt{\frac{2k_B T}{m}}\,x}{\left(\sqrt{\frac{2k_B T}{m}} + u\right)^2} \qquad (8.10)$$

where k_B is the Boltzmann constant and m the molecular mass of the carrier gas. Using the criteria that δt should be small compared to the characteristic time $1/k_{1st}$ (<10%), he defined an approximate criterion for validity of supersonic flow sampling:

$$\delta t \cdot k_{1st} = \frac{\sqrt{\frac{2k_B T}{m}} \, x k_{1st}}{\left(\sqrt{\frac{2k_B T}{m}} + u \right)^2} < 0.1 \qquad (8.11)$$

For $\delta t \cdot k_{1st} > 0.1$, Taatjes suggests that there will be a non-negligible difference between the predicted and observed radical decay rate. For a given pseudo-first-order rate, Eq. (8.11) suggests that the error made on k_{1st} depends on the distance between the sampling pinhole and the ionization source, x. For a supersonic flow of Mach number M = 3, Soorkia *et al.* [31] calculated that in order to measure rates of up to $80{,}000\,\mathrm{s}^{-1}$ with negligible error, the ionization source must be no further than 4 mm away from the pinhole. For a conical skimmer with an angle smaller than 115.2°, this distance will always be significantly greater, due to the narrowing of the skimmer near the pinhole. In the design of Lee *et al.* [32], the ionization laser is placed 130 mm away from the pinhole (Fig. 8.3), corresponding to $\delta t \cdot k_{1st} \sim 0.8$. Comparing modeled and experimental radical decays suggests that in this case rates of $25{,}000\,\mathrm{s}^{-1}$ are already underestimated by 20%.

8.3.2 *Photoionization mass spectrometry for neutral product detection*

Photoionization coupled to mass spectrometry (PIMS) has now been widely employed in order to investigate the products of radical reactions with closed-shell species at room and high temperatures and reported in review articles [41–43]. Its application to the detection of neutral products of chemical reactions in supersonic flows is somewhat underdeveloped and only three CRESU apparatuses, including two pulsed versions, have been coupled to PIMS for detection of radical–neutral reaction products (not including dimerization) [31, 32, 44]. Although electron-impact ionization can easily be coupled to the system, photoionization near the ionization threshold has the advantage of minimizing excess energy in the resulting ions.

A Vacuum Ultra-Violet (VUV) light source below (12 eV) allows probing nascent reaction products for radical–neutral and radical–radical reactions.

A version of such apparatus was coupled to synchrotron VUV radiation at the Advanced Light Source synchrotron in Berkeley, Ca [31]. The tunable (7–13 eV) radiation allows the selective detection of isomer products based on their ionization energy. The apparatus was employed to measure the low-temperature branching ratios for the C_2H reaction with C_2H_4, as well as C_3H_6 and C_4H_8 isomers [38, 39]. The following paragraphs present the results from the two pulsed Laval experiments coupled to PIMS, 118-nm laser photoionization [32] and tunable synchrotron radiation [31], compare the rate coefficients inferred from the product profiles to literature data, and discuss low-temperature isomer-resolved product branching ratios. The effects of the temporal instrument response and sampling techniques are also discussed.

8.3.2.1 *Pulsed Laval nozzle coupled to time-of-flight mass spectrometry using a pulsed laser ionization source*

The first Laval nozzle experiment for the detection of neutral products using mass spectrometry used a fixed-wavelength laser ionization at 118 nm (10.53 eV) [32]. It was employed to detect the reaction products of the C_2H radical reacting with O_2 and C_2H_2 [33, 45]. The product temporal profiles were used to determine the reaction rate coefficient at low temperature. A schematic of the pulsed Laval nozzle apparatus coupled to time-of-flight mass spectrometry is presented in Fig. 8.3 [32]. The 118-nm ionization laser pulse, 13 cm down from the sampling pinhole is the 9th harmonic of an Nd:YAG laser. It is generated by focusing the 355-nm radiation into a low-pressure Xe-Ar gas cell. The 10.5-eV radiation is sufficient to ionize most hydrocarbons with minimum fragmentation. Multiphoton ionization due to the high flux at 355 nm also leads to the detection of nitrogen molecules and other background gases.

Figure 8.9 displays a mass spectrum recorded at low temperature upon dissociation of C_2H_2 at 193 nm in the presence of molecular oxygen [45]. The peak at m/z 41 is attributed to HCCO from the $C_2H + O_2$ reaction, while that at m/z 50 is attributed to C_4H_2 from the $C_2H + C_2H_2$ reaction. The reaction of C_2H with acetylene

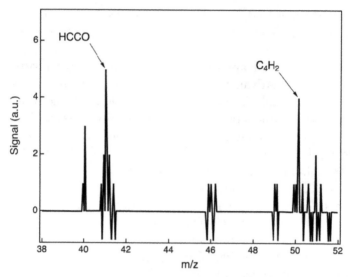

Fig. 8.9. Mass spectrum recorded at low temperature upon dissociation of C_2H_2 at 193 nm in the presence of molecular oxygen. Adapted from Ref. [45].

is expected to only lead to the formation of diacetylene by H loss [32, 33], while that with O_2 has five thermodynamically accessible pathways.

In a recent high-level theoretical study, Bowman *et al.* [46] calculated that the less exothermic HCCO + O and most exothermic HCO + CO channels will have equal branching ratios. The detection of signal at the m/z value of HCCO in Fig. 8.9 is consistent with the theoretical analysis. Lee and Leone [45] do not report the detection of HCO in the pulsed Laval experiment. The non-detection of the formyl radical in the supersonic flow may be due to its possible lower ionization cross section at 118 nm or to successive dissociation after its initial formation in a highly vibrational excited state. Its signal also overlaps with the large N_2 signal due to 355-nm multiphoton ionization. Indeed, based on Fig. 8.9 (average of 3000 laser shots), the sensitivity of the product detection may not be sufficient to detect the elusive HCO radical. Under these conditions, the product detection may not be used for branching ratio measurements nor mechanistic investigation but only for rate coefficient measurements.

Under pseudo-first-order conditions, the concentration of any reaction product P at a time t from reaction of the C_2H radical with a reactant R may be represented by

$$[P] = \frac{k_p[R][C_2H]_{t=0}}{\sum k_{1st}}(1 - e^{-\sum k_{1st} \cdot t}) \qquad (8.12)$$

where k_p is the product-specific reaction rate coefficient, $[C_2H]_{t=0}$ is the initial concentration of the C_2H radical, and $\sum k_{1st}$ is the sum of all the pseudo-first-order rates contributing to the decay of the C_2H radical. Only products with sufficiently large branching ratios are observable. The time behavior of the observed product is independent of the product channels and provides the overall pseudo-first-order decay rate of the radical.

Figure 8.10 displays the HCCO temporal profile measured at $T = 90\,K$ upon 193-nm photolysis of a C_2H_2/O_2 mixture [45]. The signal is plotted as a function of the delay time between the photolysis laser and the ionizing laser. The vertical error bars are from the average of several independent mass spectra at a specific delay time. Lee and Leone [45] do not provide any indications on the origin of the horizontal error bars, although these are likely to result from the travel time dispersion of the products between the pinhole and the ion

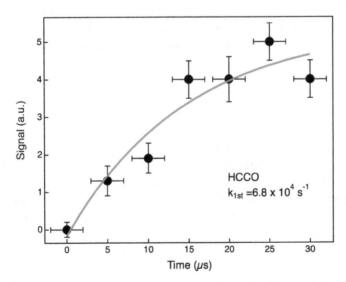

Fig. 8.10. HCCO temporal profile measured at $T = 90.4\,K$ upon 193-nm photolysis of a C_2H_2/O_2 mixture. The red line is an exponential fit to the data. Adapted from Ref. [45].

extraction region. The red line is a fit to the data using Eq. (8.12) and returning a total pseudo-first-order rate of $6.8 \times 10^4 \, \text{s}^{-1}$. Under the experimental conditions of Fig. 8.10, the $\sum k_{1st}$ term contains contributions from reactions with both O_2 and C_2H_2. The absolute value of the second-order rate coefficient for the C_2H reaction with one of the reactants is obtained by plotting the total pseudo-first-order rate as the function of the reactant number density.

Figure 8.11 displays the rate coefficients for the C_2H reaction with C_2H_2 (top) and O_2 (bottom) measured by time-of-flight mass spectrometry (filled red circles) [33, 45], chemiluminescence (filled black squares) [47], and transient IR absorption (opened blue triangles) [48, 49]. The rate coefficients measured by following the rise of the products (HCCO or C_4H_2) are found to be systematically higher than that measured by monitoring the decay of the C_2H radical. Due to the fast reaction of the C_2H radical with its precursor C_2H_2, the observed rates in the pulsed Laval nozzle flow range from $6.0 \times 10^4 \, \text{s}^{-1}$ to $15.0 \times 10^4 \, \text{s}^{-1}$. In the case of the reaction with C_2H_2, the linear fit of the pseudo-first-order rates as a function of reactant number density is found to intercept the y-axis at about $3.0 \times 10^4 \, \text{s}^{-1}$. This high value is far beyond potential systematic errors due to the mass filtering geometry established by Taatjes [30]. Without careful modeling of the experimental sampling, it is difficult to estimate the direct effect of the sampling bias on the determined second-order rate coefficient.

8.3.2.2 *Tunable photoionization at the advanced light source synchrotron*

The apparatus described above was coupled to synchrotron VUV radiation at the Advanced Light Source synchrotron in Berkeley, Ca [41]. The tunable (7–13 eV) radiation allows one to selectively detect isomer products based on their ionization energy. The apparatus was employed to measure the low-temperature branching ratios for the C_2H reaction with C_2H_4, as well as C_3H_6 and C_4H_8 isomers [38, 39]. The design uses airfoil sampling, which minimizes any timing issues when measuring kinetic profiles.

At a fixed ionization wavelength, the quadrupole mass filter provides temporal profiles of a given mass channel relative to the photolysis laser. Figure 8.12 displays the measured ion signal (open circles)

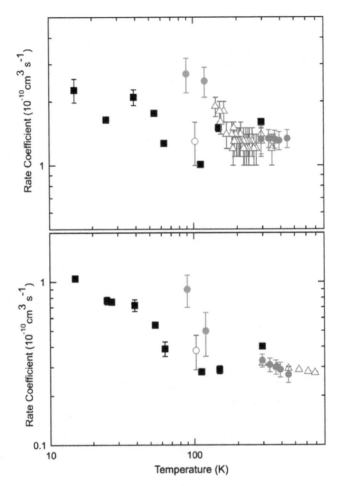

Fig. 8.11. Rate coefficients for the C_2H reaction with C_2H_2 (top) and O_2 (bottom) measured by time-of-flight mass spectrometry (filled red circles) [33, 45], chemiluminescence (filled black squares) [47], and transient IR absorption (opened blue triangles) [48, 49].

at m/z 52 from the reaction of the C_2H radical with propene together with a fit to the data [39]. The black line is the Gaussian profile, with unit area, that represents the temporal response function of the ion detector, obtained from the time profile of a photolysis fragment [38]. The data in Fig. 8.12 are then fitted with a function obtained from the convolution of an exponential rise function with the Gaussian instrument temporal response. As for the Laval flow coupled to the

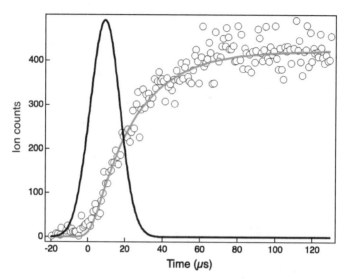

Fig. 8.12. Ion signal (open circles) at m/z 52 from the reaction of the C_2H radical with propene together with a fit to the data (red line) [39]. The black line represents the instrument temporal characteristic obtained from the time profile of photolysis fragment [38].

118-nm ToF, second-order rate coefficients are obtained by plotting the formation rate from the convoluted fit as a function of the reactant number density.

Table 8.1 displays reaction rate coefficients obtained from products detection for the C_2H reaction with small unsaturated hydrocarbons. The values are in very good agreement with low-temperature literature data, validating the airfoil design for kinetic measurements.

The main advantage of the technique is to be able to tune the ionization energy and record photoion spectra as a function of wavelength at a given mass channel. Figure 8.13 displays the photoion spectra at m/z 52 (C_4H_4) detected in Fig. 8.12 as a function of the ionizing photon energy together with a fit to the data [39]. The ionization onset observed at 9.6 eV is characteristic of the ionization energy of the C_4H_4 isomer vinylacetylene. The sharps peaks at higher energies have been observed previously and are attributed to auto-ionization states of the molecule. In order to fully confirm the detection of vinylacetylene in the reaction flow, the photoionization spectrum is fitted with the spectrum of known molecules. In this case, the photoionization spectrum can be fit with the spectrum of

Table 8.1. Reaction rate coefficients k_{2nd} obtained from products' detection for the C_2H reaction with small unsaturated hydrocarbons.

Reactant	Product Detection at 74 K $k_{2nd}(10^{-10}\ cm^3 s^{-1})$ [38, 39]	$T(K)$	C$_2$H Radical Decay $k_{2nd}(10^{-10}\ cm^3 s^{-1})$
C$_2$H$_4$	1.30($\pm$0.10)	112	1.59($\pm$0.04) [47]
		103	1.49($\pm$0.40) [50]
C$_3$H$_6$	2.10($\pm$0.20)	112	2.35($\pm$0.09) [47]
		103	2.40($\pm$0.60) [50]
C$_4$H$_8$			
1-butene	1.90($\pm$0.50)	103	2.60($\pm$0.60) [51]
Cis-2-butene	1.70($\pm$0.50)		
Trans-2-butene	2.10($\pm$0.70)		
isobutene	1.80($\pm$0.90)	103	1.40($\pm$0.30) [51]

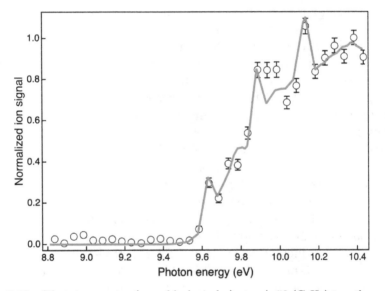

Fig. 8.13. Photoion spectra (open black circles) at m/z 52 (C_4H_4) together with a fit (red line) to the experimental data using the corrected photoionization of vinylacetylene [39].

vinylacetylene over the whole ionization range. The fit function in Fig. 8.13 is corrected by adding 4% of the propene photoion spectrum due to possible charge transfer between propene cations and neutral products [39].

The technique has been employed to confirm the product of the reaction of the C_2H radical with acetylene, ethylene, propene, as well as with butene isomers [38, 39]. For larger unsaturated hydrocarbons such as propene or butene isomers, the reaction with the C_2H radical leads to several product mass channels through H or CH_3 elimination as well as to several isomers within the same mass channel. Branching ratios between the different products may be obtained upon knowledge of the mass discrimination factor and the absolute photoionization cross section. At a given mass channel, the ion signal I is directly proportional to the sum of the ionization cross section of each isomer σ_i weighted by their molar fraction y_i according to [39].

$$I \propto \sum_i y_i \sigma_i \qquad (8.13)$$

The branching ratios between two isomer products are inferred from the ratio of the y_i. Although the knowledge of the absolute ionization cross section is the main source of error, semi-empirical methods [52] have allowed one to provide a good estimate of the branching ratios.

Figure 8.14 displays the m/z 66 ion signal as a function of photon energy for the C_2H reaction with isobutene [38]. The experimental data are fit with photoion spectra of the two C_5H_6 isomers, trans-3-penten-1-yne and 2-methyl-1-buten-3-yne. In this case, it is found that the isobutene reactant adsorbs on the reservoir laser window causing a rapid decrease of the laser fluence and radical density in the supersonic flow. As the energy is scanned linearly as a function of experimental time, the decreasing signal is taken into account by multiplying the photoion spectra of the individual isomers by an exponential function of the energy. The fit to the data reports branching ratios of 0.57 for 2-methyl-1-buten-3-yne and 0.43 for trans-3-penten-1-yne [38]. These ratios are given with an error of ± 0.15 mostly corresponding to the error made on the estimated ionization cross section. The resolution of the ionizing radiation is found insufficient to differentiate between the trans and cis stereoisomers.

Similar fits are performed for the ion signal recorded at m/z 80. In this case, the experimental spectrum can be fit with the photoionization spectrum of only one C_6H_8 isomer, 4-methyl-3-penten-1-yne.

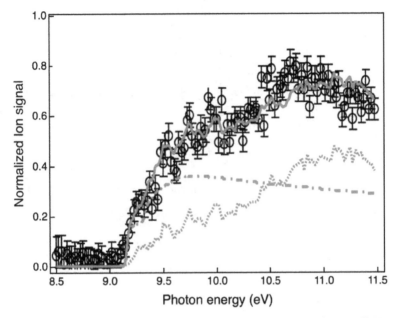

Fig. 8.14. Ion signal at m/z 66 as a function of photon energy for the C_2H reaction with isobutene (open black circles). The experimental data are fit (solid red line) with corrected photoion spectra of the two C_5H_6 isomers, trans-3-penten-1-yne (dashed and dotted red line) and 2-methyl-1-buten-3-yne (dotted red line) [38].

Branching ratios for the reaction products from the C_2H reaction with isobutene are inferred by normalizing the ion signals at each mass channel by the mass-dependent sensitivity of the ion detector and ionization cross sections of the C_5H_6 and C_6H_8 isomers. Scheme 8.1 gives an overview of the branching ratios for each isomer product measured at 79 K.

By performing similar analysis on four butene isomers and compiling results for the reactions with ethylene and propene, a generalized mechanism may be proposed. The reaction proceeds by addition of the C_2H radical onto the carbon–carbon double bond followed by loss of an H atom, or alkyl radical. Table 8.2 summarizes the H loss and alkyl loss branching ratios measured at 79 K. Based on the obtained branching ratios, the loss of a methyl or vinyl group is found to dominate that of the H atom. These reactions are therefore likely

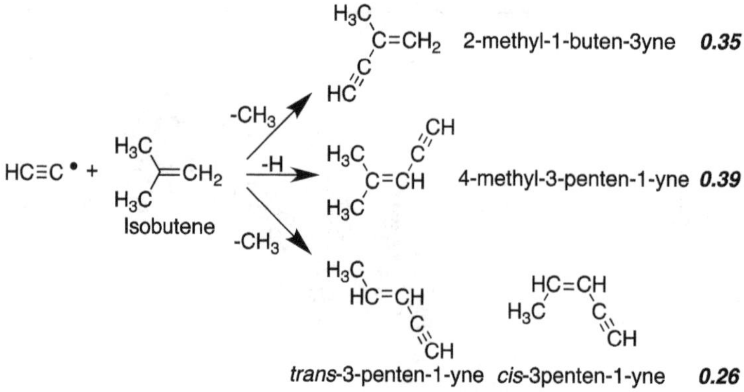

Scheme 8.1.   Products formed in the reaction of C_2H with isobutene with their respective branching ratios.

Table 8.2.   H loss and alkyl loss branching ratios measured at 79 K for the reaction of C_2H with small alkenes [38].

Reactants		Products					
		C_4 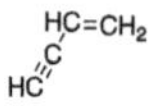	C_5 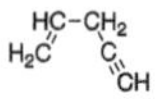	C_5 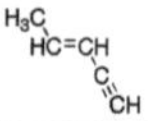	C_5 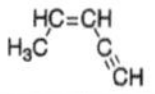	C_5 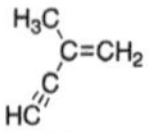	C_6 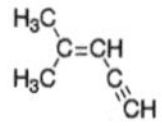
		HC=CH₂ / HC≡C	HC-CH₂ / H₂C-CH	H₃C HC=CH / C≡CH	HC=CH / H₃C C≡CH	H₃C C=CH₂ / HC≡C	H₃C C=CH / H₃C C≡CH
C_2	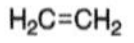 $H_2C{=}CH_2$	1					
C_3	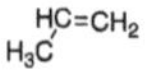 $HC{=}CH_2$ / H_3C	0.85	0.02	0.04		0.09	
C_4	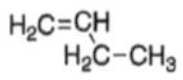 $H_2C{=}CH$ / $H_2C{-}CH_3$	0.65	0.35				
C_4	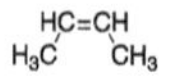 $HC{=}CH$ / H_3C CH_3				1		
C_4	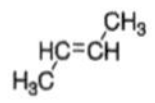 CH_3 / $HC{=}CH$ / H_3C			1			
C_4	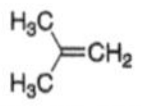 H_3C ${=}CH_2$ / H_3C			0.26		0.35	0.39

to play a significant role in the formation of alkyl-free radicals in low-temperature gas phase environments.

8.3.3   *Tunable photoionization at the SOLEIL synchrotron facility: the CRESUSOL apparatus*

A new CRESU apparatus, CRESUSOL, is under development at the Institute of Physics of Rennes and synchrotron SOLEIL Paris

(DESIRS (a French acronym, Dichroïsme Et Spectroscopie par Inter-action avec le Rayonnement Synchrotron, for Dichroism and inter-action spectroscopy with synchrotron Radiation) beamline) that couples a CRESU flow to a photoelectron–photoion coincidence (PEPICO) spectrometer in order to detect neutral–neutral reaction products including nucleation processes, and derive branching ratios at very low temperatures, down to 20 K [40].

In order to couple a CRESU device to the synchrotron radiation at SOLEIL, pumping capacities have to be reduced (pumping speed of ca. 3,000 m^3/h for molecular nitrogen) in order to fit onto a non-permanent end station on the VUV DESIRS beamline. Such a reduction of pumping speed prevents the generation of continuous supersonic flows at temperatures lower than 40 K. A pulsed uniform supersonic flow generation is needed to reach lower temperatures (<20 K) and will be implemented as a further step of development.

The CRESUSOL apparatus has been designed in two sections (see Fig. 8.15) with an upper chamber where the CRESU supersonic flows are generated and skimmed to reach a lower chamber equipped with a house-made modified Wiley-McLaren time-of-flight mass spectrom-eter [53] collecting ions and electrons particles in coincidence, the electrons serving as the start of the ion time of flight. Such a config-uration provides multiplex detection and 100% transmission for all the ions, only limited by the detection efficiency of the microchannel plates (MCP), about 60%, a clear advantage over quadrupole mass analyzers which transmit only the ions with a given mass-to-charge ratio and need to be scanned to acquire the whole mass spectrum. This is, however, at the expense of a decreased linear dynamic range of the mass analysis. Multiplex acquisition is crucial for branching ratio analysis as the m/z information is recorded as a function of other variables that also need to be scanned (photon energy and reaction time, for instance).

As explained in Section 8.3.1.1, the flange supporting the skim-mer, which is also separating the two chambers, has been designed with streamlined geometry to avoid the formation of detached shock-waves upstream of the skimmer (see Fig. 8.16). In addition, square geometry for the electrodes of the time-of-flight mass spectrometer was chosen to minimize the distance between the skimmer apex and the photoionization region to preserve the kinetic time resolution.

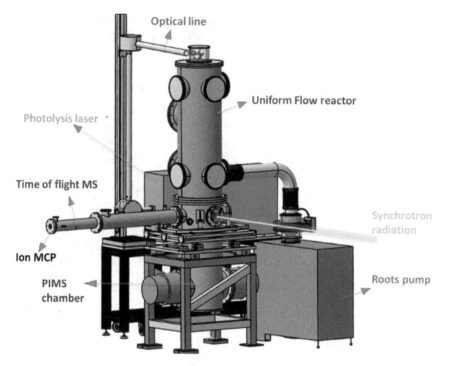

Fig. 8.15.   Outside view of the whole experimental setup [40].

A schematic diagram of the mass spectrometer is shown in
Fig. 8.17. The synchrotron radiation (SR) is centered between the
first electrode on the electron side and the one on the ion side, separated by 4.6 mm, and hits the molecular beam near the upper edge
of the extraction region, in order to minimize the distance from
the skimmer apex (~53 mm). Ions are extracted continuously by
static fields through two acceleration regions and guided toward the
200-cm free flight tube at the end of which they are collected onto
40-mm-diameter microchannel plates. Electrons are also accelerated
continuously toward the opposite side before being collected onto
25-mm-diameter microchannel plates, about ten centimeters from the
ionization zone. The electron signal is used to provide the start of
the time of flight (TOF) of the ions following a multistart/multistop
coincidence scheme [54]. The resolving power obtained experimentally, $m/\Delta m \sim 800$, where Δm is the full width at half maximum

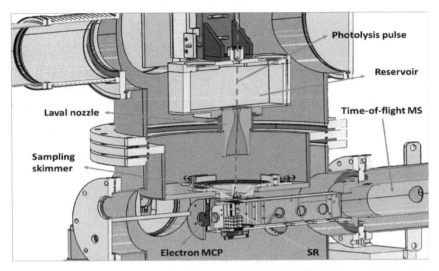

Fig. 8.16. Inside view of the lower part of the reaction chamber and top of the detection chamber equipped with the time-of-flight mass spectrometer (SR is for synchrotron radiation and MCP for microchannel plate). The ion microchannel plate detector, implemented at the end of the drift tube, is out of view [40].

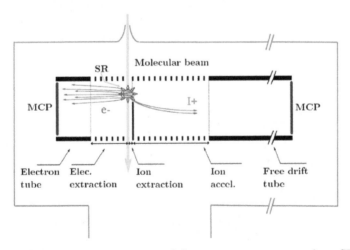

Fig. 8.17. Schematic representation of the mass spectrometer, where SR is for synchrotron radiation, MCP for microchannel plates, and I+ for positive ions (see text for details) [40].

(FWHM) is found to be close to that theoretically expected by the overall resolution of the Wiley-McLaren equation, ~ 900.

In the present configuration, a series of experiments were performed at SOLEIL to test the potential of the apparatus for studying nucleation processes as well as product detection of chemical reaction products. The formation of formic acid dimers was monitored at 52 and 74.6 K, giving access to both the kinetic eVolution of the dimer and the monomer channels [44]. The detection of C_4H_2 from the reaction $C_2H + C_2H_2$ was also performed at low temperatures.

Figure 8.18 displays the mass spectrum recorded at 74.6 K upon pulsed-photolysis dissociation of C_2H_2 at 193 nm with a repetition rate of 50 Hz. The data are integrated from 0 to 600 μs after the laser shot (corresponding to the duration of the cold uniform supersonic flow) and over a photon energy range between 10.0 eV and 11.35 eV, stopping just below the ionization energy of C_2H_2 (11.40 eV), $\Delta E = 0.05$ eV, with an acquisition time of 500 s per energy step. In addition, the false coincidence background has been subtracted. The ion signal at m/z = 50 shows the presence of C_4H_2 formed by the reaction between C_2H and C_2H_2, known to be very fast ($k \geq 10^{-10}$ cm^3 molecule^{-1}s^{-1}) down to 15 K [47].

In the present configuration of the apparatus, a large number of false coincidences significantly reduce the sensitivity of the detection,

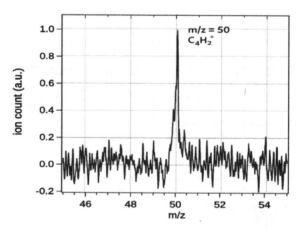

Fig. 8.18. Mass spectrum of C_4H_2 reaction product obtained from photolysis of C_2H_2 ($n_{C2H2} = 1.8 \times 10^{15}$ cm^{-3}) introduced in a 74.6-K uniform supersonic flow of nitrogen (total density $= 6.6 \times 10^{16}$ cm^{-3}).

and improvements need to be undertaken to perform identification of chemical reaction products, especially for the less favorable cases where ionization energy of the reactants is lower than the ionization energy of the expected products. In the mid-term, a position-sensitive detector (PSD) will be adapted on the ion detector side of the time-of-flight mass spectrometer and should increase by orders of magnitude the ion signal contrast by minimizing the contribution of false coincidences and background species present in the chamber. Due to the large velocity of the CRESU flow, the species arrive at a position of the detector that depends on $\sqrt{m/z}$. For a false coincidence to be counted not only does it have to arrive within the same time range as a true coincidence but also at the same position, which dramatically reduces the probability of occurrence. Furthermore, ions originating from the background gas present in the PIMS chamber will hit the center of the detector, in contrast to the ions produced in the CRESU flow that will have arrive at different positions off the center due to their higher velocity. The removal will be facilitated with the consequence being a significant increase of the signal-to-background ratio. The implementation of a temporal ion deflection coupled with a position-sensitive ion detector as proposed by Osborn and co-workers [55] will also be considered, which should allow one to essentially remove the false coincidence background, increasing the dynamic range in the PEPICO TOF mass spectrum by 2 to 3 orders of magnitude.

8.4   Detection of Products Using Chirped-Pulse Fourier Transform MicroWave (CP-FTMW) Spectroscopy

Provided they possess a permanent dipole, reaction products can also be detected using rotational spectroscopy, which has the advantage of being highly specific. For spectroscopic applications, microwave spectroscopy has been used with great success in molecular beams or free jet expansions. Such collision-free environments, however, are not suitable for performing chemical reaction kinetic studies. Furthermore, microwave spectroscopy has long suffered from a lack of sensitivity. The Chirped-Pulse Fourier Transform MicroWave (CP-FTMW) technique, invented by Pate and co-workers at the University of Virginia, has opened new perspectives by improving the rate of data acquisition by several orders of magnitude, as well as by covering a wide range of frequencies that enables simultaneous detection of multiple species [56, 57]. The principle is based on the generation of a waveform which is used to produce a short microwave pulse, typically 1 ms or less, with a frequency sweep. The target-probed molecular species absorbs at all rotational transitions within the frequency range of the chirp and is polarized by the radiation. Free induction decay (FID) of the polarization ensues and the emitted radiation is collected, down-converted, and averaged in the time domain by a high-bandwidth oscilloscope or digitizer card. The collected signal is then Fourier transformed to give the rotational spectrum at MHz resolution.

The CP-FTMW technique was coupled to a CRESU reactor for the first time in 2014 by Suits and co-workers [58] and is referred to as the CPUF (chirped pulse in uniform flow) technique (see Fig. 8.19). The CRESU flow characteristics present the advantage of a very cold and thermalized environment with very fast rotational cooling of the reaction products (for products formed rotationally

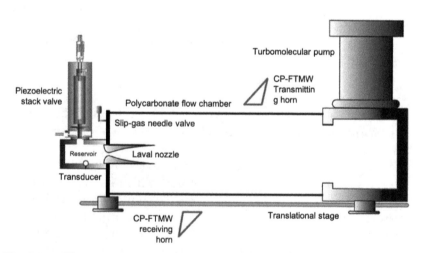

Fig. 8.19. The chirped-pulse, uniform flow (CPUF) instrument. Pulsed uniform supersonic flows are generated through de Laval nozzles using a piezoelectric stack actuator. Microwave antennas on opposite sides of the polycarbonate-made chamber transmit and receive microwave radiation. The figure is not to scale. Adapted from [58].

excited from exothermic reactions) leading to rotational state distributions at very low values of rotational energy. In addition, this cold and multiple-collision environment leads to a large increase in difference in adjacent rotational-level populations. However, FID, is significantly quenched in the relatively high-density environment of a typical CRESU experiment (a few 10^{16} cm^{-3} to 10^{18} cm^{-3}), reducing dramatically the signal-to-noise ratio by shortening the duration of the detectable signal.

The CPUF technique has been applied by Suits and co-workers to study the products formed at low temperature ($\sim$22 K) by the reactions of CN radical with acetylene (C_2H_2) [59] and propyne (C_3H_4) [60]. BrCN laser pulsed photolysis at 193 nm (10 Hz) was used to generate CN radicals in the supersonic flows. For the bimolecular reaction CN + C_2H_2, only one exit channel is expected, forming HCCCN and H atoms. The HCCCN in the v = 0 level near 27.3 and 34.4 GHz was therefore targeted and observed using short chirp bandwidths. Spectra of the J = 4 − 3 pure rotational transition were recorded (see Fig. 8.20) at different time delays, from 10 μs to 110 μs following the laser trigger and the first chirped-pulse excitation.

For the reaction of CN with propyne, the following channels are accessible:

Direct abstraction:

$$CN + CH_3CCH \rightarrow HCN + CH_2CCH \tag{8.14}$$

CN addition/methyl elimination:

$$CN + CH_3CCH \rightarrow CH_3 + HCCCN \tag{8.15}$$

CN addition/H elimination:

$$CN + CH_3CCH \rightarrow H + CH_3CCCN \tag{8.16}$$

$$CN + CH_3CCH \rightarrow H + CH_2CCHCN \tag{8.17}$$

The strongest transitions of the accessible products were targeted in two frequency regions: HCN (J = 1 − 0 transition at 88.631 GHz), HCCCN (J = 9 − 8 transition at 81.881 GHz), CH$_3$CCCN (J_K = 20_0 − 19_0 at 82.627 GHz; J_K = 21_0 − 20_0 at 86.750 GHz), and CH$_2$CCHCN ($J_{Ka,Kc}$ = $16_{0,16}$ − $15_{0,15}$ at 81.674 GHz; $J_{Ka,Kc}$ = $17_{0,17}$ − $16_{0,16}$ at 86.668 GHz). HCN, HCCCN, and CH$_3$CCCN reaction products were clearly detected, while CH$_2$CCHCN was not

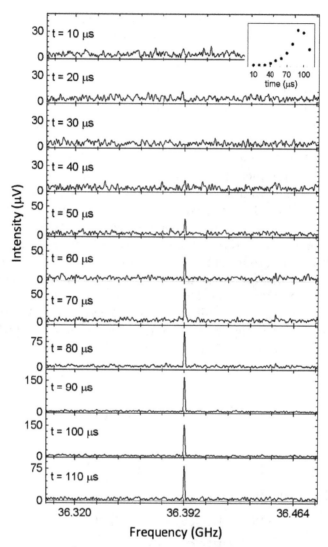

Fig. 8.20. HCCCN ($\tilde{X}^1\Sigma^+$) pure rotational spectra of the $J = 4 - 3$ rotational transition illustrating the time evolution of HCCCN produced by the bimolecular reaction of CN with C_2H_2. Each spectrum is an average of 22,000 acquisitions obtained over the frequency range 35.25–36.45 GHz using a 1-μs duration upchirp. The inset displays the time dependence of HCCCN (see text for details) [59]. Reproduced with permission from Ref. [59], Copyright 2014 AIP Publishing LLC.

observed. The product column densities were calculated from the integrated line intensities and the following branching ratios were determined assuming product thermalization: $(12\pm5)\%$ for HCN, $(66\pm4)\%$ for HCCCN, $(22\pm6)\%$ for CH_3CCCN, and $(0\pm8)\%$ for CH_2C_2HCN. The relatively small branching to the direct abstraction forming HCN was attributed to the low collision energy and the strong electrophilic interaction of CN with the propyne π electrons. These branching ratio values were found to be consistent with statistical calculations based on ab initio results at the CBS-QB3 level of theory [60].

For both sets of experiments, it is worth noting that reaction products began to appear approximately a few tens of μs after the laser trigger, which, according to the authors, is likely owing to rotational cooling of the CN prior to reaction, as it is known to be generated vibrationally cold but rotationally hot from 193-nm photolysis of BrCN [61]. Rotational and vibrational thermalization of the products may also contribute to this delay. A rise of the product signals was also observed at later times. The authors attributed it to reactions occurring in the nozzle throat that may not have reached the 22-K flow temperature and assume this does not impact the measured branching ratios.

Two new CPUF apparatuses are currently being built in Bordeaux and in Rennes with the well-identified challenge to find a way to take advantage of the relatively high pressure of the CRESU flows to ensure thermalization of the reactants and products, with the main drawback to lead to potential low FID signal intensity and duration. This quenching of the FID in the time domain is directly related to pressure broadening in the frequency domain, which contributes to the speed of the decay of the FID, known as the dephasing rate. Sims and co-workers [62] recently explored in detail this effect on the OCS + He system at room temperature. They employed a new CP-FTMW spectrometer built in the millimeter wave region to determine the product channel branching ratios with the perspective of using it in a dedicated CRESU apparatus, named CRESUCHIRP. In this study, the FID decays very rapidly at the highest pressures of the experiment (0.05–0.35 mbar), which reduces the available signal for both time domain fitting and Fourier transformation. In addition, it was noticed that the initial amplitude FID signal was also affected by the pressure. This could be explained by a reduction of the polarization of the polarizing pulse as the pressure rises. These experiments allowed one to highlight the role of the buffer gas used in kinetic experiments in dephasing of the probed transitions, leading to decay rates that correspond to pressure broadening coefficients in the frequency domain. This rate of decay of a transition has to be taken into account when comparing relative intensities, as two transitions could have identical polarizations but differing rates of decay, which would lead to different intensities in the FT spectrum.

8.5 Perspectives

Modeling gaseous environments with large networks of chemical reactions is a major challenge in various scientific fields such as combustion, environmental science, or astrophysics. Models are designed to help us understand how molecules, radicals, and particles are formed and to quantify their abundances. They are based on extensive theoretical and experimental research that provides quantitative information on the rates of the physical and chemical processes involved. Simple gas phase reactions are the building blocks of complex chemical molecules and particles forming in combustion reactors, plasmas, planetary atmospheres, and the interstellar medium. Most of these reactions can form different sets of products, including isomers, that must be unambiguously identified. Knowledge of the fraction of a reaction proceeding to each channel is crucial in modeling the chemistry of these environments. In addition, this information gives valuable insight into the fundamental mechanisms involved in the chemical reactions and especially at low temperature where quantum effects, difficult to access accurately even for molecules of few atoms, prevail.

The quest for channel-specific rate coefficients at very low temperature remains an important challenge that will undoubtedly reveal unexpected quantum effects on reactivity. Great efforts are in progress in various research groups for the development of methods to determine quantitatively the nature of the reaction products through laser or microwave spectroscopy and threshold VUV photoionization coupled to mass spectrometry. Another promising technique that has not been mentioned in this chapter is time-resolved frequency comb spectroscopy that has been developed recently in the mid-infrared range with broadband and high-resolution capabilities [63, 64]. This technique has the advantage of being sensitive enough to measure trace species in a multiplex manner. Indeed, frequency combs are coherent light sources that emit a broad spectrum consisting of discrete and evenly spaced narrow lines. Recently, the cavity-enhanced direct frequency comb spectroscopy (CE-DFCS) technique has been successfully used for kinetics and dynamics studies of OH + CO, OD + CO, and D + CO_2 by the group of Jun Ye at JILA [64, 65]. The coupling of this technique to CRESU flows is under development in the Rennes group and is expected to provide valuable data for the analyses of data from the next generation of infrared observatories like the James Webb Space Telescope (JWST).

State-to-state reactivity is even more challenging than branching ratios determination and appears to be much needed for a better understanding of chemistry in the low-pressure regions prevailing in the interstellar medium, where non-local thermal equilibrium (non-LTE) populations are very common. For instance, little information is known about the influence of reagent rotational excitation on reactivity. One reason for the scarcity of such data is the rapidity at which rotational energy transfer occurs during collisions under laboratory conditions, which hinders the quantitative measurement of the process. More generally, the reactivity of specific states, the assignment of the structure, and internal quantum states of the products of reactive or non-reactive collisions and photodissociation are great experimental challenges for the future.

References

[1] Caravan RL, Khan MAH, Zador J, Sheps L, Antonov IO, Rotavera B, Ramasesha K, Au K, Chen MW, Rosch D, Osborn DL, Fittschen C, Schoemaecker C, Duncianu M, Grira A, Dusanter S, Tomas A, Percival CJ, Shallcross DE, Taatjes CA. The reaction of hydroxyl and methylperoxy radicals is not a major source of atmospheric methanol. Nat Commun. 2018;9(4343):1–9.

[2] Heard DE. Rapid acceleration of hydrogen atom abstraction reactions of OH at very low temperatures through weakly bound complexes and tunneling. Acc Chem Res. 2018;51:2620–2627.

[3] Gómez-Martín JC, Caravan RL, Blitz MA, Heard DE, Plane JMC. Low temperature kinetics of the CH_3OH + OH reaction. J Phys Chem A. 2014;118:2693–2701.

[4] Shannon RJ, Blitz MA, Goddard A, Heard DE. Accelerated chemistry in the reaction between the hydroxyl radical and methanol at interstellar temperatures facilitated by tunnelling. Nat Chem. 2013;5:745–749.

[5] Antiñolo M, Agúndez M, Jiménez E, Ballesteros B, Canosa A, El Dib G, Albaladejo J, Cernicharo J. Reactivity of OH and CH_3OH between 22 and 64 K: modeling the gas phase production of CH_3O in Barnard 1b. Astrophys J. 2016;823:1–8.

[6] Ocaña AJ, Blázquez S, Potapov A, Ballesteros B, Canosa A, Antiñolo M, Vereecken L, Albaladejo J, Jiménez E. Gas-phase reactivity of CH_3OH toward OH at interstellar temperatures (11.7–177.5 K): experimental and theoretical study. Phys Chem Chem Phys. 2019;21:6942–6957.

[7] Lique F, Faure A. Gas-phase chemistry in space. IOP Publishing, Bristol, UK; 2019. https://iopscience.iop.org/book/978-0-7503-1425-1

[8] Seakins PW. Product branching ratios in simple gas phase reactions. Ann Rep Prog Chem Sect C Phys Chem. 2007;103:173–222.

[9] Hickson KM, Loison JC, Wakelam V. Temperature dependent product yields for the spin forbidden singlet channel of the $C(^3P)$ + C_2H_2 reaction. Chem Phys Lett. 2016;659:70–75.

[10] Nuñez-Reyes D, Hickson KM. Rate constants and H-atom product yields for the reactions of $O(^1D)$ atoms with ethane and acetylene from 50 to 296 K. J Phys Chem A. 2018;122:4696–4703.

[11] Nuñez-Reyes D, Hickson KM. Kinetic and product study of the reactions of $C(^1D)$ with CH_4 and C_2H_6 at low temperature. J Phys Chem A. 2017;121:3851–3857.

[12] Meng QY, Hickson KM, Shao KJ, Loison JC, Zhang DH. Theoretical and experimental investigations of rate coefficients of $O(^1D)$ + CH_4 at low temperature. Phys Chem Chem Phys. 2016;18:29286–29292.

[13] Hickson KM, Suleimanov YV. An experimental and theoretical investigation of the $C(^1D)$ + D_2 reaction. Phys Chem Chem Phys. 2017;19:480–486.

[14] Hickson KM, Loison JC, Guo H, Suleimanov YV. Ring-polymer molecular dynamics for the prediction of low-temperature rates: an investigation of the $C(^1D)$ + H_2 reaction. J Phys Chem Lett. 2015;6:4194–4199.

[15] Hickson KM. Low-temperature rate constants and product-branching ratios for the $C(^1D)$ + H_2O reaction. J Phys Chem A. 2019;123:5206–5213.

[16] Hickson KM, Loison JC, Nuñez-Reyes D, Mereau R. Quantum tunneling enhancement of the C + H_2O and C + D_2O reactions at low temperature. J Phys Chem Lett. 2016;7:3641–3646.

[17] Shannon RJ, Cossou C, Loison JC, Caubet P, Balucani N, Seakins PW, Wakelam V, Hickson KM. The fast $C(^3P)$ + CH_3OH reaction as an efficient loss process for gas-phase interstellar methanol. Rsc Adv. 2014;4:26342–26353.

[18] Bourgalais J, Capron M, Kailasanathan RKA, Osborn DL, Hickson KM, Loison JC, Wakelam V, Goulay F, Le Picard SD. The $C(^3P)$ + NH_3 reaction in interstellar chemistry. I. Investigation of the product formation channels. Astrophys J. 2015;812:1–12.

[19] Douglas KM, Blitz MA, Feng WH, Heard DE, Plane JMC, Rashid H, Seakins PW. Low temperature studies of the rate coefficients and branching ratios of reactive loss vs quenching for the reactions of $^1CH_2$ with C_2H_6, C_2H_4, C_2H_2. Icarus. 2019;321:752–766.

[20] Douglas K, Blitz MA, Feng WH, Heard DE, Plane JMC, Slater E, Willacy K, Seakins PW. Low temperature studies of the removal reactions of $^1CH_2$ with particular relevance to the atmosphere of Titan. Icarus. 2018;303:10–21.

[21] Bourgalais J, Spencer M, Osborn DL, Goulay F, Le Picard SD. Reactions of atomic carbon with butene isomers: implications for molecular growth in carbon-rich environments. J Phys Chem A. 2016;120:9138–9150.

[22] Hickson KM, Loison JC, Bourgalais J, Capron M, Le Picard SD, Goulay F, Wakelam V. The $C(^3P)$ + NH_3 reaction in interstellar chemistry. II. Low temperature rate constants and modeling of NH, NH_2, and NH_3 abundances in dense interstellar clouds. Astrophys J. 2015;812:1–8.

[23] Nuñez-Reyes D, Hickson KM, Larregaray P, Bonnet L, Gonzalez-Lezana T, Bhowmick S, Suleimanov YV. Experimental and theoretical study of the $O(^1D)$ + HD reaction. J Phys Chem A. 2019;123:8089–8098.

[24] Leonori F, Petrucci R, Segoloni E, Bergeat A, Hickson KM, Balucani N, Casavecchia P. Unraveling the dynamics of the $C(^3P, {}^1D)+C_2H_2$ reactions by the crossed molecular beam scattering technique. J Phys Chem A. 2008;112:1363–1379.

[25] Costes M, Halvick P, Hickson KM, Daugey N, Naulin C. Non-threshold, threshold, and nonadiabatic behavior of the key interstellar C + C_2H_2 reaction. Astrophys J. 2009;703:1179–1187.

[26] Costes M, Daugey N, Naulin C, Bergeat A, Leonori F, Segoloni E, Petrucci R, Balucani N, Casavecchia P. Crossed-beam studies on the dynamics of the $C+C_2H_2$ interstellar reaction leading to linear and cyclic C_3H+H and C_3+H_2. Faraday Discuss. 2006;133:157–176.

[27] Gu XB, Guo Y, Zhang FT, Kaiser RI. Investigating the chemical dynamics of the reaction of ground-state carbon atoms with acetylene and its isotopomers. J Phys Chem A. 2007;111:2980–2992.

[28] Bergeat A, Loison JC. Reaction of carbon atoms, C ($2p^2$, ^3P) with C_2H_2, C_2H_4 and C_6H_6: overall rate constant and relative atomic hydrogen production. Phys Chem Chem Phys. 2001;3:2038–2042.

[29] John JE, Keith TG. Gas dynamics. Upper Saddle River, NJ: Pearson; 2006.

[30] Taatjes CA. How does the molecular velocity distribution affect kinetics measurements by time-resolved mass spectrometry? Int J Chem Kinet. 2007;39:565–570.

[31] Soorkia S, Liu CL, Savee JD, Ferrell SJ, Leone SR, Wilson KR. Airfoil sampling of a pulsed Laval beam with tunable vacuum ultraviolet synchrotron ionization quadrupole mass spectrometry: application to low-temperature kinetics and product detection. Rev Sci Instrum. 2011;82:124102.

[32] Lee S, Hoobler RJ, Leone SR. A pulsed Laval nozzle apparatus with laser ionization mass spectroscopy for direct measurements of rate coefficients at low temperatures with condensable gases. Rev Sci Instrum. 2000;71:1816–1823.

[33] Lee S, Samuels DA, Hoobler RJ, Leone SR. Direct measurements of rate coefficients for the reaction of ethynyl radical (C_2H) with C_2H_2 at 90 and 120 K using a pulsed Laval nozzle apparatus. J Geophys Res. 2000;105:15085–15090.

[34] Sabbah H, Biennier L, Klippenstein SJ, Sims IR, Rowe BR. Exploring the role of PAHs in the formation of soot: pyrene dimerization. J Phys Chem Lett. 2010;1:2962–2967.

[35] Speck T, Mostefaoui TI, Travers D, Rowe BR. Pulsed injection of ions into the CRESU experiment. Int J Mass Spectrom Ion Proc. 2001;208:73–80.

[36] Maccoll JW. The conical shock wave formed by a cone moving at a high speed. Proc Royal Soc London A. 1937;159:0459–0472.

[37] Anderson JD. Modern compressible flow with historical perspective. McGraw-Hill: New York; 2004.

[38] Bouwman J, Fournier M, Sims IR, Leone SR, Wilson KR. Reaction rate and isomer-specific product branching ratios of C_2H + C_4H_8: 1-Butene, cis-2-Butene, trans-2-Butene, and isobutene at 79 K. J Phys Chem A. 2013;117:5093–5105.

[39] Bouwman J, Goulay F, Leone SR, Wilson KR. Bimolecular rate constant and product branching ratio measurements for the reaction of C_2H with ethene and propene at 79 K. J Phys Chem A. 2012;116:3907–3917.

[40] Durif O, Capron M, Messinger JP, Benidar A, Bourgalais J, Canosa A, Courbe J, Garcia GA, Gil JF, Okumura M, Rutkowski L, Sims IR, Thievin J, Le Picard SD. A new instrument for kinetics and branching ratio studies of gas phase collisional processes at very low temperatures. Rev Sci Instrum. 2021;92(014102):1–10.

[41] Osborn DL, Zou P, Johnsen H, Hayden CC, Taatjes CA, Knyazev VD, North SW, Peterka DS, Ahmed M, Leone SR. The multiplexed chemical kinetic photoionization mass spectrometer: a new approach to isomer-resolved chemical kinetics. Rev Sci Instrum. 2008;79(104103):1–10.

[42] Taatjes CA, Hansen N, Osborn DL, Kohse-Hoeinghaus K, Cool TA, Westmoreland PR. "Imaging" combustion chemistry via multiplexed synchrotron-photoionization mass spectrometry. Phys Chem Chem Phys. 2008;10:20–34.

[43] Trevitt AJ, Goulay F. Insights into gas-phase reaction mechanisms of small carbon radicals using isomer-resolved product detection. Phys Chem Chem Phys. 2016;18:5867–5882.

[44] Durif O, Benidar A, Biennier L, Canosa A, Rutkowski L, Sims IR, Nahon L, Garcia GA, Le Picard SD. In preparation.

[45] Lee S, Leone SR. Rate coefficients for the reaction of C_2H with O_2 at 90 K and 120 K using a pulsed Laval nozzle apparatus. Chem Phys Lett. 2000;329:443–449.

[46] Bowman MC, Burke AD, Turney JM, Schaefer HF. Mechanisms of the ethynyl radical reaction with molecular oxygen. J Phys Chem A. 2018;122:9498–9511.

[47] Chastaing D, James PL, Sims IR, Smith IWM. Neutral-neutral reactions at the temperatures sf interstellar clouds–rate coefficients for reactions of C_2H radicals with O_2, C_2H_2, C_2H_4 and C_3H_6 down to 15 K. Faraday Discuss. 1998;109:165–181.

[48] Opansky BJ, Leone SR. Low-temperature rate coefficients of C_2H with CH_4 and CD_4 from 154 to 359 K. J Phys Chem. 1996;100:4888–4892.

[49] Thiesemann H, Taatjes CA. Temperature dependence of the reaction $C_2H(C_2D)$ + O_2 between 295 and 700 K. Chem Phys Lett. 1997;270:580–586.

[50] Vakhtin AB, Heard DE, Smith IWM, Leone SR. Kinetics of C_2H radical reactions with ethene, propene and 1-butene measured in a pulsed Laval nozzle apparatus at T = 103 and 296 K. Chem Phys Lett. 2001;348:21–26.

[51] Nizamov B, Leone SR. Kinetics of C_2H reactions with hydrocarbons and nitriles in the 104–296 K temperature range. J Phys Chem A. 2004;108:1746–1752.

[52] Bobeldijk M, van der Zande WJ, Kistemaker PG. Simple models for the calculation of photoionization and electron impact ionization cross section of polyatomic molecules. Chem Phys. 1994;179:125–30.

[53] Wiley WC, McLaren IH. Time-of-flight mass spectrometry with improved resolution. Rev Sci Instrum. 1955;26:1150–57.

[54] Bodi A, Sztaray B, Baer T, Johnson M, Gerber T. Data acquisition schemes for continuous two-particle time-of-flight coincidence experiments. Rev Sci Instrum. 2007;78(084102):1–7.

[55] Osborn DL, Hayden CC, Hemberger P, Bodi A, Voronova K, Sztaray B. Breaking through the false coincidence barrier in electron-ion coincidence experiments. J Chem Phys. 2016;145(164202):1–8.

[56] Brown GG, Dian BC, Douglass KO, Geyer SM, Shipman ST, Pate BH. A broadband Fourier transform microwave spectrometer based on chirped pulse excitation. Rev Sci Instrum. 2008;79(053103):1–13.

[57] Park GB, Field RW. Perspective: the first ten years of broadband chirped pulse Fourier transform microwave spectroscopy. J Chem Phys. 2016;144(200901):1–10.

[58] Oldham JM, Abeysekera C, Joalland B, Zack LN, Prozument K, Sims IR, Park GB, Field RW, Suits AG. A chirped-pulse Fourier-transform microwave/pulsed uniform flow spectrometer. I. The low-temperature flow system. J Chem Phys. 2014;141(154202):1–7.

[59] Abeysekera C, Zack LN, Park GB, Joalland B, Oldham JM, Prozument K, Ariyasingha NM, Sims IR, Field RW, Suits AG. A chirped-pulse Fourier-transform microwave/pulsed uniform flow spectrometer. II. Performance and applications for reaction dynamics. J Chem Phys. 2014;141(214203):1–9.

[60] Abeysekera C, Joalland B, Ariyasingha N, Zack LN, Sims IR, Field RW, Suits AG. Product branching in the low temperature reaction of CN with propyne by chirped-pulse microwave spectroscopy in a uniform supersonic flow. J Phys Chem Lett. 2015;6:1599–604.

[61] Khachatrian A, Dagdigian PJ, Bennett DIG, Lique F, Klos J, Alexander MH. Experimental and theoretical study of rotationally inelastic collisions of $CN(A^2\Pi)$ with N_2. J Phys Chem A. 2009;113:3922–31.

[62] Hayes BM, Guillaume T, Hearne TS, Cooke I, Gupta D, Khedaoui OA, Le Picard SD, Sims IR. Design and performance of an E-band chirped pulse spectrometer for kinetics applications: OCS — He pressure broadening. J Quant Spectrosc Radiat Transfer. 2020;250:107001.

[63] Bui TQ, Bjork BJ, Changala PB, Nguyen TL, Stanton JF, Okumura M, Ye J. Direct measurements of DOCO isomers in the kinetics of OD. Science Adv. 2018;4:1–8.

[64] Bjork BJ, Bui TQ, Heckl OH, Changala PB, Spaun B, Heu P, Follman D, Deutsch C, Cole GD, Aspelmeyer M, Okumura M, Ye J. Direct frequency comb measurement of OD plus CO $\rightarrow$ DOCO kinetics. Science. 2016;354:444-8.

[65] Bui TQ, Bjork BJ, Changala PB, Heckl OH, Spaun B, Ye J. OD plus CO $\rightarrow$ D + CO_2 branching kinetics probed with time-resolved frequency comb spectroscopy. Chem Phys Lett. 2017;683:91-5.

Chapter 9

Infrared Absorption Spectroscopy in Laval Nozzle Supersonic Flows

Robert Georges[*,‡], Eszter Dudás[*,§], Nicolas Suas-David[†,¶]
and Lucile Rutkowski[*,‖]

*CNRS, IPR(Institut de Physique de Rennes)-UMR 6251,
Université de Rennes, F-35000 Rennes, France
†Leiden Observatory, Leiden Universiteit, Oort — Niels Bohrweg 2,
2333 Leiden, Netherlands
‡robert.georges@univ-rennes1.fr
§eszter.dudas@univ-rennes1.fr
¶suas@strw.leidenuniv.nl
‖lucile.rutkowski.univ-rennes1.fr

Abstract

The extraordinary cooling caused by the supersonic expansion of a polyatomic gas on its infrared footprint has been extensively employed for applications ranging from nucleation studies to post-shock spectroscopy. In this chapter, we recall the benefits of decreasing the molecular sample temperature, involving the spectral simplification and the subsequent magnification of the absorption transitions starting from the lowest rotational energy levels, the drastic line width narrowing, the supersaturation, the formation, and stabilization of weakly bounded molecular complexes. These effects are illustrated by infrared spectra of, e.g. water, carbon dioxide, methane, and more complex molecules such as trans-butadiene or naphthalene. The scope of this chapter includes different types of flows obtained using the Laval nozzle: perfectly expanded or slightly overexpanded flows produced by contoured Laval nozzles, and underexpanded flows produced by simpler conical divergent nozzles.

The various infrared spectroscopic techniques coupled with Laval nozzle flows are also reviewed. Conventional techniques like tunable diode laser absorption spectroscopy (TDLAS) and Fourier transform infrared spectroscopy (FTIR) have been intensively applied to the study of homogeneous nucleation. Other recent and more sensitive techniques such as cavity ring-down spectroscopy (CRDS), cavity enhanced absorption spectroscopy (CEAS), and optical frequency comb spectroscopy (OFCS) are now used for the measurement of reaction kinetics in the infrared, or for the non-ETL spectroscopy of molecules such as methane, which is of great astrophysical interest.

Keywords: Low-temperature spectroscopy; Molecular aggregates; Non-equilibrium spectroscopy; Hypersonic CRDS; FTIR; TDLAS.

9.1   Introduction

Laval (or *de Laval*) nozzles have long been used to produce high-speed collimated flows of several kilometers per second, propelling, for instance, civilian and military planes at supersonic or hypersonic speeds. In this case, the combustion of a propellant in a combustion chamber leads to high enthalpy gases which after expansion exit the nozzle at hypersonic velocities. Hypersonic flows, usually characterized by a Mach number >5, have also had a huge impact on the physics of, e.g. planetary space probes or vehicles during their reentry phase into terrestrial or planetary atmospheres. The atmospheric entry causes a sudden deceleration of the gas flow through the shock wave that develops in front of the probes, and a rise in the temperature of the gas layer that flows around the flying object can reach several thousand degrees. This high temperature is the cause of many complex physical phenomena, the so-called real gas effects that modify the aerodynamic properties of the gas layer, namely, dissociation, ionization, and vibrational and electronic excitation of molecules. These molecular processes favor a strong decoupling of the internal degrees of freedom of molecules and the emergence of non-equilibrium chemistry.

To reproduce hypersonic flight conditions, hypersonic and supersonic wind tunnels and chambers have been designed and built. However, and as discussed in details in Chapter 1, a major difficulty had arisen: when starting from a reservoir at room temperature, the resulting hypersonic flows at the nozzle exit are extremely cold, with a static temperature dropping far below atmospheric temperatures. Therefore, in this case, the resulting flow around a body is not really representative of hypersonic flight. The real gas (i.e. high temperature) effects are not present at all. To overcome this problem, it is necessary to heat up the gas in the reservoir at thousands of degrees, a very tricky technical problem. "Cold" hypersonic facilities have been built to study high Mach number flights from a fundamental point of view. On the contrary, they can be used as a precious tool in chemical physics and spectroscopy as shown throughout this book.

Different types of hypersonic facilities have emerged to simulate high Mach number flight regimes and to study the associated complex processes in the laboratory [1]. A large number of spectroscopic techniques have been implemented at the same time to probe and characterize the flows produced by Laval nozzles in hypersonic aerodynamic tunnels; see, for instance, the excellent summary written by Danehy *et al.* [2] Spectroscopy provides a non-invasive diagnostic which can reach a high sensitivity and is therefore well suited to probe the low density of rarefied flows. In addition, spectroscopy can be implemented in a time-resolved fashion, allowing one to follow the evolution of rapid processes. Among the techniques implemented in high-velocity flows, direct absorption spectroscopy, wavelength or frequency modulation spectroscopy, laser-induced fluorescence, coherent anti-Stokes Raman scattering, and emission spectroscopy are widely used to retrieve the gas temperature (translational, rotational, vibrational), pressure, speed, and the concentration of the different probed species, but also enthalpy and mass flux.

At the same time as high-temperature wind tunnels devoted to high-speed gas dynamics, Laval nozzle expansions have become one of the most efficient techniques to cool molecular samples, relying on the tremendous lowering of the temperature which results from the conversion of the enthalpy of the gas into kinetic energy. This approach is used in physical chemistry to depopulate the energy levels, which simplifies the molecular spectral signatures.

Spectroscopy at cold temperatures is an ideal probe of the energy levels of polyatomic molecules in order to study their internal dynamics, including the couplings that may exist between their degrees of freedom. It also provides a way to study the formation process of molecular complexes and aggregates, i.e. the first stages of condensation of a gas, which occur though weak intermolecular interactions of the van der Waals type. Laser-induced fluorescence was the first spectroscopic technique ever employed to probe the cold gas of a supersonic expansion [3, 4], and it was followed later by all kind of spectroscopic techniques, including infrared absorption spectroscopy, which is more specifically discussed in this chapter. Historically, it is the technique known as "free jets" (i.e. extremely underexpanded sonic jets) that was used first — and still is widely used — for its simplicity. It is obtained by freely expanding a gas in an expansion chamber through a sharp-edged orifice, without requiring the use of a convergent–divergent profiled nozzle. Because their streamlines strongly diverge, free jets allow one to achieve extremely low temperatures, as low as a few kelvins. The density collapse of a free jet is advantageous to "freeze" the condensation of the sample gas; therefore, free jets are ideal for the study of the properties of dimers (i.e. two atoms/molecules complexes). An outstanding synthesis of spectroscopy of polyatomic molecules complexes can be found in the review from Herman *et al.* [5] Contrarily to free jet flows, the use of the convergent–divergent profile specific to Laval nozzles cancels out, or at least strongly reduces, the divergence of the flow streamlines, which in turns allow one to tune the density of the flow and modulate the collisional processes. Combined with spectroscopy, Laval nozzles are excellent tools for studying the homogeneous nucleation of a gas and for monitoring the decoupling of the internal degrees of freedom of molecules (non-LTE[a] spectroscopy).

This chapter focuses on the combined use of infrared spectroscopy and Laval nozzles to probe the fundamental properties of molecules, nucleation processes, and gas phase kinetics. In the following, the name Laval nozzle is used in a broad understanding applied to any nozzle equipped with a diverging part, whether conical or contoured. Section 9.2 recalls the advantages linked to very low temperatures for spectroscopy. Section 9.3 compares extremely underexpanded flows (so-called free jet type) to flows produced by Laval nozzles. Section 9.4 presents a synthesis of the different infrared spectroscopic techniques coupled with flows produced by Laval nozzles and their applications to nucleation and kinetic experiments, among others. Section 9.5 presents an apparatus, named SMAUG, aimed at

---

[a]LTE: local thermal equilibrium.

performing non-LTE spectroscopy in a hypersonic flow by cavity ring-down spectroscopy. Finally, Section 9.6 presents the technique of post-shock spectroscopy producing spectra of small organic molecules at high temperature by probing the detached shock wave formed by the impact of a hypersonic Laval nozzle flow on an obstacle.

9.2 Infrared Spectroscopy at Low Temperatures

Cold environments offer a number of advantages for spectroscopy, which will be described in this section. The principles of absorption spectroscopy are quickly covered in Section 9.2.1, while the following sections focus on the different spectroscopic advantages of lowering the temperature of a gas, namely, the simplification of the rotation-vibration spectra (Section 9.2.2), the sensitivity enhancement (Section 9.2.3), the supersaturation effect (Section 9.2.4), the stabilization of van der Waals complexes (Section 9.2.5), and the line narrowing (Section 9.2.6).

9.2.1 *Infrared absorption spectroscopy*

Infrared absorption spectroscopy targets rovibrational transitions of the molecular structures, providing an unambiguous absorption fingerprint for each species. Generally, measuring the infrared spectrum of a molecular sample can have two types of objectives. When the sample composition is known, it provides valuable information about the energy levels of the molecular system. On the contrary, when the transitions are well documented, it is then possible to achieve quantitative spectroscopy. This allows obtaining the absolute concentrations of the sample compounds and the simultaneous retrieval of the sample thermodynamic conditions.

Infrared absorption spectroscopy generally relies on the measurement of an intensity spectrum I fundamentally ruled by the Beer-Lambert law:

$$I(\tilde{\nu}) = I_0(\tilde{\nu})e^{-\alpha(\tilde{\nu}) \cdot L} \tag{9.1}$$

where I_0 is the power spectral density of the light source, $\tilde{\nu}$ is the optical wavenumber, α is the absorption coefficient of the gaseous sample, and L is the interaction length between the species and the optical beam. Most approaches require measuring the two spectra $I(\tilde{\nu})$ and $I_0(\tilde{\nu})$ in order to retrieve the absorbance spectrum $\alpha(\tilde{\nu}) \cdot L$ of the measured molecular sample. If the absorption lines are well resolved and if the light source spectrum is sufficiently smooth and stable, then I_0 can be extrapolated from the baseline of $I(\tilde{\nu})$. When the spectrum is measured using cavity-enhanced techniques, then Eq. (9.1) has to be modified to consider the cavity response. In the particular case of cavity ring-down spectroscopy (CRDS), the cavity decay time which is measured is directly proportional to the absorption coefficient α; see Section 9.4 for more details about the CRDS technique.

The absorption coefficient is related to the absorption cross section σ, usually expressed in cm^2 molecule^{-1}, through the relationship $\alpha(\tilde{\nu}) = \sigma(\tilde{\nu})n$, where n is the molecular concentration. The infrared absorption spectrum of a gas is composed of many individual molecular rotation-vibration lines k, so that

$$\alpha(\tilde{\nu}) = \sum_k \alpha_k(\tilde{\nu}, T) = \sum_k \sigma_k(\tilde{\nu}, T)n \qquad (9.2)$$

where the absorption cross section $\sigma_k(\tilde{\nu}, T)$ of the molecular line k is related to its integrated absorption cross section $\bar{\sigma}_k(T)$ under the normalized line profile $g(\tilde{\nu} - \tilde{\nu}_k)$ centered on the wavenumber $\tilde{\nu}_k$:

$$\sigma_k(\tilde{\nu}, T) = \bar{\sigma}_k(T)g(\tilde{\nu} - \tilde{\nu}_k) \qquad (9.3)$$

The collisional broadening of the lines is negligible due to the very low static pressure (typically a few Pa) characterizing most supersonic jet expansions, so that only the thermal broadening by the Doppler effect needs to be considered. The thermal motion of the molecules leads to the Gaussian line profile, g_D,

$$g_D(\tilde{\nu} - \tilde{\nu}_k) = \sqrt{\frac{ln2}{\pi}} \frac{1}{\gamma_D} \exp\left[-\ln 2 \left(\frac{\tilde{\nu} - \tilde{\nu}_k}{\gamma_D}\right)^2\right] \qquad (9.4)$$

where γ_D is the half-width at half-maximum (HWHM) of the molecular line. When performing spectroscopy at room temperature of heavy molecules, the frequency separating adjacent transitions can reach values similar to the thermal width of the lines, leading to an unresolved absorption spectrum difficult to assign.

9.2.2   *Spectral simplification*

The depopulation of high energy levels induced by the drop of temperature involves an extreme simplification of the rotation-vibration spectrum of molecules having many degrees of internal freedom. This technique is nowadays associated with high-resolution spectroscopy to probe molecules as heavy as polycyclic aromatic hydrocarbon molecules [6], such as pyrene [7] or naphthalene [8], to suppress the contribution of transitions of high rotational quantum numbers, which prevent precise determination of the origin of the bands and blur the identification of the different hot bands[b] of the spectrum. The spectral simplification induced by the low temperature reveals the position of the Q branches of the different hot bands, as can be seen in Fig. 9.1 for the case of the naphthalene molecule. Lowering the temperature drastically modifies the intensities of the rovibrational lines of an infrared spectrum, as can be seen on the *trans*-butadiene spectrum in Fig. 9.2 [9].

9.2.3   *Absorption cross section vs temperature*

The integrated absorption cross section $\bar{\sigma}_k$ over wavenumber, expressed in cm molecule^{-1}, is highly sensitive to temperature. It is obtained from the absorption coefficient integrated over the profile of the molecular line k:

$$\bar{\sigma}_k(T) = \frac{1}{n} \int_0^\infty \alpha_k(\tilde{\nu}, T) d\tilde{\nu} \tag{9.5}$$

The integrated absorption cross section $\bar{\sigma}_k$ can be calculated at any temperature from the absorption cross section $\bar{\sigma}_k(T_0)$ listed at the reference temperature $T_0 = 296\,\mathrm{K}$ in spectroscopic molecular databases (e.g. HITRAN, Ref. [11]), using the following relation:

$$\bar{\sigma}_k(T) = \bar{\sigma}_k(T_0)\frac{Q(T_0)}{Q(T)}\exp\left(\frac{E''}{k_B T_0} - \frac{E''}{k_B T}\right)\left[\frac{1 - \exp\left(-\frac{\tilde{\nu}_k}{k_B T}\right)}{1 - \exp\left(-\frac{\tilde{\nu}_k}{k_B T_0}\right)}\right] \tag{9.6}$$

---

[b]A hot band is a vibrational band starting from an excited vibrational level whereas a cold band starts from the vibrational ground state. The number of hot bands observed on a spectrum increases with the population of excited vibrational states, therefore with the temperature.

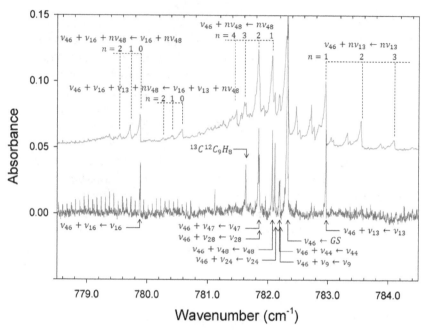

Fig. 9.1. High-resolution $(0.005\,\mathrm{cm}^{-1})$ jet spectrum of naphthalene (lower spectrum, rotational temperature 25 K) compared to a room-temperature spectrum (upper spectrum [10]) recorded at the same resolution. Note that the absorbance of the room-temperature spectrum was divided by 10 and shifted upward to facilitate comparison. The cold data were obtained using a free jet with the far-infrared beamline AILES at the SOLEIL synchrotron [8]. An identification of the main hot bands is provided in the figure. The series of resolved rovibrational lines observed in the low wavenumber side belongs to the P-branch of the cold band $\nu_{46} \leftarrow \mathrm{GS}$.

where Q is the total partition function (increasing function with T) and E'' is the energy of the lower level of the transition corresponding to the line.

Following Eq. (9.6), the lowering of the temperature of a gas induces an increase of the infrared absorption cross section of the low-J levels, where J is the total angular momentum of the molecule. Figure 9.3 illustrates this temperature effect on selected transitions, $R(0)$, $R(1)$, $R(10)$, $R(20)$, $R(30)$, of the $3 \leftarrow 0$ cold band of carbon monoxide centered on $6354\,\mathrm{cm}^{-1}$. The intensity of the low-J absorption lines is magnified at low temperatures, leading to a better signal-to-noise ratio of the spectrum. In the case of carbon monoxide, lowering the gas temperature from 300 to 10 K multiplies the

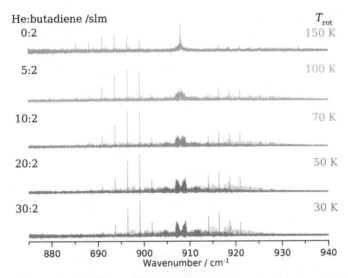

Fig. 9.2.   Jet spectrum of *trans*-butadiene (10-atom molecule) recorded with the Jet-AILES setup at the SOLEIL synchrotron [9]. The temperature indicated in each panel was retrieved from the relative intensity of the rovibrational lines.

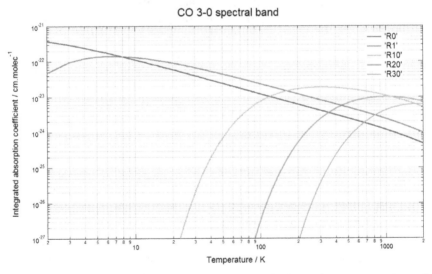

Fig. 9.3.   Temperature evolution of the integrated absorption cross sections of five rovibrational lines of CO.

integrated absorption cross section of low-J transitions by a factor of ~10, while the high-J transitions drop by several orders of magnitude. Note that the magnitude of the intensity discrepancy between the low- and high-J transitions is linked to the evolution of the partition function with temperature, therefore it is going to increase with the number of internal degrees of freedom of the considered molecule.

9.2.4   *Supersaturation*

One of the most interesting advantages of the free jet for spectroscopy is undoubtedly linked to its extremely fast cooling rate, of the order of -10^8 K/s for a pinhole expansion. This ultrafast expansion brings the gas into a highly supersaturated state, far from the thermodynamic equilibrium. The sample remains gaseous although well below its sublimation point; that is, its partial pressure remains thousands of times higher than the saturated vapor pressure at the temperature reached. This state of high supersaturation is eminently interesting for absorption spectroscopy techniques since it makes it possible to probe densities which can never be reached at equilibrium in a cryogenic cooling cell at the same temperature. Note that the collisional cooling cells are based on the same principle of sympathetic cooling, where a polyatomic gas is cooled by colliding with a cold buffer gas, generally helium [12] (see Chapter 1). The reader is referred to Chapter 6 for a detailed description of the nucleation processes taking place in supersonic expansions.

9.2.5   *Stabilization of van der Waals complexes*

The formation of small, weakly bonded molecular complexes takes place under the combined action of the very low temperature and three-body collisions. The third body is necessary to stabilize the complex before it spontaneously dissociates by removing the binding energy. The formation of complexes involving one (homo-complexes) or several types of molecules (hetero-complexes) depends on the temperature of the jet and the number of collisions occurring between the molecules. In the case of a free jet expansion, the stabilization of the formed complex starts close to the nozzle exit, where the number of collisions is significantly high (low Knudsen number).

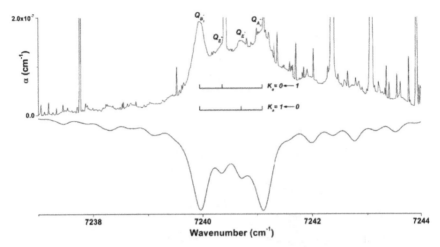

Fig. 9.4. Cluster spectroscopy obtained in a free jet: cavity ring-down spectrum of the water dimer formed at $-10\,\mathrm{K}$ in a slit-jet expansion of argon. The narrower spectral features correspond to water monomer transitions. Reproduced with permission from Ref. [13], Copyright 2015 ACS Publications.

Once the complex is formed, the number of collisions collapses quickly (high Knudsen number), blocking a further growth of the complexes. Experimental conditions can be tuned to limit the growing process to complexes as small as dimers, for example, the dimer of water (see Fig. 9.4) [13]. One of the main difficulties is to generate a sufficiently high number of dimers for their detection while limiting their growth. The high sensitivity specific to spectroscopic techniques based on high finesse optical cavities is a real advantage because it allows limiting the sample pressure which triggers the formation of large clusters. Conversely, large aggregates can be formed provided that the stagnation pressure is significantly increased, from hundreds of mbar up to several bars, then increasing the number of collisions during the first phase of expansion (see Fig. 9.5).

9.2.6 *Effect of the thermal motion on Doppler line broadening*

The absorption line narrowing resulting from the reduced thermal agitation (i.e. translational temperature) of very cold gases is another

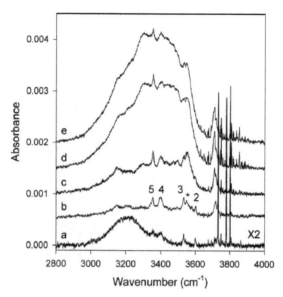

Fig. 9.5.   FTIR spectrum of water aggregates formed in a slit-jet expansion of argon. Broader absorption features correspond to large aggregates. Numbered absorption peaks are attributed to moderate size clusters, from dimer to pentamer, while narrower spectral features at $-3800\,\mathrm{cm}^{-1}$ correspond to monomer water lines. Reproduced with permission from Ref. [14], Copyright 2009 AIP Publishing.

advantage of supersonic cooling (see Fig. 9.6). The reduction of line broadening associated with the Doppler effect allows one to perform high-resolution spectroscopy, allowing one to resolve the complex spectrum of, e.g. heavy molecules. The Doppler line broadening, defined in Eq. (9.4), provides a way to estimate the translational temperature of the gas flow in the expansion via the following relation:

$$\gamma_D(HWHM) = 3.581 \times 10^{-7} \tilde{\nu}_k \sqrt{\frac{T_T}{\mathcal{M}}} \qquad (9.7)$$

where T_T is the translational temperature; $\tilde{\nu}_k$ is the wavenumber of the transition expressed in cm^{-1}; and $\mathcal{M}$ is the molar mass of the gas in grams. This approach is a convenient way to track the presence of condensation of the sample or, in the case of free jet expansions, to assess the cooling efficiency linked to the type of carrier gas used (see Fig. 9.6). However, it assumes that other potential sources of

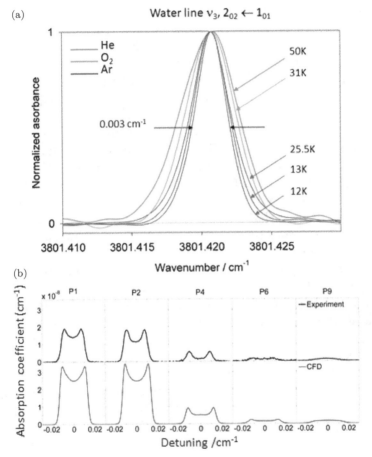

Fig. 9.6. (a) Water absorption line recorded by FT spectroscopy in a slit-jet expansion of argon, helium, or dioxygen admixed with various amounts of water vapor. The temperature values are retrieved from the absorption line widths. (b) CO absorption line profiles recorded probing by CRDS as an axisymmetric free jet perpendicular to its axis [17]. Reproduced with permission from Ref. [17], Copyright 2016 Elsevier B.V.

line broadening remain negligible. In particular, it does not apply in the presence of important line broadening linked to the internal dynamics of molecules, such as vibrational predissociation characterizing weakly bound complexes such as the water dimer [13] or internal vibrational redistribution of energy, which in case of large molecules typically broadens the absorption lines to a point where the rotational structure of the band is no longer observable. Collisional broadening can be excluded as the pressure drops rapidly downstream of the nozzle down to less than 1 Pa, typically.

It is of course advisable to adapt the instrumental function to the width of the observed transitions. With a spectral finesse of about 10^{-4} cm^{-1}, distributed feedback laser diodes are well adapted to the low Doppler widths associated with very low temperatures [15]. High-resolution Fourier transform spectroscopy is also well suited provided that the instrumental resolution and the apodization function [16]. are set properly to not mask the Doppler broadening. Moreover, to measure the translational temperature of the isentropic core, it is preferable to use mainly the absorption lines with low J values, which remain less affected than the lines with high J value by the absorption due to the boundary layers and the background static gas filling the vacuum chamber.

Figure 9.6(a) depicts the water absorption line recorded by Fourier transform spectroscopy in a slit-jet expansion of argon, helium, or dioxygen. The translational temperature is in this case retrieved from the Doppler line broadening, and it appears that both the mass and the specific heat ratio of the carrier gas have an influence on the achieved temperature. A diatomic (O_2) or a light (He) carrier gas is less efficient to achieve very low translational temperatures. Condensation of water is triggered beyond a certain concentration of vapor which is monitored using the line broadening induced by the heat of condensation released in the flow [26].

However, line-of-sight absorption spectroscopy can come up against the convective Doppler effect, which in some cases makes impossible the retrieval of the translational temperature from the absorption line profile. The convective Doppler effect comes from the global convective motion of the gas which creates a non-zero velocity component along the probe axis, which in turns induces a frequency shift of the measured absorption line. This effect shifts, broadens, and distorts the transitions, especially in the case of an axisymmetric free jet whose divergence leads to a significant transverse velocity distribution. This is at the origin of highly specific line profiles [17, 18], such as the ones shown in Fig. 9.6(b). In this case, the CO absorption line profiles were obtained probing an axisymmetric free jet perpendicular to its axis. The atypical shape of the lines is caused by the high transverse speed of the gas along the line of sight.

9.3   Underexpanded Sonic Jets vs Laval Nozzle Flows

Reaching low temperatures to perform cold spectroscopy has been achieved extensively using extremely underexpanded sonic jets, usually designed as "free jet" type expansions. As will be explained at the end of the following subsection, the term "free jet" is widely used to denote a supersonic jet exiting a sharp-edged orifice where the gas velocity is sonic and its pressure is far greater than the vacuum chamber pressure. This is the reason why a "free jet" is also called an underexpanded sonic jet (see Chapter 1). The properties of these flows will be recalled to better grasp the interest and the complementarity of Laval nozzle flows.

9.3.1   *Underexpanded sonic jets*

The temperature of a gas can easily be lowered down to a few kelvins by adiabatically expanding through a simple orifice separating two volumes set at different pressures, provided that the gas flow breaks the speed of sound to reach a supersonic regime. In a free jet expansion, very high Mach numbers can be achieved, meaning that almost all the stagnation enthalpy of the gas is converted into kinetic energy, which leads to a drastic drop in temperature. A translational temperature[c] as low as 6×10^{-3} K has been reported in a free jet of helium [20], corresponding to a Mach number above 140.

The terminology "free jet" must be clarified at this stage. The term "free jet" has been widely used in chemical physics and spectroscopy to designate a jet of high Mach number expanding freely from a sharp-edged orifice. As discussed at length in Chapter 1, it should, however, be emphasized that a free jet is by definition a flow which is not constrained by physical walls. Thus, a flow generated by any type of nozzle is formally a free jet. What has become customary to call a free jet, especially in spectroscopy, is a jet produced by a nozzle having no divergent section. In this case, the pressure at the nozzle exit far exceeds the background pressure, leading then to an

---

[c]The concept of translational temperature breaks down in this case (see Chapter 1). Two parallel and transverse Maxwell-Boltzmann velocity distributions associated with two different temperatures are required to model the molecular motion in the center of mass frame [19].

extremely "underexpanded" jet [21]. Following the laws of physics of compressible fluids, the existence of a throat is necessary for the gas to reach the speed of sound. In the particular case of a "free jet", the throat is located at the nozzle exit, and the gaseous expanding region located downstream forms a divergent section limited by the jet boundaries. For these reasons, a supersonic jet exiting a sharp-edged orifice should be designated as an underexpanded "sonic" jet, but we will be using "free jet" throughout this chapter.

To reach a supersonic regime, a flow has to meet two conditions. The first one is the chocked flow condition: the speed of sound has to be reached at the ejection port, which then acts as a sonic throat characterized by a Mach number equal to 1. The second condition ensues from the mass conservation law: in a supersonic regime, the density drop is extremely rapid and cannot be compensated by the velocity increase alone to maintain a constant mass flow ($\dot{m} = \rho A V$). In this case, the density (ρ) decrease must be associated with an increase of both the flow velocity (V) and area (A). Therefore, the divergent area of the supersonic nozzles (e.g. Laval or conical nozzles) is due to the mass conservation. In a free jet, the expansion of the gas is not constrained by any divergent duct, but it opens directly from the throat into a large volume allowing for further acceleration beyond the speed of sound. The Mach number of the flow is imposed by both the geometry of the orifice and the specific heat ratio of the gas. In spectroscopic studies, a distinction is made between planar free jets associated with a slit-type aperture [22–24] and axisymmetric free jets associated with circular orifices [20].

To obtain a sonic throat, it is necessary that the pressure of the gas at the nozzle exit (p_e) reaches a critical value p^* given by the isentropic chocking pressure ratio $\frac{p^*}{p_0} = (\frac{2}{\gamma+1})^{\frac{\gamma}{\gamma-1}}$, where p_0 is the stagnation pressure and γ denotes the specific heat ratio of the gas. Thus, a supersonic flow is forming provided that $p^*/p_0 \sim 0.5$, or equivalently that the exit pressure ratio $p_0/p_e \sim 2$ with $p_e = p^*$. This condition is easily satisfied. For example, the gas which escapes from a bottle of champagne is supersonic [25], as well as the air jet pushed out of the cavity formed when a pebble hits the surface of the water [26].

A free jet is by nature underexpanded as the critical pressure (p^*) is much higher than the backpressure (p_b), i.e. the pressure of the vacuum chamber. Free jets used for spectroscopic purposes are characterized by very high exit pressure ratios, generally exceeding 1000 and they are therefore part of the extremely underexpanded jets. Note that the total pressure ratio p_0/p_b, popularly known as the nozzle pressure ratio (NPR), is more widely used than the exit pressure ratio p_0/p_e.

The structure of an underexpanded supersonic flow discharging from a sonic nozzle into a low-pressure volume is far from trivial [21, 27, 28]. Two expansion effects combine to form a strong expansion fan at the nozzle outlet edge. The first is linked to the underexpanded state of the flow at the throat, while the second is due to the relaxation felt by the flow when it enters a large volume. When passing the expansion fan, the flow exiting the nozzle is suddenly deflected away from the edge aperture and its pressure drops well below the backpressure. As this pressure cannot physically reach a zero value, the flow is recompressed by an oblique shock referred to as a barrel shock (see Fig. 9.7), which redirects the gas toward the axis. The oblique shocks on either side of the axis converge downstream up to intersect on the flow axis. In case of moderate NPR, the angle of intersection is smaller than a critical value so that a regular reflection occurs and a single reflected shock follows in the wake of

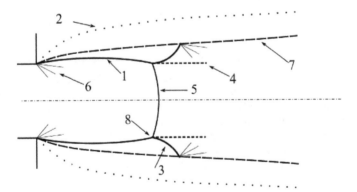

Fig. 9.7. Structure of underexpanded sonic jet (axisymmetric free jet): (1) oblique shock (barrel shock); (2) jet boundary; (3) reflected shock; (4) slip line separating two zones of same pressure but of different speeds; (5) normal shock or Mach disk; (6) group of expansion waves generated at the nozzle; (7) line of pressure equilibrium from which the expansion waves reflect as compression waves; and (8) triple point.

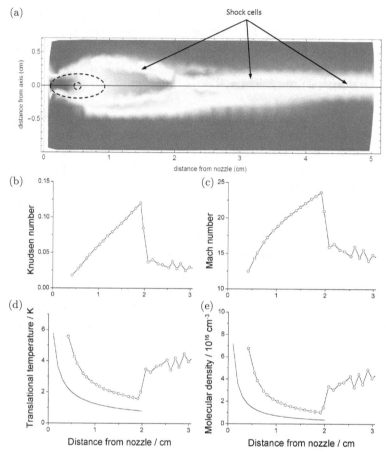

Fig. 9.8. Flow structure of an Ar expansion from a high-pressure reservoir into a low-backpressure volume (NPR = 1570) through an 18-μm-thick slit. (a) Section view of a Pitot mapping of the flow, (b) Knudsen number, (c) Mach number, (d) temperature, and (e) density along the symmetry plane. The open circles show the measured data, and the solid lines in (d) and (e) correspond to the isentropic model based on a Mach number distribution given by the correlation from Ref. [31]. Reproduced with permission from Ref. [44], Copyright 2017 American Chemical Society.

the intersection point [29]. For high NPR expansions, often used for spectroscopic purposes, the angle of intersection exceeds the critical value and the intersection point becomes a normal shock known as a Mach stem, or Mach disk in the case of axisymmetric flows [30]. The region comprising the area between the expansion fan forming at the nozzle outlet and the oblique shock waves intersection is called a shock cell in jet literature. An underexpanded jet tries to equilibrate its internal pressure with the background pressure through a succession of shock cells (see Fig. 9.8). After each crossing of a shock wave intersection, the flow is slowed down — to a subsonic speed for a normal intersection — and compressed to a pressure higher than the background pressure. Then, it is reaccelerated downstream up to a supersonic speed through a new set of expansion waves similar to the ones at the nozzle exit, consequently forming a new shock cell. It is usually the first shock cell, i.e. the so-called zone of silence located between the outlet of the nozzle and the Mach disk, which is probed for spectroscopic purposes.

9.3.2 *Laval nozzle flows*

9.3.2.1 *Convergent–divergent nozzles*

Several types of convergent–divergent nozzles can generate supersonic flows. A convergent–divergent nozzle with straight walls can deliver a supersonic Mach number, but with a flow not unidirectional. Such nozzles were coupled to infrared detection to follow the condensation of gases either in the divergent portion [32–35] or downstream from the nozzle exit [36–39]. On the contrary, a nozzle with a contoured divergent portion, called *Laval* (or *de Laval*), generates supersonic Mach numbers and a flow parallel to the nozzle axis. Two types of Laval nozzles should be distinguished. There are nozzles which are designed to operate in the so-called "adapted" or "perfectly expanded" regime. Under certain conditions of mass flow, stagnation, and backpressures, they will produce a uniform flow downstream of the divergent portion. Such adapted flow is free of gradients along its axis and is not limited downstream by the presence of normal (Mach disk) or oblique shock waves which would modify the speed and thermodynamic parameters of the gas. This type of flow has become commonly used to measure gas reaction kinetics at low temperature; this method is known under its French acronym "CRESU", which stands for Cinétique de Réaction en Ecoulement Supersonique Uniforme [40]. The Computational Fluid Dynamics (CFD) simulation of such a flow is shown in Fig. 9.9.

Some Laval nozzles can be operated in a not fully adapted regime. The profile of their divergent portion is calculated to generate a flow

(a)   **Temperature /K**

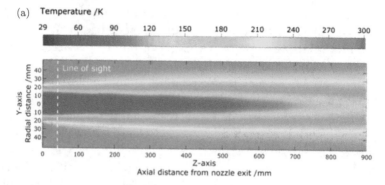

(b)

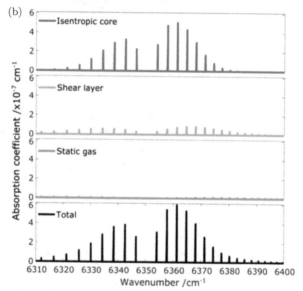

(c)

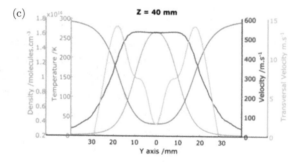

Fig. 9.9.   (*Continued*)

parallel to the nozzle axis, and is an efficient way to avoid the convective Doppler effect responsible for the broadening of absorption lines [39, 41]. Unlike CRESU flows characterized by a constant Mach number, these flows have an increasing Mach number and are limited downstream by the development of oblique shock waves intersecting on the flow axis and reflected on the jet boundary to form shock cells.

Flows produced by Laval nozzles are fundamentally different from free jets because the evolution of their Mach number is smoother, as imposed by the geometry of the diverging part of the nozzle. An initially confined supersonic flow consequently leads to a lower cooling rate, which in turn reduces the "thermal soaking" specific to free jets, freezes the thermodynamic parameters, and prevents some internal degrees of freedom of polyatomic molecules to relax. In the divergent duct of a Laval nozzle, the thermodynamic gradients are reduced along the axis. In particular, the density of a flow produced by a Laval nozzle decreases much less quickly than a free jet flow, which maintains a higher frequency of collisions between molecules themselves and between molecules and carrier gas. In addition, by properly designing the divergent part of the nozzle, the isentropic flow can be preserved over large distances along its axis, up to several tens of centimeters, before being perturbed by the intersection of two oblique shock waves (regular reflection) or by the formation of a normal shock wave (Mach reflection or Mach disk), as described in [42]. In the case of a divergent duct specifically designed to work in adapted conditions, the presence of such shock waves can be totally eliminated and the flow is therefore called "uniform". The position of the first shock wave from the nozzle exit, and more generally the spatial extension of the shock cells, depends on the NPR. For a free jet, the Mach disk is usually relatively close to the gas exhaust, typically

---

Fig. 9.9. (a) Temperature CFD simulation of an argon CRESU flow produced using a Laval nozzle whose stagnation temperature and stagnation pressure are 295 K and 23.24 hPa, respectively. The jet is characterized by a Mach number of 5.12, a temperature of 30.5 K, a static pressure of 0.082 hPa, and a density of $1.9\,10^{16}\,\mathrm{cm}^{-3}$. (b) Calculated absorption spectra of CO (0.5% in concentration) relative to the different gas layers (isentropic core, shear layer, static gas) simulated at a distance from the nozzle of 40 mm. (c) Density, temperature, axial, and radial velocities along the Y-axis at the same position. Note the very weak radial velocity.

few tens of millimeters [43]. The time of flight of molecules moving in the silence zone between the throat and the Mach disk — that we will call "hydrodynamic time" — is therefore reduced to a few tens of microseconds typically, [44]. which is very short to study relaxation phenomena. In contrast, the hydrodynamic time can reach about one millisecond for a flow generated by an adapted Laval nozzle. A high collision frequency combined with significant hydrodynamic time is responsible for a large number of 2- and 3-body collisions. This property has been widely exploited to trigger the formation of molecular aggregates at very low temperatures [33, 35, 45], using infrared absorption spectroscopy among other techniques to monitor this process.

9.3.2.2 *Perfectly expanded flows probed by infrared spectroscopy*

Unlike laser-induced fluorescence [46] or Raman spectroscopy [47], capable of probing a supersonic flow over a very limited spatial area, infrared spectroscopy delivers an absorption signal integrated along a line of sight. Unfortunately, the spectrum resulting from this line of sight technique is a superposition of different gas layers which might interfere to degrade the absorption signal coming from the isentropic core. Indeed, to reach the isentropic core, the infrared beam must cross first the static gas of the vacuum chamber (background pressure) and second the surrounding shear layers (see Fig. 9.9). The recorded infrared spectrum is thus made up of the superposition of the contributions relating to these different gas layers. It is possible to deconvolve this integrated signal by using tomographic absorption spectroscopy [48] or well-known inversion techniques such as Abel inversion in the case of radially symmetric flows [18, 49]. However, using one of these techniques increases the measurement time because the laser beam (which probes the jet perpendicularly) must be scanned across the jet section. An approach circumventing this issue involves penetrating the boundary layers with a profiled tube to avoid their contribution [50]. Nevertheless, the lower density and higher temperature associated with these outer layers can be neglected most of the time by focusing on the lowest J value transitions.

In a CRESU flow, the adapted regime of the nozzle imposes a constant static pressure throughout the flow field, strictly equal to the background pressure prevailing in the expansion chamber. The static pressure remains *a fortiori* constant along the line of sight through all the gaseous layers crossed by the optical beam. By virtue of the ideal gas law, it follows that the density of the flow varies inversely with temperature (see Fig. 9.9(c)). This property is very interesting for absorption spectroscopy as it minimizes the warmer parasitic contributions of the boundary layers and static residual gas. Figure 9.9 illustrates this property of adapted supersonic flows. It was made from a simulation of a uniform supersonic flow of argon at 30.5 K seeded with a small amount of CO. It shows that the density of the gas gradually drops as the gas heats up from the core to the outer layers of the flow. The spectra showed in this figure were simulated following an approach based on the thermodynamic parameters of the flow calculated by CFD [17]. It can be seen that the spectrum relative to the isentropic core is dominated by low-J rotation-vibration transitions, while the spectra relative to the boundary layers and the static gas are dominated by higher-J values.

It can also be seen that the contribution to the total intensity of the hot layers remains moderate and that it does not significantly interfere with the absorption signal from the cold layers. In addition, coupling an IR source with a high-finesse cavity enhances the sensitivity by increasing the effective absorption length by several orders of magnitude: in this way, the cavity ring-down spectroscopy technique can lead to an effective absorption length of several hundred meters with an isentropic core diameter of only few millimeters [39].

9.3.2.3   *Perfectly expanded flows and capillary injection*

The formation of molecular aggregates in adapted Laval nozzles disrupts the operating conditions for which the nozzles are designed, *cf.* Chapter 6. Beyond 1% of admixed polyatomic molecules, the specific heat ratio and the viscosity will start to differ significantly from the one of the monoatomic gas alone. Furthermore, the heat of condensation will raise the temperature of the flow and increase its speed so that the Laval nozzle will not work anymore in adapted conditions to produce a uniform flow. In particular, the formation of nanometric-size aggregates made of up to several million molecules requires a very large injection of monomers which will break the uniformity of the supersonic flow. This problem is well known in wind tunnels and is referred to as condensation shock [51].

To work around this problem, it is possible to use a capillary injector centered on the nozzle throat. We showed that the polyatomic molecule gas injected by the capillary expands to adapt its pressure to the one of the nozzle, but remains confined in the central part of the divergent portion, as well as downstream from the nozzle exit (see Fig. 9.10) [37].

The molecular diffusion is too slow for the molecules to migrate into the surrounding monoatomic gas (as discussed in Chapter 1),

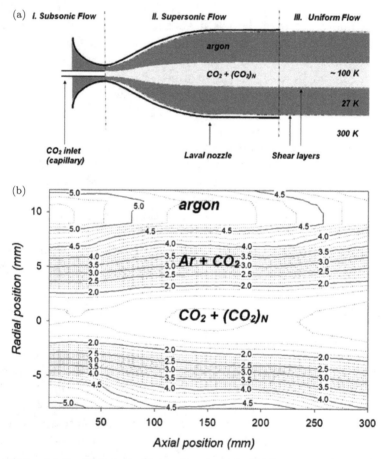

Fig. 9.10. (a) Capillary injection in a Laval nozzle. (b) Pitot map of the flow field downstream from the Laval nozzle exhaust. The nozzle is supplied with argon, while pure carbon dioxide is injected by the capillary. Argon and CO_2 only mix in the shear layer.

and an intense condensation process takes place under the effect of the temperature drop induced by the adiabatic expansion. In the same way, thermal diffusion is too low compared to the gas velocity of several hundred meters per second. The heat released during the molecular condensation process then remains trapped in the central part of the flow, without however impacting the peripheral carrier gas flow, but leading to a severe temperature increase, up to ∼100 K, of the central condensing flow. Simultaneously, the condensation is responsible for a significant acceleration of about 100 m/s of the condensing flow (see Fig. 9.11). A combined FTIR and Rayleigh

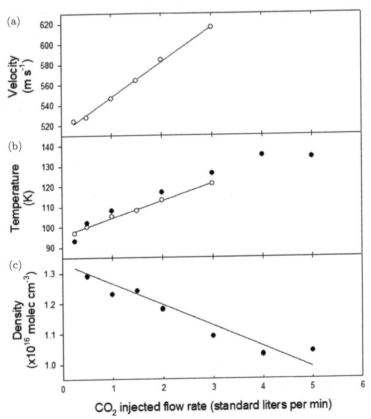

Fig. 9.11. Evolution of (a) the velocity, (b) the translational [white circles] and rotational [black circles] temperatures, and (c) the flow density of the central part of the flow as a function of the amount of CO_2 injected through a capillary, see text for details.

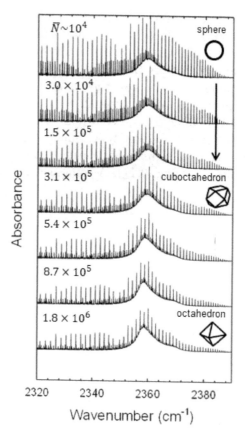

Fig. 9.12. High-resolution (0.005 cm^{-1}) spectra of CO_2 aggregates formed by capillary injection [52]. The sharp absorption features are the absorption lines of the CO_2 monomer. $\bar{N}$ denotes the average number of molecules per aggregate, determined by Rayleigh scattering.

scattering approach [37, 52] revealed condensation rates of 10 to 20% of the injected carbon dioxide, leading to aggregates composed of up to several million molecules with a radius extending up to 26 nm (see Figs. 9.12 and 9.13). Figure 9.12 shows the high-resolution absorption spectrum of CO_2 aggregates obtained with a capillary injection. The high resolution is required to distinguish resolved monomers from the broader aggregate contributions. As the change in amplitude of the monomer lines shows, the number of uncondensed molecules decreases from the top to the bottom of the figure.

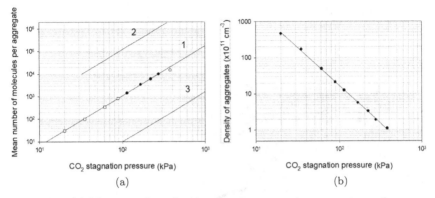

Fig. 9.13. (a) Mean number of molecules per aggregate versus stagnation pressure. (1–2) Aggregates formed by capillary injection with a low- and high-background Laval jet pressure, respectively; (3) aggregates formed in a free jet expansion. (b) Evolution of the density of aggregates formed by capillary injection versus stagnation pressure, in the case of a low-background Laval jet pressure.

9.3.2.4 *Non-adapted Laval nozzle*

As will be described in Section 9.4, a non-adapted Laval nozzle offers interesting properties for non-LTE (non-local thermodynamic equilibrium) absorption spectroscopy. Starting from a very high stagnation temperature ($>1000\,\mathrm{K}$), the Laval nozzle developed in Rennes and equipping the SMAUG setup [39] produces a slightly overexpanded jet (with $p_e \approx p_b/5$). The CFD simulation of Fig. 9.14(a) shows that a regular intersection forms 60 millimeters downstream from the nozzle exit (4). The jet further expands downstream from the nozzle exit to reach a static pressure which is even lower than the chamber pressure. This induces a continuous increase of the Mach number along the jet axis between the nozzle exit and the shock wave intersection (Fig. 9.14(c)), leading to a continuous decrease in temperature on the jet axis (Fig. 9.14(b)).

The CFD simulations, shown in Fig. 9.15, reveal that the concentration of absorbing molecules remains significantly higher in the isentropic core than in the hottest parts of the boundary layers. The boundary layers are responsible for the peaks of the density and temperature profiles. The contribution of the boundary layers to the absorption spectrum is thus negligible, i.e. a non-adapted Laval nozzle is suitable for absorption spectroscopy across the flow axis.

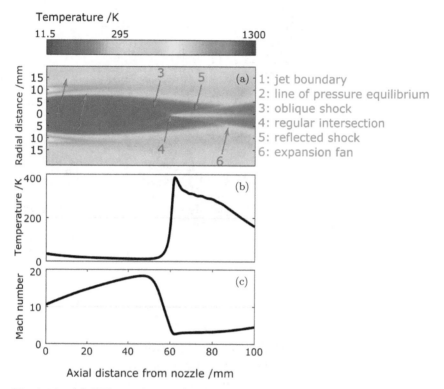

Fig. 9.14. (a) CFD simulation of a hypersonic jet generated by a slightly overexpanded Laval nozzle supplied by argon initially heated at 1300 K. (b) Temperature on the jet axis. (c) Mach number on the jet axis.

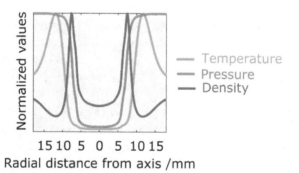

Fig. 9.15. Normalized radial temperature (green curve), pressure (red curve), and molecular density (blue curve) obtained from the CFD simulations of a hypersonic jet of argon 20 mm from the nozzle exit (stagnation temperature: 1300 K; stagnation pressure: 1100 Torr; residual pressure: 0.5 Torr). The blue zone is the cold flow region. Note the mismatch of the temperature and density profiles.

9.4 Infrared Absorption Spectroscopy Techniques Associated with Laval Nozzle Flows

The spectroscopic technique employed to probe the narrow cold core of a Laval flow should have the following characteristics:

- *Intensity precision*: Absorption spectroscopy is fundamentally ruled by the Beer law, as given in Eq. (9.1). Measuring the intensity depletion due to absorption gives access to the absolute species concentrations provided that both the absorption cross section and the interaction length between the light beam and the sample are known, which is one of the main attractions of absorption spectroscopy.

- *Sensitivity*: Laval supersonic expansions have a sub-Torr pressure and the fraction of species of interest is usually a few percent at maximum.

- *Frequency precision*: Low-temperature spectroscopy offers a unique chance to determine precisely the frequencies of the vibrational modes usually with a precision set by the spectrometer.

- *High spectral resolution*: The low rotational temperatures characterizing the Laval flows lead to narrow absorption lines (few $10^{-3}\,\mathrm{cm}^{-1}$ levels for the FWHM of monomer species); therefore, retrieving undistorted lineshapes requires high-resolution spectrometers.

- *Broad spectral coverage*: Measuring entire absorption bands or absorption contributions of different species in a single measurement is always an advantage when trying to untangle the different molecular processes occurring in the flows (e.g. nucleation or chemical reaction).

- *Robustness to mechanical vibrations*: Generating the Laval flows necessitates rather large pumping capacities, which brings mechanical noises to the walls of the flow chambers. The spectroscopic technique should be as immune as possible to this noise source.

Several approaches have been used over time: tunable diode laser absorption spectroscopy (TDLAS), Fourier transform spectroscopy based on incoherent sources (FTIR), cavity ring-down spectroscopy (CRDS), and optical frequency comb spectroscopy (OFCS). New developments aimed at combining broad spectral coverage with sensitivity are under development and will be quickly scanned.

9.4.1 *Tunable Diode Laser Absorption Spectroscopy*

Tunable Diode Laser Absorption Spectroscopy (TDLAS) relies on the use of a laser diode combined with either a simple photodetector (if the diode is single mode) or a monochromator (if the diode is multimode). It has been used extensively in Laval flows in a transverse fashion, where the optical beam crosses the supersonic jet perpendicularly [35, 53]. The main advantage of this simple optical scheme is that it only requires windows on each side of the flow chamber to allow the optical beam to propagate through the cold sample. However, the single-pass configuration still limits the sensitivity of TDLAS, despite the spectral brightness of the diode laser source, which is much higher than the one of broadband, incoherent light sources. This sensitivity limitation can be improved by coupling TDLAS with lock-in detection strategies. The resulting spectrum covers a narrow band set by the laser diode tunability, which limits its application to the detection of monomer species. Despite this limitation, the first works relying on TDLAS in Laval flows have been focusing on the investigation of nucleation [32, 35, 54, 55]. In the work of Wu and Laguna [32], TDLAS was used to monitor the concentration drop of the monomer of SF_6 in the infrared around $16\,\mu m$.

Further spectroscopic works, aimed at retrieving the low-temperature absorption line profile, were performed two decades later on CO_2 and N_2O [53, 56], retrieving the temperature difference between the flow core and the boundary layers, the velocity distribution of the molecule, and the translational velocity from the Voigt profile of the recorded transitions.

9.4.1.1 *To characterize condensation in flows*

As nucleation occurs in the gas sample when it flows from the nozzle throat downstream, measuring the absorption spectrum of a monomer species (taking part to the nucleation process or not) at different distances from the nozzle gives access to the flow conditions during nucleation. When nucleation occurs, one expects a drop of monomer species, an increase of temperature (nucleation is exothermic), a variation of the pressure, and a significant increase of the gas velocity. Therefore, measuring the depletion of the monomer species, provided that this depletion is caused by nucleation, gives an indirect picture of the nucleation process.

Most of the applications of TDLAS in Laval flows concerned clustering flows. Molecules contained in the flow offer a sensitive probe to temperature and pressure conditions. Retrieving temperature from an absorption spectrum requires that two transitions belonging to the same vibrational band of the same species are measured simultaneously: the absolute absorption strength gives access to the concentration and the temperature [35]. However, in most cases, TDLAS fails to provide enough information to know which size and density of clusters are produced. This limitation has led to the combining of TDLAS with mass spectrometry or X-ray scattering techniques in order to retrieve simultaneously the flow temperature, monomer concentration, and the apparition of larger size clusters; see Chapter 6 for details.

9.4.1.2 *Wavelength modulation spectroscopy based on TDLAS*

Wavelength modulation spectroscopy relies on a periodic tuning of the laser diode wavelength coupled to a lock-in detection to improve the detection sensitivity of the spectrometer. The modulation allows performing the measurement at a known frequency higher than the most intense noise frequencies (usually in the 10–1000 Hz range). It is an efficient noise reduction technique especially when used in harsh environments where mechanical vibrations are the main source of noise. The full spectrum is obtained in the same way as regular TDLAS, and each spectral point is measured by demodulation of the signal at the first or second harmonic of the modulation frequency. This leads to absorption signals that have the shape given by the absolute value of the first or second derivative of the initial absorption profile. This approach was successfully used for detection of the temperature and the concentrations of H_2O and NH_3 in a supersonic flow [57].

9.4.2 *Fourier Transform Infrared Spectroscopy*

Fourier Transform Infrared Spectroscopy (FTIR) has been the workhorse of physical chemistry for many decades. It relies on incoherent light sources, typically lamps displaying a broad spectrum, analyzed using a Michelson interferometer. It provides a parallel measurement of the entire spectrum, allowing simultaneous identification of multiple species and recording of an entire absorption band of the molecules of interest. Because of its high spectral versatility, FTIR is the ideal method to perform spectroscopy in the far infrared range, in the so-called fingerprint region where each molecule has a unique absorption pattern. However, it is intrinsically limited in sensitivity by the poor spectral brightness of incoherent sources, the low molecular density, and the small diameter of the Laval flows (typically at the cm scale). Similarly to TDLAS experiments, FTIR in Laval flows has been used extensively to study nucleation processes, more precisely to measure the spectra of molecular clusters [33, 34, 38, 45, 52, 58]. Performing spectroscopy with a satisfying signal-to-noise ($\gtrsim 10$) ratio still requires trade-offs between acquisition time, number of averaging, resolution, simplicity, and interaction length. Different approaches have been tried, each of them achieving different spectral performances.

9.4.2.1 *Low-resolution spectroscopy of molecular clusters*

A first approach relies on a reduction of the resolution of the FTIR in order to increase the power density per spectral element and shorten the total acquisition time (inversely proportional to the spectral resolution in case of FTIR). The measurement is then performed in a single path scheme, where the collimated optical beam is going through two windows placed across the Laval flow before entering the interferometer. This simple scheme allows one to measure the absorption spectrum of clusters. Molecular clusters are weakly bonded which implies broad absorption features and releases the need for high spectral resolution. However, the absorption modes of clusters, especially those of small clusters, usually lie close to the monomer transition frequencies, or even overlap with the monomer absorption bands, as shown in Fig. 9.12. When the resolution of the spectrum is too poor to untangle the spectral signature of the clusters from the monomers, it is very challenging to dissociate them. Using this approach, Lippe *et al.* retrieved convolved spectra containing the monomer and cluster absorption feature and attempted to subtract the monomer contribution to isolate the cluster contributions [38]. The size of the probed cluster was determined in parallel using mass spectrometry. This approach requires an accurate knowledge of the flow conditions regarding the monomer thermodynamic conditions, in order to subtract the correct monomer absorption spectrum.

9.4.2.2 *Increasing the sensitivity: probing the flow along its axis*

A second approach to increase the absorption sensitivity of FTIR measurements is to increase the interaction length of the optical beam with the Laval flow by probing the flow along the nozzle axis, as depicted in Fig. 9.16 [36]. The incoherent source is placed behind the reservoir. The optical beam enters first the reservoir through a window, and propagates through the Laval nozzle and in the subsequent supersonic flow. The light beam is collected to the FTIR via another window, placed across the Laval flow. This allows the interaction length to be increased to a few tens of cm, at the price of two drawbacks: the optical beam probes the high-density gas in the reservoir and the exit window placed downstream produces a shock wave which warms up the neighboring gas back to the reservoir temperature. These two effects have to be considered to be able to isolate the part of the absorption spectrum which belongs to the cold Laval flow. Another noticeable effect is the Doppler shift affecting all the absorption lines measured using this method: contrary to a transverse measurement where the velocity component collinear with the

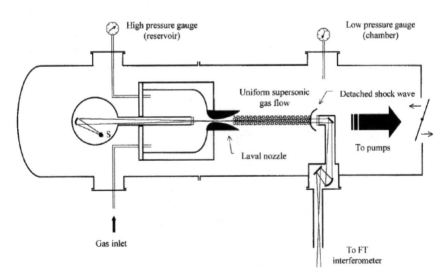

Fig. 9.16. Schematic of the FTIR probe set collinearly to the Laval flow. Reproduced with permission from Ref. [36], Copyright 2000 Academic Press.

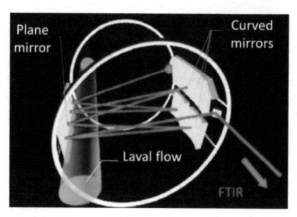

Fig. 9.17. Schematic of the multipass cell increasing the interaction length between the FTIR and the flow (here in an 8-pass configuration); by courtesy of A. Bonnamy [59].

optical beam is null, an optical probing along the stream implies a supersonic average velocity.

9.4.2.3 *Increasing the sensitivity using multipass cells*

Measuring the absorption spectrum over a path length longer than a single path without suffering from the spectral distortion due to the collinear probing scheme can be achieved using a multipass cell, such as the one depicted in Fig. 9.17 [59]. To successfully probe the core of the Laval flow multiple times, it is necessary to carefully design the cell pattern in order to match the 1 cm diameter of the flow. Using the design shown in Fig. 9.17, Bonnamy *et al.* recorded the spectra of CO_2 clusters (which can be seen in Fig. 9.12) with a resolution high enough to distinguish without any confusion the cluster from the monomer contributions [52]. This multipass cell geometry allowed multiplying up to 24 times the interaction length with the Laval flow. Increasing this performance again could be possible using a cell design close to the one proposed by, for instance, Tuzson *et al.* [60].

9.4.3 *Cavity Ring-Down Spectroscopy*

Pursuing the goal of increasing the sensitivity led to the application of laser-based cavity-enhanced techniques to the detection in Laval flows. In their simplest form, an optical cavity involves two high reflectivity mirrors placed face to face. When a light beam enters the cavity with the proper shape and frequency, it constructively interferes with its many intra-cavity reflections leading to effective propagation length inside the cavity that can exceed several kilometers. By comparison, state-of-the-art FTIR spectroscopy measurements are limited to 10^{-28} cm/molecule, while Cavity Ring-Down Spectroscopy (CRDS) operating in static gas cell conditions can routinely detect intensities in the range of 10^{-32} cm/molecule. Among the different cavity techniques, CRDS offers an appealing combination of high sensitivity and robustness, which makes it the technique of choice in noisy environments. It relies on the measurement of the photon lifetime of the cavity *vs* the light frequency. This ring-down time is set by the cavity length and losses arising from the cavity mirror coatings and the possible absorption of a gas contained between the two mirrors. Retrieving the ring-down time is performed by turning off the laser light when its frequency matches one of the cavity resonances and measuring the intensity decay at the cavity output; see Kassi *et al.* [61] for a typical CRDS layout. This measurement gives an absolute quantification of the absorption spectrum, without requiring prior knowledge of the mirror reflectivity or interaction length.

9.4.3.1 *Application to non-equilibrium spectroscopy*

CRDS has been applied to detection in Laval flows to perform non-LTE spectroscopy, as discussed in Section 4. The cavity involved two curved mirrors with a high reflectivity placed across the Laval flow, as described in Fig. 9.18 [39]. The mirrors are used directly at the entrance and exit windows in the flow chamber and they are mounted on bellows to isolate them as much as possible from the chamber vibrations. Their reflective surfaces are protected from the potential pollution induced by adsorption of the chamber gases (which would decrease their reflectivity) using a continuous purge of pure N_2. The purge gas flows inside pipes coming close to the Laval flow without disturbing it. The cavity length is swept periodically using a piezo transducer mounted on one of the cavity mirrors to match periodically a cavity resonance to the laser diode frequency. When the laser frequency is coupled into a cavity resonance mode, the laser is shut off using an intensity-threshold detector. At the same time as the decay recording, the laser frequency is calibrated on the fly using a

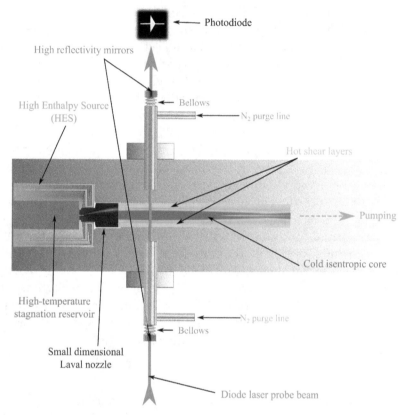

Fig. 9.18. Schematic of the SMAUG experimental setup showing the main components (from left to right): the HES terminated by the Laval nozzle and the CRDS diagnostic probing the hypersonic flow [39].

high finesse wavemeter (precision of $0.002\,\text{cm}^{-1}$) to reconstruct the frequency scale of the spectrum.

The flow mixture involves typically 90% of the buffer gas and 10% of the sample molecule, which is originally contained in a high-pressure reservoir at room temperature. In the non-LTE experiment, the reservoir gas is brought to high temperature by going through a porous graphite rod. The rod is kept at a high temperature by applying $\sim$100-A intensity current through it and relying on the Joule effect to warm up the carbon semiconductor. A Laval nozzle is fixed at the rod exit so that the gas contained in the core of the rod (up to $\sim$2000 K in LTE conditions) expands in a uniform fashion. This expansion relaxes the rotational degrees of freedom of the molecular motion much more efficiently than the vibrational ones, bringing the molecular sample in a very strong out-of-equilibrium thermodynamic state [39].

The diameter of the isentropic core produced by such a Laval nozzle is about 10 mm, which yields an effective absorption length of about 450 m when combined to a typical ring-down time of 120 μs.

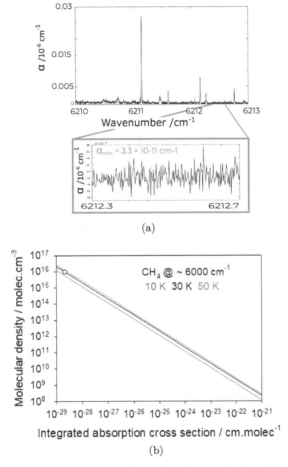

Fig. 9.19. (a) Limit of sensitivity of the CRDS setup deduced from a non-LTE spectrum of methane recorded in the 1.6 μm range [39]. (b) Integrated absorption cross sections which can be measured with the same setup. The circle corresponds to one of the weakest transitions (3.7×10^{-29} molecules/cm) measured [39]. As the amplitude of the absorption lines increases with decreasing temperature, slightly lower absolute intensities can be measured as the jet temperature is lowered.

Depending on the set of mirrors used, ring-down times as high as $260\,\mu s$ can be achieved in the 1.5–1.7 micrometer range, equivalent to an absorption length of about 1 km long through a cold flow of 1 cm in diameter. A minimum absorption length coefficient α_{min} of $3.3 \times 10^{-11}\,\text{cm}^{-1}$ has been deduced from the root mean square of the CRD spectrum baseline (see Fig. 9.19). Thus, for example, a 30-K supersonic flow of argon admixed with 9.5×10^{15} molecules of methane per cm^3 allows the detection of absolute intensities (i.e. integrated absorption cross section) as low as $\sim 4 \times 10^{-29}$ cm/molecule. Conversely, molecular concentration as low as $\sim 10^9\,\text{cm}^{-3}$ can be detected provided that intense transitions can be probed (see Fig. 9.19).

9.4.3.2 *Application to reaction kinetics*

Reaction kinetic studies have long been the main interest of the uniform flows through the CRESU approach (see previous chapters). This wall-less configuration limits the unwanted condensation and reaction of the studied mixture especially at very low temperature. Thus, the CRESU technique has been very successful in a probing reaction of astrophysical interest such as encountered in the interstellar medium or planetary atmospheres [62].

In a CRESU experiment, a reactant is expanded along with a precursor carrying a reactive species to study. This species, which can be a radical, is then produced by pulsed laser photolysis injected collinearly to the uniform flow. It can also be an ion either produced by photoionization or by an electron gun, or directly drifted into the flow from a selective ion source. Thus, the isentropic core holds a reactive medium allowing the study of the reaction rate only limited by the expansion length defining the maximum time for the reaction that can be investigated. Then, a mass spectrometer or a laser probe is set as far as possible from the output of the nozzle to quantify the depletion/production of one or more species according to the fire of the photolysis probe which triggers the reaction (Fig. 9.20).

For decades, reaction rates have been recorded by laser-induced fluorescence (LIF). While being a very sensitive approach, it remains limited by the low number of accessible species. The CRESU approach also benefited from the development of broadband millimeter and microwave spectrometers [64]. More recently, a new setup

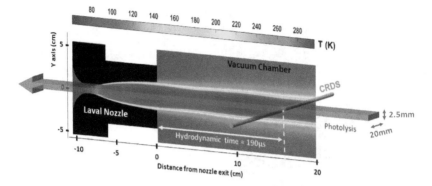

Fig. 9.20. Schematic diagram of the main components of the UF-CRDS setup. The temperature distribution of the uniform flow of N_2 has been calculated through a computational fluid dynamics software (OpenFOAM). The photolysis probe travels backward along the axis of the flow. The target species is probed perpendicularly at 15 cm from the output of the Laval nozzle [63]. Reproduced with permission from Ref. [63], Copyright 2019 AIP Publishing.

named UF-CRDS (Uniform Flow–Cavity Ring-Down Spectrometer) extended the accessible wavelength region to the infrared using a high-sensitivity cavity ring-down spectrometer [63]. The exponential decay of the empty high finesse cavity is most of the time similar to the hydrodynamic time of the uniform flow (between 500 and $1000 \, \mu s$), making this coupling particularly suitable for kinetic studies.

Thus, instead of performing multiple measures corresponding to different reaction times (time delays between the photolysis probe and the measurement), the UF-CRDS records the whole reaction during a unique non-exponential decay due to the density variation of the probed species as described by the SKaR (Simultaneous Kinetics and Ring-Down) method [65] (Fig. 9.21).

The uniform flow produced by the UF-CRDS, as per other recent setups, is pulsed through a high-throughput piezo stack valve to reduce gas consumption and pump capacity. The flow shows similar characteristics to its continuous counterpart, that is to say, a temperature down to 10 K and a length of several tens of centimeters (see Chapter 2 for a complete description of systems based on pulsed Laval nozzles). The CRDS is currently operated in the near infrared (1.2 to $1.4 \, \mu m$) using an external cavity diode laser (ECDL) and distributed feedback diodes (DFB) and can reach a minimum absorption coefficient of $2 \times 10^{-9} \, cm^{-1}$ by averaging only three ring downs.

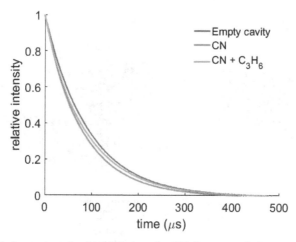

Fig. 9.21.   Informative plot highlighting the SKaR approach from three intensity decays: the empty cavity (blue) and in the presence of stable (red) and changing (green) concentration of CN due to its reaction with propene [63]. Reproduced with permission from Ref. [63], Copyright 2019 AIP Publishing.

This wavelength range could be extended to the mid-infrared to scan strong fundamental vibration transitions to increase the number of detectable species and in particular the ubiquitous complex hydrocarbons like polycyclic aromatic hydrocarbons (PAHs) for astrophysics studies.

The UF-CRDS approach has been demonstrated through the reaction kinetics of CN (v = 1) with propene at $70 \, K$ [63]. The high sensitivity of the CRDS technique is a strong advantage when aimed at detecting small product densities. However, covering a broad spectral range is challenging as each wavelength is probed sequentially and thus requires a prohibitive experimental time.

9.4.4   *Toward cavity-enhanced absorption spectroscopy in Laval nozzle flows*

Several technical outlooks are arising concerning spectroscopy in Laval flows. These involve using cavity enhancement in a different way than CRDS: when CRDS relies on shutting off the incoming light to measure ring-down events in a row, cavity-enhanced absorption spectroscopy (CEAS) directly measures the entire spectrum transmitted through the cavity in a parallel fashion.

9.4.4.1 *Broadband cavity-enhanced absorption spectroscopy*

The first implementation of this technique in a supersonic expansion is currently under development at the Institute of Physics of Rennes and it involves using a supercontinuum source as the probe and an FTIR to analyze the cavity transmission [66] (see Fig. 9.22). This combination is called broadband Fourier transform CEAS (BB-FT-CEAS). The scientific objective is to perform the spectroscopy of cold molecular ions and ionic van der Waals complexes of astrophysical interest.

The plasma source involves a discharge taking place in a gas sample of acetylene diluted in a buffer of argon or helium. The ion produced in the plasma then enters a planar Laval nozzle (10 cm long) and flows with the supersonic expansion afterward. In order to perform spectroscopy of the ions, the expansion is placed at the center of an enhancement cavity with a finesse of few thousands. The light source is a laser supercontinuum emitting from 400 to 2400 nm. The supercontinuum beam is fiber coupled to enter the supersonic chamber before propagating in open space to be mode matched to the cavity. The absorption spectrum of the sample is measured in transmission using a commercial high-resolution FTIR (model Bruker IFS 120HR). In order to not saturate the FTIR detector, a spectral filter is placed before the cavity which only transmits the spectral range of interest (620–1120 nm).

The main advantage of this configuration is that it does not require any active stabilization of the light source and cavity system:

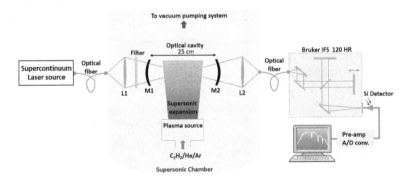

Fig. 9.22. Schematic of the implementation of BB-CEAS in a Laval flow. The supercontinuum source is coupled to the enhancement cavity centered on the flow and the transmitted light is analyzed using an FTIR. $L_{1,2}$, mode matching lenses; $M_{1,2}$, high-reflectivity mirrors of the cavity.

since the spectrum of the light source is a continuum, there is always part of the spectrum matching the frequencies of the cavity modes. Laser supercontinuum is also an attractive light source for FTIR as it solves most of the collimation and alignment issues encountered using incoherent lamps. One of the challenges associated lies in the rather large noise of the intensity spectrum of supercontinuum sources, which makes mandatory long averaging times. The cavity also spectrally filters the supercontinuum light (i.e. only the light matching the resonance modes of the cavity is transmitted) which will limit the final spectral resolution of the measurement since recording a spectrum with a resolution better than the cavity-free spectral range would mostly bring noise in the final spectrum.

9.4.4.2 *Optical frequency comb spectroscopy*

Optical frequency combs are the spectral signature of femtosecond mode locked lasers and provide a broad spectrum discretized at the harmonics of the laser repetition rate [67]. These features apply ideally in cold spectroscopy and in multispecies detection. Pioneering work has been performed by the Ye group at JILA [18], where a near-infrared frequency comb source was used to probe acetylene supersonically cooled using a Laval flow. The gas sample was probed after the nozzle and the absorption signal were enhanced using a Fabry-Perot cavity. This work focused on the lineshape of the acetylene absorption, which showed that despite the use of a Laval nozzle, the gas flow was far from being uniform in velocity. More recently, the technique was pushed to the mid-infrared using a buffer gas cooling cell [68] to perform cold spectroscopy of naphthalene, adamantane, and hexamethylenetetramine. The same approach was also used to perform for the first time the rotationally resolved spectroscopy of fullerene around $8.84\,\mu$m [69].

9.5 Non-LTE Spectroscopy in Laval Nozzles

9.5.1 *Rotationally cold hot bands*

Low-pressure jets are characterized by a strong non-LTE low-collisional regime. The molecular internal degrees of freedom are inherently decoupled, and several temperatures are necessary to describe both the velocity and the population distributions on the

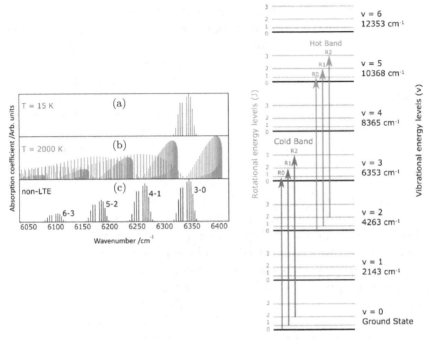

Fig. 9.23.   Carbon monoxide absorption spectrum simulated in LTE conditions at (a) 15 K and (b) 2000 K, and (c) in non-LTE conditions with $T_V = 2000$ K and $T_R = 15$ K. The right panel shows a schematic of the rotation-vibration transitions of CO (only the $3 \leftarrow 0$ cold band and $5 \leftarrow 2$ hot band are reproduced with permission from Ref. [39], Copyright 2020 AIP Publishing.

different rotational and vibrational energy levels of the molecule. The translational (T_T), rotational (T_R), and vibrational (T_V) temperatures can take very different values in low-pressure supersonic expansions. As we will see, such a strong non-LTE regime is of great interest to interpret the complex pattern of highly excited vibrational states characterizing polyatomic molecules.

As illustrated in Fig. 9.23, which represents the simple case of a diatomic molecule (carbon monoxide), high temperature is responsible for the presence of numerous hot bands — vibrational transitions between two excited vibrational states — which are superimposed on their associated cold band — the transition starting from the vibrational ground state (GS). Regarding carbon monoxide at 2000 K, this means that 21.4%, 4.6%, and 1% of the population are on the $v = 1, 2$, and 3 vibrational state, respectively. The absorption spectrum around 1.6 μm is therefore composed of the cold band $3 \leftarrow 0$

connecting the GS to $v = 3$, and the hot bands $4 \leftarrow 1$, $5 \leftarrow 2$, $6 \leftarrow 3$ involving the upper vibrational states $v = 4$, 5, and 6, respectively (note that $v = 6$ is located at the energy of $12353 \, \text{cm}^{-1}$, which would require a laser emitting at 800 nm to be probed from the GS). Each vibrational band is associated with its set of rotational transitions involving rotational quantum number $J > 50$ at 2000 K. As can be seen from panel (b) in Fig. 9.23, the resulting infrared spectrum is crowded by a number of overlapping rovibrational transitions. The situation is of course much more complex for molecules which have more vibrational modes than CO, hence the interpretation of their infrared spectrum becomes rapidly inextricable as temperature increases! One solution would be to cool the gas sample down to a very low temperature to limit the number of rovibrational transitions to low-J values. LTE spectroscopy at a low temperature would lead, however, to a complete depopulation of the excited vibrational states and a loss of information regarding the highly excited ones, as can be seen in panel (a) of Fig. 9.23. Decoupling the vibrational and rotational temperatures, so that $T_V \gg T_R \geq T_T$, provides a mean to observe rotationally cold hot bands. This in turns allows easy access to the highly excited vibrational states of rotationally cooled polyatomic molecules by infrared spectroscopy, as depicted in panel (c) of Fig. 9.23.

9.5.2 *Hypersonic jet generated by a high-temperature contoured Laval nozzle (SMAUG device)*

The device used in Rennes to produce non-LTE spectra was described in Section 9.3.3. It is called spectroscopy of molecules accelerated in uniform gas flow, abbreviated SMAUG [39], a nod to the dragon character imagined by J.R.R. Tolkien. The gas is preheated to a very high temperature ($>1000 \, \text{K}$) and then accelerated using a Laval graphite nozzle to hypersonic Mach numbers between 10 and 18.3 (see Fig. 9.14). The jet is probed transversely by CRDS (see Fig. 9.18). The jet is not uniform along the axis in the strict sense since the nozzle is not perfectly adapted, but what matters is that its radial uniformity is good, which is the case if we disregard boundary layers whose impact is low (see Figs. 9.15 and 9.24). The nozzle profile has been designed to meet the following two criteria: (1) a sufficiently low density to prevent the vibrational relaxation of the

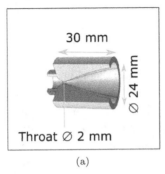

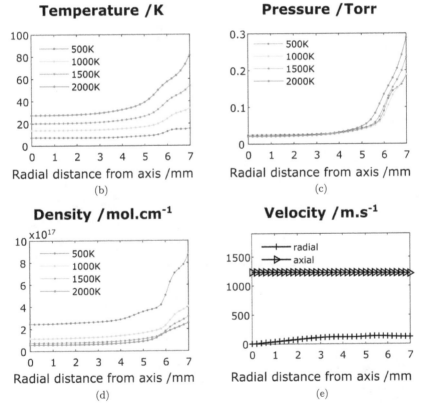

Fig. 9.24.   Radial profiles extracted from CFD simulations of the hypersonic flow generated by (a) a small-dimension Laval nozzle withstanding high temperatures. (b) temperature, (c) static pressure, (d) argon density, (e) radial and axial velocities. Radial profiles are taken 20 mm downstream from the nozzle exit. The stagnation temperature and pressure have been fixed to 1300 K and 1100 Torr, respectively.

molecules and (2) a reduced jet divergence (radial velocity) so as not to modify the absorption line profile by convective Doppler effect, as is the case when using axisymmetric free jets, which broaden the lines and deform their profile [17, 18].

The SMAUG setup was applied to the non-LTE spectroscopy of methane [39] (see also next section). An efficient rotational cooling of about 30 K was obtained with only little vibrational relaxation, revealing many hot bands, as can be seen in Fig. 9.25, which compares the non-LTE spectrum of methane to an LTE spectrum recorded at 964 K [70] and an LTE spectrum recorded at 81 K [71].

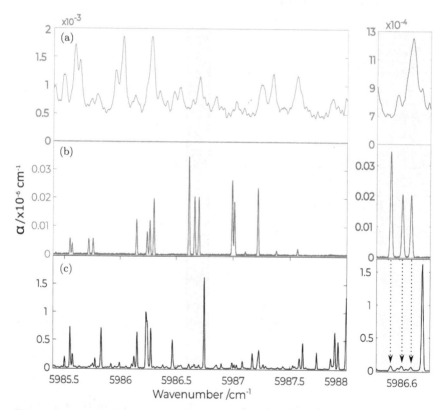

Fig. 9.25. Absorption spectra of methane [39]. (a) LTE spectrum at 964 K, (b) LTE spectrum at 81 K, and (c) hypersonic jet CRDS non-LTE spectrum recorded with SMAUG. The inserts on the right shows enlargements of the yellow areas in panels (a–c).

The hypersonic jet spectrum exhibits remarkably narrow absorption lines of methane with an FWHM of $\sim 0.01\,\mathrm{cm}^{-1}$, surprisingly above the expected value of $0.006\,\mathrm{cm}^{-1}$ estimated from the rotational temperature of $30\,\mathrm{K}$ obtained from the intensity of the lines. A radial velocity of $\sim 500\,\mathrm{m.s}^{-1}$ is necessary to explain such a line broadening, which is not compatible with the radial flow velocity simulated by CFD of $130\,\mathrm{m.s}^{-1}$ associated with our mini graphite Laval nozzle (see Fig. 9.24) [39]. The presence of vortices forming at the nozzle exit and developing downstream, giving rise to a turbulent shear layer [72], could lead to Doppler-broadened lines. Such a Doppler effect induced by turbulences is well known in astrophysics [73] and could explain our observations (see Chapter 1 where turbulence and eddies are discussed).

9.5.3 *Non-LTE spectroscopy of methane*

Methane is a spherical top molecule that has four vibrational modes, three of which are degenerated. Bending modes ν_2 and ν_4 have a frequency half that of the elongation modes ν_1 and ν_3, such as $\nu_1 \approx \nu_3 \approx 2\nu_2 \approx 2\nu_4$. This yields that the vibrational states of methane form groups of energy levels, called polyads (P_n), which are characterized by neighboring energies. If we exclude the fundamental vibrational state P_0, the polyad of lower energy is the dyad (P_1) composed of states ν_4 and ν_2, followed by the pentad (P_2) composed of the five states $2\nu_4$, $\nu_2 + \nu_4$, ν_1, ν_3 and $2\nu_2$, by the octad (P_3, 8 states), the tetradecad (P_4, 14 states), etc. The energy structure of the two first polyads is depicted in Fig. 9.27. The number of interactions between states therefore increases with the order of the polyad which makes the modeling of the infrared spectrum of the molecule complex, particularly at high temperature. Non-LTE approaches help in analyzing the complex spectroscopy of hot methane by isolating the contribution of the different hot bands of the spectrum. It makes it possible to access the information on the vibrational states of high-order polyads (icosad P_5, triacontad P_6), which are poorly known, although they contribute significantly to the high-temperature infrared spectrum of methane. Figure 9.26 shows a non-LTE CRDS spectrum of methane recorded in the $6000\,\mathrm{cm}^{-1}$ region in which one can observe cold band transitions ($P_4 \leftarrow P_0$) but also hot bands $P_5 \leftarrow P_1$, $P_6 \leftarrow P_2$, starting from the dyad and

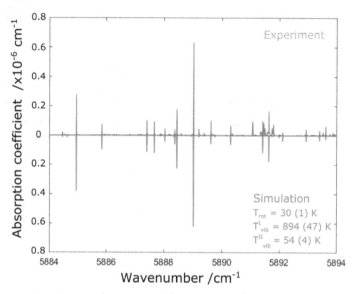

Fig. 9.26. Upper trace: Non-LTE CRD spectrum of methane. Lower trace: Simulation based on the TheoReTS spectroscopic database (unpublished results).

the pentad and ending, respectively, on the icosad (P_5) and the tricontad (P_6). The observed spectrum is compared to a simulation based on the TheoReTS *ab initio* spectroscopic database [74]. The missing lines in the simulation arise from an inaccurate modeling of the upper polyads P_5 and P_6, which could be improved on the basis of the non-LTE spectra.

As expected, the analysis of the recorded non-LTE spectrum revealed a decoupling between rotational and vibrational degrees of freedom. A rotational temperature $T_R \approx 30\,K$ was extracted from the intensity of lines belonging to a same vibrational band. More surprising was the evidence of the existence of two vibrational temperatures of very different values: $T_V^I = 894\,K$ and $T_V^{II} = 54\,K$. The higher vibrational temperature (T_V^I) is needed to describe the distribution of the population on the different polyads (see Fig. 9.27). In other words, the vibrational population of a given polyad does not relax efficiently on the lower polyads, leading to a high vibrational temperature of 894 K. On the contrary, the lower pseudo vibrational temperature (T_V^{II}) is associated with the distribution of vibrational states belonging to a given polyad. This pseudo temperature is calculated by considering the first vibrational state of each polyad as the

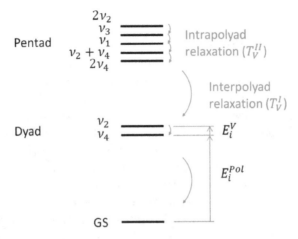

Fig. 9.27.   Diagram of the first two polyads of the methane molecule.

origin of the energies (see Eq. (9.8)). The relaxation between these relatively close states is more efficient, leading to a temperature of 54 K, closer to the rotational one.

Equation (9.5) is only valid for LTE conditions and needs to be adapted to simulate a non-LTE spectrum from a spectroscopic database. To do so, it is necessary to separate the population contributions relative to the decoupled temperatures T_R, T_V^I, and T_V^{II}. This leads to Eq. (9.8), where Q_0, Q_V, and Q_R are the total, the vibrational, and the rotational partition functions, respectively. E_i^{pol} is the energy of the lowest energy level of a given polyad. As is schematically shown on Fig. 9.27, the lowest energy level of the dyad is the ν_4 level, the lower levels of the other polyads being the ν_4 overtones of energy $n\nu_4$. In this model, E_i^V is measured from $n\nu_4$ rather than from the ground state. Including the rotational energy E_i^R, the total energy E_i of a rovibrational state is then given by $E_i = E_i^{pol} + E_i^V + E_i^R$.

$$\bar{\sigma}_{ij}(T_R, T_V^I, T_V^{II}) = \bar{\sigma}_{ij}(T_0)\frac{Q_0(T_0)}{Q_R(T_R) \times Q_V}$$

$$\times \exp\left[\frac{E_i}{kT_0} - \frac{1}{k}\left(\frac{E_i^{pol}}{T_V^I} + \frac{E_i^V}{T_V^{II}} + \frac{E_i^R}{T_R}\right)\right] \quad (9.8)$$

Well suited to rarefied flows, the CRDS technique shows a high potential for diagnosing non-LTE regimes in hypersonic flows.

9.6   Post-Shock CRDS

Infrared spectroscopy is an efficient and non-intrusive way to remotely probe extremely hot gaseous media, including combustion gases, flames [75], shock tubes [76], or exotic atmospheres of hot planets [77]. The extraction of information from infrared spectra regarding the temperature, pressure, or density of these media encounters a major problem of interpretation. Their complexity increases with temperature because the thermal population of the high-energy rotation-vibration states of the molecules increases, which leads to a drastic increase of the density of absorption or emission infrared lines. New theoretical approaches have emerged to produce databases [74, 78–81] compiling millions of rotation-vibration transitions, with the calculation of synthetic spectra of various molecules at high temperatures (up to 3000 K typically). Laboratory infrared spectra, however, remain essential because they constitute the basic material from which models are built and/or (in)validated.

The experimental devices dedicated to the production of infrared spectra at high temperature are subject to severe constraints. The optical elements that make up the optical cells and spectrometers (mirrors, optical windows, detectors, etc.) must be protected or cooled to avoid rapid damage. The thermal expansion of materials must be considered to limit mechanical stresses and avoid possible leaks. The withstanding of materials to temperature is of course a crucial parameter, so heat-resistant materials such as alumina or tantalum must be used. It is also necessary to make sure that the probed gas flow does not undergo a strong thermal gradient which would complexify the interpretation of the collected data. Finally, it is necessary to eliminate or limit the gradual dissociation of molecules under the effect of temperature. The "post-shock CRDS" technique that has been developed very recently in Rennes allows one to overcome many of these inconveniences. Its principle consists of slowing down the gas expelled by the hypersonic nozzle so that its kinetic energy is converted back into internal energy and its temperature returns to a value close to its stagnation temperature. In practice, it suffices to insert a plane obstacle perpendicular to the flow. A detached shock wave forms in front of the obstacle through which the flow becomes suddenly subsonic. The experimental setup is described in Fig. 9.28. The shock layer located between the shock

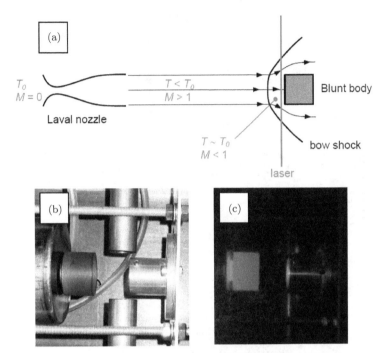

Fig. 9.28. (a) Schematic of the shock layer and the probe laser for absorption spectroscopy. (b) Photograph of the experimental device. The Laval graphite nozzle is on the left and the obstacle impacted by the flow is on the right. The tubes at the top and bottom are fed by a continuous flow of N_2 to eliminate absorption due to static gas from the expansion chamber. (c) Image taken in working condition. The nozzle temperature exceeds 1000 K. The glow at the level of the metallic obstacle is a reflection of the radiation emitted by the nozzle.

wave and the material wall is characterized by a low speed, a high temperature close to the stagnation temperature T_0 (which is strictly equal to T_0 on the plate obstacle where the gas speed is zero), and a much higher pressure and density than upstream of the shock wave, the values of which can be easily calculated from Rankine-Hugoniot formulas relating to the normal shock wave [27]. The molecules, initially in a state of strong non-LTE upstream of the shock, regain a state of thermodynamic equilibrium downstream of the shock under the effect of the multiple collisions that they undergo.

The shock layer is probed by CRDS. In Fig. 9.29, the post-shock spectrum of the methane molecule (estimated temperature around 1400 K) is compared to a non-LTE spectrum obtained by probing

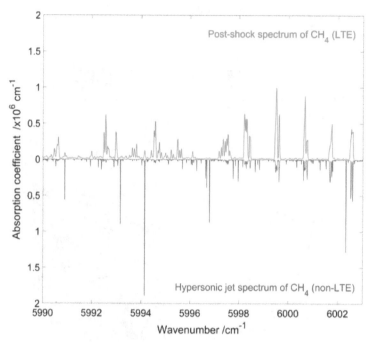

Fig. 9.29. Post-shock CRD spectrum of LTE methane obtained by probing the subsonic gas located downstream from a normal shock wave. This spectrum is compared to the non-LTE jet CRD spectrum obtained by probing the hypersonic gas upstream from the normal shock (unpublished results).

the jet downstream of the shock wave. Because it probes a flowing gas, this approach avoids the delicate heating of a confined optical cell. The increase in density caused by the shock is an advantage for absorption spectroscopy. By nature, the flow provides continuous sample turnover and maintains a constant column density of absorbent molecules in the infrared laser beam. The actual length of gas probed is approximately 2 cm, which leads to a laser cavity filling rate of 2/80 and therefore to an effective absorption path length of approximately 1.5 km for a decay time of 200 μs.

9.7 Conclusion

Laval nozzles generate high-speed flows of cold molecules which are of the greatest interest for physical chemistry applications. They allow one to probe molecules at temperatures relevant for astrophysical and fundamental applications. Infrared absorption spectroscopy gives both a powerful diagnostic tool to characterize the flow conditions and a mean to detect with a high resolution the energy levels of molecular structure growing in complexity. This chapter has focused mostly on this last aspect.

The Laval flows are characterized by relatively high densities sustained on a large length, which are particularly adapted to the study of condensation processes and properties of molecular complexes but also for reaction kinetics recently performed in the infrared range. This rather high density favors spectroscopy, while the boundary layers of the expansion contribute to the absorption spectrum only marginally (and for high-J transitions). These properties allow one to perform spectroscopy without requiring inversion techniques which are both complex and costly in acquisition time.

Among the spectroscopic techniques which have been used to probe the Laval flows, FTIR and TDLAS are the most conventional, while cavity-enhanced approaches such as the CRDS, CEAS, and OFCS are recent additions. These latter techniques offer high resolution and high sensitivity, at the level of several hundreds of meters to few kilometers in the case of CRDS. They open up new perspectives for cold and non-LTE molecular physics which are of interest for fundamental spectroscopy and astrochemistry but also for the aerospace and hypersonic research fields.

Acknowledgments

We warmly thank Dr. Samir Kassi (LIPhy Grenoble) for having designed and installed the CRDS spectrometer in Rennes, and for the wise counsel he regularly provides us. We are very grateful to Prof. Vinayak Kulkarni (IIT Guwahati) for his invaluable help during the design of the hypersonic nozzle which equips the SMAUG device. Thank you very much to our colleague Dr. Abdessamad Benidar for his expertise in computational fluid dynamics and for having carried out a large number of calculations for our various studies. Robert Georges thanks Prof. Michel Herman (ULB Brussels) and Dr. Andrei Vigasin (Obukhov Institute of Atmospheric Physics, Moscow) for their constant and friendly support.

References

[1] Chazot O, Panerai F. High-enthalpy facilities and plasma wind tunnels for aerothermodynamics ground testing. In: Hypersonic nonequilibrium flows: fundamentals and recent advances. AIAA, Inc., Reston; 2015. pp. 329–42.

[2] Danehy PM, Weisberger J, Johansen C, Reese D, Fahringer T, Parziale NJ, Dedic C, Estevadeordal J, Cruden BA. Non-intrusive measurement techniques for flow characterization of hypersonic wind tunnels. In: Lectures series STO-AVT-325 flow characterization and modeling of hypersonic wind tunnels. Sint-Genesius-Rode, Belgium; 2018 Nov.

[3] Smalley RE, Wharton L, Levy DH. Molecular optical spectroscopy with supersonic beams and jets. Acc Chem Res. 1977;10(4):139–45.

[4] Levy DH. The spectroscopy of very cold gases. Science. 1981; 214(4518):263–9.

[5] Herman M, Földes T, Didriche K, Lauzin C, Vanfleteren T. Overtone spectroscopy of molecular complexes containing small polyatomic molecules. Int Rev Phys Chem. 2016;35(2):243–95.

[6] Huneycutt AJ, Casaes RN, McCall BJ, Chung C-Y, Lee Y-P, Saykally RJ. Infrared cavity ringdown spectroscopy of jet-cooled polycyclic aromatic hydrocarbons. ChemPhysChem. 2004;5(3):321–6.

[7] Brumfield BE, Stewart JT, McCall BJ. Extending the limits of rotationally resolved absorption spectroscopy: pyrene. J Phys Chem Lett. 2012;3(15):1985–8.

[8] Pirali O, Goubet M, Huet TR, Georges R, Soulard P, Asselin P, Courbe J, Roy P, Vervloet M. The far infrared spectrum of naphthalene characterized by high resolution synchrotron FTIR spectroscopy and anharmonic DFT calculations. Phys Chem Chem Phys. 2013;15(25):10141.

[9] Martin-Drumel M-A, Porterfield JP, Goubet M, Asselin P, Georges R, Soulard P, Nava M, Bryan Changala P, Billinghurst B, Pirali O, McCarthy MC, Baraban JH. Synchrotron-based high resolution far-infrared spectroscopy of *trans*-butadiene. J Phys Chem A. 2020; 124(12):2427–35.

[10] Pirali O, Vervloet M, Mulas G, Malloci G, Joblin C. High-resolution infrared absorption spectroscopy of thermally excited naphthalene. Measurements and calculations of anharmonic parameters and vibrational interactions. Phys Chem Chem Phys. 2009;11(18):3443.

[11] Gordon IE, Rothman LS, Hill C, Kochanov RV, Tan Y, Bernath PF, Birk M, Boudon V, Campargue A, Chance KV, Drouin BJ, Flaud J-M, Gamache RR, Hodges JT, Jacquemart D, Perevalov VI, Perrin A,

Shine KP, Smith M-H, Tennyson J, Toon GC, Tran H, Tyuterev VG, Barbe A, Császár AG, Devi VM, Furtenbacher T, Harrison JJ, Hartmann J-M, Jolly A, Johnson TJ, Karman T, Kleiner I, Kyuberis AA, Loos J, Lyulin OM, Massie ST, Mikhailenko SN, Moazzen-Ahmadi N, Müller HSP, Naumenko OV, Nikitin AV, Polyansky OL, Rey M, Rotger M, Sharpe SW, Sungh K, Starikova E, Tashkun SA, Vander Auwera J, Wagner G, Wilzewski J, Wcisło P, Yu S, Zak EJ. The HITRAN2016 molecular spectroscopic database. J Quant Spectrosc Rad Trans. 2017;203:3–69.

[12] Willey DR, Crownover RL, Bittner DN, De Lucia FC. Very low temperature spectroscopy: the pressure broadening coefficients for CO–He between 4.3 and 1.7 K. J Chem Phys. 1988;89(4):1923–8.

[13] Suas-David N, Vanfleteren T, Földes T, Kassi S, Georges R, Herman M. The water dimer investigated in the 2OH spectral range using cavity ring-down spectroscopy. J Phys Chem A. 2015;119(39): 10022–34.

[14] Moudens A, Georges R, Goubet M, Makarewicz J, Lokshtanov SE, Vigasin AA. Direct absorption spectroscopy of water clusters formed in a continuous slit nozzle expansion. J Chem Phys. 2009;131(20):204312.

[15] Louviot M, Suas-David N, Boudon V, Georges R, Rey M, Kassi S. Strong thermal nonequilibrium in hypersonic CO and CH_4 probed by CRDS. J Chem Phys. 2015;142(21):214305.

[16] Griffiths PR, De Haseth JA. Fourier transform infrared spectrometry. John Wiley & Sons, Hoboken; 2007.

[17] Suas-David N, Kulkarni V, Benidar A, Kassi S, Georges R. Line shape in a free-jet hypersonic expansion investigated by cavity ring-down spectroscopy and computational fluid dynamics. Chem Phys Lett. 2016;659:209–15.

[18] Thorpe MJ, Adler F, Cossel KC, de Miranda MHG, Ye J. Tomography of a supersonically cooled molecular jet using cavity-enhanced direct frequency comb spectroscopy. Chem Phys Lett. 2009;468(1):1–8.

[19] Toennies JP, Winkelmann K. Theoretical studies of highly expanded free jets: Influence of quantum effects and a realistic intermolecular potential. J Chem Phys. 1977;66(9):3965–79.

[20] Campargue R. Progress in overexpanded supersonic jets and skimmed molecular beams in free-jet zones of silence. J Phys Chem. 1984;88(20):4466–74.

[21] Franquet E, Perrier V, Gibout S, Bruel P. Free underexpanded jets in a quiescent medium: a review. Prog Aerosp Sci. 2015;77:25–53.

[22] Dupeyrat G. Two-and three-dimensional aspects of a freejet issuing from a long rectangular slit. Rarefied gas Dyn. 1981;74:812–9. https://arc.aiaa.org/doi/book/10.2514/4.865480

[23] Amjrav A, Even U, Jortner J. Absorption spectroscopy of ultracold large molecules in planar supersonic expansions. Chem Phys Lett. 1981;83(1):1–4.

[24] Putignano M, Welsch CP. Numerical study on the generation of a planar supersonic gas-jet. Nucl Instrum Methods Phys Res A. 2012;667:44–52.

[25] Liger-Belair G, Cordier D, Georges R. Under-expanded supersonic CO_2 freezing jets during champagne cork popping. Sci Adv. 2019;5(9):eaav5528.

[26] Gekle S, Peters IR, Gordillo JM, van der Meer D, Lohse D. Supersonic air flow due to solid-liquid impact. Phys Rev Lett. 2010;104(2):024501.

[27] Rathakrishnan E. Applied gas dynamics. John Wiley & Sons, Chichester; 2019. https://onlinelibrary.wiley.com/doi/pdf/10.1002/978111 9500377.fmatter

[28] Menon N, Skews BW. Shock wave configurations and flow structures in non-axisymmetric underexpanded sonic jets. Shock Waves. 2010;20(3):175–90.

[29] Hartigan P, Foster J, Frank A, Hansen E, Yirak K, Liao AS, Graham P, Wilde B, Blue B, Martinez D, Rosen P, Farley D, Paguio R. When shock waves collide. ApJ. 2016;823(2):148.

[30] Velikorodny A, Kudriakov S. Numerical study of the near-field of highly underexpanded turbulent gas jets. Int J Hydrog Energy. 2012;37(22):17390–99.

[31] Scoles G. Atomic and molecular beam methods. Oxford University Press, New York-Oxford; 1988. vol 1.

[32] Wu BJC, Laguna GA. Gasdynamic and infrared spectroscopic measurements in a condensing flow of a sulfur hexafluoride–argon mixture. J Chem Phys. 1979;71(7):2991.

[33] Tanimura S, Okada Y, Kuga Y, Takeuchi K. Determination of the rotational temperature of UF_6 and pressure in a supersonic expansion by use of the Q branch under the influence of pressure broadening. Chem Phys Lett. 1995;232(1–2):176–80.

[34] Tanimura S, Okada Y, Takeuchi K. FTIR spectroscopy of UF_6 clustering in a supersonic Laval nozzle. J Phys Chem. 1996;100(8):2842–48.

[35] Tanimura S, Zvinevich Y, Wyslouzil BE, Zahniser M, Shorter J, Nelson D, McManus B. Temperature and gas-phase composition measurements in supersonic flows using tunable diode laser absorption spectroscopy: the effect of condensation on the boundary-layer thickness. J Chem Phys. 2005;122(19):194304.

[36] Benidar A, Georges R, Le Doucen R, Boissoles J, Hamon S, Canosa A, Rowe BR. Uniform supersonic expansion for FTIR absorption spectroscopy: the ν_5 band of $(NO)_2$ at 26 K. J Mol Spectrosc. 2000;199(1):92–9.

[37] Bonnamy A, Georges R, Benidar A, Boissoles J, Canosa A, Rowe BR. Infrared spectroscopy of $(CO_2)_N$ nanoparticles ($30 < N < 14500$) flowing in a uniform supersonic expansion. J Chem Phys. 2003;118(8):3612–21.

[38] Lippe M, Szczepaniak U, Hou G-L, Chakrabarty S, Ferreiro JJ, Chasovskikh E, Signorell R. Infrared spectroscopy and mass spectrometry of CO_2 clusters during nucleation and growth. J Phys Chem A. 2019;123(12):2426–37.

[39] Dudás E, Suas-David N, Brahmachary S, Kulkarni V, Benidar A, Kassi S, Charles C, Georges R. High-temperature hypersonic Laval nozzle for non-LTE cavity ringdown spectroscopy. J Chem Phys. 2020;152(13):134201.

[40] Dupeyrat G, Marquette JB, Rowe BR. Design and testing of axisymmetric nozzles for ion-molecule reaction studies between $20\,K$ and $160\,K$. Phys Fluids. 1985;28(5):1273–9.

[41] Ferrer R, Barzakh A, Bastin B, Beerwerth R, Block M, Creemers P, Grawe H, de Groote R, Delahaye P, Fléchard X, Franchoo S, Fritzsche S, Gaffney LP, Ghys L, Gins W, Granados C, Heinke R, Hijazi L, Huyse M, Kron T, Kudryavtsev Yu, Laatiaoui M, Lecesne N, Loiselet M, Lutton F, Moore ID, Martínez Y, Mogilevskiy E, Naubereit P, Piot J, Raeder S, Rothe S, Savajols H, Sels S, Sonnenschein V, Thomas J-C, Traykov E, Van Beveren C, Van den Bergh P, Van Duppen P, Wendt K, Zadvornaya A. Towards high-resolution laser ionization spectroscopy of the heaviest elements in supersonic gas jet expansion. Nature Commun. 2017;8:14520.

[42] Addy AL. Effects of axisymmetric sonic nozzle geometry on Mach disk characteristics. AIAA J. 1981;19(1):121–2.

[43] Driftmyer RT. A correlation of freejet data. AIAA J. 1972;10(8):1093–5.

[44] Georges R, Michaut X, Moudens A, Goubet M, Pirali O, Soulard P, Asselin P, Huet TR, Roy P, Fournier M, Vigasin A. Nuclear spin symmetry conservation in $^1H_2^{16}O$ investigated by direct absorption FTIR spectroscopy of water vapor cooled down in supersonic expansion. J Phys Chem A. 2017;121(40):7455–68.

[45] Laksmono H, Tanimura S, Allen HC, Wilemski G, Zahniser MS, Shorter JH, Nelson DD, McManus JB, Wyslouzil BE. Monomer, clusters, liquid: an integrated spectroscopic study of methanol condensation. Phys Chem Chem Phys. 2011;13(13):5855.

[46] Creasey DJ, Halford-Maw PA, Heard DE, Pilling MJ, Whitaker BJ. Implementation and initial deployment of a field instrument for measurement of OH and HO_2 in the troposphere by laser-induced fluorescence. J Chem Soc Faraday Trans. 1997;93(16):2907–13.

[47] Tejeda G, Maté B, Fernández-Sánchez JM, Montero S. Temperature and density mapping of supersonic jet expansions using linear raman spectroscopy. Phys Rev Lett. 1996;76(1):34–37.

[48] Cai W, Kaminski CF. Tomographic absorption spectroscopy for the study of gas dynamics and reactive flows. Prog Energy Combust Sci. 2017;59:1–31.

[49] Hickstein DD, Gibson ST, Yurchak R, Das DD, Ryazanov M. A direct comparison of high-speed methods for the numerical Abel transform. Rev Sci Instrum. 2019;90(6):065115.

[50] Mohamed A, Rosier B, Sagnier P, Henry D, Louvet Y, Bize D. Application of infrared diode laser absorptio n spectroscopy to the F4 high enthalpy wind tunnel. Aerosp Sci Technol. 1998;2(4):241–50.

[51] Wegener PP, Mack LM. Condensation in supersonic and hypersonic wind tunnels. Adv Appl Mech. 1958;5:307–447.

[52] Bonnamy A, Georges R, Hugo E, Signorell R. IR signature of $(CO_2)_N$ clusters: size, shape and structural effects. Phys Chem Chem Phys. 2005;7(5):963.

[53] Baldacchini G, Chakraborti PK, D'Amato F. Infrared diode laser absorption features of N_2O and CO_2 in a laval nozzle. Appl Phys B. 1992;55(1):92–101.

[54] Fisher SS. SF_6 condensation in supersonic nozzle expansions. Phys Fluids. 1979;22(7):1261.

[55] Okada Y, Tanimura S, Okamura H, Suda A, Tashiro H, Takeuchi K. Vibrational spectroscopy and predissociation of UF_6 clusters in a supersonic Laval nozzle. J Mol Struct. 1997;410–411:299–304.

[56] Baldacchini G, Chakraborti PK, D'Amato F. Lineshape in a laval molecular beam and pressure broadening of N_2O transition lines. J Quant Spectrosc Rad Transfer. 1993;49(4):439–47.

[57] Makowiecki AS, Hayden TR, Nakles MR, Pilgram NH, MacDonald NA, Hargus WA, Rieker GB. Wavelength modulation spectroscopy for measurements of temperature and species concentration downstream from a supersonic nozzle. In: Proceeding 53rd AIAA/SAE/ASEE Joint Propulsion Conference; Atlanta, USA; 2017 Jul.

[58] Bonnamy A, Georges R, Benidar A, Boissoles J, Canosa A, Rowe BR. Infrared spectroscopy of $(CO_2)_N$ nanoparticles (30 < N < 14500) flowing in a uniform supersonic expansion. J Chem Phys. 2003;118(8):3612–21.

[59] Bonnamy A. Spectroscopie infrarouge de molécules en écoulements supersoniques. Applications aux nanoparticules de CO_2 et aux molécules organiques volatiles [PhD Thesis]. Université de Rennes 1; 2002.

[60] Tuzson B, Mangold M, Looser H, Manninen A, Emmenegger L. Compact multipass optical cell for laser spectroscopy. Opt Lett. 2013;38(3):257–9.

[61] Kassi S, Campargue A. Cavity ring down spectroscopy with $5 \times 10^{-13}\,\text{cm}^{-1}$ sensitivity. J Chem Phys. 2012;137(23):234201.

[62] Cooke IR, Sims IR. Experimental studies of gas-phase reactivity in relation to complex organic molecules in star-forming regions. ACS Earth Space Chem. 2019;3(7):1109–34.

[63] Suas-David N, Thawoos S, Suits AG. A uniform flow–cavity ring-down spectrometer (UF-CRDS): a new setup for spectroscopy and kinetics at low temperature. J Chem Phys. 2019;151(24):244202.

[64] Abeysekera C, Joalland B, Ariyasingha N, Zack LN, Sims IR, Field RW, Suits AG. Product branching in the low temperature reaction of CN with propyne by chirped-pulse microwave spectroscopy in a uniform supersonic flow. J Phys Chem Lett. 2015;6(9):1599–604.

[65] Brown SS, Ravishankara AR, Stark H. Simultaneous kinetics and ring-down: rate coefficients from single cavity loss temporal profiles. J Phys Chem A. 2000;104(30):7044–52.

[66] Bejjani R, Benidar A, Georges R. Fourier transform incoherent broadband cavity enhanced absorption spectroscopy developed for the study of cold astrophysical anions in a planar Laval nozzle expansion. In: Proceeding 2020 International Symposium on Molecular Spectroscopy; Urbana Champaign, USA; 2020 Jun.

[67] Adler F, Thorpe MJ, Cossel KC, Ye J. Cavity-enhanced direct frequency comb spectroscopy: technology and applications. Annu Rev Anal Chem. 2010;3(1):175–205.

[68] Spaun B, Bryan Changala P, Patterson D, Bjork BJ, Heckl OH, Doyle JM, Ye J. Continuous probe of cold complex molecules with infrared frequency comb spectroscopy. Nature. 2016;533(7604):517–20.

[69] Bryan Changala P, Weichman ML, Lee KF, Fermann ME, Ye J. Rovibrational quantum state resolution of the C_{60} fullerene. Science. 2019;363(6422):49–54.

[70] Ghysels M, Vasilchenko S, Mondelain D, Béguier S, Kassi S, Campargue A. Laser absorption spectroscopy of methane at $1000\,\text{K}$ near $1.7\,\mu\text{m}$: a validation test of the spectroscopic databases. J Quant Spectrosc Rad Transfer. 2018;215:59–70.

[71] Kassi S, Gao B, Romanini D, Campargue A. The near-infrared (1.30–$1.70\,\mu\text{m}$) absorption spectrum of methane down to $77\,\text{K}$. Phys Chem Chem Phys. 2008;10(30):4410.

[72] Ball CG, Fellouah H, Pollard A. The flow field in turbulent round free jets. Prog Aerosp Sci. 2012;50:1–26.

[73] Lazarian A, Pogosyan D. Studying turbulence from doppler-broadened absorption lines: statistics of logarithms of intensity. ApJ. 2008;686(1):350–62.

[74] Rey M, Nikitin AV, Babikov YL, Tyuterev VG. TheoReTS — an information system for theoretical spectra based on variational predictions from molecular potential energy and dipole moment surfaces. J Mol Spectrosc. 2016;327:138–58.

[75] Rutkowski L, Foltynowicz A, Schmidt FM, Johansson AC, Khodabakhsh A, Kyuberis AA, Zobov NF, Polyansky OL, Yurchenko SN, Tennyson J. An experimental water line list at 1950 K in the 6250–6670 cm^{-1} region. J Quant Spectrosc Rad Trans. 2018;205:213–9.

[76] Strand CL, Ding Y, Johnson SE, Hanson RK. Measurement of the mid-infrared absorption spectra of ethylene (C_2H_4) and other molecules at high temperatures and pressures. J Quant Spectrosc Rad Transfer. 2019;222–223:122–9.

[77] Tinetti G, Encrenaz T, Coustenis A. Spectroscopy of planetary atmospheres in our Galaxy. Astron Astrophys Rev. 2013;21(1):63.

[78] Rothman LS, Gordon IE, Barber RJ, Dothe H, Gamache RR, Goldman A, Perevalov VI, Tashkun SA, Tennyson J. HITEMP, the high-temperature molecular spectroscopic database. J Quant Spectrosc Rad Transfer. 2010;111(15):2139–50.

[79] Tashkun SA, Perevalov VI. CDSD-4000: high-resolution, high-temperature carbon dioxide spectroscopic databank. J Quant Spectrosc Rad Transfer. 2011;112(9):1403–10.

[80] Ba YA, Wenger C, Surleau R, Boudon V, Rotger M, Daumont L, Bonhommeau DA, Tyuterev VG, Dubernet M-L. MeCaSDa and ECaSDa: methane and ethene calculated spectroscopic databases for the virtual atomic and molecular data centre. J Quant Spectrosc Rad Transfer. 2013;130:62–8.

[81] Tennyson J, Yurchenko SN, Al-Refaie AF, Barton EJ, Chubb KL, Coles PA, Diamantopoulou S, Gorman MN, Hill C, Lam AZ, Lodi L, McKemmish LK, Na Y, Owens A, Polyansky OL, Rivlin T, Sousa-Silva C, Underwood DS, Yachmenev A, Zak E. The ExoMol database: molecular line lists for exoplanet and other hot atmospheres. J Mol Spectrosc. 2016;327:73–94.

Chapter 10

Cold Chemistry and Beyond: The Astrochemical Context

Eric Herbst

Departments of Chemistry and Astronomy,
University of Virginia,
22904 Charlottesville, VA, USA
eh2ef@virginia.edu

Abstract

Our knowledge of the gas phase chemistry of the interstellar medium in our galaxy and many others owes a major debt to the many laboratory and theoretical scientists who studied critical classes of ion–neutral and neutral–neutral reactions to improve our understanding of interstellar chemistry. Of the different types of apparatuses utilized, the CRESU (*in English: reaction kinetics in uniform supersonic flow*) has played the dominant role, especially but not solely at temperatures near 10 K, a temperature that pertains to the coldest regions of interstellar clouds. In these cold sources, many of the gas phase molecules are very unsaturated organic species, known to astronomers as "carbon chains". In addition to these ultracold regions, interstellar clouds are also the birthplaces of new stars and planets, and the chemistry in these regions, which is very different from that of colder regions, leads to the synthesis of terrestrial-like organic molecules that may predate the formation of life. In this chapter, we interweave the story of CRESU experiments with that of gas phase astrochemistry, involving both the historical context and our current state of knowledge. We end with a hint of what the future might bring.

Keywords: Interstellar clouds; Star formation; Molecular concentrations in space; Dust particles; Ion-neutral reactions; Neutral-neutral reactions.

10.1   Introduction

Astrochemistry is currently a major research topic, encompassing the significant role that the study of molecules throughout the universe plays in increasing our understanding of both chemistry and astronomy. The subject contains a number of specific fields including the observation of molecules in space, mainly via the use of radio and infrared telescopes, both laboratory and theoretical studies of chemical reactions and molecular spectra, and the construction of large chemical simulations with thousands of chemical reactions to calculate molecular concentrations ("abundances") [1]. These calculated concentrations can be compared with molecular spectroscopic observations with the goals of understanding the physical conditions of the sources where the molecules in space are found and the lifetimes of these sources, which range from thousands to billions of years [2]. In addition to their astronomical role as probes of sources, molecules are of interest in themselves since they are often exotic by terrestrial standards, and are often synthesized by chemical processes not previously studied in the terrestrial laboratory [2]. Thus, the study of astrochemistry is of assistance both to astronomers and to chemists.

Although molecules are found in both stellar and non-stellar sources in our galaxy and others, most attention has been paid to so-called interstellar clouds, which are large objects comprising both gas phase species and tiny dust particles, with molecules detected in both of these phases. Interstellar clouds in our galaxy are part of a rather complex region known as the interstellar medium, or ISM for short [1]. These interstellar clouds come in a variety of densities, with the denser regions richer in their chemistry. These regions have densities in the vicinity of 10^4 cm^{-3}, mostly in the form of hydrogen (see the discussion below). The dense clouds are also the only known sources of new generations of stars and planets, which form in a complex evolutionary process, in which portions of the clouds, often at temperatures as low as 10 K and known as cold cores, collapse, first isothermally and subsequently adiabatically, leading to the formation of denser regions of infalling material known as "hot corinos" or "hot cores". Both of these objects collapse onto stars in the process of formation, known as "protostars". These regions differ in their mass and brightness as well as the final star or stars produced: low-mass stars such as our sun derive from hot corinos and high-mass stars from hot cores. In the remainder of this chapter, we mostly use the term "hot core" for both.

Temperatures of hot cores exceed 100 K, above which temperature mantles of molecular ices, surrounding dust particles and composed mainly of water ice, desorb into the gas phase. While the collapse and desorption are occurring, the infalling material can lead to the production of disk-shaped objects, especially for low-mass sources. As the protostar becomes a full-fledged star, the disk surrounding it becomes a protoplanetary disk, in which the dust particles coalesce to eventually form comets, meteors, asteroids, planetesimals, and eventually planets. Each evolutionary stage has its own chemistry [3].

The overall process of star formation, from cold cores in dense clouds to new planets and stars, is cyclical in the sense that the new solar-type systems age and the central star itself goes through evolutionary stages over billions of years, after which a thick atmosphere of gas and dust is blown out into space, forming a rather wispy diffuse cloud, which under the influence of gravity eventually forms a dense cloud, and the process of star formation then begins anew. High-mass stars are formed from more massive cold regions known as "infrared dark" clouds [4]. Unlike cold cores, which block out visible and ultraviolet radiation but can be penetrated by longer wavelength radiation, infrared dark clouds are mainly, as their name suggests, dark in the infrared. After these regions pass through a hot core phase, further evolutionary stages lead to objects of ionized matter known as HII regions, which surround the hottest of stars, which can end their lives in supernova bursts [5].

10.2 Molecules in Space

Over the last 50 years, more than 200 different molecules, excluding isotopomers, have been discovered in space, mostly in the gas phase. A full list can be found in Woon [6] along with the date and place of first observation. These molecules can be divided into a number of types, which have a rather complex relationship with the evolutionary stages of star formation. The names of the types, with selected examples, are shown in Table 10.1. Figure 10.1 shows a view of the

Table 10.1.   Types of interstellar molecules with examples.

Simple: H_2, CO, OH, CN, H_2O, H_3^+, H_2D^+, HCN, HNC, NH_3
Carbon-Chains: CCCS, HNCCC, C_6H, C_6H^-, HC_6CN
COMs: CH_3OH, CH_3OCH_3, $HCOOCH_3$, C_2H_5CN, C_3H_7CN, CH_3COCH_3
Fullerenes: C_{60}, C_{60}^+, C_{70}
Aromatics & Rings: C_6H_6, C_6H_5CN, $C_{10}H_7CN$, c-C_5H_5CN
Inorganic Species: PN, MgNC, HCP, NaCN, TiO_2
Ice Mantles: H_2O, CO_2, CO, CH_3OH, CH_4, NH_3, OCN^-
Dust Cores: silicates, amorphous carbon

Molecules and Low-Mass Star Formation

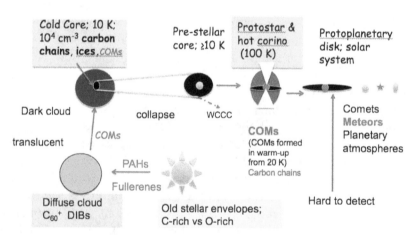

Fig. 10.1.   The evolutionary stages of low-mass stellar and planetary formation along with the classes of molecules detected in each stage. Molecular classes in boldface are dominant. Terms not yet defined will be discussed throughout the chapter.

various evolutionary stages starting from old low-mass stars with expanding atmospheres to new stars and planets with the types of molecules detected in each stage.

Simple molecules tend to be rather ubiquitous, small, and abundant. Molecular hydrogen is the most abundant molecule in the universe, which is not surprising since hydrogen is the most abundant element. It is formed on the surfaces of dust grains [7]. Carbon monoxide and water are tied for second at a fractional abundance of 10^{-4} with respect to hydrogen in dense clouds if one counts the ice phase as well. The H_3^+ ion is listed since it is a very abundant ion with a fractional abundance of up to 10^{-7} in dense clouds and its presence indicates that interstellar clouds are in reality low-density plasmas, where the state of ionization is non-thermal. The problem of thermodynamic state with regard to kinetics is discussed in detail in Chapter 1.

More generally, in interstellar clouds, a uniform temperature often does not exist except in the highest density regions, and each individual mode (e.g. fine structure, rotation, vibration, translation) possesses a unique temperature or non-thermal distribution, with translation the highest. Astronomers often refer to the temperature of the translational mode as "kinetic temperature" or simply "temperature". Where a uniform temperature exists over at least a small spatial range, astronomers use the term "local thermodynamic equilibrium" or LTE. This term is also defined kinetically to refer to the situation when the radiant energy absorbed by a molecule is distributed across other molecules by inelastic collisions before it is reradiated by emission. The density at which this occurs for rotational levels is known as the critical density (see Eq. (10.32)). In the interstellar medium, rates of chemical reactions can depend strongly upon the departure from LTE, although this departure is rarely treated in chemical simulations unless there is a strong and obvious effect, such as the dependence of the rate of the $N^+ + H_2$ reaction on the ortho and para nuclear spin-rotation states of H_2 [7a] and the effect of ortho/para states on the rate of the reaction between H_2D^+ and H_2, which helps determine the deuterium fractionation between H_3^+ and H_2D^+ [7b].

The isotopomer H_2D^+ is listed because, under certain low-temperature conditions, its abundance almost reaches the abundance of H_3^+, despite the fact that the elemental D/H ratio is 10^{-5}, indicating a huge low-temperature isotope effect [8]. The radicals OH and CN make the list because of their importance as reactants in neutral–neutral reactions, as discussed below. Finally, HNC is listed because its low-temperature abundance is equal to that of HCN, despite its high energy, indicating that the chemistry is kinetically based and that the HNC/HCN ratio is not in local thermodynamic equilibrium [9], which is true for most isomeric pairs, and which is why chemical simulations involving isomers are based on kinetics.

Carbon-chain molecules are highly unsaturated, often linear, and rather exotic species, found mainly in cold cores, although recent observations indicate that they can exist at higher temperatures if they are not reactive with H_2 [10]. One set of examples is the cyanopolyyne family ($HC_{2n}CN$), of which the largest detected in cold cores is $HC_{10}CN$ [10a]. Negative ions are also detected in cold cores, especially the $C_{2n}H^-$ family, with the larger ions having higher abundance ratios with respect to their neutral counterparts [11]. There is also evidence for small carbon chains such as $CCCH^+$ in more diffuse gas and in so-called photon-dominated regions, which are located near bright stars [12].

"Complex" organic molecules, abbreviated as COMs, are more terrestrial-like organic molecules ranging in size from six atoms to 13 atoms [13]. Although now observed in a wide variety of sources in the galaxy, they are still mostly associated with hot corinos and hot cores, with temperatures in the 100–300 K range. These species are partially to totally saturated and contain oxygen or nitrogen in addition to carbon and hydrogen. Examples are simple alcohols (methanol, ethanol), the ester methyl formate, dimethyl ether, glycolaldehyde, acetone, ethyl cyanide, and propyl cyanide. Interestingly, the nitrogen-containing species are often spatially distinct from the oxygen-containing COMs. Although it is difficult to detect gaseous COMs in protoplanetary disks, a few have been detected, notably methanol, which is also detected in cold cores.

Fullerenes, aromatics, and rings have all been detected, some quite recently. The fullerenes — C_{60} and its cation as well as C_{70} — have been detected in unusual sources, C_{60} and C_{70} by four infrared-active fundamental transitions in star clusters, planetary nebulae, and reflection nebulae [14], and C_{60}^+ by low-frequency electronic transitions in the near infrared in diffuse clouds against the light of background stars [15]. In addition to C_{60}^+, large numbers of unidentified broad spectral features in similar sources, named diffuse interstellar bands (DIBs), have remained unidentified for 100 years [16], but are thought to arise from organic molecules. Three aromatic species have been detected, the simplest, benzene, in the atmosphere of an old carbon-rich star, and more recently, the CN derivatives of benzene ($c\text{-}C_6H_5CN$) and probably naphthalene ($c\text{-}C_{10}H_7CN$) in the best-known cold core TMC-1 [17]. The non-aromatic ring $c\text{-}C_5H_5CN$ has also been observed in this source [18]. In the detection of the nitriles,

CN enhances the dipole moment of the parent species, making the detection by rotational emission possible at sufficiently high densities. A number of infrared emission bands seen in many sources have been attributed to polycyclic aromatic hydrocarbons (PAHs), but no individual species have been detected unambiguously, although it is estimated that 10% of the carbon in the galaxy is in the form of PAHs [19]. One possible source of PAHs lies in the atmospheres of old stars, but if this is correct, the PAHs would have to survive transit in space to reach "nearby" dense clouds.

Inorganic species, especially those containing metallic elements, tend to be found in circumstellar regions, especially surrounding older stars.

Molecules in and on ice mantles in cold clouds and stellar envelopes are detected using infrared absorption with broad features [20]. The largest such molecule is methanol ice; observing larger molecules is expected to be difficult for a number of reasons including the lack of laboratory spectral data and spectral congestion with the already observed features. The dust particles themselves are thought to be formed in the atmospheres of old stars, and consist of either silicates or amorphous carbon depending on the elemental abundances in the star. The dust-to-gas ratio in terms of mass is approximately 0.01 but depends upon environment.

10.3 Mechanisms for the Synthesis of Interstellar Molecules

Depending upon the density and temperature of the source, different types of chemistry can play major and minor roles in producing interstellar molecules. The chemistry can be divided into gas phase and grain surface mechanisms; each of these mechanisms to be plausible must yield observable quantities of gas phase species, since a vast majority of observed molecules reside in the gas. In this review, we will focus on the gas phase synthesis of molecules in cold cores and to a much lesser extent, hot cores. *Specific examples of important reactions will most frequently involve CRESU studies if available.* Databases of all types of gas phase reactions important in astrochemistry can be found on two websites: the KIDA database (https://kida.astrochem-tools.org/) and the UMIST database (http://udfa.ajmarkwick.net/.) The evaluation procedure for inclusion in KIDA is discussed in Section 10.4. The astrochemical databases also list codes for chemical simulations of assorted astrochemical sources. A large list of neutral–neutral gas phase reactions can be found on the NIST Chemical Kinetics Database (https://kinetics.nist.gov/kinetics/index.jsp), which is not related to astrochemistry. A similar ion-molecule database, An Index of the Literature for Bimolecular Gas-Phase Cation-Molecule Reaction Kinetics (https://trs.jpl.nasa.gov/handle/2014/7981), can be found on the JPL website.

10.3.1 *Gas phase ion–molecule chemistry*

Of the two major mechanisms, gas phase chemistry was investigated first, starting from the earliest days of astrochemistry, when polyatomic molecules were first observed in dense interstellar clouds in the early 1970s [21, 22]. Given the low temperatures of dense clouds, early investigators concentrated on ion–molecule reactions because it was known even then that these reactions are rapid at low temperatures if exothermic and if they do not possess activation energy barriers for at least one exit channel. Indeed, the few measurements made as a function of temperature over a limited range showed that the rate coefficients of exothermic ion–molecule reactions with non-polar neutral species such as H_2 are often totally independent of temperature, with rate coefficients $k\,(\mathrm{cm^3\,s^{-1}})$ closely approximated by the Langevin formula

$$k_L = 2e\sqrt{\alpha/\mu} \approx 10^{-9} \tag{10.1}$$

with e the electrostatic unit of charge, α the polarizability in $\mathrm{cm^3}$, and μ the reduced mass in gm [22]. The rate coefficients of ion–molecule reactions with a polar neutral k_D could be fit reasonably to the formulae of Su and Chesnavich: [23]

$$x = \mu_D/(2\alpha kT)^{1/2} \tag{10.2}$$

$$k_D = k_L(0.4767x + 0.6200) \tag{10.3}$$

or

$$k_D = k_L(((x + 0.5090)^2/10.256) + 0.9754) \tag{10.4}$$

where k is the Boltzmann constant and μ_D is the dipole moment (D). The first equation for k_D pertains if $x \geq 2$, while the second pertains if $x < 2$. At temperatures significantly under room temperature, $k_D \propto T^{-1/2}$.

Of course, at the time, well before the advent of the CRESU technique, measurements at very low temperature were uncommon. With the CRESU technique, starting in the 1980s, the situation improved dramatically [24], and gave an impetus to more detailed theories (ACCSA and SACM) capable of calculating rate coefficients for individual rotational states [25, 26]. The history and development of the CRESU technique are presented in detail in Chapters 1 and 2.

In early chemical simulations of interstellar chemistry (1970–1980), most neutral–neutral reactions were assumed to have barriers and therefore to be unimportant at low temperatures, although such reactions involving an atomic reactant were occasionally included in the models of the day [22]. It was known from earlier ultraviolet measurements of diffuse clouds that molecular hydrogen was the most abundant molecular species, and it was assumed that atomic helium was also quite abundant [27]. It had already been shown theoretically that H_2 was formed on the surfaces of interstellar grains since the gas phase processes known were too slow to explain the abundance of H_2, especially in dense clouds where virtually all hydrogen was assumed to be molecular [7, 22].

Ion–molecule chemistry is initiated principally in dense clouds by cosmic rays, which are high-energy nuclei, mainly protons, that travel at speeds near the speed of light and travel great distances even through dense clouds. Their bombardment of H_2 and He leads to H_2^+ and He^+ as well as H and H^+ [22]. The H_2^+ is quickly converted to H_3^+ by reaction with H_2 [21, 22, 28]:

$$H_2 + H_2^+ \rightarrow H_3^+ + H \qquad (10.5)$$

while He^+ reacts slowly with H_2 and, like H_3^+, is available to react with a wide variety of other neutral species. Reactions of these two dominant ions with CO occur, after CO is itself synthesized by atom–diatom reactions [22]. The reaction of CO with H_3^+ produces the HCO^+ ion, important as the molecular ion first observed by radioastronomy [29], which helped to secure the role of ion–molecule reactions in low-temperature chemistry in cold cores. The reaction of CO with He^+ produces C^+ which, along with neutral atomic carbon, initiates the synthesis of carbon-chain species, starting with one-carbon

species. These reactions have been studied with the CRESU appara-
tus down to temperatures of $8\,\mathrm{K}$ for $\mathrm{He^+}$-CO and $30\,\mathrm{K}$ for $\mathrm{H_3^+}$-CO
[30, 31].

10.3.2 *Ion–molecule formation of carbon chains*

The chemistry below is focused on cold cores of dark clouds, although
some of it pertains to more diffuse clouds. In modern time-dependent
simulations of cold core chemistry, elemental carbon, which is mainly
in the form of $\mathrm{C^+}$ at the earliest of times, is converted into neutral
C at intermediate times, and, if the elemental C/O abundance ratio
is below unity, which is common in the galaxy, is eventually mainly
converted into CO, although there is still enough neutral and ionic
atomic carbon to continue carbon-chain development. Figure 10.2
shows this time dependence.

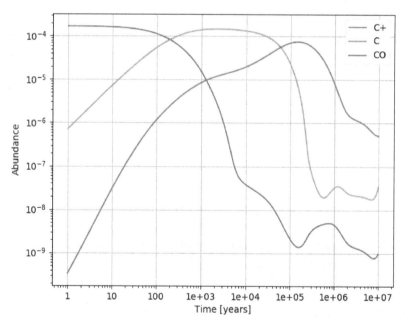

Fig. 10.2. Time dependence of gaseous fractional abundances for $\mathrm{C^+}$, C, and
CO in a cold core, as calculated with a gas-grain code. The inclusion of grain
chemistry becomes important at long times when gaseous CO begins to accrete
onto the dust.

One initial reaction in the production of carbon chains is the radiative association between C^+ and H_2 [32]:

$$C^+ + H_2 \rightarrow CH_2^+ + h\nu \qquad (10.6)$$

Radiative association is a low-density process in which an intermediate complex is formed, which then can be stabilized by emission of a photon. Although radiative association is inefficient unless the complex is a large one, which can live for a long time, the reaction rate with H_2 as a reactant can often be competitive, especially at low temperatures such as $10\,\mathrm{K}$. Note that the more normal reaction

$$C^+ + H_2 \rightarrow CH^+ + H \qquad (10.7)$$

is endothermic by $0.4\,\mathrm{eV}$, and can only occur in excited regions such as shocks [33]. Although most instances of radiative association are thought to occur via vibrational photons, the $C^+ + H_2$ association occurs via emission from excited electronic states of CH_2^+ which emit to the ground state via a complex process [34].

It is difficult to study radiative association in the laboratory because unless the density is lower than approximately $10^{11-12}\,\mathrm{cm}^{-3}$, the three-body regime dominates in which the complex is stabilized by collisions with the bath gas. The few ion-neutral radiative systems studied directly to date have been undertaken with ion traps of various designs although there is some ambiguity in most of the results [35]. To the best of our knowledge, the $C^+ + H_2$ association has only been studied theoretically or in the three-body limit with either He or H_2 as the bath gas. But, a to an order of magnitude to the rate coefficient for the radiative mechanism to an order of magnitude can be obtained from the ternary regime, unless it is saturated, which occurs at densities such that collisional stabilization of the complex is faster than the complex redissociation rate. One requires knowledge of the density and both the collisional stabilization rate coefficient and the emission rate of photons with sufficient energy to stabilize the complex. Expressions for complex stabilization by vibrational photons and electronic photons for the special case of CH_2^+ are given by Herbst *et al.* [34] and Herbst [36]. A standard stabilization rate for vibrational photons is $\approx 10^{2-3}\,\mathrm{s}^{-1}$. For He as the bath gas, the collisional stabilization rate coefficient often utilized is $10^{-10}\,\mathrm{cm}^3\,\mathrm{s}^{-1}$, whereas that for H_2 is ten times larger [35]. The result

Table 10.2.　Radiative association rate coefficients $(\mathrm{cm^3\,s^{-1}})$.

System	Temperature	$k_{ra}\,(\mathrm{cm^3\,s^{-1}})$	Ref.
$C^+ + H_2$	13 K	2×10^{-15}	CRESU [37]
	10 K	8×10^{-16}	Theory [38]
	80 K	7×10^{-16}	Ion trap [35]
	80 K	5×10^{-16}	SIFT [35]
	90 K	2×10^{-16}	Afterglow [35]
$CH_3^+ + H_2$	23 K	3×10^{-13}	CRESU [39]
	13 K	1×10^{-13}	Penning [40]
	20 K	1×10^{-13}	Theory [41]
	80 K	6×10^{-15}	Ion trap [35]
	80 K	6×10^{-14}	Theory [41]

for radiative association is at worst an order of magnitude in uncertainty. In addition, a plot of the effective two-body rate coefficient in the three-body regime can be extrapolated to the low-pressure limit to yield an upper limit to the radiative association rate coefficient, as has been done and explained by Gerlich and Horning [35].

It is interesting to see how the CRESU-based results compare with previous estimates of two important radiative association rate coefficients. Table 10.2 contains results for the radiative association rate coefficients of $C^+ + H_2$ and $CH_3^+ + H_2$ as obtained at assorted temperatures, mainly indirectly via three-body collisions at one density, as performed here, or by effective rate coefficients extrapolated to low pressure, as done by others [35]. There are theoretical rate coefficients listed as well. Let us start with the $C^+ + H_2$ radiative association. The lowest temperature experimental value, at $\approx 13\,\mathrm{K}$, is derived from the ternary CRESU value with He bath gas [37] to be $2 \times 10^{-15}\,\mathrm{cm^3\,s^{-1}}$, which is only somewhat higher than the theoretical value of $8 \times 10^{-16}\,\mathrm{cm^3\,s^{-1}}$ at 10 K [38]. Other tabulated values using various techniques at temperatures from 80–90 K, range from $2 \times 10^{-16}\,\mathrm{cm^3\,s^{-1}}$ to $7 \times 10^{-16}\,\mathrm{cm^3\,s^{-1}}$, while higher temperature CRESU results show little temperature dependence from the 13 K value. Although these rate coefficients are small compared with the collisional rate coefficient, the overall rates are competitive.

In addition to the $C^+ + H_2$ radiative association, carbon-chain formation can be initiated via competitive reactions between neutral atomic carbon and H_3^+:

$$C + H_3^+ \rightarrow CH^+ + H_2 \tag{10.8a}$$

and

$$C + H_3^+ \rightarrow CH_2^+ + H \tag{10.8b}$$

although these reactions have not been studied with the CRESU apparatus but with the use of merged beams [42].

Once CH_2^+ is formed, it can react rapidly with H_2 to produce CH_3^+ +H. The methyl ion is not reactive with H_2 in the normal sense, but undergoes a radiative association reaction with H_2 to form CH_5^+. This process has been studied in a number of ways. Experimental and theoretical values in the 13–23 K temperature range and at 80 K are shown in Table 10.2. The value derived from the ternary CRESU measurement at 23 K [39] is higher by a factor of three than the direct Penning trap result at 13 K [40] and the theoretical value at 20 K [41]. This is reasonable agreement. Moreover, these values are sufficiently large that the rate of the $CH_3^+ + H_2$ reaction to form CH_5^+ at low temperatures is a dominant loss process for CH_3^+.

The measured 80 K ion trap value is significantly lower, although the theoretical value at the same temperature is an order of magnitude higher, and quite close to the lower-temperature results. Theory shows that little temperature dependence is expected at the lowest temperatures, although there is a strong inverse temperature dependence according to theory at higher temperatures [43].

We now move on to larger species. Most of the reactions discussed in the following paragraph can be found in the UMIST and/or KIDA databases. Let us start with CH_5^+. The protonated methane ion can be destroyed by reactions with CO to form CH_4 and by dissociative recombination with electrons. Dissociative recombination reactions normally lead to a variety of products, and in this case the neutral products include CH, CH_2, CH_3, and CH_4 with H and /or H_2. The dominant neutral product is CH_3, which can likely react with C^+ to form the ions C_2H^+ or $C_2H_2^+$. These ions can react with H_2 to form more saturated ions through $C_2H_4^+$, but these ions then undergo dissociative recombination to form neutrals such as C_2H, C_2H_2, and C_2H_3. Gaseous acetylene cannot be detected in space via radio astronomy, because it is non-polar, but it has been detected in absorption in the source IRc2 in the infrared through a bending transition with the SOFIA airborne observatory [44]. Among the interesting observations are those of the *ortho* and *para* forms of acetylene, which are clearly separate spatially.

The carbon-chain neutral species grow in size as these processes continue. For example, C^+ can react with C_2H_2 to form three-carbon ions and H/H_2. As hydrocarbon ions increase in number of carbon atoms, the likelihood of barrierless hydrogenation with H_2 decreases, and the result is likely to be very unsaturated species, such as the cyanopolyyne family $HC_{2n}CN$ and the radical family C_nH. Moreover, there are also so-called condensation reactions between hydrocarbon ions and neutrals, which can produce larger unsaturated ions, such as [45]

$$C_2H_2^+ + C_2H_2 \rightarrow C_4H_3^+ + H \qquad (10.9)$$

which then undergo dissociative recombination with electrons to form species such as C_4H and C_4H_2 (diacetylene). A gas phase model containing carbon chains with up to 9 carbon atoms illustrates these processes [46].

A new carbon-chain molecule recently detected in the well-known cold core TMC-1 is HC_4NC, the isocyanide isomer of HC_4CN, which is thought to be formed by dissociative recombination from two larger ions: HC_4CNH^+ and HC_4NCH^+ [47]. The cyanopolyyne $HC_{10}CN$, originally thought to have been found in TMC-1, an observation which was later refuted, has now been observed definitively.

As the cold core begins to heat up to more than 20–30 K and the core is more aptly called a pre-stellar core, as shown in Fig. 10.1, another phase of gas phase chemistry known as WCCC (warm carbon-chain chemistry) sometimes occurs [48]. This stage starts when volatile hydrocarbons such as methane formed earlier on 10 K grains start to desorb. Ion–molecule reactions such as

$$C^+ + CH_4 \rightarrow C_2H_3^+ + H; C_2H_2^+ + H_2 \qquad (10.10)$$

then begin a second stage of carbon-chain formation. Finally, neutral–neutral chemistry plays a competitive role in the formation of carbon chains as well as COMs, as will be discussed later in the chapter.

10.3.3   *Other positive ion–molecule syntheses*

10.3.3.1   *Ammonia synthesis*

The synthesis of ammonia starts with N^+ and N. Although the reaction of N^+ and H_2

$$N^+ + H_2 \rightarrow NH^+ + H \qquad (10.11)$$

is either slightly endothermic or has a small barrier, it occurs rapidly enough at 10 K to initiate the formation of ammonia. A detailed CRESU study by Marquette *et al.* from 8 K to 70 K allows a determination of the barrier as a function of the lowest two rotational states of H_2 [49]. The reaction between N and H_3^+, on the contrary, does not occur. The NH^+ ion reacts with H_2 to form NH_2^+, which then reacts again with H_2 to form NH_3^+. The ammonia cation undergoes what was assumed initially to be a rather unusual reaction at the time it was studied [50]. It is rather slow at 300 K because of a barrier to produce NH_4^+, decreases further as the temperature is lowered, but then reaches a minimum at 100 K and then starts to increase as the temperature decreases further. By 10 K, the rate coefficient is up to $10^{-12}\,\mathrm{cm^3\,s^{-1}}$, which is sufficiently rapid to synthesize NH_4^+. The unusual temperature dependence of the reaction has been ascribed to a potential minimum leading to complex formation which allows efficient tunneling under the transition state of the activation energy barrier. A more recent explanation of this unusual temperature dependence for a CRESU study of $CN + C_2H_6$ [51] replaces the intermediate complex with a "pre-exponential" complex associated with a loose transition state, and does not include tunneling [52]. However, tunneling is indicated in the earlier studied reaction since replacing the tunneling H with a deuterium atom likely negates the process. A tunneling mechanism has also been suggested for the reaction between OH and CH_3OH, studied with the CRESU, although the mechanism for this reaction is still unclear (see Section 10.4.3 below). Once NH_4^+ is produced, it undergoes dissociative recombination to form NH_3 and H.

10.3.3.2   *Water synthesis*

Gaseous water is produced by a number of syntheses starting with neutral and singly ionized oxygen atoms. Neutral O reacts with H_3^+

to produce OH^+ and OH_2^+ [53]:

$$O + H_3^+ \rightarrow OH^+ + H_2; OH_2^+ + H \qquad (10.12)$$

while O^+ reacts with H_2

$$O^+ + H_2 \rightarrow OH^+ + H \qquad (10.13)$$

The ions OH^+ and OH_2^+ react with H_2 once or twice to produce H_3O^+, a relatively stable ion with a structure and inversion motion similar to ammonia. Interestingly, the three ions have all been detected. Although none is detected in cold cores, the ions are seen in hot core sources as well as photon-dominated regions (PDRs). The H_3O^+ ion does react dissociatively with electrons, but the dominant heavy product is OH, while the water channel has a branching fraction of around 10%. Although water is seen in the gas phase, it has a much higher abundance in the ice mantles of interstellar grains, where it is the dominant species [54].

10.3.4 *Negative ions and attachment processes*

Negative ions are observed in cold cores, especially the well-studied TMC-1. They are also detected in the carbon-rich star IRC+10216. The best-known anions are carbon chains of the $C_{2n}H^-$ family, ranging up to C_8H^- [11]. There are a variety of differing views on how these ions are synthesized, the simplest of which is the low-density radiative attachment [55, 56]:

$$X + e^- \rightarrow X^- + h\upsilon \qquad (10.14)$$

where X is normally a species with a high electron affinity. Recent theoretical work shows that this mechanism is inefficient for C_4H^-, but more efficient for larger anions [56]. Although several theoretical approaches to radiative attachment have been reported, including a long-lived intermediate complex and an intermediate dipolar species, no direct experiments have been undertaken in which emitted photons have been detected given the need for very low densities. CRESU studies with electrons and neutral species have been undertaken for a variety of neutrals including PAHs [57], which are thought by some to be negatively charged in cold cores. However, electrons are not well thermalized in CRESU experiments (see Chapter 4).

The rate coefficient derived for anthracene vapor in the range 48–300 K, 1–$3 \times 10^{-9} \mathrm{cm}^3 \mathrm{s}^{-1}$, is likely the result of collisional stabilization rather than radiative attachment given bath nitrogen densities of $\approx 10^{16} \mathrm{cm}^{-3}$. Another possibility is the formation of cluster ions. Some of the measured rate coefficients are probably near the collisionally saturated bimolecular values of $\approx 10^{-7} \mathrm{cm}^3 \mathrm{s}^{-1}$, which can be obtained by a Langevin-like capture approach [22] or by the assumption of s-wave capture [55]. Unlike the non-saturated cases of radiative association, discussed above, saturated high-density bimolecular rate coefficients cannot easily be converted to low-density radiative rate coefficients, needed for interstellar chemical models. For rate coefficients in the ternary range, conversion to a radiative rate coefficient would require actual three-body rate coefficients, which do not appear to be available.

Simple theoretical models for radiative attachment, with the assumption of a long-lived complex and the emission of vibrational photons, do a reasonable job of reproducing the abundance ratios of the $C_{2n}H$ anions to their neutral precursors for $n = 3$ and 4 but not $n = 2$ in TMC-1 [56]; however, this theoretical approach has been criticized as simplistic [58].

10.3.5 *Negative ion reactions*

Negative ion–molecule reactions do not play a major role in most of the standard sources studied in astrochemistry, yet can affect the outcome of most chemical simulations. Negative ions can react with neutral species via associative detachment reactions such as [59, 60]

$$C + C_nH^- \rightarrow C_{n+1}H + e \qquad (10.15)$$

$$H + C_n^- \rightarrow C_nH + e \qquad (10.16)$$

$$H + C_3N^- \rightarrow HC_3N + e \qquad (10.17)$$

to produce carbon-chain neutrals. The reverse reactions, discussed in Chapter 4, and known as dissociative attachment, can play a role in the formation of negative ions if they are exothermic. When negative ions are included in the chemistry of the dark cloud cold core TMC-1, the abundances of the carbon-chain molecular families C_n, C_nH, C_nH_2, C_nN, and $HC_{2n}CN$ are enhanced [61]. The role of negative ions in the production of much larger carbon chains in the outer dusty envelope of the carbon star IRC + 10216 is thought to involve negative ion–molecule reactions such as the radiative association [62]

$$C_8 + C_8^- \rightarrow C_{16}^- + h\upsilon \qquad (10.18)$$

followed by associative detachment with atomic hydrogen to form $C_{16}H$. Models of IRC + 10216 show that, in the inner envelope, the high density and temperature allow for thermodynamics to determine the molecular abundances, but in the outer stages the material cools and grows more rarefied, reaching and passing through the physical conditions of dense interstellar clouds, where the large abundance of carbon species promotes a rich chemistry [63].

Negative ion chemistry is also important in the chemistry of the upper atmosphere (the ionosphere) of Titan, the large moon of Saturn. Recent studies with the Cassini plasma spectrometer combined with photochemical models reveal negative ions of the types C_nH^- and $C_{n-1}N^-$ (n = 2–6) [63a]. For example, the reaction [64]

$$CN^- + HC_3N \rightarrow C_3N^- + HCN \qquad (10.19)$$

is thought to be responsible for most of the C_3N^- production. CRESU studies have been undertaken for reactions of importance in the Titan atmosphere including (10.19), studied down to 50 K, and a variety of reactions between the anions $C_{2n+1}N^-$ and the polar neutrals HC_3N and $HCOOH$ [65].

10.4 Neutral–Neutral Chemistry

Although radical–neutral reactions were not ignored during the first twenty years of astrochemical simulations, they played second fiddle to ion–molecule chemistry. The breakthrough came in the early 1990s with initial experiments by Haider and Husain [66] on reactions between carbon atoms and alkynes and alkenes, which showed that very large rate coefficients of 10^{-9}–10^{-10} cm^3 s^{-1} of astrochemical importance could occur at 300 K. Initial theoretical work was done by Clary *et al.* and by Woon and Herbst [67, 68]. Some of the earliest CRESU studies of astrochemical interest were reported at around the same time: examples are $CN+C_2H_2$ and C_2H_4, which show an inverse temperature dependence down to 25 K, and $CN+C_2H_6$, which shows a U-shaped temperature dependence similar to that discussed above for $NH_3^+ + H_2$ [51, 52].

Other reactions with larger hydrocarbons such as allene and methyl acetylene were also studied with the CRESU apparatus and show little temperature dependence [69]. The rate coefficients at the lowest temperatures are quite large, and clearly important in cold cores. The problem with these and other early experiments using the CRESU apparatus regarding astrochemistry was the frequent inability to measure products directly and unambiguously, an inability that has now been partially corrected. For example, a more recent CRESU experiment on CN and propyne uses chirped-pulse microwave spectroscopy to determine the product branching [70]. When products have not been measured unambiguously in CRESU experiments, theoretical treatments and databases that include other types of experiments can often be utilized for this purpose. Some examples include the flowing afterglow [35], merged beams [53], ion-cyclotron resonance, and crossed-beam devices [71]. A crossed-beam apparatus has often been used to determine products, but at very high, non-thermal energies. The values currently used for rate coefficients with measured or estimated temperature ranges as well as products if known for all classes of reactions important in astrochemistry can be found in two principal databases cited above: the udfa (UMIST) database (http://udfa.ajmarkwick.net/) and the more recent KIDA database (https://kida.astrochem-tools.org/), both of which have available websites and list references when available. A check next to a reaction in KIDA means that a panel of experts has evaluated the temperature range, rate coefficient, and possibly the products. Rate coefficients are normally tabulated in terms of the modified Arrhenius formula:

$$k(T) = \alpha(T)(T/300)^{\beta} e^{-\gamma/T}$$

10.4.1 *Carbon-chain formation*

The reaction between CN and acetylene

$$CN + C_2H_2 \rightarrow HC_3N + H \tag{10.20}$$

is now thought to be competitive with ion–molecule syntheses in the production of the carbon-chain cyanoacetylene, the simplest polyyne in cold cores. In fact, one school of thought suggests that the reaction is more important than several ion–molecule processes, based on the observed unequal abundances of ^{13}C isotopomers of cyanoacetylene [72]. Neutral–neutral reactions can also presumably form larger cyanopolyynes by reactions with larger unsaturated hydrocarbons. CRESU studies of the reactions of CN with a number of other hydrocarbons, including allene, methyl acetylene, 1-propyne, propene, and propane, have been reported to be rapid at low temperatures [51, 69], with unsaturated hydrocarbons probably leading to unsaturated nitriles.

In general, rapid radical–neutral reactions at low temperature are often competitive with ion–molecule syntheses of carbon-chain species in cold cores. Some examples of important low-temperature reactions involve carbon atoms and small hydrocarbon radicals such as CH_n with hydrocarbons. These include the following:

$$C + C_2H_2 \rightarrow c - C_3H + H; \, 1 - C_3H + H \quad (10.21)$$

$$C + C_2H_2 \rightarrow C_3 + H_2 \quad (10.22)$$

$$CH + C_2H_2 \rightarrow (c/1-) \, C_3H_2 + H \quad (10.23)$$

$$C_4H + C_2H_2 \rightarrow C_6H_2 + H \quad (10.24)$$

Reactions (10.21) and (10.22) have been studied both theoretically [73] and by a number of experiments at assorted energies, some with products [74]. Other reactions between carbonaceous radicals (including C) and assorted hydrocarbons have been studied at low temperatures [75], but the products, as those for (10.24), are sometimes only inferred or found in databases with less than definitive values. Neutral–neutral reactions that lead to carbon-chain species are also important in models of the carbon star IRC + 10216, which does not have large numbers of ions in its envelope, as well as the atmosphere of Titan.

10.4.2 The formation of COMs

As astronomers began to understand the evolutionary stages of star formation, especially regions such as hot corinos and hot cores, the chemistry occurring at higher temperatures than 10 K gained interest, with the detection of large abundances of COMs in the gas phase.

The simplest COM, methanol, had also been detected weakly in cold gaseous environments, and a suggestion of an ion–molecule synthesis consisting of two reactions was proposed starting with radiative association to form protonated methanol followed by dissociative recombination:

$$CH_3^+ + H_2O \rightarrow CH_3OH_2^+ + h\upsilon \qquad (10.25)$$

$$CH_3OH_2^+ + e \rightarrow CH_3OH + H \qquad (10.26)$$

This process was studied experimentally and found to be untenable. The radiative association was studied in an ion trap and found to be too slow to measure [76], and the dissociative recombination, measured with a storage ring, showed methanol to be a minor product channel [77]. A second measurement of reaction (10.25) might be in order given that it was reported originally in a non-refereed source. A synthesis on the surface of dust particles was later proposed and duplicated in the laboratory. In this synthesis, CO on the dust surface reacts with successive hydrogen atoms landing on the surface to form methanol, which desorbs into the gas [78]:

$$CO \rightarrow HCO \rightarrow H_2CO \rightarrow H_3CO/H_2COH \rightarrow CH_3OH \uparrow \qquad (10.27)$$

This sequence, followed by standard ionic destruction reactions, reproduces the observed fractional abundance for methanol of 10^{-9} in cold cores. To explain the production of COM abundances in hot cores via gas phase reactions, it was originally assumed that the physical evolution from cold core to hot core occurred instantaneously. An ion–molecule chemistry based heavily on the precursor molecule CH_3^+ was considered. Eventually it was realized that a more reasonable assumption was the slow evolution of cold cores into prestellar cores and finally into protostars and hot corinos/cores during which time the chemistry occurs, analogous to the laboratory procedure known as temperature-programmed desorption. Inclusion of gas phase processes in this physical evolution was able to explain a few of the abundances of COMs larger than methanol, but was unable to explain most of them [79].

In the first decade of this century, the idea was formulated and worked out that the major chemical processes leading to COMs during the warm up to the hot core stage are diffusive reactions on the surfaces of dust particles between reactive radicals, themselves produced by photodissociation processes with photons generated by cosmic ray bombardment in the gas. Once produced on granular surfaces, the COMs eventually desorb into the gas as the temperature exceeds a threshold of approximately $100\,\mathrm{K}$ [79, 80]. The process of diffusion cannot occur at $10\,\mathrm{K}$ because the radicals are too massive to diffuse below $20\,\mathrm{K}$. The theory became the favored theory because it did explain the abundances of hot core COMs in the gas.

10.4.3 *Formation of cold COMs*

But, astronomy is a fast-changing subject, especially with the rapid improvement of telescopes, and less than a decade after the publication of the grain reaction hypothesis, several of the most abundant COMs, namely, dimethyl ether and methyl formate, were found in several cold core regions at a temperature of $10\,\mathrm{K}$ [81, 82]. The diffusive reaction theory could not reproduce these observational results because the radicals could not diffuse at such low temperatures. At present, the problem of low-temperature COMs has not been resolved, with both gas phase and revised grain surface mechanisms suggested. Here, we emphasize the gas phase mechanism since it is related to CRESU measurements.

The gas phase processes discussed up to now in cold regions tend to produce unsaturated carbon chains rather than partially saturated COMs. An interesting type of chemistry that might be applicable to form COMs under cold conditions has been used to explain some of the chemistry occurring in the upper portions of the atmosphere of Titan. This chemistry is composed of neutral radical–radical radiative association reactions, which are likely to have no activation energy barriers [83]. Moreover, for the Titan chemistry the reactions must have very large rate coefficients, near the collisional limit of $\approx 10^{-10}\,\mathrm{cm^3\,s^{-1}}$ or even larger. To the best of our knowledge, no such process has been measured in the laboratory, nor is it obvious that radiative association rates can be so large.

In the cold interstellar medium, radiative association reactions were originally limited to ion–molecule systems with few exceptions. However, the neutral–neutral analog eventually was used in studies of the chemistry of cold cores. One of the specific processes suggested by several groups to explain the abundance of dimethyl ether at $10\,\mathrm{K}$ cold cores is the radiative association of the radicals CH_3 and

CH$_3$O [82]:

$$CH_3 + CH_3O \rightarrow CH_3OCH_3 + h\upsilon \qquad (10.28)$$

Calculations in our group are in progress, and it does appear that the rate coefficient is almost as large as the collisional rate coefficient at 10 K. The reasons are as follows: First, the complex formed initially has a deep potential well, which is deeper than the systems of ion–molecule collisions previously studied. This deep well enhances the reaction rate coefficient because it allows the complex to live for a longer period of time. Second, the radiative relaxation rate to stabilize the complex is relatively large. Finally, radiative association rate coefficients at 10 K are at close to their maximum values. But, considerable gas phase abundances of the reacting radicals are also necessary for the reactions to be competitive in chemical simulations. The methyl radical can be formed from the dissociative recombination of CH$_5^+$ and other reactions. But, what about methoxy?

One way to form the methoxy radical in cold clouds is the reaction between OH and CH$_3$OH [81, 84]:

$$OH + CH_3OH \rightarrow CH_3O(CH_2OH) + H_2O \qquad (10.29)$$

which has been studied with the CRESU technique down to low temperatures by three groups [84, 85]. The CH$_3$O product was detected using laser-induced fluorescence, and it was formed with a rate coefficient which was the same as that measured for the removal of OH [84]. The two product channels shown above dominate at different temperatures; higher temperatures favor hydroxymethyl and lower temperatures favor methoxy [84]. The CRESU experiment to temperatures as low as 22 K allowed a reasonable extrapolation of the rate coefficient down to 10 K, which was then used in a chemical simulation of the cold core B1-b to show that the reaction was a major contributor to the production of methoxy [85]. There is now a newer experiment in which a minimum temperature of 11.7 K was achieved [85a].

The dependence of the rate coefficient on temperature is of the "U type" in which the rate coefficient is a minimum at some intermediate temperature and increases as temperature increases and decreases. Detailed reviews discussing this and other U-type systems have been written by Cooke & Sims [52] and Heard [85b]. But, in this case, below a certain temperature, the reaction increases dramatically by more than two orders of magnitude in rate as the temperature is decreased below 100 K, reaching very high values near the collisional limit. If the standard explanation applies, the lifetime of a weak complex increases as the temperature decreases, allowing more time for tunneling under the transition state. But, an alternative explanation involves a loose transition state in the entrance channel along with a "pre-reactive" complex, similar to the explanation for $CN + C_2H_6$ [52]. Other alternative explanations have been suggested such as the system reaching a high-pressure limit, in which the three-body process saturates.

Nevertheless, the experimental data for $OH + CH_3OH$ at low temperatures suggest a tunneling mechanism well below the high-pressure limit, indicating that the data for methoxy production can be used for astrochemical purposes [84]. In a review mainly of $OH + CH_3OH$, Canosa [85c] noted that, at very low temperatures, the pressure dependence of this reaction remains uncertain although it appears to be almost pressure independent, while calculations suggest that these experimental results are consistent with the low-pressure limit. This reaction is also discussed in both Chapter 5 (Hickson and Heard) and Chapter 11 (Halvick and Stoecklin).

As mentioned above, the methoxy radical can then undergo radiative association to form dimethyl ether. A different radiative association between the methyl and hydroxymethyl radicals could possibly lead to the formation of ethanol, but the production of hydroxymethyl by the reaction of OH and CH_3OH does not occur until warm temperatures, on the way toward star formation.

The formation of dimethyl ether can lead to the subsequent synthesis of methyl formate, another important COM [82]. Dimethyl ether can react with halogen atoms or the CN radical to produce CH_3OCH_2, which can subsequently react with O atoms to produce methyl formate [82]:

$$O + CH_3OCH_2 \rightarrow HCOOCH_3 + H \qquad (10.30)$$

The entire synthesis is shown in Fig. 10.3. Additional evidence in favor of this reaction sequence is the overlap of spatial distributions of dimethyl ether and methyl formate in interstellar clouds. Inclusion of this sequence of reactions into a chemical simulation network for a cold core did not quite produce enough dimethyl ether and methyl

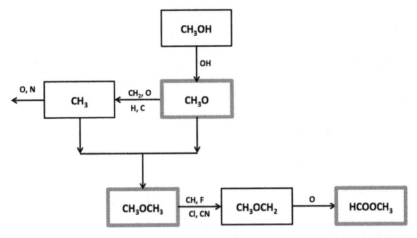

Fig. 10.3. A possible low-temperature route to the production of the COMs dimethyl ether and methyl formate using neutral–neutral chemistry and starting from the OH + CH₃OH reaction. Reproduced with permission from Ref. [82], Copyright 2015 Oxford University Press.

formate, although a lower value was used for the rate coefficient of the radiative association reaction that produces dimethyl ether than now appears to be the correct value [82].

Other reactions involving OH studied at low temperatures with the CRESU apparatus include the reactants formaldehyde, ethanol, acetone, methyl formate, and dimethyl ether [85b, 86]. Some of these produce COMs, while some produce radicals that might lead to new COMs. These reactions tend to show varieties on the U-type rate coefficient vs temperature pattern, which involve moderate to strong increases at low temperature. For some of the reactions, detailed theoretical calculations were also undertaken to help analyze the data and yield more information. The OH + ethanol reaction [87] has been measured down to 21 K. With respect to the rate coefficient at 300 K, the value at 21 K is larger by two orders of magnitude, indicating a similarity in mechanism to the methanol reaction. At higher temperatures, a variety of products has been detected, although the major product is the radical CH_3CH_2O. In addition to higher-temperature studies, the OH + acetone reaction [88] has been studied from 64.4 K down to 11.7 K. The value for the estimated rate coefficient at 10 K is 1.8×10^{-10} cm³ s⁻¹ and the suggested products are $CH_3(CO)CH_2$ and water, which are the known products at higher temperatures.

The reaction between OH and formaldehyde [89] was studied between 22 K and 107 K. The rate coefficient was extrapolated to 10 K to obtain a value of 2.6×10^{-10} cm^3 s^{-1}. The reaction was included in a chemical simulation and appears to be the dominant mechanism to produce HCO at that temperature. The reaction between OH and methyl formate was studied between 24.4 and 64.2 K and found to be quite rapid, and important as a destruction mechanism for methyl formate (HCOOCH$_3$) [90].

The radical CN also plays a role in the chemistry of cold COMs albeit a smaller one than OH. A CRESU study of the reaction between CN and methylamine from 23 K to 297 K has been reported. Used along with theoretical calculations, it was determined that the major products are cyanamide, NH$_2$CN, and methyl cyanamide, CH$_3$NHCN [91]. A more recent calculation, however, indicates that the dominant product is produced by hydrogen abstraction [91a]. In an earlier study using two techniques, Sims *et al.* studied reactions between CN and assorted hydrocarbons, with results from a CRESU apparatus down to 25 K [51]. An addition–elimination product channel was inferred for unsaturated hydrocarbons in which the CN replaces an H atom. Reactions with methane and ethane were inferred with some additional evidence to lead directly to methyl and C$_2$H$_5$ radicals. Although some of the products are doubtless closer to carbon chains than to COMs, there are exceptions, such as the almost saturated C$_2$H$_5$, which can then react to form larger species.

Can cold COMs be made by ion–molecule chemistry? One possibility would be to return to early semi-quantitative treatments of the chemistry of CH$_3^+$ in which this ion reacts with polyatomic neutral species via normal and radiative association reactions to produce large ionic species, which then undergo dissociative recombination reactions to form neutral species. At that time, no real distinction was made between cold and hot regions, and some of the chemistry today, minus the radiative association, would be considered to be more appropriate to hot cores. A glance at Fig. 10.2 from the review article of Smith [92] shows some of the many pathways envisaged. Consider, for example, the reaction with NH$_3$, which was thought to lead to CH$_3$NH$_2$, or the reaction with CH$_3$OH which was thought to lead to both ethanol and dimethyl ether. In each case, an efficient ion–molecule radiative association must occur even if there are competitive pathways, which, although possible, is unusual. Another problem with this approach is that it does not take into account the branching fractions of dissociative recombination reactions, which are now known to favor three-body product channels to a great extent. The failure of the radiative association between $CH_3^+ + H_2O$ followed by dissociative recombination to produce methanol illustrates the problems.

An interesting suggestion was made recently that if there is enough ammonia present in the gas, ion-neutral syntheses could produce large abundances of COMs in addition to carbon-chain species [93]. The mechanism works as follows. Instead of undergoing dissociative recombination with electrons, a protonated ion can react with NH_3. In this case, the protonated ion normally only transfers its proton to ammonia because ammonia has a very high proton affinity. For example, consider

$$CH_3OHCH_3^+ + NH_3 \rightarrow CH_3OCH_3 + NH_4^+ \qquad (10.31)$$

which can be compared with the dissociative recombination, which produces only a 7% channel for dimethyl ether. The result could be a more efficient method of making, in this case, dimethyl ether and, in general, could convert many protonated COMs into the neutral species. But, it turns out that there is insufficient ammonia to dominate the dissociative recombination at $10\,K$ [94]. So, it seems that neutral–neutral reactions will dominate gas phase approaches to form cold COMs. But, what about the competition with grain-surface approaches? To generalize, at the present time, grain-based syntheses that do not involve surface diffusion but do involve desorption in a non-thermal manner following surface reactions seem to be somewhat more successful than gas phase syntheses of COMs in explaining their abundance in cold cores, but the matter has not been settled definitively, and both mechanisms may play a role [82, 95]. After all, the methanol that starts the gas phase synthesis of dimethyl ether and methyl formate in Fig. 10.3 is produced on grain surfaces.

Another process involving grains that may play a role in the formation of cold COMs is known as radiolysis. In the interstellar context, the term "radiolysis" pertains to the bombardment of ice mantles of cold grains by cosmic rays, mainly protons. Numerous experiments have been done showing that high-energy protons (or electrons) bombarding cold ices containing carbonaceous species leads to COMs (see, e.g. [96]). Of course, the chemistry is very complex, but recent studies have made reasonable approximations of the sequences of reactions that lead from proton bombardment to synthetic processes [97, 98]. These processes have been added to chemical simulations and appear to form certain COMs such as methyl formate much more rapidly than other processes, although the same is not true for dimethyl ether.

10.5   Aromatics, Rings, and Fullerenes

In addition to carbon chains and COMs, reactions with CN lead to ringed species, including aromatics. As discussed above, other than benzene, no pure individual aromatic species has been observed in the interstellar medium, although there is strong evidence for aromatic infrared features in many sources. There are many instances in spectroscopy in which the addition of atoms or molecules or their reactions with hard-to-detect species can make the altered species more detectable. Since aromatic species tend to have small or zero dipole moments, detection by the standard technique of rotational spectroscopy is very difficult and has so far not been successful. But, what if CN radicals displace a hydrogen atom from an aromatic or ringed species, causing or enlarging the dipole moment and making the product molecule more detectable? This supposition is actually more accurate for absorption spectroscopy than the more normal emission features seen in the interstellar medium because the molecules are typically detected following collisional excitation, and if dipole moments are too large, the emission rate downward will be larger than the collisional rate upward, and the emission features will be sub-thermal and weak. Nevertheless, this effect is likely to be small at low emission frequencies, as can be seen from the following simplified equation:

$$k_{coll} n > A \qquad (10.32)$$

where A is the Einstein A coefficient for a specific emission rotational transition, k_{coll} is the bimolecular rate coefficient for the inelastic rotational collision upward for the same transition, and n is the so-called critical density, typically that of H_2 in cold cloud cores. The inequality in the direction indicated is needed for strong rotational emission. Since the Einstein A coefficient is proportional to the frequency cubed as well as the dipole of the emitting species squared, the inequality is obeyed for larger molecules with lower-frequency rotational transitions as long as the density is large enough. If the inequality is not obeyed, transitions can still be detected but they will be non-thermal and weak.

Now, let us consider the benzene molecule, which has no dipole moment. The reaction between CN and C_6H_6

$$CN + C_6H_6 \rightarrow C_6H_5CN + H \tag{10.33}$$

has recently been measured in a CRESU apparatus with product detection by pulsed laser-induced fluorescence [99] down to a temperature of 15 K with a rate coefficient of $(3.6–5.4) \times 10^{-10}$ cm^3 s^{-1} that decreases little up to the maximum temperature studied of 295 K. The product molecule, benzonitrile, has a well-known rotational spectrum, which can be used to detect it. A successful detection in the cold core TMC-1 at 10 K can act as a proxy for the identification of benzene in this source. The predicted thermal emission spectrum is weak despite the fact that inequality (10.32) is obeyed even with a dipole moment of 4.5 Debye. To observe benzonitrile more strongly, a relatively new technique known as stacking, in which a number of rotational transitions are added in intensity in velocity space, was used. This technique succeeded in detecting benzonitrile in the cold core TMC-1 [17]. Even more astounding is the likely detection of the naphthalene proxies 1- and 2-cyanonaphthalene in the same source [100]. In this case, stacking appears to have been successful even though individual lines could not be detected. Another ringed but non-aromatic molecule detected in TMC-1 is cyanocyclopentadiene (c-C_5H_5)CN [101]. The chemistry of these aromatic proxies is still quite uncertain.

10.5.1 *Fullerenes*

The fullerenes were discovered in the laboratory in 1985 [102] and, many years later, seen in interstellar sources such as planetary nebulae and reflection nebulae [14]. They are nowhere near as abundant and widespread as PAHs. A number of suggestions as to how they are synthesized have been made. An early scheme was developed by Bettens and Herbst [103] based loosely on laboratory work in which the fullerenes are formed from linear carbon chains through a sequence of ringed species. The idea is that linear chains are first converted to monocyclic rings by spontaneous isomerization at a number of carbon atoms large enough so that the process of ring closure has no barrier. The monocyclic rings then condense into tricyclic rings, and eventually into fullerenes with the help of He^+ ion reactions. The main growth mechanism is based on the reactions of hydrocarbons with C^+ and the reactions of hydrocarbon ions with neutral C. A joint CRESU-SIFT study of an example reaction — $C^+ + c - C_6H_{12}$ — has been undertaken over a wide temperature range but could not be used to confirm the growth mechanism since the products are the results of charge transfer or dissociative charge transfer [104].

This synthesis was utilized to determine the abundances of fullerenes and other non-hydrogenated species through 64 carbon atoms in dense clouds and diffuse clouds [105, 106] via extension of then current gas phase networks used in simulations. In diffuse clouds, it was necessary to include a change in the physical conditions of the cloud in which initially dense matter was dispersed into the current diffuse conditions. The diffuse cloud model was also directed at DIBs through 64 carbon atoms. The fullerene cation C_{60}^+, detected in diffuse clouds, can be reproduced by this dispersive model, both with a purely gas phase simulation and also a gas-grain one.

A more recent "top-down" model for the formulation of fullerenes is based on the evolution of PAHs in objects such as a reflection nebula, which is an interstellar cloud on which a nearby star shines light which reflects back to the observer [107, 108]. The PAH considered at the start of the model is $C_{66}H_{20}$, which is bombarded by irradiation of ultraviolet photons from the nearby star and becomes fully dehydrogenated while it shrinks to C_{60}. The shrinking determines the overall rate of the process, in which the initial PAH is fully converted to C_{60} in 10^5 yr. Lowering the number of carbons in the initial PAHs down to 60 will continually speed up the process of C_{60} formation, while even smaller PAHs will just be photodissociated.

10.6   The Future

Astrochemistry is entering a new age driven by new and more powerful telescopes, allowing detection of weaker sources and, perhaps more importantly, showing dramatic increases in spatial resolution. Sources treated as homogeneous in the past are now known to be inhomogeneous in both physical and chemical aspects. As a simple example, if species A and B, previously included in a chemical simulation that treats a source as homogeneous in space, are now observed to have spatial distributions that only partially overlap, how do we treat them? Is their spatial distinctiveness due to different physical conditions, which may both be time dependent? Or is there some chemical basis, which of course could depend upon local changes in physical conditions? But, even if two objects appear to be co-spatial along the plane of the sky, they need not be co-spatial in a direction toward the observer. Here, though, one can get information from the Doppler effect, which distinguishes objects by their relative velocity with respect to what astronomers call the "local standard of rest". The existence of more than one narrow Doppler effect or a broad Doppler effect along the line of sight indicates that the matter is not homogeneous in this direction. The three dimensions are now referred to as "data cubes".

The chemistry of inhomogeneous objects with little to no symmetry remains an unsolved problem. Sources with some symmetry have already been treated as inhomogeneous, the longest known of which are so-called PDRs where the acronym stands alternatively for "photon-dominated regions" or "photodissociation" regions [109–111]. These are regions of larger clouds that lie relatively close to an external star, and exhibit differences in molecular abundances as a function of one-dimensional distance from the star. The one-dimensional chemistry has been modeled for some time, but more recently its time dependence has been looked at carefully as well [112]. The study of protoplanetary discs, a stage of intense current interest because the disks are precursors of new planetary systems, can be looked at simply as two-dimensional objects, which viewed from above or below appear to be circular, but in which the distribution of matter becomes cooler and spreads out perpendicularly to the circular plane as a function of distance away from the central star or young stellar object [113]. Chemical simulations of protoplanetary disks, including the gas and dust phases, have been undertaken for some time [114]. It remains difficult to find COMs in these sources, although gaseous methanol has recently been detected with the use of stacking [115].

Although earlier stages of disks are now observed surrounding protostars, and are classified as Class 0 and Class 1 [116], the transition from these stages through the millions of years required for them to become protoplanetary disks is very difficult to treat along with the chemistry because the physical changes starting from earlier stages require the solution of hydrodynamic equations in tandem with solutions of chemical kinetic equations. Moreover, since protoplanetary disks are thought to be the hosts for planet formation, models of the condensation of dust particles into much larger macroscopic objects need to be connected with the chemistry as protoplanetary disks evolve into planetary systems. Such models are just starting to appear and should yield the initial organic chemical inventory of planets.

Finally, astronomy is not always a subject in which time scales are orders of magnitude longer than terrestrial. One of the best-known "giant" molecular clouds, the Orion Nebula, in which in addition to many star-forming regions, there are also many full-fledged stars, appears to have undergone an explosion about 500 years ago, with matter hurtling away from a central object! At much shorter time scales, molecules have been detected in some of the strangest of objects, such as novae, which can be thought of as short-lived recurring explosions. The chemistry of such high-temperature objects may take less than a year to come to completion.

What will the impact be on chemical networks for this burst of activity? It is clear that more binary reactions will be needed at all temperatures. To improve our knowledge of the synthesis of carbon chains and COMs at low temperatures, more attention will have to be paid to the production of molecules somewhat larger than currently handled for both of these classes of molecules. *It is likely that the observation of more COMs in cold cores will drive a renaissance of low-temperature gas phase chemistry.* The synthesis of somewhat larger molecules than currently observed at low temperatures should be emphasized. For example, because methanol is found to be reasonably abundant, ethanol should be looked at more carefully; because methyl formate and dimethyl ether are reasonably abundant, ethyl formate and methyl-ethyl ether should be searched for. An understanding of the synthesis of these molecules at low temperatures will require more laboratory kinetic data. The spectra are certainly available, and stacking techniques are quite powerful for observing weak interstellar signals. The chemistry that synthesizes aromatic species in cold cores is also in dire need of a better understanding.

Are these molecules formed in situ, or do they come from another type of source?

In addition to low temperature kinetic data, more rate coefficients at intermediate and high temperatures will certainly be needed. The highest temperatures in the networks will have to be raised to at least 1000 K, and possibly much higher to study hot regions close to stars and short-lived sources such as shocks [33]. At these high temperatures, one will have to be careful to use detailed balancing to obtain "backwards" endothermic reactions because these endothermic reactions may become important (see Chapter 1). But, in addition, high densities will require the inclusion of ternary reactions in both the three-body range and the saturated limit, as in novae, for example. The current KIDA network does utilize detailed balancing to obtain rate coefficients for selected endothermic binary reactions, and at least one paper reports ternary reaction rates using detailed balancing [117]. Nevertheless, there is obviously much more to be done!

Acknowledgment

E. H. wishes to acknowledge the support of the National Science Foundation (US) via grant AST 19-06489.

References

[1] Tielens AGGM. The physics and chemistry of the interstellar medium. Cambridge University Press, Cambridge; 2010. https://www.cambridge.org/core/books/physics-and-chemistry-of-the-interstellar-medium/B71008A57D96A2875E851DAEA70B16AE

[2] Wakelam V, Smith IWM, Herbst E, Troe J, Geppert W, Linnartz H, Oberg K, Roueff E, Agundez M, Pernot P, Cuppen HM, Loison JC, Talbi D. Reaction networks for interstellar chemical modelling. Space Sci Rev. 2010;156:13–72.

[3] Herbst E. Three milieux for interstellar chemistry. PCCP. 2013;16:3344–3359.

[4] McGuire C, Fuller GA, Peretto N, Zhang Q, Traficante A, Avison A, Jimenez-Serra I. The structure and early evolution of massive star forming regions. Substructure in the infrared dark cloud SDC13. Astron Astrophys. 2016;594(A118):13.

[5] Churchwell E, Sievers A, Thum C. A millimeter survey of ultra-compact HII regions and associated molecular clouds. Astron Astrophys. 2010;513(A9):13.

[6] Woon DE. The astrochymist; 2020. Available from: http://www.astrochymist.org/

[7] Cazaux S, Tielens AGGM. H_2 formation on grain surfaces. Astrophys J. 2004;604(1):222–37.

[7a] Grozdanov TP, McCarroll R, Roueff E. Reactions of $N^+(^3P)$ ions with H_2 and HD molecules at low temperatures. Astron Astrophys. 2016;589:A105.

[7b] Gerlich D, Herbst E, Roueff E. H_3^+ + HD $\leftrightarrow$ H_2D^+ + H_2: low-temperature laboratory measurements and interstellar implications. Planet Sp Sci. 2002;50:1275–85.

[8] Albertsson T, Semenov DA, Vasyunin AI, Henning TH, Herbst E. New extended deuterium fractionation model: assessment at dense ISM conditions and sensitivity analysis. Astrophys J Supp. 2013;207(2):29.

[9] Graninger DM, Herbst E, Oberg KI, Vasyunin AI. The HNC/HCN ratio in star-forming regions. Astrophys J. 2014;787(74):11.

[10] Taniguchi K, Herbst E, Caselli P, Paulive A, Maffucci DM, Saito M. Cyanopolyyne chemistry around massive young stellar objects. Astrophys J. 2019;881(1):13.

[10a] Loomis RA, Burkhardt AM, Shingledecker CN, Charnley SB, Cordiner MA, Herbst E, Kalenskii S, Kelvin Lee KL, Willis ER, Xue C, Remijan AJ, McCarthy MC, McGuire BA. An investigation of spectral line stacking techniques and application to the detection of $HC_{11}N$. Nature Astron. 2021; 5:188–196.

[11] Agúndez M, Cernicharo J, Guélin M, Gerin M, McCarthy MC, Thaddeus P. Search for anions in molecular sources: C_4H^- detection in L1527. Astron Astrophys. 2008;478(1):L19–L22.

[12] Gupta H, Lee K, McCarthy MC. Detection of $CCCH^+$ toward W49N: elucidating the molecular complexity of the diffuse interstellar gas. 74th Int Symp Mol Spectr. 2019;WG02. https://www.ideals.illinois.edu/handle/2142/104259

[13] Herbst E, van Dishoeck EF. Complex organic interstellar molecules. Ann Rev Astron Astrophys. 2009;47(1):427–80.

[14] Cami J, Bernard-Salas J, Peeters E, Malek SE. Detection of C_{60} and C_{70} in a young planetary Nebula. Science. 2010;329(5996):1180–1182.

[15] Cordiner MA, Linnartz H, Cox NLJ, Cami J, Najarro F, Proffitt CR, Lallement R, Ehrenfreund P, Foing BH, Gull TR, Sarre PJ, Charnley SB. Confirming interterlar C_{60}^+ using the hubble space telescope. Astrophys J Lett. 2019;875(L28):7.

[16] Sonnentrucker P, Cami J, Ehrenfreund P, Foing B. The diffuse interstellar bands at 5797, 6379 and 6613 angstroms. Ionization properties of the carriers. Astron Astrophys. 1997;327:1215–21.

[17] McGuire BM, Burkhardt AM, Kalenskii S, Shingledecker CN, Remijan AJ, Herbst E, McCarthy MC. Detection of the aromatic molecule benzonitrile (c-C_5H_5CN) in the interstellar medium. Science. 2018;359(6372):202–5.

[18] McCarthy MC, Lee KLK, Loomis RA, Burkhardt AM, Shingledecker CN, Charnley SB, Cordiner MA, Herbst E, Kalenskii S, Willis ER, Xue C, Remijan AJ, McGuire BA. Interstellar detection of the highly polar five-membered ring cyanocyclopentadiene. Nature Astronomy. 2021;5:176–180.

[19] Allamandola LJ, Hudgins DM, Sandford SA. Modeling the unidentified infrared emission with combinations of polycyclic aromatic hydrocarbons. Astrophys J. 1999;511(2):L115–L119.

[20] Oberg KI, Boogert ACA, Pontoppidan KM, Blake GA, Evans NJ, Lahuis F, van Dishoeck EF. The c2d spitzer spectroscopic survey of ices around low-mass young stellar objects. III. CH_4. Astrophys J. 2008;678(2):1032–41.

[21] Watson WD. The rate of formation of interstellar molecules by ion-molecule reactions. Astrophys J. 1973;183:L17–L20.

[22] Herbst E, Klemperer W. The formation and depletion of molecules in dense interstellar clouds. Astrophys J. 1973;185:505–34.

[23] Su T, Chesnavich WJ. Parametrization of the ion-polar molecule collision rate constant by trajectory calculations. J Chem Phys. 1982;76:5183–5185.

[24] Rowe BR, Dupeyrat G, Marquette JB, Gaucherel P. Study of the reactions $N_2^+ + 2N_2 \rightarrow N_4^+ + N_2$ and $O_2^+ + 2O_2 \rightarrow O_4^+ + O_2$ from 20 to 160 K by the CRESU technique. J Chem Phys. 1984;80:4915–4921.

[25] Rebrion C, Marquette JB, Rowe BR, Clary DC. Low temperature reactions of He^+ and C^+ with HCl, SO_2 and H_2S. Chem Phys Lett. 1988;143:130–34.

[26] Troe J. Statistical adiabatic channel model of ion-neutral dipole capture rate constants. Chem Phys Lett. 1985;122(5):423–430.

[27] Wagenblast R. Interpretation of the level population distribution of highly rotationally excited H_2 molecules in diffuse clouds. Mon Not R Astr Soc. 1992;259(1):155–65.

[28] Solomon PM, Werner MW. Low energy cosmic rays and the abundance of atomic hydrogen in dark clouds. Astrophys J. 1971;165:41–9.

[29] Snyder LE, Hollis JM. HCN, X-ogen (HCO^+), and U90.66 emission spectra from L134. Astrophys J. 1976;204:L139–L142.

[30] Rowe B, Marquette J, Dupeyrat G, Ferguson E. Reactions of He$^+$ and N$^+$ ions with several molecules at 8 K. Chem Phys Lett. 1985;113:403–6.

[31] Marquette J, Rebrion C, Rowe B. Proton transfer reactions of H$_3{}^+$ with molecular neutrals at 30 K. Astron Astrophys. 1989;213:L29–L32.

[32] Black JH, Dalgarno A. The formation of CH in interstellar clouds. Astrophys Lett. 1973;15:79–82.

[33] Elitzur MD, Watson WD. Interstellar shocks and molecular CH$^+$ in diffuse clouds. Astrophys J. 1980;236:172–81.

[34] Herbst E, Schubert JG, Certain PR. The radiative association of CH$_2{}^+$. Astrophys J. 1977;213:696–704.

[35] Gerlich D, Horning S. Experimental investigations of radiative association processes as related to interstellar chemistry. Chem Rev. 1992;92:1509–39.

[36] Herbst E. An approach to the estimation of polyatomic vibrational radiative relaxation rates. Chem Phys. 1982;65(2):185–95.

[37] Rowe BR, Canosa A, Le Page V. FALP and CRESU studies of ionic reactions. Int J Mass Spectrom Ion Proc. 1995;149–150:573–596.

[38] Herbst E. A reinvestigation of the rate of the C$^+$ + H$_2$ radiative association reaction. Astrophys J. 1982;252:810–813.

[39] Speck T, Mostefaoui TL, Travers D, Rowe BR. Pulsed injection of ions into the CRESU experiment. Int J Mass Spectrom. 2001;208: 73–80.

[40] Barlow SE, Dunn GH, Schauer M. Radiative association of CH$_3^+$ and H$_2$ at 13 K. Phys Rev Lett. 1984;52:902.

[41] Herbst E. An update of and suggested increase in calculated radiative association rate coefficients. Astrophys J. 1985;291:226–9.

[42] Vissapagada S, Buzard CF, Miller KA, O'Connor AP, de Rusette N, Urbain X, Savin DW. Recommended thermal rate coefficients for the C + H$_3^+$ reactions and some astrochemical implications. Astrophys J. 2016;832(1):6.

[43] Herbst E. An additional uncertainty in calculated radiative association rates of molecular formation at low temperatures. Astrophys J. 1980;241:197–9.

[44] Rangwala N, Colgan SWJ, Le Gal R, Acharyya K, Huang X, Lee TJ, Herbst E, deWitt C, Richter M, Boogert A, McKelvey M. High spectral resolution SOFIA/EXES observations of C$_2$H$_2$ toward Orion IRc2. Astrophys J. 2018;856(1):12.

[45] Myher JM, Harrison AJ. Ion-molecule reactions in acetylene and acetylene-methane mixtures. Can J Chem. 1968;46:1755–62.

[46] Herbst E, Leung CM. Gas phase production of complex hydrocarbons, cyanopolyynes, and related compounds in dense interstellar clouds. Astrophys J Suppl Ser. 1989;69:271–99.

[47] Xue C, Willis ER, Loomis RA, Lee KLK, Burkhardt AM, Shingledecker CN, Charnley SB, Cordiner MA, Kalenskii S, McCarthy MC, Remijan AJ, McGuire BA. Detection of HC_4NC and an investigation of isocyanopolyyne chemistry in TMC-1 conditions. Astrophys J Lett. 2020;900(1):L9.

[48] Araki M, Takano S, Sakai N, Yamamoto S, Oyama T, Kuze N, Tsukiyama K. Long carbon chains in the warm carbon-chain- chemistry source L1527: first detection of C_7H in molecular clouds. Astrophys J. 2017;847(1):7.

[49] Marquette JB, Rowe BR, Dupeyrat G, Roueff E. CRESU study of the reaction $N^+ + H_2$ yields $NH^+ + H$ between 8 and 70 K and interstellar chemistry implications. Astron Astrophys. 1985;147(1):115–120.

[50] Herbst E, DeFrees DJ, Talbi D, Pauzat F, Koch W, McLean AD. Calculations on the rate of the ion-molecule reaction between NH_3^+ and H_2. J Chem Phys. 1991;94:7842–9.

[51] Sims IR, Queffelec J-L, Travers D, Rowe BR, Herbert LB, Karthauser J, Smith IWM. Rate constants for the reaction of CN with hydrocarbons at low and ultra-low temperatures. Chem Phys Lett. 1993;211:461–8.

[52] Cooke IR, Sims IR. Experimental studies of gas-phase reactivity in relation to complex organic molecules in star-forming regions. ACS Earth Space Chem. 2019;3:1109–34.

[53] de Ruette N, Miller KA, O'Connor AP, Urbain X, Buzard CF, Vissapragada S, Savin DW. Merged-beams reaction studies of $O + H_3^+$. Astrophys J. 2016;816(1):11.

[54] van Dishoeck EF, Herbst E, Neufeld DA. Interstellar water chemistry: from laboratory to observations. Chem Rev. 2013;113(12):9043–85.

[55] Herbst E, Osamura Y. Calculations on the formation rates and mechanisms for C_nH anions in the interstellar and circumstellar media. Astrophys J. 2008;679(2):1670–79.

[56] Gianturco FA, Grassi T, Wester R. Modelling the role of electron attachment rates on column density ratios for C_nH^-/C_nH (n = 4, 6, 8) in dense molecular clouds. J Phys B. 2016;49(20):204003.

[57] Moustefaoui T, Rebrion-Rowe C, Le Garrec J-L, Rowe BR, Brian A, Mitchell J. Low temperature electron attachment to polycyclic aromatic hydrocarbons. Faraday Discuss. 1998;109:71–82.

[58] Douguet N, Fonseca dos Santos S, Raoult M, Dulieu O, Orel AE, Kokoouline V. Theory of radiative electron attachment to molecules: benchmark study of CN^-. Phys Rev A. 2013;88(5):052710.

[59] Jerosimic SVJ, Gianturco FA, Wester R. Associative detachment (AD) paths for H and CN$^-$ in the gas phase: astrophysical implications. Phys Chem Chem Phys. 2018;20:5490–500.

[60] Snow TP, Stepanovic M, Betts NB, Eichelberger BR, Martinez O, Bierbaum V. Formation of gas-phase glycine and cyanoacteylene via associative detachment reactions. Astrobio. 2009;9(10):10001–1005.

[61] Walsh C, Harada N, Herbst E, Millar TJ. The effects of molecular anions on the chemistry of dark clouds. Astrophys J. 2009;700(1): 752–61.

[62] Millar TJ, Herbst E, Bettens RPA. Large molecules in the envelope surrounding IRC+10216. Mon Not R Astr Soc. 2000;316(1):195–203.

[63] Agúndez M, Cernicharo J, Pardo JR, Fonfría Expósito JP, Guélin M, Tenenbaum ED, Ziurys LM, Apponi AJ. Understanding the chemical complexity in circumstellar envelopes of C-Rich AGB stars: the case of IRC+10216. Astrophys Sp Sci. 2008;313(1–3):229–33.

[63a] Dubois D, Carrasco N, Bourgalais J, Vettier L, Desai RT, Wellbrock A, Coates AJ. Nitrogen-containing anions and tholin growth in titan's ionosphere: implications for *Cassini* CAPS-ELS observations. Astrophys J. 2019;872(L31):6.

[64] Bienner L, Carles S, Cordier D, Guillemin J-C, Le Picard SD, Faure A. Low temperature reaction kinetics of CN$^-$ + HC$_3$N and implications for the growth of anions in titan's atmosphere. Icarus. 2014;227:123–31.

[65] Joalland B, Jamal-Eddine N, Klos J, Lique F, Trolez Y, Guillemin J-C, Carles S, Bienner L. Low-temperature reactivity of C$_{2n+1}$N$^-$ anions with polar molecules. J Phys Chem Lett. 2016;7:2957–61.

[66] Haider N, Husain D. Absolute rate data for the reactions of ground-state atomic carbon, C[2p^2(^{3}P$_J$)], with alkenes. J Chem Soc Faraday Trans. 1993;89:7–14.

[67] Clary DC, Haider N, Husain D, Kabir M. Interstellar carbon chemistry: reaction rates of neutral atomic carbon with organic molecules. Astrophys J. 1994;422:416–22.

[68] Woon DE, Herbst E. The rate of the reaction between CN and C$_2$H$_2$ at interstellar temperatures. Astrophys J. 1997;477:204–8.

[69] Carty D, Le Page V, Sims IR, Smith IWM. Low temperature rate coefficients for the reactions of CN and C$_2$H radicals with allene (CH$_2$ = C = CH$_2$) and methyl acetylene (CH$_3$CCH). Chem Phys Lett. 2001;344:310–16.

[70] Abeysekera C, Joalland B, Ariyasingha N, Zack LN, Sims IR, Field RW, Suits AG. Product branching in the low temperature reaction of CN with propyne by chirped-pulse microwave spectroscopy in a uniform supersonic flow. J Phys Chem Lett. 2015;6:1599–604.

[71] Morales SB, Bennett CJ, Le Picard SD, Canosa A, Sims IR, Sun BJ, Chen PH, Chang AHH, Kislov VV, Mebel AM, Gu X, Zhang F, Maksyutenko P, Kaiser RI. Crossed molecular beam, low-temperature kinetics, and theoretical investigation of the reaction of the cyano radical (CN) with 1,3-butadiene (C_4H_6): a route to complex nitrogen-bearing molecules in low-temperature extraterrestrial environments. Astrophys J. 2011;742:10.

[72] Takano S, Masuda A, Hirahara Y, Suzuki H, Ohishi M, Ishikawa S-I, Kaifu N, Kasai Y, Kawaguchi K, Wilson TL. Observations of ^{13}C iosotopomers of HC_3N and HC_5N in TMC-1: evidence for isotopic fractionation. Astron Astrophys. 1998;329:1156–69.

[73] Mebel AM, Kislov VV, Hayashi M. Prediction of product branching ratios in the $C(^3P) + C_2H_2 \rightarrow 1 - C_3H_2 + H/c - C_3H_2 + H/C_3 + H_2$ reaction using ab initio coupled cluster calculations extrapolated to the complete basis set combined with Rice-Ramsperger-Kassel-Marcus and radiationless transition theories. J Chem Phys. 2007;126(20):11.

[74] Hickson KM, Loison J-C, Wakelam V. Temperature-dependent product yields for the spin-forbidden singlet channel of the reaction $C(^3P) + C_2H_2$. Chem Phys Lett. 2016;659:70–5.

[75] Chastaing D, James PL, Sims IR, Smith IWM. Neutral-neutral reactions at the temperatures of interstellar clouds: rate coefficients for reactions of atomic carbon (^3P) with O_2, C_2H_2, C_2H_4 and C_3H_6 down to 15 K. Phys Chem Chem Phys. 1999;1:2247–56.

[76] Luca A, Voulot D, Gerlich D. WDS02. Proceedings of contribution papers Part II; 2002. p. 294.

[77] Geppert WD, Hamberg M, Thomas RD, Osterdahl F, Hellberg F, Zhaunerchyk V, Ehlerding A, Millar TJ, Roberts H, Semaniak J, Af Ugglas M, Kallberg A, Simonsson A, Kaminska M, Larsson M. Dissociative recombination of protonated methanol. Faraday Discuss. 2006;133:177–90.

[78] Watanabe N, Kouchi A. Efficient formation of formaldehyde and methanol by the addition of H atoms to CO in H_2O-CO ice at 10 K. Astrophys J. 2002;571(2):L173–L176.

[79] Garrod RT, Herbst E. Formation of methyl formate and other organic species in the warm-up phase of hot molecular cores. Astron Astrophys. 2006;457(3):927–36.

[80] Garrod RT, Widicus Weaver SL, Herbst E. Complex chemistry in star-forming regions: an expanded gas-grain warm-up chemical model. Astrophys J. 2008;682:283–302.

[81] Cernicharo J, Marcelino N, Roueff E, Gerin M, Jiménez-Escobar A, Muñoz Caro GM. Discovery of the methoxy radical, CH_3O, toward

B1: dust grain and gas-phase chemistry in cold dark clouds. Astrophys J. 2012;759(2):4.

[82] Balucani N, Ceccarelli C, Taquet V. Formation of complex organic molecules in cold objects: the role of gas-phase reactions. Mon Not R Astron Soc. 2015;449:L16–L20.

[83] Vuitton V, Yelle RV, Lavvas P, Klippenstein SJ. Rapid association reactions at low pressure. Astrophys J. 2012;744(1):7.

[84] Shannon RJ, Blitz MA, Goddard A, Heard DE. Accelerated chemistry I. the reaction between the hydroxyl radical and methanol at interstellar temperatures facilitated by tunneling. Nature Chem. 2013;5(9):745–9.

[85] Antiñolo M, Agúndez M, Jiménez E, Ballesteros B, Canosa A, El Dib G, Albaladejo J, Cernicharo J. Reactivity of OH and CH_3OH between 22 and 64 K: modeling the gas phase production of CH_3O in Barnard 1b. Astrophys J. 2016;823(25):8.

[85a] Ocaña AJ, Blázquez S, Potapov A, Ballesteros B, Canosa A, Antiñolo M, Vereecken L, Albaladejo J, Jiménez E. Gas-phase reactivity of CH_3OH toward OH at interstellar temperatures (11.7–177.5 K): experimental and theoretical study. Phys Chem Chem Phys. 2019;21:6942.

[85b] Heard DE. Rapid acceleration of hydrogen atom abstraction reactions of OH at very low temperatures through weakly bound complexes and tunneling. Acc Chem Res. 2018;51:2620–27.

[85c] Canosa A. Gas phase reaction kinetics of complex organic molecules at temperatures of the interstellar medium: the OH + CH_3OH case, in laboratory astrophysics: from observations to interpretation. In: Salama F, Linnartz H, Editors. IAU Proceedings Symposium, N° 350, 2019. Cambridge University Press; 2020. vol 15, pp. 35–40.

[86] Shannon RJ, Caravan RL, Blitz MA, Heard DE. A combined experimental and theoretical study of reactions between the hydroxyl radical and oxygenated hydrocarbons relevant to astrochemical environments. Phys Chem Chem Phys. 2014;16:3466–78.

[87] Ocaña AJ, Blázquez S, Ballesteros B, Canosa A, Antiñolo M, Albaladejo J, Jiménez E. Gas phase kinetics of the OH + CH_3CH_2OH reactions at temperatures of the interstellar medium (T = 21–107 K). Phys Chem Chem Phys. 2018;20:5865–73.

[88] Blázquez S, González A, García-Sáez M, Antiñolo M, Bergeat A, Caralp F, Mereau R, Canosa A, Ballesteros B, Albaladejo J, Jiménez E. Experimental and theoretical investigation on the OH + $CH_3(CO)CH_3$ reaction at interstellar temperatures (T = 11.7–64.4 K). ACS Earth Space Chem. 2019;3(9):1873–83.

[89] Ocaña A, Jiménez E, Ballesteros B, Canosa A, Antiñolo M, Albaladejo J, Agúndez M, Cernicharo J, Zanchet A, Del Mazo P, Roncero O, Aguado A. Is the gas-phase $OH + H_2CO$ reaction a source of HCO in interstellar cold dark clouds? A kinetic, dynamic, and modeling study. Astrophys J. 2017;850:28.

[90] Jiménez E, Antiñolo M, Ballesteros B, Canosa A, Albaladejo J. First evidence of the dramatic enhancement of the reactivity of methyl formate $(HC(O)OOCH_3)$ with OH at temperatures of the interstellar medium: a gas phase kinetic study between 22 and 64 K. Phys Chem Chem Phys. 2016;18:2183–91.

[91] Sleiman C, El Dib G, Rosi M, Skouteris D, Balucani N, Canosa A. Low temperature kinetics and theoretical studies of the reaction $CN + CH_3NH_2$: a potential source of cyanamide and methyl cyanamide in the interstellar medium. Phys Chem Chem Phys. 2018;20:5478–89.

[91a] Sleiman C, El Dib G, Talbi D, Canosa A. Gas phase reactivity of the CN radical with methyl amines at low temperatures (23–297 K): a combined experimental and theoretical investigation. Earth Space Chem. 2018;2:1047–57.

[92] Smith D. The ion chemistry of interstellar clouds. Chem Rev. 1992;92:1473–85.

[93] Taquet V, Wirstrom ES, Charnley SB. Formation and recondensation of complex organic molecules during protostellar luminosity outbursts. Astrophys J. 2016;821(1):12.

[94] Skouteris D, Balucani N, Ceccarelli C, Faginas Lago N, Codella C, Falcinellli S, Rosi M. Interstellar dimethyl ether gas-phase formation: a quantum chemistry and kinetics study. Mon Not R Astr Soc. 2019;482:3567–75.

[95] Jin M, Garrod RT. Formation of complex organic molecules in cold interstellar environments through nondiffusive grain-surface and ice-mantle chemistry. Astrophys. J Suppl. Ser. 2020;249(2):26.

[96] Zhu C, Turner AM, Meinert C, Kaiser RI. On the production of polyols and hydroxycarboxylic acids in interstellar analogous ices of methanol. Astrophys J. 2020;889:134.

[97] Shingledecker CN, Tennis J, Le Gal R, Herbst E. On cosmic-ray-driven grain chemistry in cold core models. Astrophys J. 2018;861(1):15.

[98] Shingledecker CN, Herbst E. A general method for the inclusion of radiation chemistry in astrochemical models. Phys Chem Chem Phys. 2018;20:5359–67.

[99] Cooke IR, Gupta D, Messinger JP, Sims IR. Benzonitrile as a proxy for benzene in the cold ISM: low-temperature rate coefficients for $CN + C_6H_6$. Astrophys J Lett. 2020;891(2):L41.

[100] McGuire BA, Loomis RA, Burkhardt AM, Lee KLK, Shingledecker CN, Charnley SB, Cooke IR, Cordiner MA, Herbst E, Kalenskii S, Siebert MA, Willis ER, Xue C, Remijan AJ, McCarthy MC. Discovery of the interstellar polycyclic aromatic hydrocarbons 1- and 2-cyanonaphthalene. Science. 2021;371:1265.

[101] McCarthy MC, Lee KLK, Loomis RA, Burkhardt AM, Shingledecker CN, Charnley SB, Cordiner MA, Herbst E, Kalenskii S, Willis ER, Xue C, Remijan AJ, McGuire BA. Interstellar detection of the highly polar five-membered ring cyanocyclopentadiene. Nature Astronomy. 2021;5:176–180.

[102] Kroto HW, Heath JR, Obrien SC, Curl RF, Smalley RE. C_{60}: Buckminsterfullerene. Nature. 1985;318(6042):162–3.

[103] Bettens RPA, Herbst E. The interstellar gas phase production of highly complex hydrocarbons: construction of a model. Int J Mass Spectrom Ion Proc. 1995;149–150:321–43.

[104] Rebrion C, Marquette JB, Rowe BR, Adams NG, Smith D. Low-temperature reactions of some atomic ions with molecules of large quadrupole moment: C_6F_6 and c-C_6H_{12}. Chem Phys Lett. 1987;136(6):495–9.

[105] Bettens RPA, Herbst E. The abundance of very large hydrocarbons and carbon clusters in the diffuse interstellar medium. Astrophys J. 1996;468:686–93.

[106] Bettens RPA, Herbst E. The formation of large hydrocarbons and carbon clusters in dense interstellar clouds. Astrophys J. 1997;478(2):585–93.

[107] Berné O, Montillaud J, Joblin C. Top-down formation of fullerenes in the interstellar medium. Astron Astrophys. 2015;577:9.

[108] Berné O, Montillaud J, Joblin C. Top-down formation of fullerenes in the interstellar medium (Corrigendum). Astron Astrophys. 2016;588:1.

[109] Tielens AGGM, Hollenbach D. Photodissociation regions. I. Basic model. Astrophys J. 1985;291:722–46.

[110] Tielens AGGM, Hollenbach D. Photodissociation regions II. A model for the orion photodissociation region. Astrophys J. 1985;291:747–54.

[111] Le Petit F, Nehme C, Le Bourlot J, Roueff E. A model for atomic and molecular interstellar gas: the Meudon PDR code. Astrophys J Suppl Ser. 2006;164:506–29.

[112] Le Gal R, Herbst E, Dufour G, Gratier P, Ruaud M, Vidal THG, Wakelam V. A new study of the chemical structure of the Horsehead nebula: the influence of grain-surface chemistry. Astron Astrophys. 2017;605:A88.

[113] Bergin EA, Aikawa Y, Blake GA, van Dishoeck EF. The chemical evolution of protoplanetary disks. In: Reipurth VB, Jewitt D, Keil K, Editors. Protostars and planets. Tucson: University of Arizona Press; 2007. pp. 751–766.

[114] Aikawa Y, Furuya K, Hincelin U, Herbst E. Multiple paths of deuterium fractionation in protoplanetary disks. Astrophys J. 2018;855:21.

[115] Walsh C, Loomis RA, Oberg KI, Kama M, van't Hoff MLR, Millar TJ, Aikawa Y, Herbst E, Widicus Weaver SL, Nomura H. First detection of gas-phase methanol in a protoplanetary disk. Astrophys J. 2016;823(1):L10.

[116] Enoch ML, Corder St, Dunham MM, Duchene G. Disk and envelope structure in class 0 protostars. I. The resolved massive disk in serpens firs 1. Astrophys J. 2009;707(1):103–13.

[117] Willacy K, Klar HH, Millar TJ, Henning Th. Gas and grain chemistry in a protoplanetary disk. Astron Astrophys. 1998;338:995–1005.

Chapter 11

Theoretical Rate Constants

Philippe Halvick* and Thierry Stoecklin[†]

*Université de Bordeaux, Institut des Sciences Moléculaires,
CNRS UMR 5255, 33405 Talence, France*
*philippe.halvick@u-bordeaux.fr
†thierry.stoecklin@u-bordeaux.fr

Abstract

A review of selected theoretical works devoted to the understanding of
the rate constants measured by the flow supersonic reactor experiments
is presented. This review focuses on the kinetics of reactive and inelastic
neutral–neutral collisions at low temperature. Neutral–neutral reactions
can show puzzling non-Arrhenius temperature dependence, while most of
ion–neutral collisions are well explained by a long-range capture model.
First, a short summary of the adiabatic capture method and the variational transition state theory is presented. Then, a variety of reaction
mechanisms are detailed, involving spin-orbit effects, short-range potential effects, and quantum tunneling effects, along with the theoretical
methods necessary to unveil these mechanisms. In the second part of
this chapter, a similar analysis is applied to purely inelastic collisions.
The different models available for describing specifically electronic, spin-
orbit, vibrational, and rotational transitions are reviewed and compared
to the results of Close Coupling calculations.

Keywords: Reaction rate constants; Low temperature; Fast reactions;
Adiabatic capture; Transition state theory; Semi-classical dynamics;
Inelastic collisions; Electronic quenching; Vibrational quenching; Rota-
tional quenching; Spin-orbit relaxation.

11.1   Reaction Rate Constants

11.1.1   *Introduction*

The flow supersonic reactor experiments, by providing numerous measurements of low-temperature rate constants on a variety of exothermic neutral–neutral reactions, have been a major challenge for theory. These reactions may have a low-energy barrier or no barrier at all. They are rapid, especially at low temperature, and their important role in low-temperature environments, such as planetary atmospheres, protostellar disks, and dense interstellar clouds, is nowadays widely recognized. There is an intense scientific interest in this chemistry which leads to the formation of the large inventory of molecules, including complex organic molecules (COMs) and prebiotic molecules [1] which have been observed in cold and dense interstellar environments. For most of these reactions, at least one of the reactants is an open-shell species such as an atom or a radical. At low temperature, it is essentially the long-range interaction between reactants, which determines the magnitude of the cross section and the rate constant. As the temperature rises, short-range effects such as change of the electronic configuration usually associated with a potential barrier, steric repulsion, and chemical bonding become progressively more important.

While it has long been recognized that the rate constants of exothermic reactions with no activation energy can rise drastically as temperature decreases, recent experiments have unveiled a new class of exothermic reactions which are fast at low temperature despite a substantial energy barrier [2]. This is in contrast with the commonly accepted principle that almost no reactive process is possible if the collision energy is significantly smaller than the activation barrier. These reactions are essentially the reactions of OH with COMs, such as formaldehyde, methanol, ether, and aldehyde groups [3–11]. The reaction mechanism proceeds in two steps. First, a van der Waals structure stabilized by a strong long-range dipole–dipole interaction is formed. Then, hydrogen abstraction, allowed by tunneling through the reaction barrier, leads to the products. The two important physicochemical properties which foster such reactions are the large dipole of the COM and the light mass of H.

The potential energy surface (PES) is the fundamental ingredient for any calculation of the reaction rate constants. Theoretical methods can be classified according to the amount of knowledge they require on the PES. Capture theories need only the values of a few electric multipole moments and polarizability properties of the reactants from which a realistic approximation of the long-range interaction energy can be set up. Statistical theories require potential energy data only in a small region around every bottleneck existing along the reaction path in order to evaluate more or less qualitatively the energy levels of the internal motions. Quasi-classical trajectories (QCTs) or quantum dynamics need the global PES, i.e. the mathematical model of the PES in all the regions of the coordinate space which are expected to be visited by the classical trajectories or populated by the scattering wave function at the highest energy considered. If only a small subset of the nuclear coordinates is directly involved in the bond breaking or bond forming of the reactive process, then a strategy of reduced dimensionality can be used [12, 13]. The coordinate space is reduced by transforming the spectator coordinates in frozen coordinates or relaxed coordinates.

It is nowadays easy to compute global PES with high-level ab initio methods for system with a few free degrees of freedom, namely, small polyatomic systems comprising 3 or 4 atoms. However, for larger systems, calculation of a global PES, or even a part of it, becomes rapidly out of reach since each supplementary nuclear coordinate corresponds to an increase by about a factor 10 of the demand for computational resources. Indeed, the short-range interaction is usually very anisotropic, with high values of the first and second derivatives of the potential energy with respect to the internuclear coordinates. Consequently, a high density of ab initio points is required for a realistic model of PES. With the QCT method, it is possible to compute the potential energy on the fly instead of building a global PES. However, in that case, the accuracy of the ab initio calculations has to be limited in order to keep the computational demand below a reasonable limit since classical trajectories usually need a large number of potential energy calculations in order to be propagated accurately.

Obviously, methods using global PES are the most accurate ones, but can be applied only to small systems. Methods using only a small part of the PES are approximate, but can be applied to large systems. The challenge is to control the quality of the approximations made so that realistic results can be obtained even for large polyatomic molecules. This is essentially the playing field of the statistical methods and, to a lesser extent, of the QCT and quantum dynamics within the reduced dimensionality approach. In this chapter, we summarize the methods most frequently used to reproduce the rate constants measured by CRESU (French acronym for Reaction Kinetics in a Uniform Supersonic Flow), namely, the adiabatic capture and the variational transition state theory. Then, we discuss a selection of theoretical studies of neutral–neutral reactions, focusing on the spin-orbit effects, the short-range potential effects, and the tunneling effects, all observed in CRESU rate constants.

11.1.2 *Adiabatic capture*

A method which combines rotationally adiabatic capture theory and centrifugal sudden approximation (ACCSA) was proposed [14–17] to treat the case of exothermic reactions, which are governed by long-range attractive intermolecular forces. The PES of such reactions is assumed to not present any energy barriers which could impede the close encounter of the reactants. This is mostly the case when at least one of the reactants is a radical or an ion. These reactions are fast, with large rate constants usually close to the Langevin limit. It is therefore reasonable to assume a classical capture approximation. However, the reaction can occur if there is enough energy in the radial motion to overcome the centrifugal rotational barrier. Furthermore, it is not necessary to solve the Schrödinger equation over the whole reaction path, but only in the long-range part of the entrance channel to the reaction, where the interaction potential can be described in terms of multipolar moments and polarizabilities of the reactants.

In the following paragraphs, we outline the main features of the ACCSA method in the case of the reaction between an atom and a linear molecule for which the vibration motion is neglected, and therefore reduced to a linear rigid rotor. The collisional system is described by the vector $\mathbf{R}$ which lies between the atom and the center of mass of the rigid rotor, and the rotation angle θ of the rigid rotor. The interaction potential is defined in a 2-dimensional space, spanned by the body-fixed coordinates $R = \|\mathbf{R}\|$ and θ. The Hamiltonian, written in the body-fixed coordinates and using atomic units, is as follows:

$$\hat{H} = -\frac{1}{2\mu R}\frac{\partial^2}{\partial R^2}R + B\hat{j}^2 + \frac{|\hat{J} - \hat{j}|^2}{2\mu R^2} + V(R,\theta) \qquad (11.1)$$

where $\hat{j}$ is the rotational angular momentum operator of the rigid rotor, $\hat{J}$ is the total angular momentum operator of the atom–molecule system, B is the rotational constant of the rigid rotor, $V(R,\theta)$ is the interaction potential, and μ is the reduced mass of the collisional system. A basis set of body-fixed wave functions which are eigenfunctions of the operators $\hat{J}, \hat{j}$ and $\hat{\Omega}$ is used to solve the Hamiltonian H. $\hat{\Omega}$ is the projection of $\hat{J}$ or $\hat{j}$ on an arbitrary chosen body-fixed axis, usually the axis defined by the vector **R**. The centrifugal potential is given by the term $|\hat{J} - \hat{j}|^2/|2\mu R^2$ which couples the basis set wave functions with different values of Ω. In the centrifugal sudden approximation (CSA), it is assumed that the latter coupling matrix elements can be neglected, therefore reducing the representation of the centrifugal potential to a diagonal matrix with the elements:

$$\frac{J(J+1) + j(j+1) - 2\Omega^2}{2\mu R^2} \qquad (11.2)$$

The CSA is known to give reasonable results for averaged quantities such as integral cross sections or rate constants. The sum of the centrifugal and interaction potentials

$$\frac{J(J+1) + j(j+1) - 2\Omega^2}{2\mu R^2} + V(R,\theta) \qquad (11.3)$$

is diagonalized, giving a set of rotationally adiabatic potential energy curves denoted by $\varepsilon_{Jj\Omega}(R)$. Other approximations of the rotational motion have been also employed with the adiabatic capture. Applying the infinite order sudden approximation (IOSA) gives the ACIOSA method [14]. Combining IOSA for one reactant and CSA for the other reactant gives the partial centrifugal sudden approximation (ACPCSA) [14, 18].

The capture approximation assumes that the chemical reaction will happen if the reactants have enough energy to pass over the centrifugal barrier. This approximation is reliable in the case of strongly exothermic reactions, since once the reactants have formed an intermediate complex, it is very unlikely that the complex will be able to decompose back to the reactants. Indeed, for the intermediate complex, the reactants' channel is now behind a potential energy barrier, which can be passed only if the major part of the available kinetic energy is deposited on the reactants' recoil mode of motion.

For given values of J, j, Ω, and the initial translational energy E_t, the reaction probability $P_{Jj\Omega}$ is 0 or 1 depending on if there is enough translational energy to pass the centrifugal barrier:

$$E_t \geq \varepsilon_{Jj\Omega}(R) \Rightarrow P_{Jj\Omega} = 1 \qquad (11.4)$$

$$E_t < \varepsilon_{Jj\Omega}(R) \Rightarrow P_{Jj\Omega} = 0 \qquad (11.5)$$

The first inequality must hold for any value of R, except at short range where the interaction potential $V(R,\theta)$ becomes strongly repulsive. Conversely, the second inequality needs to hold only in one point at least. The reaction cross section is then obtained from the partial wave expansion

$$\sigma_j(E_t) = \frac{\pi}{k^2(2j+1)} \sum_{J=0}^{J_{max}(j,\Omega)} \sum_{\Omega=-\min(J,j)}^{\min(J,j)} (2J+1)P_{Jj\Omega} \qquad (11.6)$$

where $k^2 = 2\mu E_t$. The rate constant is obtained by the Maxwell-Boltzmann integral on the initial values of j and E_t:

$$k(T) = \left[\frac{8}{\pi\mu(k_BT)^3} \right]^{1/2} \sum_j Q_j(T) \int_0^\infty \sigma_j(E_t)e^{-E_t/k_BT}E_t dE_t$$

$$(11.7)$$

$$Q_j(T) = \frac{(2j+1)e^{-Bj(j+1)/k_B^T}}{\sum_j (2j+1)e^{-Bj(j+1)/k_BT}} \qquad (11.8)$$

where $Q_j(T)$ is the rotational population, k_B is the Boltzmann constant, and T the temperature.

The same theory can be also applied to reactions involving two linear rigid rotors, as well as symmetric or asymmetric or spherical top rigid rotors in collision with an atom or another rigid rotor [19]. For example, let us just write the Hamiltonian in the case of two linear rigid rotors. There are four body-fixed coordinates, denoted by

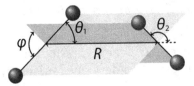

Fig. 11.1. Body-fixed coordinates for a system of two diatomic molecules.

R, θ_1, θ_2, and φ as defined by Fig. 11.1. The body-fixed Hamiltonian is then written as follows:

$$\hat{H} = -\frac{1}{2\mu R}\frac{\partial^2}{\partial R^2}R + B_1\hat{j}_1^2 + B_2\hat{j}_2^2 + \frac{|\hat{J} - \hat{j}|^2}{2\mu R^2} + V(R, \theta_1, \theta_2, \varphi)$$

$$(11.9)$$

where $\hat{j} = \hat{j}_1 + \hat{j}_2$ is the sum of the rotational angular momentum operators of both rigid rotors and B_1 and B_2 are the rotational constants.

The ACCSA has been applied successfully in many studies not only for ion–polar systems [15, 20], but also for molecule–molecule reactions involving both ions and neutrals. A variety of other theoretical treatments based on capture theory have been developed and applied successfully on ion–polar molecule reactions [21–23], among which the statistical adiabatic channel model (SACM) [24, 25] is quite close to ACCSA. Indeed, considering a simple isotropic interaction potential such as

$$V(R) = -\frac{C}{R^n}(n > 2)$$

$$(11.10)$$

for the reaction between an atom and a linear rotor molecule, both methods give the same result [26].

Capture theories based on the use of long-range interaction are generally expected to provide an upper bound to the true rate constant. However, for neutral–neutral reactions, the rate constants calculated by capture theory are seldom in agreement with the experimentally observed ones. Besides the neglect of the short-range interaction, this poor performance might come from the difficulties in building an accurate model of the PES. The long-range potential is weaker than for ion–neutral reactions and it is the sum of a variety of contributions such as dispersion, induction, and electrostatics [27], none of which could be neglected.

11.1.3　*Variational transition state theory*

The transition state theory (TST) and its extension specially designed for unimolecular reaction, the Rice-Ramsperger-Kassel-Marcus (RRKM) theory, were developed originally for reactions with a potential energy barrier. TST is based on the reaction bottleneck which must be identified in the reaction path, namely, the transition state (TS). The latter is defined as a hypersurface in the phase space that divides reactants from products. The fundamental assumption is that the TS corresponds to a configuration of no return. In other words, once the system has passed the TS, then the probability of the reaction is equal to 1. If we transpose this assumption to classical dynamics, it means that all trajectories cross the transition-state dividing surface only once. The saddle point associated with an energy barrier is the highest energy point on the minimum energy path (MEP) from reactants to products, and therefore it appears as a good choice for the location of the TS. TST computes the rate of crossing at the TS by counting the number of reactants passing toward products, per unit time. If the condition of no return is not satisfied, then TST provides an upper bound to the rate. Or, conversely, the TS corresponds to the location on the reaction path where the crossing rate is minimal [28]. This generalized definition of the TS allows one to apply TST to barrierless reactions [29–31]. In practice, this requirement is satisfied by the variational minimization of the calculated rate constant with respect to the curvilinear coordinate of the reaction. This procedure is referred to as variational TST (VTST). Various implementations of VTST have been developed, depending on if the reaction coordinate is defined as the distance between the two atoms involved in bond breaking, or between the centers of mass of the two fragments, or as the parametric coordinate along the MEP [32].

The TST rate constant for unimolecular dissociation or bimolecular reaction at a given total energy E and total angular momentum J is given by

$$k_J(E) = \frac{N_J^\dagger(E)}{h\rho_J(E)} \tag{11.11}$$

where $\rho_J(E)$ is the density of states of the reactants at given E and J (per unit energy per unit volume for bimolecular reactions or per unit energy for unimolecular dissociation), while $N_J^\dagger(E)$ is the flux integral in the product direction through the dividing surface [33–35]. The latter quantity corresponds also to the number of quantum states on the dividing surface [33] for a given J and with an energy lower than or equal to E. The rate constant in a microcanonical ensemble is given by

$$k(E) = \frac{\sum_J (2J+1)\rho_J(E) k_J(E)}{\rho(E)} \tag{11.12}$$

where

$$\rho(E) = \sum_J (2J+1)\rho_J(E) \tag{11.13}$$

Let $N^\dagger(E)$ be the number of internal quantum states of the TS with an energy lower than or equal to E, defined by

$$N^\dagger(E) = \sum_J (2J+1) N_J^\dagger(E) \tag{11.14}$$

then

$$k(E) = \frac{N^\dagger(E)}{h\rho(E)} \tag{11.15}$$

Finally, the rate constant in a canonical ensemble is given by

$$k(T) = \frac{1}{Q(T)} \int_0^\infty k(E)\rho(E) e^{-\beta E} \mathrm{d}E \tag{11.16}$$

where $Q(T)$ is the partition function of reactants given by

$$Q(T) = \int_0^\infty \rho(E) e^{-\beta E} \mathrm{d}E \tag{11.17}$$

and

$$\beta = \frac{1}{k_{\mathrm{B}} T} \tag{11.18}$$

If we define the partition function of the TS by

$$Q^\dagger(T) = \int_0^\infty \rho^\dagger(E) e^{-\beta E} \mathrm{d}E \tag{11.19}$$

then the canonical rate constant is given by

$$k(T) = \frac{1}{\beta h} \frac{Q^\dagger(T)}{Q(T)} e^{-\beta V^\dagger} \tag{11.20}$$

where $V^\dagger$ is the potential energy of the TS defined with respect to the ground state energy of the reactants. VTST can be implemented with any of the three formulations of TST defined above. We distinguish first the fixed-J microcanonical VTST (μJ-VTST) based on the E and J resolved TST rate constant (Eq. (11.11)), second the microcanonical VTST (μ-VTST) based on the E resolved TST rate constant (Eq. (11.15)), and third the canonical VTST (C-VTST) based on Eq. (11.20). Since the location of the TS is determined by the minimum of $N_J^\dagger(E)$ in the μJ-VTST implementation, then a specific TS has to be sought for each value of E and J. In the μ-VTST implementation, it is for each value of E.

In some cases, the VTST rate constant has more than one minimum as a function of the reaction coordinate. In terms of classical trajectories, this means that some trajectories reflected by a second TS are expected to recross the first TS previously crossed. For such reactions, a simple estimate of the μ-VTST rate constant is provided by the unified statistical theory [34]. In the case of two TSs along the reaction path, let us denote by $N_1^\dagger(E)$ and $N_2^\dagger(E)$ the microcanonical number of states at each TS, and by $N_{max}^\dagger(E)$ the maximum of the number of states in the region between both TSs. Then, the two-transition state μ-VTST rate constant is calculated with an effective number of states given by

$$N^\dagger(E) = \left(\frac{1}{N_1^\dagger(E)} + \frac{1}{N_2^\dagger(E)} - \frac{1}{N_{max}^\dagger(E)} \right)^{-1} \qquad (11.21)$$

This can be applied also to the μJ-VTST rate constant calculation by considering the microcanonical numbers of states for every given value of J. If $N_{max}^\dagger(E)$ can be assumed much larger than $N_1^\dagger(E)$ and $N_2^\dagger(E)$, then

$$N^\dagger(E) \cong \left(\frac{1}{N_1^\dagger(E)} + \frac{1}{N_2^\dagger(E)} \right)^{-1} \qquad (11.22)$$

In the case of a reaction with two or more product channels, we have several independent reaction paths which act in parallel. For example, let us assume that the reactants overcome a first bottleneck and form an intermediate complex, which in turn dissociates toward several products. Each dissociation pathway (labeled by i) has its own bottleneck and then its own number of states $N_{2i}^{\dagger}(E)$ at its own TS. The effective global number of states is then

$$N^{\dagger}(E) \cong \left(\frac{1}{N_1^{\dagger}(E)} + \frac{1}{\sum_i N_{2i}^{\dagger}(E)} \right)^{-1} \qquad (11.23)$$

The canonical unified statistical (CUS) method [36] provides a direct formulation of the rate constant for a canonical ensemble. In the case of a two TSs reaction, the rate constant is

$$k^{\text{CUS}}(T) = \left(\frac{1}{k_1(T)} + \frac{1}{k_2(T)} - \frac{1}{k_{max}(T)} \right)^{-1} \qquad (11.24)$$

In the case of several parallel independent reaction paths, as described just above, the CUS rate constant is replaced by the competitive CUS (CCUS) rate constant [36]:

$$k^{\text{CCUS}}(T) = \left(\frac{1}{k_1(T)} + \frac{1}{\sum_i k_{2i}(T)} - \frac{1}{k_{max}(T)} \right)^{-1} \qquad (11.25)$$

11.1.4 *Spin-orbit effects*

The fine structure is an important issue in the low-temperature reactivity of exothermic reactions without a barrier. Within the approximation where only the long-range forces are considered, the theory is based on the coupling of the electronic, rotational, and electronic spin angular momenta of the reactants as well as the orbital angular momenta. The Hamiltonian, including rotational and electronic angular momenta, and spin-orbit coupling is diagonalized for a set of values of the intermolecular separation R in order to obtain the long-range adiabatic potential energy curves. Every potential curve correlates with a state of the reactants and has its particular shape and therefore its particular reaction rate constant. Depending on temperature, the thermal population of the energy levels of reactants will then control the reactivity. Depending on the symmetry properties of the adiabatic states, crossings and avoided crossings are observed [37].

The calculation of rate constants for the $O(^3P) + OH(X^2\Pi)$ reaction was done with the adiabatic capture approximation, using ACIOSA [37]. Only the two lowest-order terms of the electrostatic interaction were retained: dipole–quadrupole and quadrupole–quadrupole. The matrix representation of these operators was obtained in the fine structure basis set expansion, yielding an 18×18 potential matrix. The diagonalization of the latter matrix gives rise to 18 doubly degenerate potential energy surfaces correlating to the 36 fine structure states of the reactants $O(^3P) + OH(X^2\Pi)$. In the fine structure basis set expansion, the spin-orbit operator is easily represented as a diagonal matrix by using the fine structure splitting of reactants. This approximation is valid only for the long-range region. The dispersion interaction and the coupling between the electronic and rotational angular momenta of OH were neglected. A good agreement was obtained with the experimental rate constant. Furthermore, the latter authors investigated the effects of nonadiabatic couplings by solving the radial Hamiltonian within the IOSA and found them to be of minor importance, therefore supporting the validity of the adiabatic approximation.

The rate constant of the $Al + O_2$ reaction was calculated with the ACCSA method and a long-range potential including the quadrupole–quadrupole and dispersion contributions [38, 39]. The matrix elements of this potential were evaluated in the spin-orbit basis set, using the same procedure as Graff and Wagner for the $O + OH$ reaction, but with the dispersion interaction. The temperature dependence of the ACCSA rate constant shows a good agreement with the experimental one (see Fig. 11.2). Indeed, negative temperature dependence is shown, i.e. the rate constantly decreases as the temperature rises, with a slope close to the experimental one for the temperature range 20–400 K. However, the ACCSA rate constant is about 50% larger than the experimental one. Conversely, if the spin-orbit coupling is ignored, then a positive temperature dependence of the rate constant is observed, therefore demonstrating that the negative temperature dependence of the rate constant is due to the open-shell nature of the reactants. The effects of the fine structure have also proven to be essential in the low-temperature rate constants of reactions $C + NO$ [40, 41], $C + O_2$ [41], $Si + O_2$, $Si + NO$ [42], $B + O_2$ [43], $O + OH$, $S + OH$, $N + OH$ [44].

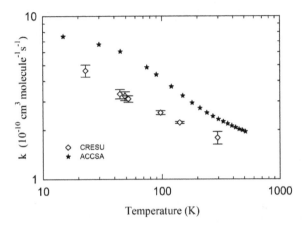

Fig. 11.2. Measured and calculated rate constants for the Al+O$_2$ reaction. CRESU experimental rate constants and ACCSA rate constants are from Le Picard *et al.* [38].

11.1.5 *Short-range potential effects*

For many neutral–neutral reactions, the adiabatic capture theory has provided rate constants showing strong discrepancies with the experimental measurements. One of the most extensively investigated reactions is the CN + O$_2$ reaction for which the adiabatic capture rate constant has been calculated with three different approximations of the reactants rotational motion, namely, ACIOSA, ACPCSA, and global CSA [45]. With an interaction potential including only the dipole–quadrupole contribution, all three variants of capture theory give a negative temperature dependence in the range 30–200 K, and then a weakly positive temperature dependence in the range 200–500 K. But, if the interaction potential includes both the dipole–quadrupole and the dispersion contributions, then the rate constants rise steeply in the temperature range 1–5 K and continue to rise monotonously in the range 5–500 K, but more gradually. In contrast, all experimental rate constants in the same temperature range show a negativetemperature dependence [46–48]. Indeed, since both CN and O$_2$ are radicals, their intermolecular potential is expected to be monotonically attractive and then a negative temperature dependence of the reactive rate constant is also expected. Consequently, rather than using only a long-range model, Klippenstein and Kim [47]

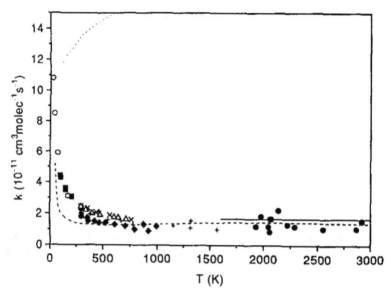

Fig. 11.3. Calculated and measured rate constant for the reaction $CN+O_2$. The dashed line represents the C-VTST results [47], while the dotted line represents the global CSA results [45]. Experimental data are from (o) Sims *et al.* [49], (■) Sims and Smith [50], (×) Sims and Smith [51], (Δ) Durant and Tully [52], (♦) Atakan *et al.* [53], (+) Balla and Casleton [54], (•) Burmeister *et al.* [55], and (solid line) Davidson *et al.* [56] Reproduced with permission from Ref. [47], Copyright 1993 AIP Publishing.

performed ab initio calculations of the middle- and short-range interactions. They employed the C-VTST method to calculate the reaction rate constant and obtained a qualitative agreement with the experiment (see Fig. 11.3). The C-VTST rate constant reproduced quite well the increase which is observed experimentally as the temperature falls from 3000 to 50 K.

This result demonstrates that explicitly including ab initio short-range interactions is a key factor for a realistic numerical simulation of the $CN+O_2$ reaction. Indeed, the latter reaction proceeds through the formation of the NC–OO complex. In the path from the reactants up to the NC–OO complex, there is the important step where the CO bond formation is accompanied by the weakening of the O_2 bond. This effect must be properly included in the interaction energy model in order to calculate a realistic temperature dependence of the rate constant.

A further enlightening example is provided by the reaction $CN+C_2H_2$. Liao and Herbst [57] have calculated the rate constant of this reaction with the ACCSA and SACM methods. They used an interaction potential comprising a dipole-quadrupole term, an isotropic induction term, and an isotropic dispersion term [27]. The calculated rate constants were shown to increase as the temperature rises from 25 K up to 298 K. This is in qualitative agreement with the experimental rate constant [58], which increases slowly between 20 and 50 K. But, there is a large discrepancy in the range 50–700 K where the experimental rate constant decreases quickly. An attempt to check if a different angular dependence of the interaction potential could significantly change the calculated rate constant was unsuccessful. Liao and Herbst assumed that the capture theory based on long-range potential is valid only at very low temperature, pointing out that the short-range potential effects must be included in the calculations since at higher temperatures, and hence on average higher collision energies, the bottleneck for capture moves to closer separation of the reactants. Consequently, Woon and Herbst [59] investigated the minimum energy path, using the coupled cluster method CCSD(T). They found that the reaction $CN + C_2H_2$ proceeds through the intermediate complex C_3H_2N and then dissociates to the products $H + HC_3N$ by passing an exit barrier which occurs at 9.7 kcal/mol below the energy of the ground state of reactants. Such an exit channel barrier tends to reduce the reaction rate when the sum of the potential energy barrier and centrifugal barrier becomes larger than the energy of the reactants, which occurs progressively as the temperature increases. The rate constant was calculated with a phase-space approach conserving the total angular momentum and combined with an RRKM calculation of the C_3H_2N dissociation rate constant toward the products $H + HC_3N$ [59]. At low temperature, the phase-space rate constant is very close to the ACCSA rate constant. But, above 100 K, in contrast with the ACCSA rate constant, the phase-space rate constant shows a negative temperature dependence in agreement with the experimental one. However, the agreement being rather qualitative, the method used by Woon and Herbst had room for improvement.

The ACPCSA has been applied to reactions involving a polar linear molecule and a polar symmetric top: $CN + NH_3$, $CH + NH_3$ and $OH + PBr_3$ [19]. The long-range interaction energy between reactants was considered as the sum of dipole–dipole, induction, and dispersion contributions. At 300 K, the ACPCSA rate constants for the reactions CH + NH$_3$ and OH + PBr$_3$ are about two times larger than the experimental ones, and ten times larger for the reaction CN + NH$_3$ [51, 60, 61]. Considering that the dipole moment and isotropic polarizabilities of CH and CN are quite close, this difference between the reaction rate constants of NH$_3$ with CH and CN comes as a surprise. Furthermore, the experimental rate constant of the CH + NH$_3$ reaction exhibits a steeper negative temperature dependence than the ACPCSA one.

Sims *et al.* [46] reported a low-temperature experimental rate constant for the CN + NH$_3$ reaction, in the range 25–295 K. A particularly strong negative temperature dependence of the rate constant was observed. As a matter of fact, the rate constant is multiplied by ~17 when the temperature is lowered from 298 K to 25 K. Talbi and Smith [62] investigated the PES of the latter reaction and found that only HCN + NH$_2$ can be produced at low temperature, since no low-energy pathway to NCNH$_2$ + H was found. The reaction proceeds via several steps: first, an intermediate complex NCNH$_3$, then a submerged barrier which is the consequence of the abstraction of one H atom by the CN molecule, therefore leading a second intermediate complex, NCH-NH$_2$, which in turn dissociates to the HCN + NH$_2$ products. A schematic minimum energy path is presented in Fig. 11.4. They computed the reaction rate constant by using the μJ-VTST method combined with the procedure for two TSs as defined by Eq. (11.22). The outer TS, which corresponds to the long-range centrifugal barrier, was treated within the approach described by Georgievskii and Klippenstein [63]. The inner TS is close to the saddle point of the submerged barrier. A good agreement with the experimental rate constant was found [46, 51].

The calculation of separate rate constants for each of the two TSs provides a physical insight into the reaction process. The capture rate constant, calculated with the outer TS, remains almost constant on a large range of temperature, from 25 K up to 298 K, while the abstraction rate constant, calculated with the inner TS, is much larger at 25 K than the capture rate constant, and then sharply decreases as temperature rises and becomes much lower than the capture rate (see Fig. 11.5).

The calculation of separate rate constants for each of the two TSs provides a physical insight into the reaction process. The capture rate

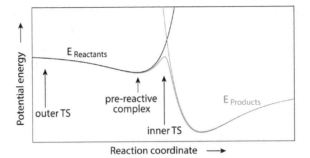

Fig. 11.4. Schematic minimum energy path for an exoergic reaction with a barrier resulting from a change in the electronic configuration. $E_{Reactants}$ and $E_{Products}$ are the diabatic energies of the electronic configurations of reactants and products, respectively. Diagonalization of the electronic Hamiltonian couples both configurations and gives the adiabatic energy (red curve). Depending on the position of the crossing point, on the intensity of the electronic coupling and on the vibrational zero-point energy, the barrier can be submerged or not.

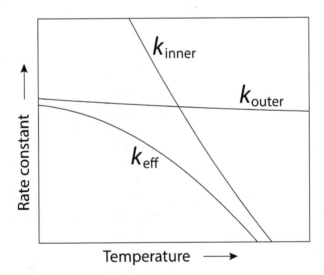

Fig. 11.5. Kinetic scheme for the two TSs case. k_{inner} and k_{outer} are the rate constants calculated with a single TS, the inner and outer TS, respectively. k_{eff} is the rate constant resulting from the two-TS calculation.

constant, calculated with the outer TS, remains almost constant on a large range of temperature, from 25 K up to 298 K, while the abstraction rate constant, calculated with the inner TS, is much larger at 25 K than the capture rate constant, and then sharply decreases as temperature rises and becomes much lower than the capture rate (see Fig. 11.5). Indeed, as the temperature increases, so does the relative orbital angular momentum of the reactants. The orbital rotational energy, added to the potential energy of the saddle point, prevents the reactants from getting close enough for bond formation.

The reaction of addition of OH to C_2H_4 provides another example of the effect of a submerged barrier on the temperature dependence of the reaction rate constant. The main contribution to the long-range potential energy is the interaction between the dipole of OH and the quadrupole of C_2H_4. The formation of a chemical bond between the oxygen atom of OH and a carbon atom of C_2H_4 is accompanied by the breaking of the π bond of C_2. The consequence of this electronic rearrangement is a potential energy barrier which separates the two potential wells associated with the van der Waals structure and the C_2H_4OH adduct. Two TSs can be identified. The inner TS is located in the neighborhood of the latter barrier, while the outer TS is located at the large separation between the reactants and the van der Waals structure.

A μJ-VTST treatment of both inner and outer TSs has been performed by Greenwald *et al.* [64], using the approximation defined by Eq. (11.22) adapted to the case where the total angular momentum J is conserved. A remarkably good agreement between the calculated and the experimental rate constants is observed (see Fig. 11.6), both for the temperature and pressure dependences. The relative contributions of each TS have been evaluated. The rate constant calculated with only the outer TS is almost independent of the temperature, while the rate constant of the inner TS shows a very fast negative temperature dependence. Both rate constants have the same value around 130 K. However, the total rate constant starts to decrease well before the temperature of 130 K is reached as a consequence of the interaction between both TSs implemented in the two-TS μJ-VTST treatment. The inner and outer TSs act as a series of bottlenecks to reaction. The outer TS is the dominant bottleneck of the reaction at low temperature, but the inner TS becomes progressively the dominant bottleneck as the temperature rises. The J-conserving model is an important ingredient for the accuracy of the calculation since it is the addition of the orbital rotational energy and potential energy at the location of the inner saddle point which provides the bottleneck to the reaction for medium and high temperature.

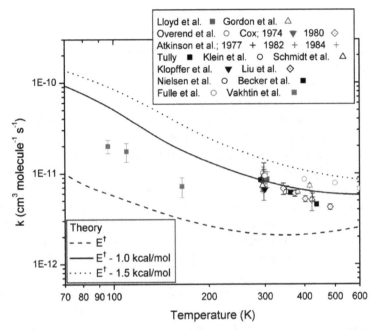

Fig. 11.6. Experimental and theoretical rate constants for the addition of OH to C_2H_4. Experimental data are from Lloyd *et al.* [65], Gordon *et al.* [66], Overend *et al.* [67], Cox [68], Cox *et al.* [69], Atkinson *et al.* [70–72], Tully [73, 74], Klein *et al.* [75], Schmidt *et al.* [76], Klopffer *et al.* [77], Liu *et al.* [78, 79], Nielsen *et al.* [80], Becker *et al.* [81], Fulle *et al.* [82], Vakhtin *et al.* [83] Theoretical rate constants are reproduced with permission from Ref. [64], Copyright 2005 American Chemical Society.

In summary, the two-TS picture has proven to be essential to the quantitative understanding of the dynamics of reactions where a submerged barrier reflects a change in the electronic configuration. This was also observed in the addition of OH to isoprene [84], the reactions of oxygen atoms with alkenes [85], the reaction CN + C_2H_6 [86], the reaction CN + HC_3N [87], the reaction of OH with methanol investigated with the CCUS method [88], or the μJ-VTST method [4].

11.1.6 *Tunneling effects*

There are essentially two possible exothermic channels for the reaction of OH with acetone $CH_3C(O)CH_3$: (i) the abstraction of an

H atom leading to water and acetonyl radical $CH_3C(O)CH_2$ and (ii) the addition of OH leading to acetic acid CH_3COOH and methyl radical. It is now experimentally well established that the second channel can be neglected at room temperature [89]. The reaction mechanism has been investigated by MP2 and CCSD(T) calculations [90]. Two distinct pathways leading to water and acetonyl radical, with the same energy profile shown in Fig. 11.7, proceed in three steps: (i) the formation of a hydrogen-bonded intermediate complex, (ii) the transition state corresponding to the formation of a chemical bond between the O atom of OH and an H atom of acetone, and (iii) the formation of a second hydrogen-bonded intermediate complex between H_2O and acetonyl radical which then dissociates easily since the reaction is exothermic. Both pathways correspond to the schematic minimum energy path presented in Fig. 11.4. There is a large uncertainty in the activation energy, since calculations have given $16.7\,kJ\,mol^{-1}$ [90], $14.8\,kJ\,mol^{-1}$ [91], $6.1\,kJ\,mol^{-1}$ [92]. The reaction is strongly exothermic, since products are $65.6\,kJ\,mol^{-1}$ below reactants. Caralp *et al.* [93] performed an RRKM-master equation [94] calculation of the reaction rate constant in which only the first H-bonded potential well and the abstraction saddle point were included.

The kinetic model is based on the competition between the processes affecting the steady state concentration of the H-bonded complex: reactants capture, collisional stabilization, redissociation of the reactants, and transition state crossing with or without tunneling (see Fig. 11.7). Rate constants for each of the two distinct pathways have been calculated. Let us denote by k_a and k_b the reaction rate constants associated to the lowest and highest activation energy pathway, respectively. As expected, both rate constants show a positive temperature dependence. The reaction rate constant is the sum $k_a + k_b$. At low temperature, k_a dominates, and at high temperature, above 330 K, k_b dominates. The slope of k_b is steeper than the one of k_a, and therefore, the sum $k_a + k_b$ shows an unusual temperature dependence: it increases slowly below room temperature, and more rapidly above room temperature. The calculated reaction rate constant in the range 200–700 K is in perfect agreement with the experimental results, therefore confirming the validity of the kinetic model. Furthermore, if tunneling is not included in the model, then the calculated reaction rate constant is much smaller than the experimental one. In order to confirm the effect of tunneling, the rate constant of

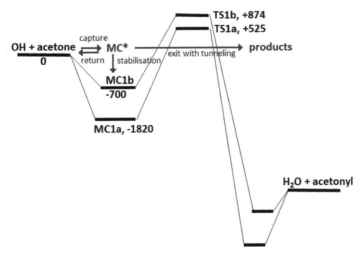

Fig. 11.7. Schematic energy profile and kinetic model for an H-abstraction reaction of OH with acetone. The pre-reactive complex MC* can either be stabilized by a collision with the bath gas, redissociate back to reactants, or form products by surmounting or tunneling through the abstraction energy barrier. Energy of stationary points is given in cm^{-1}. Reproduced with permission from Ref. [95], Copyright 2019 American Chemical Society.

the reaction of OH with deuterated acetone $CD_3C(O)CD_3$ has been also calculated and shows the same temperature dependence and is also in good agreement with the experiments.

Subsequent studies of the OH + acetone reaction, both theoretical and experimental, have been reported by Shannon *et al.* [3, 96] in a lower-temperature range, 63–148 K. A large negative temperature dependence of the rate constant was measured, along with an increase with pressure, in contrast with the positive temperature dependence observed above 200 K. The rate constant was found to increase by a factor of 334, while the temperature was decreasing from 295 K down to 68 K. This finding is rather unexpected since the lowest ab initio activation energy is 6.1 kJ mol^{-1}, which corresponds to a temperature $T = E/k_B = 733$ K. Calculations of the rate constant were done with the master equation solver for multi-energy well (MESMER) software package [97]. The kinetic model was the same as the one used previously by Caralp *et al.* [93] However, it was a semi-empirical calculation since several parameters were optimized by a weighted least-squares process in order to reproduce at best the experimental rate constant. The calculated rate constant was found to be very sensitive to the energies of the two most important stationary points of the PES: the hydrogen-bonded intermediate complex and the abstraction transition state. Below 200 K, The MESMER calculated rate constant rises rapidly as the temperature decreases. Conversely, if the kinetic model does not include the tunneling effect, then the calculated rate constant decreases rapidly and becomes several orders of magnitude smaller than the experimental one. The pressure dependence is also well described, owing to the dominant role played by collisional stabilization of the hydrogen-bonded complex. As the temperature lowers, the rate of dissociation toward reactant becomes negligible compared to the rate of complex formation. The complex has a longer lifetime and its concentration may become significant. This enables stabilization by collisions with the bath gas and facilitates product formation by tunneling. In summary, a tunneling effect is required in order to calculate a rate constant that is in good agreement with the experimental results.

Further progress toward lower temperature for the OH + acetone reaction has been reported recently [95]. The reaction rate constant was measured in the temperature range 11.7–64.4 K and a strong negative temperature dependence was observed, in agreement with the previous results of Shannon *et al.* [3, 96] Blázquez *et al.* [95] computed the rate constant using the same method as in their previous work [93], based on E and J resolved RRKM and master equation theories. Below 30 K, the calculated rate constant is in good agreement with the experimental result, while in the range 30–65 K, the calculated rate constant is about one order of magnitude lower than the experimental data. This qualitative agreement validates the treatment of the competition between the dissociation of the complex back to the reactants and its reactivity via tunneling through a barrier to the products, H_2O + acetonyl.

The OH + methanol reaction, for which there is a large body of experimental and theoretical works, exhibits two distinct reaction pathways with similar energy profiles, as shown by Fig. 11.8. A prereactive complex is formed by the hydrogen bonding of OH onto the methyl or the hydroxyl group of CH_3OH. In the first case, H atom abstraction leads to the products CH_2OH + H_2O (reaction R1), while in the second case, it leads to CH_3O + H_2O (reaction R2). In both case, there is a potential energy barrier.

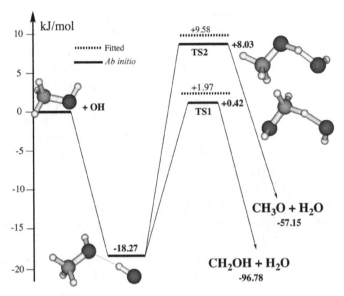

Fig. 11.8. Schematic energy profile for the OH+methanol reaction. Reproduced with permission from Ref. [98], Copyright 2019 AIP Publishing.

Gao *et al.* [88] have calculated the reaction rate constants with the VTST and CCUS theories, along with the small-curvature tunneling approximation [99] in the temperature range 30–2000 K. Their best estimates of the 0-K enthalpy of the pre-reactive complex are -4.91 and -5.04 kcal mol^{-1} for R1 and R2, respectively, and for the barriers they are 0.54 and 0.72 kcal mol^{-1}, respectively. They have calculated the rate constants within two kinetic models: low-pressure limit (LPL-CCUS) and high-pressure limit (HPL-CCUS). The high-pressure limit implies a larger tunneling effect since the stabilized pre-reactive complex is populated at low energies and then hydrogen can tunnel from more energy levels than in the low-pressure limit where tunneling occurs only at energies above the zero-point energy (ZPE) of reactants since it is assumed that no collisional stabilization occurs at low pressure. It is necessary to consider two bottlenecks: the capture transition state between reactants and pre-reactive complex, and the energy barrier between the pre-reactive complex and the products. The reaction rate constant at low temperature is controlled by the capture bottleneck. LPL-CCUS and HPL-CCUS rate constants are in good agreement with the experiments for T $\geq$ 200 K, with a positive temperature dependence since in that range of temperature, the dominant bottleneck is the abstraction barrier. Below 200 K, only the HPL-CCUS rate constant shows a strong negative temperature dependence. The tunneling effect plays an important role and the long-range centrifugal barrier becomes the dominant bottleneck. A large kinetic isotope effect (KIE) is observed, clearly validating the crucial role of the tunneling effect. The HPL-CCUS rate constant is in good agreement with the experiments [4, 5, 7, 11], but this agreement is misleading since all experiments below 200 K have been performed at low pressure, from ∼0.1 up to 3 Torr. Furthermore, no KIE experimental data are available.

Observing that Gao *et al.* [88] have reported only the low- and high-pressure limits, Nguyen *et al.* [98] have used the master equation in order to calculate the rate constant for any pressure. They also based their calculations on highly accurate extrapolated energies computed by using the composite method mHEAT [100] which combines the results of several ab initio methods, from SCF up to CCSDT(Q). Since the barrier height is the most important parameter, it was adjusted, within the plausible error bar for the mHEAT method, in order to reproduce the experimental rate constant at room temperature. The calculated reaction rate constant is in good agreement with all experimental results. Above 200 K, the rate constant is almost independent of the pressure and shows a positive temperature dependence. Below 200 K, the rate constant shows a strong negative temperature dependence and a positive pressure dependence, shown in Fig. 11.9. This can be explained by the stabilization of the pre-reactive complex, both by the collisions with the bath gas and the lowering of temperature which prevents the back-dissociation toward reactants. The tunneling probability is therefore enhanced by the increased lifetime of the complex. However, these theoretical results have been obtained for a much higher pressure than in the experiments. The experimental rate constant of Gómez Martin *et al.* [11] was reproduced with a pressure of 3000 Torr and those of Antiñolo *et al.* [7] with a pressure of 200 Torr (Fig. 11.9). The pressure appears here as an external parameter allowing one to adjust the master-equation rate constant on the experimental results. At low pressure, the latter rate constant becomes pressure independent, but it is then one order of magnitude lower than the experimental rate constant.

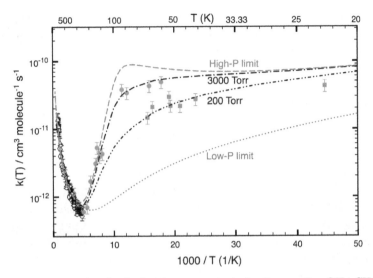

Fig. 11.9. Measured and calculated rate constants for the reaction $OH+CH_3OH$. Experimental data are from (solid red circles) Shannon *et al.* [5] and Gómez Martin *et al.* [101] with $P_{exptl.} < 2$ Torr, (solid pink squares) Antiñolo *et al.* [7] with $P_{exptl.} < 0.4$ Torr. Reproduced with permission from Ref. [98], Copyright 2019 AIP Publishing.

Shortly after the work of Nguyen *et al.* [98], a complete set of measured rate constants in the range 11.7–177.5 K was reported [4], along with a calculation of the pressure and temperature dependences of the reaction rate constant with the μJ-VTST method. A good agreement between theory and experiment is observed in the temperature range studied. However, as in the calculations of Nguyen *et al.*, this agreement is obtained between the rate constant calculated in the HPL and the low-pressure experimental data. The rate constant calculated in the LPL is much too small in the temperature range 30–130 K. The theoretical investigation confirms the previous modelizations of the reaction kinetics. Above 200 K, an Arrhenius behavior is observed, and the reaction is controlled by the abstraction barrier. Below 100 K, the rate constant is well described by the capture theory and shows a moderate negative temperature dependence. The reaction is controlled by the capture transition state. In the range 100–200 K, a sudden transition between the capture and the Arrhenius behavior is observed, and this induces a strong negative temperature dependence.

Further physical insight into the OH + methanol reaction was gained by two recent investigations of the reaction dynamics. In contrast with the kinetic approaches, the time-dependent simulations of the reaction dynamics allow one to observe all the features of a reaction, such as the atomic and molecular motions, the distribution of available energy among the various modes of molecular motions, and the typical time of reaction steps. Using Ring polymer molecular dynamics (RPMD) [102–105] and TST, del Mazo-Sevillano *et al.* [106] have studied the reactions of OH with formaldehyde and methanol in the temperature range 20–1200 K.

The RPMD approach is based on the isomorphism between the statistical properties of a quantum system and the classical statistical mechanics of a fictitious ring polymer made up of several copies of the original system linked by harmonic oscillators [107]. It allows one to approximate the quantum time-dependent dynamics of a molecular system by propagating all the copies in an extended phase space by classical dynamics. It preserves the ZPE along the reaction path and provides an exact solution for tunneling through a parabolic barrier [103]. The RPMD rate theory has been applied to many reactions and the comparison with accurate or approximate quantum rate constants has shown a good agreement [103, 104, 108–118].

The $OH + H_2CO$ and $OH + CH_3OH$ rate constants computed with the RPMD rate theory are in good agreement with the experimental results and the previous calculations, as illustrated in Fig. 11.10 for $OH + CH_3OH$. It should be noted that the RPMD rate theory is applied here to an isolated system and therefore produces zero pressure rate constants. The rate constants of both reactions show a strongly negative temperature dependence below 300 K and a positive temperature dependence above 300 K. The minimum of both rate constants is then found around 300 K. While in the previous works [4, 88, 98], the negative temperature dependence for the reaction $OH + CH_3OH$ was obtained under HPL condition, the same dependence is obtained by RPMD with a zero-pressure condition, which is not too much different than the low pressure typical of the CRESU experiments (see Fig. 11.10). The RPMD trajectories show a long-lived pre-reactive complex which is trapped in the effective potential well resulting from the attractive dipole–dipole interaction and the repulsive energy of the relative rotational motion. A transfer of the collision energy toward the other mode of intermolecular

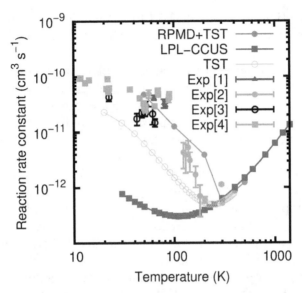

Fig. 11.10. Calculated and measured rate constant for the reaction OH + CH₃OH. The [1–4] experimental results are from Shannon *et al.* [5], Gómez-Martin *et al.* [11], Antiñolo *et al.* [7], and Ocaña *et al.* [4], respectively. LPL–CCUS data are from Gao *et al.* [88] TST refers to the TST results obtained in the LPL by Ocaña *et al.* [4]. Reproduced with permission from Ref. [106], Copyright 2019 American Chemical Society.

motions happens, essentially driven by the dipole–dipole interaction which reorients the reactants, therefore inducing some rotational excitation of the reactants. It is then unlikely for the molecular system to get enough recoil energy to back-dissociate. In a statistical point of view, we can say that the pre-reactive complex has a large density of states, while only a small number of states allow back-dissociation.

In the RPMD study, pre-reactive complexes with a lifetime over 100 ns are observed, therefore long enough for formation of the products by tunneling. This trapping effect is increased by the lowering of the collision energy and also by quantum effects, such as the conservation of the ZPE by the semi-classical RPMD. Indeed, at very low collision energy, in the region of the centrifugal barrier, the vibrational energy of the ground state is much larger than the rotational and translational energies. In a classical simulation of the dynamics, we observe a transfer of vibrational energy toward the other mode of motions, the so-called ZPE leakage. This can increase the recoil energy therefore decreasing the lifetime of the pre-reactive complex.

Owing to the necessity of using a buffer gas in flow supersonic reactor experiments, there is never a zero pressure in such experiments. It is then necessary to investigate the pressure effect on the long-lived pre-reactive complexes predicted by RPMD studies. However, the zero-pressure RPMD rate constants for the reaction of OH with formaldehyde and methanol are close to the experimental data [106], suggesting that pressure effects are not important for the reactivity in these cases. A preliminary study of such effects has been reported by Naumkin *et al.* [119] This study is based on the assumption that in a CRESU experiment, the He buffer gas atoms are much more abundant than the COM (here formaldehyde or methanol) which in turn are more abundant than the radical OH. At low temperature, owing to the high density of He buffer gas, complexes like He–COM and He–OH are expected to be easily formed, however with a weak bond. Naumkin *et al.* have performed QCT calculations of the cross sections for the possible inelastic and reactive channels of the He–H_2CO + OH collision and found that the major channel is the reaction leading to the H_2CO–OH + He products with a much larger cross section than other channels. Furthermore, QCT calculations of the collision He + H_2CO–OH give only very small cross sections for all inelastic and reactive channels, owing to the weak long-range interaction between reactants. While these findings suggest that pressure should have an effect on the OH + H_2CO reaction, a quantitative result cannot be obtained unless a kinetic simulation involving the rate constants of all major reactive processes, along with the density of the buffer gas and the reactants, is performed.

Besides the H-abstraction reactions of OH with acetone or methanol, a few other reactions with similar energy profile have been investigated both experimentally and theoretically, such as OH with methyl formate [120] or with HNO_3 [121]. In summary, even if it is in contradiction with the usual expectation that reactions with a significant barrier above the ZPE of reactants cannot occur at low temperatures, it has been demonstrated, in the case of hydrogen abstraction, that such reactions can occur very rapidly, even at low temperatures typical of interstellar media, when the hydrogen atom tunneling effect is enhanced by a pre-reactive complex [2]. These findings are important since they open new routes to the formation of complex organic molecules in the interstellar media.

11.2 Inelastic Collisions

Very early in the development of the CRESU technique, inelastic collisions were studied by this method. While small and intermediate case molecules do not undergo radiationless transitions under isolated conditions, they do undergo collision-induced radiationless transitions with a varying degree of efficiency depending upon the nature of the collision partner and the internal energy mode transferred.

Vibrational quenching was the first kind of inelastic collision studied in CRESU in 1996 [122] while very few other works of this type have thereafter been performed. Two years later, Spin-Orbit relaxation of atoms was considered and has remained since then a subject of strong interest. The first electronic quenching studies are more recent (2014) and the most recent works are those devoted to rotational quenching. The scale of energy spanned by each of these processes varies from a few tenths of cm^{-1} up to several tens of thousands cm^{-1}. It then allows checking very different regions of the PES describing a given inelastic collision. In what follows, we decided to not use this historic order but rather to use a decreasing energy scale of the transitions and will then consider successively electronic, vibrational, rotational, and spin-orbit transitions. Each of these kinds of transitions will be described in the following four sections.

11.2.1 *Electronic quenching*

The first experiments were performed in the 80s in a static bulb at $300 \, K$ by monitoring the decrease of the fluorescence quenching of the excited electronic state as a function of the collider pressure in order to reveal collision-induced electronic quenching. Following supersonic jets, experiments confirmed the shorter lifetimes due to "collision-induced" electronic quenching as in the case of static bulb experiments at moderate pressures. These experiments indicated that collision-induced vibrational relaxation is very inefficient compared to collision-induced electronic quenching and it was soon realized that it was possible to correlate electronic quenching rates with molecular properties such as polarizability α_M, permanent dipole moment μ_M, ionization potential I_M, and intermolecular potential well depth ε. A general trend of increasing cross section was observed with increasing dipole moment μ_M, quadrupole moment θ_M, and polarizability α_M, while the opposite trend is observed for the ionization potential I_M.

The first models were all based on the hypothesis of the formation of an intermediate collision complex A*-M between the collider M and the electronically excited state A* of the A molecule. The early model of Lin *et al.* [123] makes the quenching cross section σ depend upon the well depth ε of the metastable complex and the temperature T, as given by the following:

$$\ln \sigma = C + \beta \left(\frac{\varepsilon}{k_B} \right)^{\frac{1}{2}} \qquad (11.26)$$

where C is a constant and β is a temperature-dependent factor of A* and independent of M. This simple model was relatively successful for fitting the existing data at 300 K.

11.2.1.1 *The capture model*

The subsequent capture model of Holterman *et al.* [124] focused on the formation of the collision complex in the long-range part of the potential and takes into account its explicit form. It assumes that the probability for relaxation is unity once the A*-M complex is formed and considers that the duration of the collision is unimportant since the induced transition is completed before the dissociation of the complex. In these conditions, the electronic relaxation cross section is simply taken to be equal to the capture cross section which was presented in detail in the first section of this chapter. A given radial dependence of the intermolecular potential is then associated with a given temperature dependence. This capture model gives a better agreement with all the available experimental data at 300 K than the previous one and is also able to predict the temperature dependence of the electronic relaxation cross section at temperatures below 300 K. Conversely, it is unable to predict the variety of temperature dependences above 300 K which were measured for a few systems.

11.2.1.2 *The Harpoon model*

It was found that the quenching cross section for species having a positive electron affinity is of the order gas kinetic or greater and exhibits little or no temperature or rotational dependence. Conversely, colliders with relatively large negative electron affinities are often seen to have cross sections that are near zero at room temperature and can increase dramatically at very high temperatures without showing a marked dependence in the initial rotational state. These features inspired the Harpoon mechanism developed for collisions above 300 K, which are of interest in combustion and aerothermodynamic applications. It is appropriate for collision partners having stable negative ions, which are predicted to have large quenching cross sections. The model assumes [125] that the quenching of an electronically excited molecule A* colliding with any collider M proceeds in two steps:

$$A^* + M \rightarrow A^+M^- \rightarrow A + M \tag{11.27}$$

During the first step, electron transfer occurs from A* to the collider M, followed by ion–ion recombination to form an A*-M covalent pair, the initial electron jump occurring at the intersection r_c between the A*-M covalent and A$^+$-M$^-$ ionic surfaces. The ionic surface in turn makes an inner crossing with the A-M covalent surface on the repulsive wall. The crossings of the ionic surface with the A*-M and A-M surfaces act within this model as efficient entrance and exit channels for electronic quenching. The simplest approach neglects the angular dependence of the ionic and covalent curves which are written like

$$\varphi_{cov}(r) = [E(A^*) - E(A)] + V_{cov}(r) \tag{11.28}$$

where the first term on the right-hand side of this expression defining the covalent potential includes both the electronic and vibrational excitation energies and $V_{cov}(r)$ includes the short- and long-range contributions, and

$$\varphi_{Ionic}(r) = I(A) + E_a(M) + E_{Vib}(M^-) - \frac{e^2}{r}$$
$$- \frac{e^2[\alpha(A^+) + \alpha(M^-)]}{2r^4} + V_{Ionic}(r) \tag{11.29}$$

where $I(\mathrm{A})$ is the ionization potential of A, $E_a(\mathrm{M})$ the electroaffinity of M, $E_{Vib}(\mathrm{M}^-)$ the vibrational energy of M^-, and α stands for a molecular polarizability while e is the electron charge. For each couple of vibrational energies of the colliders A and M, one then obtains a different value of r_c. The probability of remaining on the A^*-M or A^+-M^- surface at a given r_c is taken from the Landau-Zener theory. For the electronic relaxation cross section, one obtains:

$$\sigma = Ce^{-\frac{B}{T}} \tag{11.30}$$

where B and C are constants. This model which can in principle also describe any activated quenching process was found to capture the temperature dependence, but tends to overpredict the magnitude of the quenching cross section for the simplest collision partners (e.g. atomics and homonuclear diatomics).

11.2.1.3   *General model*

A general form able to describe both the lowest and highest temperature regimes is then a combination of the Capture and Harpoon models.

$$\sigma = C_0 + C_1 \left(\frac{300}{T}\right)^{C_2} + C_3 e^{-\frac{C_4}{T}} \tag{11.31}$$

11.2.1.4   *Near-resonant energy transfer (NRET)*

An alternative mode of non-radiative electronic energy transfer can happen for a quencher having an electronic excited M^* state (acceptor) lower in energy than the initial excited electronic state of the molecule A^* (donor),

$$\mathrm{A}^* + \mathrm{M} \rightarrow \mathrm{A} + \mathrm{M}^* \tag{11.32}$$

This electronic excitation transfer mechanism, first proposed by Förster [126], revisited by Breckenridge *et al.* [127] and Scholes [128] and since then by many other authors, arises from the coulombic interaction between the electronic states of the donor A^* and the acceptor M. The standard multipole–multipole expansion of the coulombic interaction between the electronic states of the donor and the acceptor is used. For instance, the first non-zero term for two interacting neutral polar molecules is the dipole-dipole interaction while for non-polar molecules higher orders need to be considered. The NRET cross section is obtained from the Fermi golden rule involving the square of the electronic matrix elements of the coulombic potential and a Franck-Condon overlap factor for the vibrational part of the integral.

One speaks about near-resonant energy transfer as the efficiency of energy transfer is expected to increase when the energy defect (the net amount of internal energy which must be dissipated to relative translational energy in the collision) drops off to zero. The combination of final rovibrational states of the electronically relaxed molecule and of the electronically excited quencher leading to the smallest energy defect is then expected to give the largest NRET cross section. Transfer of energy to the acceptor leads to vibrational relaxation and subsequent acceptor fluorescence that is spectrally shifted from the donor fluorescence. In practice, the efficiency of energy transfer is obtained by comparing the fluorescence emitted from the donor and acceptor. A $T^{-2/3}$ temperature dependence of the thermally averaged NRET cross section is predicted from a capture cross section model for the quenching cross sections, due to dipole–dipole interaction in r^{-3}. If only one of the two molecules or none is polar, the interaction potential is in r^{-6} and the temperature dependence is $T^{-1/3}$.

11.2.1.5 *Fluorescence quenching of $NO(A^2\Sigma^+, \nu' = 0)$ by collisions with four different colliders*

To our knowledge, only two CRESU studies have been performed up to now by the same team of North and Bowersox. In the first one [129], they measured the fluorescence quenching of $NO(A^2\Sigma^+, \nu' = 0)$ by collisions with two different colliders, $NO(X^2\Pi)$ and $O_2(X^3\Sigma_g^-)$ between 34 and 109 K, and in the second [130] with C_6H_6 and C_6F_6 between 145 and 300 K. These new measurements complement those of Settersten *et al.* [131] obtained in a cryogenically cooled gas flow cell for the quenching by NO, N_2, O_2, and CO between 125 and 294 K and those of Settersten *et al.* [132] between 294 and 1300 K in a heated cell. In this latter study, the observed increase in the quenching cross sections by NO and O_2 at high temperatures is moderate and consistent with a harpoon mechanism, while the present thermally averaged cross sections for quenching by NO and O_2 between 34 and 109 K and those of Settersten *et al.* [131] for quenching by NO,

O_2, and CO from 125 to 294 K are found to decrease as temperature increases. These measurements are then consistent with the mechanism of a collision-complex formation at low temperature (below 300 K). The general functional forms defined in Section 11.1.3 appear to provide an excellent representation of all the available data.

More precisely, the fitting of their results for the quenching by NO and O_2 gives C_2 exponents (respectively, 0.22 and 0.29) in equation (11.31) which are consistent with a quenching process that is mediated by complex formation dominated by dispersion forces, in accord with the leading term in a multipole expansion of electrostatic forces. As a matter of fact, when induced dipole–dipole or dispersion forces dominate the intermolecular interaction, the expected quenching cross section dependence on temperature is $T^{-1/3}$. Instead, quenching by CO is characterized by a significantly stronger temperature dependence that is consistent with complex formation dominated by dipole–dipole forces (close to $T^{-2/3}$) even though dispersion forces are predicted to be dominant for thermally averaged molecular orientations.

On the contrary, the thermally averaged cross section for quenching by N_2 is found to remain small and constant to within the experimental uncertainty for $T < 300$ K while it was previously measured to monotonically increase above 300 K. This latter behavior is captured well by the harpoon quenching calculations of Paul and co-workers [125], which included multiple curve crossings from excited vibrational levels to account for the temperature dependence up to 1900 K. The low-temperature limit given by the harpoon mechanism is zero, while the current data, however, indicate that although apparently improbable, collisions with N_2 at low temperature do have a non-zero finite quenching cross section.

11.2.1.6 *Fluorescence quenching of $NO(A^2\Sigma^+, \nu' = 0)$ by collisions with C_6H_6 and C_6F_6*

The second study performed by Winner *et al.* [130] was dedicated to the fluorescence quenching of $NO(A^2\Sigma^+, \nu' = 0)$ by collisions with C_6H_6 and C_6F_6 between 145 and 300 K. There was only one measurement available at 300 K for C_6H_6 and the two quenchers were expected to both follow the harpoon mechanism as the many other colliders of NO studied previously had. The choice of these two new colliders was motivated by the fact that the harpoon mechanism requires that the quenching partner have a sufficiently high electron affinity $(> -1\,\text{eV})$ to ensure an efficient electron transfer from NO to the quencher. The electron affinity of C_6F_6 ($1.2\,\text{eV}$) differs significantly from the one of C_6H_6 ($-1.5\,\text{eV}$) and then offers an interesting possibility of comparison. The harpoon electron exchange mechanism predicts that species with negative EA values like C_6H_6 has negligible quenching cross sections, whereas C_6H_6 is found to have a relatively large cross section.

Furthermore, the NRET model predicts a larger quenching cross section for C_6H_6 than for C_6F_6 in agreement with experiment, while the harpoon mechanism would predict a larger quenching cross section for C_6F_6, due to its larger EA. This result suggests the NRET mechanism may better explain the quenching behavior of these two species. It requires the collisional quencher to have an electronic excited state lower in energy than the NO A state. It is indeed the case for C_6H_6 and C_6F_6 which A states energies lay at $4.9\,\text{eV}$ and $4.8\,\text{eV}$, while the one of NO is $5.45\,\text{eV}$. This mechanism has a predicted temperature dependence of $T^{-1/3}$ for C_6H_6 and C_6F_6, since the quenching cross section is expected to be proportional to the cross section for the formation of an orbiting complex, which depends on dispersion forces for these two non-polar quenchers.

In conclusion collisional electronic quenching is today described only by semi-empirical models, and there are no systems as yet for which quantum calculations were tried because of the difficulties of describing exactly the full process of collisional electronic quenching.

11.2.2 *Vibrational relaxation*

Buffer gas cooling of molecules is a universal technique, easy to implement and quite efficient to cool down rotational excitation. Conversely, collisional vibrational relaxation is known to be a very inefficient process. It is usually at least three or four orders of magnitude smaller and even up to ten orders of magnitude smaller than rotational relaxation. Also, vibrational buffer gas cooling of molecules was for long considered to be extremely inefficient. Very recently, it was however demonstrated experimentally [133] and theoretically [134] that it may be efficient at very low collision energy for molecular ions colliding with ultracold atoms. On the theoretical side, converse to electronic quenching, this process is furthermore amenable to theoretical simulations. Exact quantum Close Coupling calculations [135] are nowadays straightforward for non-reactive atom–diatom systems. They also become possible for non-reactive diatom–diatom [136], atom–triatom [137, 138], and diatom–triatom systems [139]. Two kinds of mechanism make it possible to obtain vibrational relaxation for a given excited vibrational level. For non-reactive atom–diatom collisions, only vibration to translation (V-T) transfer is possible for any excited vibrational level at any temperature and is, as discussed before, inefficient for most systems. For molecule–molecule inelastic collisions, vibration to vibration (V-V) transfer is another possible mechanism:

$$A(\nu) + B(\nu' = 0) \rightarrow A(\nu - 1) + B(\nu' = 1) \qquad (11.33)$$

which is a lot more efficient. Because of vibrational anharmonicity and as a function of the choice of A, B, and ν, this process can be endothermic and then becomes impossible at low temperature. This is the case for vibrational self-relaxation which is for this reason interesting to study in CRESU experiments as V-V relaxation should drop off at low temperature.

11.2.2.1   *Vibrational self-relaxation of NO($X^2\Pi, \nu = 3$)*

Vibrational self-relaxation was studied in CRESU by James *et al.* [140] for the NO($X^2\Pi$, $\nu = 3$) molecule between 7 and 85 K. The authors had previously obtained results for this process between 77 and 195 K for $4 \leq \nu \leq 10$ and found that the rates were similar to those for self-relaxation of CO from high vibrational levels. Interestingly, the vibrational self-relaxation of NO($\nu = 1$) which can be written

$$NO(\nu = 1) + NO(\nu = 0) \rightarrow 2NO(\nu = 0) \qquad (11.34)$$

and must occur by V-T energy transfer was found to be quite large at room temperature, $k(v = 1) = 7.7 \times 10^{-14} \, cm^3 molecule^{-1}s^{-1}$, while for CO it is found to be $< 10^{-20} \, cm^3 molecule^{-1}s^{-1}$. This unusual behavior was the subject of many theoretical studies and Nikitin and Umanski [141] proposed that V-T relaxation occurs by an electronically nonadiabatic mechanism assuming that during collisions, the system undergoes a transition between different vibronic states at geometries where the energy splitting between different electronic potential energy surfaces matches the separation of the $\nu = 1$ and $\nu = 0$ vibrational levels. Another explanation was proposed based on the fact that the well associated with the formation of the $(NO)_2$ dimer is unusually deep ($710\,\mathrm{cm}^{-1}$) and directional for what should be a van der Waals potential. This well is then expected to favor V-V energy transfer. As such weak intermolecular attractions are likely to exert their greatest effect on collisional processes at low temperatures, the authors chose to investigate vibrational self-relaxation of the $\nu = 3$ state of NO below $80\,\mathrm{K}$. This process which is a good example of V-V transfer at $300\,\mathrm{K}$ is endothermic by $55.9\,\mathrm{cm}^{-1}$, which corresponds to a temperature of $80\,\mathrm{K}$. Below this temperature, only V-T transfer inside the intermediate well is then possible and one would expect a drop of the rate constant. The authors found instead that the rate constants show a marked negative temperature dependence, reaching a value of $2.6 \times 10^{-10}\,\mathrm{cm}^3\mathrm{molecule}^{-1}\mathrm{s}^{-1}$ at $7\,\mathrm{K}$ which suggests a collisional deactivation efficiency approaching unity. They propose that V-T energy transfer is facilitated by the transient formation of $(NO)_2$ complexes which last long enough for vibrational predissociation within the complexes to compete with redissociation to $NO(\nu = 3) + NO(\nu = 0)$.

11.2.2.2 *Vibrational relaxation of highly excited toluene in collisions with He, Ar, and N_2 as a function of temperature*

Collisional relaxation of the highly excited aromatic molecules is a subject of much interest. In the case of such big molecules, the very numerous rovibronic states are forming a quasi-continuum; also, no information is gained at the state-to-state level, nor does one learn from which modes energy is removed. The system is modeled as an excited molecule in the presence of a heat bath and one measures instead macroscopic or bulk observables defined below (see [142]). The collisional relaxation is experimentally followed by observing their time-resolved infrared fluorescence. As time increases, the excited molecule relaxes and its distribution of internal energy broadens. As a function of time and for all energies less than that initially deposited in the molecule, it is then possible to follow the variation of a mean vibrational energy content $\langle E \rangle$ averaged both on the thermal collision energy distribution and on the internal energy distribution. In other words, if E is the internal energy of the molecule and $\mathcal{P}(E, t)$ the probability that internal energy shall be E at time t, the mean vibrational energy content is as follows:

$$\langle E \rangle = \int E \cdot P(E, t) dE \qquad (11.35)$$

One also obtains the corresponding mean energy removed per collision, $\langle \Delta E \rangle$ over the distributions of internal energies and of ΔE values,

$$\langle \Delta E \rangle = \int <\Delta E> \cdot P(E, t) dE \qquad (11.36)$$

where the microscopic average $<\Delta E>$ is defined by the following:

$$<\Delta E> = \int E \cdot P(E, E') dE \qquad (11.37)$$

and $P(E, E')$ is the probability, per collision and per unit energy, that if the internal energy of the subject molecule before the collision was E, it shall be E' after the collision. These experiments then ultimately provide the variation of $\langle \Delta E \rangle$ as a function of $\langle E \rangle$.

On the theoretical side, only QCT calculations using approximate model potentials have been performed to model these complicated systems. The microscopic probability $P(E, E')$ is obtained straightforwardly as a function of E' from which the variation of $\langle \Delta E \rangle$ as a function of $\langle E \rangle$ is easily computed. A good agreement between theory and experiment was obtained for the related system of the vibrational relaxation of benzene by collisions with noble gases [143]. In the case of the vibrational relaxation of highly excited toluene in collisions with He, Ar, and N_2 studied by the team of Sims [144] and detailed by him in this book in Chapter 7, there is no QCT study available as yet. The main objective of their work was to obtain the dependence of the results in the collision energy or temperature, which has never been theoretically considered up to now and will motivate new theoretical studies.

11.2.3 *Rotational relaxation*

The last ten years have seen tremendous progress in the crossed beam studies allowing one for the first time to measure state-to-state rotational excitation cross sections at very low collision energy. The agreement obtained between theory and experiment is impressive [145–147], while this technique allows one only to measure relative magnitudes in contrast with the CRESU experiments coupled with the IRVUVDR (infrared vacuum-ultraviolet double resonance) technique, which yields absolute values of the state-to-state rate constants. The recent study of the rotational relaxation of CO in collision with Ar [148] is the first example of comparison between CRESU experimental state-to-state rotational rate constants absolute values and Close Coupling calculations. The agreement is again excellent between theory and experiment. An important issue in such studies is to know whether the state-to-state mechanism for rotational relaxation within different vibrational levels (i.e. without vibrational relaxation) is identical, or at least nearly so. This assumption relies on the fact that van der Waals wells are usually quite shallow and small compared to the diatomic vibrational spacing. In other words, the intermolecular potential is not strong enough to couple different vibrational levels. Consequently, vibrational quenching is negligible compared to rotational quenching for most neutral van der Waals complexes. This is indeed the case of Ar-CO as can be seen in Fig. 11.11 which shows the rotational and vibrational quenching rate constants from the second excited vibrational state of CO in the rotational states $J = 0$, 1, 4, and 6.

The vibrational quenching rate constants are 6 to 8 orders of magnitude smaller than the rotational quenching rate constants. These results demonstrate that for Ar-CO the state-to-state mechanism for rotational relaxation within different vibrational levels is essentially identical.

11.2.4 *Spin-orbit relaxation*

Fine structure transitions in atoms induced by collisions with atoms or molecules have been the subject of many experimental and theoretical studies because of their interest in many different fields

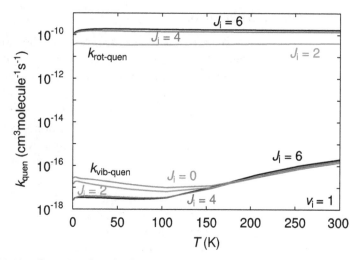

Fig. 11.11. Rotational and vibrational quenching rate constants of CO in collision with Ar [148]. Reproduced with permission from Ref. [148], Copyright 2017 Elsevier B.V.

ranging from chemical lasers to combustion, atmospheric, or interstellar chemistry. In the latter domain, the rate constants are needed at very low temperature and CRESU experiments are the tool of choice for these studies. On the theoretical side, the formalism of the quantum Close Coupling method has been available very early to compute the collision dynamics for any possible fine structure of an atom colliding with an atom or a diatom [149–154]. However, the calculation of the full fine-structure-dependent PES with a good accuracy is still a difficult task limited to a few systems. Also, most of the older studies used model potentials. The CRESU study of Le Picard et al. [155] dedicated to the intramultiplet transitions of $C(^3P)$ and $Si(^3P)$ in collisions with He was one of the first to also present a comparison with dynamics calculations on a full ab initio PES. The results of the dynamics calculations were found to be very sensitive to the quality of the ab initio PES. The agreement between theory and experiment was not satisfactory for the relaxation of $C(^3P)$, while it was found to be better for the relaxation of $Si(^3P)$ except for the lowest temperatures and only when using the highest level ab initio PES. A very recent theoretical study with a new high-quality PES dedicated to the relaxation of Si [156] quantitatively improved the agreement between theory and experiment at low temperature.

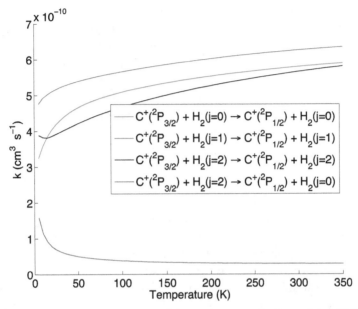

Fig. 11.12. Temperature variation of the spin-orbit quenching of the $C^+(^2P)$ ion by collisions with H_2 (j = 0, 1, 2) [157].

Some other purely theoretical studies were dedicated to the relaxation of atoms by collision with H_2, which is the most abundant molecule in the interstellar medium. The relaxation of C^+, which was studied long ago using a model potential, was recently revisited using a more elaborate ab initio potential (see Fig. 11.12). The new rate constants [157] differed from the previous one at the most by about a factor of 1.5.

To conclude this section dedicated to collisional spin-orbit relaxation, the dynamics calculations are nowadays relatively straightforward to perform using homemade or available codes like Hibridon [158], while their expected accuracy is still very much dependent on the quality of the ab initio PES.

11.3 Future Perspectives

Recent years have seen a lively interest in COMs and prebiotic molecules in the interstellar medium, raising questions of the mechanisms of formation and destruction of these species. A few of them have been investigated both by the CRESU experiments and theoretical works. However, COMs are large molecules and consequently theory faces a fundamental challenge: the molecular complexity. The size of the Hilbert space grows exponentially with the size of the molecules. Statistical theories are the customary way to circumvent this major obstacle, since they avoid all dynamics by an equilibrium assumption. However, their accuracy is limited by the statistical assumptions. In the TST, the phase space is assumed to be populated statistically which means that the population density over the whole surface defining the transition state is uniform. In the phase space theory, the probability of formation of any given product of a collision is assumed to be proportional to the ratio of the phase space available to that product divided by the total phase space available with conservation of energy and total angular momentum. These prerequisites are questionable, especially in the case of molecular systems at very low temperature when only a small number of quantum states can be populated and when intramolecular vibrational redistribution is not expected to occur fast enough. Even though the statistical theories are able to reproduce satisfactorily the experimental rate constants for most neutral–neutral reactions, they do it in some cases for pressures which are several orders of magnitude larger than in the CRESU experiments. More realistic statistical theories are then needed.

While it is relatively easy to calculate the pressure dependence of the reaction rate constants with the statistical theories, this is much more difficult with theories based on dynamics. In the case of reactions involving a pre-reactive complex and a submerged barrier, statistical theory can reproduce the experimental pressure dependence with a good or qualitative agreement. But, for the reactions driven by a pre-reactive complex and quantum tunneling through a barrier, and that are fast reactions at low temperature despite the barrier, the agreement between the calculated rate constant and the experimental one is obtained only for high pressures, much larger than in the experiment. The statistical theory can then produce a realistic rate constant only if the pre-reactive complex is submitted to the effect of a high pressure. This means that for statistical theory, these reactions are fast only if collisions with the bath gas populate all the available states of the pre-reactive complex at the gas temperature. This is a clear shortcoming which should be investigated and corrected. Now, if we turn to the theories involving dynamics, such as QCT, semi-classical, or quantum dynamics, the simulation of the reaction is much more demanding, since many different collisional processes might have to be included in the simulation. These processes are the addition reactions of the bath gas with the reactants which can produce new species in the gas flow, and then the reactions between these new species and also the reactions involving a new species and one of the original reactants. The collision between the bath gas and the pre-reactive complex should also be considered. The simulation can be simplified by taking into account only the faster reactions. Since bath gases are usually noble gases or nonreacting closed-shell molecules, they can only form van der Waals complexes with the original reactants. Such complexes are weak and can be stabilized or destroyed by subsequent collisions with the bath gas, depending on the temperature and the weakness of the van der Waals bond.

Since quantum dynamics is out of reach for large molecules, only semi-classical dynamics can provide a way to avoid statistical assumptions and to include quantum effects. RPMD has shown to be successful in the calculation of rate constants of reactions driven both by quantum tunneling and by a long-living pre-reactive complex for which the conservation of ZPE is essential. However, the computational demand increases as the temperature decreases. At very low temperature, RPMD requires far more computational resources than a standard QCT calculation, about several hundreds of times [105]. Promising new semi-classical methods, such as the quantized Hamilton dynamics approach [159] or the mixed quantum/classical theory [160], are worthy of further development and might offer a good alternative to the purely quantum approach.

In a distant future, quantum computation is expected to resolve the issue of molecular complexity. Feynman pointed out in 1982 [161] that since the Hilbert space grows exponentially with the size of quantum systems, then only a controllable quantum system would be able to simulate accurately the evolution of quantum systems. Since then, considerable progress has been made in the quantum algorithms for quantum chemistry [162]. However, today, the advent of quantum computers is held back by the difficulties associated with the decoherence issue.

Measuring state-to-state reactive rate constants remains still quite challenging because of the weakness of the signal to be measured. However, the recent progress made in the field suggests that it will become possible for more and more systems. For the same reasons, the possibility of measuring rotationally selected transitions between different vibrational states is a challenging perspective for purely inelastic collisions.

References

[1] Sandford SA, Nuevo SM, Bera PP, Lee TJ. Prebiotic astrochemistry and the formation of molecules of astrobiological interest in interstellar clouds and protostellar disks. Chem Rev. 2020;120:4616–59.

[2] Heard DE. Rapid acceleration of hydrogen atom abstraction reactions of OH at very low temperatures through weakly bound complexes and tunneling. Acc Chem Res. 2018;51:2620–27.

[3] Shannon RJ, Taylor S, Goddard A, Blitz MA, Heard DE. Observation of a large negative temperature dependence for rate coefficients of reactions of OH with oxygenated volatile organic compounds studied at 86–112 K. Phys Chem Chem Phys. 2010;12:13511–4.

[4] Ocaña AJ, Blázquez S, Potapov A, Ballesteros B, Canosa A, Antiñolo M, Vereecken L, Albaladejo J, Jiménez E. Gas-phase reactivity of CH_3OH toward OH at interstellar temperatures (11.7–177.5 K): experimental and theoretical study. Phys Chem Chem Phys. 2019;21:6942–57.

[5] Shannon RJ, Blitz MA, Goddard A, Heard DE. Accelerated chemistry in the reaction between the hydroxyl radical and methanol at interstellar temperatures facilitated by tunnelling. Nat Chem. 2013;5:745–9.

[6] Caravan RL, Shannon RJ, Lewis T, Blitz MA, Heard DE. Measurements of rate coefficients for reactions of OH with ethanol and propan-2-ol at very low temperatures. J Phys Chem A. 2015;119:7130–37.

[7] Antiñolo M, Agúndez M, Jiménez E, Ballesteros B, Canosa A, El Dib G, Albaladejo J, Cernicharo J. Reactivity of OH and CH_3OH between 22 and 64 K: modeling the gas phase production of CH_3O in Barnard 1b. Astrophys J. 2016;823:25.

[8] Jiménez E, Antiñolo M, Ballesteros B, Canosa A, Albaladejo J. First evidence of the dramatic enhancement of the reactivity of methyl formate HC(O)OCH$_3$) with OH at temperatures of the interstellar medium: a gas-phase kinetic study between 22 K and 64 K. Phys Chem Chem Phys. 2016;18:2183–91.

[9] Ocaña AJ, Jiménez E, Ballesteros B, Canosa A, Antiñolo M, Albaladejo J, Agúndez M, Cernicharo J, Zanchet A, del Mazo P, Roncero O, Aguado A. Is the gas-phase OH+H$_2$CO reaction a source of HCO in interstellar cold dark clouds? A kinetic, dynamic, and modeling study. Astrophys J. 2017;850:28.

[10] Ocaña AJ, Blázquez S, Ballesteros B, Canosa A, Antiñolo M, Albaladejo J, Jiménez E. Gas phase kinetics of the OH + CH$_3$CH$_2$OH reaction at temperatures of the interstellar medium (T = 21−107 K). Phys Chem Chem Phys. 2018;20:5865.

[11] Gómez Martín JC, Caravan RL, Blitz MA, Heard DE, Plane JMC. Low temperature kinetics of the CH$_3$OH + OH reaction. J Phys Chem A. 2014;118:2693–701.

[12] Takayanagi T. Reduced-dimensionality quantum reactive scattering calculations of the C(^{3}P) + C$_2$H$_2$ reaction on a new potential energy surface. Chem Phys. 2005;312:61–7.

[13] Clary DC, Buonomo E, Sims IR, Smith IWM, Geppert WD, Naulin C, Costes M, Cartechini L, Casavecchia P. C+C$_2$H$_2$: a key reaction in interstellar chemistry. J Phys Chem A. 2002;106:5541–52.

[14] Clary DC. Rates of chemical reactions dominated by long-range intermolecular forces. Mol Phys. 1984;53:3–21.

[15] Clary DC. Fast chemical reactions: theory challenges experiment. Ann Rev Phys Chem. 1990;41:61–90.

[16] Clary DC, Henshaw JP. Chemical reactions dominated by long-range intermolecular forces. Faraday Discuss Chem Soc. 1987;84:333–349.

[17] Dateo CE, Clary DC. Rate constant calculations on the C$^+$+HCl reaction. J Chem Phys. 1989;90:7216–28.

[18] Stoecklin T, Clary DC. Fast reactions between diatomic and polyatomic molecules. J Phys Chem. 1992;96:7346–51.

[19] Stoecklin T, Clary DC. Fast reactions between a linear molecule and a polar symmetric top. J Mol Struct (Theochem). 1995;341:53–61.

[20] González AI, Clary DC, Yáñez M. Calculations of rate constants for reactions of first and second row cations. Theo Chem Acc. 1997;98: 33–41.

[21] Bates DR, Morgan WL. Adiabatic invariance treatment of hitting collisions between ions and symmetrical top dipolar molecules. J Chem Phys. 1987;87:2611–16.

[22] Sakimoto K. On the capture rate constant of collisions between ions and symmetric-top molecules. Chem Phys Lett. 1985;116:86–88.

[23] Turulski J, Niedzielski J. Effect of quantization of rotational energy on the ion-molecule capture. Chem Phys. 1989;137:191–5.

[24] Troe J. Relation between potential and rate parameters for reactions on attractive potential energy surfaces. Application to the reaction $HO + O \leftrightarrows HO_2^* \to H + O_2$. J Phys Chem. 1986;90:3485–92.

[25] Troe J. Statistical adiabatic channel model for ion–molecule capture processes. II. Analytical treatment of ion–dipole capture. J Chem Phys. 1996;105:6249–62.

[26] Ramillon M, McCarroll R. Adiabatic capture models for fast chemical reactions. J Chem Phys. 1994;101:8697–99.

[27] Buckingham AD. Basic theory of intermolecular forces: application to small molecules. In: Pullman B, Editors. Intermolecular interactions: from diatomics to biopolymers (Perspectives in quantum chemistry & biochemistry). John Wiley & Sons, New York; 1978, pp. 1–67.

[28] Truhlar DG, Garrett BC. Variational transition-state theory. Acc Chem Res. 1980;13:440–48.

[29] Wardlaw DM, Marcus RA. RRKM reaction rate theory for transition states of any looseness. Chem Phys Lett. 1984;110:230–34.

[30] Wardlaw DM, Marcus RA. Unimolecular reaction rate theory for transition states of partial looseness. II. Implementation and analysis with applications to NO_2 and C_2H_6 dissociations. J Chem Phys. 1985;83:3462–3480.

[31] Wardlaw DM Marcus RA. Unimolecular reaction rate theory for transition states of any looseness. 3. Application to methyl radical recombination. J Phys Chem. 1986;90:5383–93.

[32] Isaacson AD, Truhlar DG. Polyatomic canonical variational theory for chemical reaction rates. Separable-mode formalism with application to $OH+H_2 \to H_2O + H$. J Chem Phys. 1982;76:1380–91.

[33] Miller WH. Unified statistical model for complex and direct reaction mechanisms. J Chem Phys. 1976;65:2216–23.

[34] Miller WH. Reaction path Hamiltonian for polyatomic systems: further developments and application. In: Truhlar DG, Editor. Potential energy surfaces and dynamics calculations. New York: Plenum; 1981. pp. 265–86.

[35] Garrett BC, Truhlar DG. Criterion of minimum state density in the transition state theory of bimolecular reactions. J Chem Phys. 1979;70:1593–8.

[36] Hu W-P, Truhlar DG. Factors affecting competitive ion-molecule reactions: $ClO^- + C_2H_5Cl$ and C_2D_5Cl via E2 and S_N2 channels. J Am Chem Soc. 1996;118:860–69.

[37] Graff MM, Wagner AF. Theoretical studies of fine-structure effects and long-range forces: potential-energy surfaces and reactivity of $O(^3P) + OH(X^2\Pi)$. J Chem Phys. 1990;92:2423–39.

[38] Le Picard SD, Canosa A, Travers D, Chastaing D, Rowe BR, Stoecklin T. Experimental and theoretical kinetics for the reaction of Al with O_2 at temperatures between 23 and 295 K. J Phys Chem A. 1997;101:9988–92.

[39] Reignier D, Stoecklin T, Le Picard SD, Canosa A, Rowe BR. Rate constant calculations for atom-diatom reaction involving an open-shell atom and a molecule in a Σ electronic state. Application to the reaction $Al(^2P_{1/2,3/2}) + O_2(X^3\Sigma_g^-) \to AlO(X^2\Sigma^+) + O(^3P_{2,1,0})$. J Chem Soc Faraday Trans. 1998;94:1681–6.

[40] Beghin A, Stoecklin T, Rayez JC. Rate constant calculations for atom-diatom reactions involving an open shell atom and a molecule in a Π electronic state: application to the $C(^3P) + NO(X^2\Pi)$ reaction. Chem Phys. 1995;195:259–70.

[41] Geppert WD, Reignier D, Stoecklin T, Naulin C, Costes M, Chastaing D, Le Picard SD, Sims IR, Smith IWM. Comparison of the cross sections and thermal rate constants for the reactions of $C(^3P_J)$ atoms with O_2 and NO. Phys Chem Phys Chem. 2000;2:2873–81.

[42] Le Picard SD, Canosa A, Reignier D, Stoecklin T. A comparative study of the reactivity of the silicon atom $Si(^3P_J)$ towards O_2 and NO molecules at very low temperature. Phys Chem Phys Chem. 2002;4:3659–64.

[43] Le Picard SD, Canosa A, Geppert W, Stoecklin T. Experimental and theoretical temperature dependence of the rate coefficient of the $B(^2P_{1/2,3/2}) + O_2(X^3\Sigma_g^-)$ reaction in the [24–295 K] temperature range. Chem Phys Lett. 2004;385:502–6.

[44] Stoecklin T, Bussery-Honvault B, Honvault P, Dayou F. Asymptotic potentials and rate constants in the adiabatic capture centrifugal sudden approximation for $X + OH(X^2\Pi) \to OX + H(^2S)$ reactions where $X = O(^3P)$, $S(^3P)$ or $N(^4S)$. Comput Theo Chem. 2012;990:39–46.

[45] Stoecklin T, Dateo CE, Clary DC. Rate constant calculations on fast diatom-diatom reactions. J Chem Soc Faraday Trans. 1991;87:1667–79.

[46] Sims IR, Queffelec J-L, Defrance A, Rebrion-Rowe C, Travers D, Bocherel P, Rowe BR, Smith IWM. Ultralow temperature kinetics of neutral–neutral reactions. The technique and results for the reactions $CN+O_2$ down to 13 K and $CN+NH_3$ down to 25 K. J Chem Phys. 1994;100:4229–41.

[47] Klippenstein SJ, Kim Y-W. Variational statistical study of the $CN+O_2$ reaction employing ab initio determined properties for the transition state. J Chem Phys. 1993;99:5790–9.

[48] Rowe BR, Canosa A, Sims IR. Rate coefficients for interstellar gas-phase chemistry. J Chem Soc Faraday Trans. 1993;89:2193–8.

[49] Sims IR, Queffelec JL, Defrance A, Rebrion-Rowe C, Travers D, Rowe BR, Smith IWM. Ultra-low temperature kinetics of neutral-neutral reactions: the reaction $CN+O_2$ down to 26 K. J Chem Phys. 1992;97:8798–800.

[50] Sims IR, Smith IWM. Rate constants for the radical-radical reaction between CN and O_2 at temperatures down to 99 K. Chem Phys Lett. 1988;151:481–4.

[51] Sims IR, Smith IWM. Pulsed laser photolysis–laser-induced fluorescence measurements on the kinetics of $CN(\nu = 0)$ and $CN(\nu = 1)$ with O_2, NH_3 and NO between 294 and 761 K. J Chem Soc Faraday Trans II. 1988;84:527–39.

[52] Durant Jr. JL, Tully FP. Kinetic study of the reaction between CN and O_2 from 295 to 710 K. Chem Phys Lett. 1989;154:568–72.

[53] Atakan B, Jacobs A, Wahl M, Weller R, Wolfrum J. Kinetic studies of the gas-phase reactions of CN with O_2 and H_2 from 294 to 1000 K. Chem Phys Lett. 1989;154:449–53.

[54] Balla RJ, Casleton KH. Kinetic study of the reactions of cyanyl radical with oxygen and carbon dioxide from 292 to 1500 K using high-temperature photochemistry. J Phys Chem. 1991;95:2344–51.

[55] Burmeister M, Gulati SK, Natarajan K, Thielen K, Mozzhukin E, Roth P. High temperature rate coefficient for the reaction $CN + O_2 \rightarrow NCO + O$ using different CN-sources. Symp (Int) Combust. 1988;22:1083–92.

[56] Davidson DF, Dean AJ, DiRosa MD, Hanson MK. Shock tube measurements of the reactions of CN with O and O_2. Int J Chem Kinet. 1991;23:1035.

[57] Liao Q, Herbst E. Capture calculations for the rates of important neutral-neutral reactions in dense interstellar clouds: $C+C_2H_2$ and $CN+C_2H_2$. Astrophys J. 1995;444:694–701.

[58] Sims IR, Queffelec J-L, Travers D, Rowe BR, Herbert LB, Karthäuser J, Smith IWM. Rate constants for the reactions of CN with hydrocarbons at low and ultra-low temperatures. Chem Phys Lett. 1993;211:461–8.

[59] Woon DE, Herbst E. The rate of the reaction between CN and C_2H_2 at interstellar temperatures. Astrophys J. 1997;477:204–8.

[60] Becker KH, Engelhardt B, Geiger H, Kurtenbach R, Wiesen P. Temperature dependence of the reactions of CH radicals with NO,

NH$_3$ and N$_2$O in the range 200–1300 K. Chem Phys Lett. 1993;210: 135–40.

[61] Jourdain JL, Le Bras G, Combourieu J. Kinetic study by electron paramagnetic resonance and mass spectrometry of the elementary reactions of phosphorus tribromide with hydrogen, oxygen, and hydroxyl radicals. J Phys Chem. 1982;86:4170–75.

[62] Talbi D, Smith IWM. A theoretical analysis of the reaction between CN radicals and NH$_3$. Phys Chem Chem Phys. 2009;11:8477–8483.

[63] Georgievskii Y, Klippenstein SJ. Long-range transition state theory. J Chem Phys. 2005;122:194103.

[64] Greenwald EE, North SW, Georgievskii Y, Klippenstein SJ. A two transition state model for radical-molecule reactions: a case study of the addition of OH to C$_2$H$_4$. J Chem Phys A. 2005;109:6031–44.

[65] Lloyd AC, Darnall KR, Winer AM, Pitts Jr. JN. Relative rate constants for reaction of the hydroxyl radical with a series of alkanes, alkenes, and aromatic hydrocarbons. J Phys Chem. 1976;80:789–794.

[66] Gordon S, Mulac WA. Reaction of the OH(X$^2\Pi$) radical produced by pulse radiolysis of water vapor. International Journal of Chemical Kinetics: Symposium. 1975;1(S):289–299. https://www.worldcat.org/title/proceedings-of-the-symposium-on-chemical-kinetics-data-for-the-upper-and-lower-atmosphere-held-at-warrenton-virginia-september-15-18-1974/oclc/1718129; Proceeding Symposium Chemical Kinetics Data Upper Lower Atmosphere; 1974. p. 289.

[67] Overend R, Paraskevopoulos G. Rates of OH radical reactions. III. The reaction OH+C$_2$H$_4$+M at 296 K. J Chem Phys. 1977;67:674–9.

[68] Cox RA. Photolysis of gaseous nitrous acid: a technique for obtaining kinetic data on atmospheric photooxidation reactions. Proceeding Symposium Chemical Kinetics Data Upper Lower Atmosphere; 1974. p. 379.

[69] Cox RA, Derwent RG, Williams MR. Atmospheric photooxidation reactions. Rates, reactivity, and mechanism for reaction of organic compounds with hydroxyl radicals. Environ Sci Technol. 1980;14: 57–61.

[70] Atkinson R, Perry RA, Pitts Jr. JN. Rate constants for the reaction of OH radicals with ethylene over the temperature range 299–425 K. J Chem Phys. 1977;66:1197.

[71] Atkinson R, Aschmann SM, Winer AM, Pitts Jr. JN. Rate constants for the reaction of OH radicals with a series of alkanes and alkenes at 299±2 K. Int J Chem Kinet. 1982;14:507–16.

[72] Atkinson R, Aschmann SM. Rate constants for the reaction of OH radicals with a series of alkenes and dialkenes at 295 ± 1 K. Int J Chem Kinet. 1984;16:1175–86.

[73] Tully FP. Laser photolysis/laser-induced fluorescence study of the reaction of hydroxyl radical with ethylene. Chem Phys Lett. 1983;96:148–53.

[74] Tully FP. Hydrogen-atom abstraction from alkenes by OH, ethene and 1-butene. Chem Phys Lett. 1988;143:510–14.

[75] Klein T, Barnes I, Becker KH, Fink EH, Zabel F. Pressure dependence of the rate constants for the reactions of ethene and propene with hydroxyl radicals at 295 K. J Phys Chem. 1984;88:5020–25.

[76] Schmidt V, Zhu GY, Becker KH, Fink EH. Study of OH reactions at high pressures by excimer laser photolysis — dye laser fluorescence. Ber Bunsenges Phys Chem. 1985;89:321–2.

[77] Klopffer VW, Frank R, Kohl E-G, Haag F. Quantitative Erfassung der photochemischen transformationsprozeese in der Troposphare. Chem Ztg. 1986;110:57–62.

[78] Liu A, Mulac WA, Jonah CD. Pulse radiolysis study of the reaction of OH radicals with C_2H_4 over the temperature range 343–1173 K. Int J Chem Kinet. 1987;19:25–34.

[79] Liu A, Mulac WA, Jonah CD. Kinetic isotope effects in the gas-phase reaction of hydroxyl radicals with ethylene in the temperature range 343–1173 K and 1-atm pressure. J Phys Chem. 1988;92:3828–33.

[80] Nielsen OJ, Jorgensen O, Donlon M, Sidebottom HW, O'Farrell DJ, Treacy T. Rate constants for the gas-phase reactions of OH radicals with nitroethene, 3-nitropropene and 1-nitrocyclohexene at 298 K and 1 atm. Chem Phys Lett. 1990;168:319–23.

[81] Becker KH, Geiger H, Wiesen P. Kinetic study of the OH radical chain in the reaction system OH + C_2H_4+ NO + air. Chem Phys Lett. 1991;184:256–61.

[82] Fulle D, Hamann HF, Hippler H, Jansch CP. The high pressure range of the addition of OH to C_2H_2 and C_2H_4. Ber Bunsenges Phys Chem. 1997;101:1433–42.

[83] Vakhtin AB, Murphy JE, Leone SR. Low-temperature kinetics of reactions of OH radical with ethene, propene, and 1-butene. J Phys Chem A. 2003;107:10055–62.

[84] Greenwald EE, North SW, Georgievskii Y, Klippenstein SJ. A two transition state model for radical-molecule reactions: application to isomeric branching in the OH-isoprene reaction. J Chem Phys A. 2007;111:5582–92.

[85] Sabbah H, Biennier L, Sims IR, Georgievskii Y, Klippenstein SJ, Smith IWM. Understanding reactivity at very low temperatures: the reactions of oxygen atoms with alkenes. Science. 2007;317:102–105.

[86] Georgievskii Y, Klippenstein SJ. Strange kinetics of the C_2H_6 + CN reaction explained. J Phys Chem A. 2007;111:3802–11.

[87] Cheikh Sid Ely S, Morales SB, Guillemin J-C, Klippenstein SJ, Sims IR. Low temperature rate coefficients for the reaction CN + HC$_3$N. J Phys Chem A. 2013;117:12155–64.

[88] Gao LG, Zheng J, Fernández-Ramos A, Truhlar DG, Xu X. Kinetics of the methanol reaction with OH at interstellar, atmospheric, and combustion temperatures. J Am Chem Soc. 2018;140:2906–18.

[89] Talukdar RK, Gierczak T, McCabe DC, Ravishankara AR. Reaction of hydroxyl radical with acetone. 2. Products and reaction mechanism. J Phys Chem. 2003;107:5021–32.

[90] Hénon E, Canneaux S, Bohr F, Dóbé S. Features of the potential energy surface for the reaction of OH radical with acetone. Phys Chem Phys Chem. 2003;5:333–41.

[91] Vanderberk S, Vereecken L, Peeters J. The acetic acid forming channel in the acetone + OH reaction: a combined experimental and theoretical investigation. Phys Chem Chem Phys. 2002;4:461–6.

[92] Yamada T, Taylor PH, Goumri A, Marshall P. The reaction of OH with acetone and acetone-d$_6$ from 298 to 832 K: rate coefficients and mechanism. J Chem Phys. 2003;119:10600–6.

[93] Caralp F, Forst W, Hénon E, Bergeat A, Bohr F. Tunneling in the reaction of acetone with OH. Phys Chem Chem Phys. 2006;8:1072–8.

[94] Forst W. Unimolecular reactions: a concise introduction. Cambridge: Cambridge University Press; 2003.

[95] Blázquez S, González D, Garcia-Sáez A, Antiñolo M, Bergeat A, Caralp F, Mereau R, Canosa A, Ballesteros B, Albaladejo J, Jiménez E. Experimental and theoretical investigation on the OH + CH$_3$C(O)CH$_3$ reaction at interstellar temperatures (T = 11.7–64.4 K). ACS Earth Space Chem. 2019;3:1873–83.

[96] Shannon RJ, Caravan RL, Blitz MA, Heard DE. A combined experimental and theoretical study of reactions between the hydroxyl radical and oxygenated hydrocarbons relevant to astrochemical environments. Phys Chem Chem Phys. 2014;16:3466–78.

[97] Glowacki DR, Liang C-H, Morley C, Pilling MJ. MESMER: an open-source master equation solver for multi-energy well reactions. J Chem Phys A. 2012;116:9545–60.

[98] Nguyen TL, Ruscic B, Stanton JF. A master equation simulation for the ·OH + CH$_3$OH reaction. J Chem Phys. 2019;150:084105.

[99] Liu Y-P, Lynch GC, Truong TN, Lu D-H, Truhlar DG, Garrett BC. Molecular modeling of the kinetic isotope effect for the [1,5] Sigmatropic rearrangement of cis-1,3-Pentadiene. J Am Chem Soc. 1993;115:2408–15.

[100] Nguyen TL, McCaslin L, McCarthy MC, Stanton JF. Thermal unimolecular decomposition of syn-CH_3CHOO: a kinetic study. J Chem Phys. 2016;145:131102.

[101] Gómez Martin JC, Caravan RL, Blitz MA, Heard DE, Plane JMC. Low temperature kinetics of the CH_3OH + OH reaction. J Phys Chem A. 2014;118:2693–701.

[102] Craig IR, Manolopoulos DE. Quantum statistics and classical mechanics: real time correlation functions from ring polymer molecular dynamics. J Chem Phys. 2004;121:3368–73.

[103] Craig IR, Manolopoulos DE. Chemical reaction rates from ring polymer molecular dynamics. J Chem Phys. 2005;122:084106.

[104] Craig IR, Manolopoulos DE. A refined ring polymer molecular dynamics theory of chemical reaction rates. J Chem Phys. 2005; 123:034102.

[105] Suleimanov YV, Aoiz FJ, Guo H. Chemical reaction rate coefficients from ring polymer molecular dynamics: theory and practical applications. J Phys Chem A. 2016;120:8488–502.

[106] del Mazo-Sevillano P, Aguado A, Jiménez E, Suleimanov YV, Roncero O. Quantum roaming in the complex-forming mechanism of the reactions of OH with formaldehyde and methanol at low temperature and zero pressure: a ring polymer molecular dynamics approach. J Phys Chem Lett. 2019;10:1900–907.

[107] Chandler D, Wolynes PG. Exploiting the isomorphism between quantum theory and classical statistical mechanics of polyatomic fluids. J Chem Phys. 1981;74:4078–95.

[108] González-Lavado E, Corchado JC, Suleimanov YV, Green WH, Espinosa-Garcia J. Theoretical kinetics study of the $O(^3P)$ + CH_4/CD_4 hydrogen abstraction reaction: the role of anharmonicity, recrossing effects, and quantum mechanical tunneling. J Phys Chem A. 2014;118:3243–52.

[109] Pérez de Tudela R, Suleimanov YV, Richardson JO, Sáez-Rábanos V, Green WH, Aoiz FJ. Stress test for quantum dynamics approximations: deep tunneling in the muonium exchange reaction D + HMu → DMu + H. J Phys Chem Lett. 2014;5:4219–24.

[110] Suleimanov YV, Collepardo-Guevara R, Manolopoulos DE. Bimolecular reaction rates from ring polymer molecular dynamics: application to H + CH_4 → H_2+ CH_3. J Chem Phys. 2011;134:044131.

[111] Li Y, Suleimanov YV, Yang M, Green WH, Guo H. Ring polymer molecular dynamics calculations of thermal rate constants for the $O(^3P)$ + CH_4 → OH + CH_3 reaction: contributions of quantum effects. J Phys Chem Lett. 2013;4:48–52.

[112] Collepardo-Guevara R, Suleimanov YV, Manolopoulos DE. Bimolecular reaction rates from ring polymer molecular dynamics. J Chem Phys. 2009;130:174713.

[113] Pérez de Tudela R, Aoiz FJ, Suleimanov YV, Manolopoulos DE. Chemical reaction rates from ring polymer molecular dynamics: zero point energy conservation in Mu + H$_2$ to MuH + H. J Phys Chem Lett. 2012;3:493–7.

[114] Suleimanov YV, Pérez de Tudela R, Jambrina PG, Castillo JF, Sáez-Rábanos V, Manolopoulos DE, Aoiz FJ. A ring polymer molecular dynamics study of the isotopologues of the H + H$_2$ reaction. Phys Chem Chem Phys. 2013;15:3655–65.

[115] Li Y, Suleimanov YV, Guo H. Ring polymer molecular dynamics rate coefficient calculations for insertion reactions: X + H$_2$ → HX + H (X = N, O). J Phys Chem Lett. 2014;5:700–705.

[116] Suleimanov YV, Kong WJ, Guo H, Green WH. Ring polymer molecular dynamics: rate coefficient calculations for energetically symmetric (near thermoneutral) insertion reactions (X + H$_2$) → HX + H(X = C(^{1}D), S(^{1}D)). J Chem Phys. 2014;141:244103.

[117] Li Y, Suleimanov YV, Li J, Green WH, Guo H. Rate coefficients and kinetic isotope effects of the X + CH$_4$ → CH$_3$+ HX (X = H, D, Mu) reactions from ring polymer molecular dynamics. J Chem Phys. 2013;138:094307.

[118] Suleimanov YV. Surface diffusion of hydrogen on Ni(100) from ring polymer molecular dynamics. J Phys Chem C. 2012;116:11141–11153.

[119] Naumkin F, del Mazo-Sevillano P, Aguado A, Suleimanov YV, Roncero O. Zero- and high-pressure mechanisms in the complex forming reactions of OH with methanol and formaldehyde at low temperatures. ACS Earth Space Chem. 2019;3:1158–69.

[120] Wu J, Ning H, Ma L, Ren W. Pressure-dependent kinetics of methyl formate reactions with OH at combustion, atmospheric and interstellar temperatures. Phys Chem Chem Phys. 2018;20:26190–26199.

[121] Nguyen TL, Stanton JF. Pressure-dependent rate constant caused by tunneling effects: OH + HNO$_3$ as an example. J Phys Chem Lett. 2020;11:3712–7.

[122] Herbert LB, Sims IR, Smith IWM, Stewart DWA, Symonds AC, Canosa A, Rowe BR. Rate constants for the relaxation of CH(X^{2}Π, ν = 1) by CO and N$_2$ at temperatures from 23 to 584 K. J Phys Chem. 1996;100:14928–35.

[123] Lin HM, Seaver M, Tang KY, Knight AEW, Parmenter CS. The role of intermolecular potential well depths in collision-induced state changes. J Chem Phys. 1979;70:5442–57.

[124] Holtermann DL, Lee EKC, Nanes R. Rate of collision-induced electronic relaxation of single rotational levels of $SO_2(\tilde{A}^1A_2)$: quenching mechanism by collision complex formation. J Chem Phys. 1982;77:5327–39.

[125] Paul PH, Gray JA, Durant Jr. JL, Thoman Jr. JW. A model for temperature-dependent collisional quenching of $NO(A^2\Sigma^+)$. Appl Phys B. 1993;57:249–59.

[126] Förster T. 10th spiers memorial lecture, transfer mechanisms of electronic excitation. Discuss Faraday Soc. 1959;27:7–17.

[127] Breckenridge WH, Blickensderfer RP, Fitzpatrick J, Oba D. Near resonant electronic energy transfer: initial rotational state populations of NO $\{A^2\Sigma^+, \nu' = 0, 1)$ produced by energy transfer from $Zn(^1P_1)$. J Chem Phys. 1979;70:4751–60.

[128] Scholes GD. Long-range resonant energy transfer in molecular systems. Annu Rev Phys Chem. 2003;54:57–87.

[129] Sánchez-González R, Eveland WD, West NA, Mai CLN, Bowersox RDW, North SW. Low-temperature collisional quenching of NO $A^2\Sigma^+(\nu' = 0)$ by $NO(X^2\Pi)$ and O_2 between 34 and 109 K. J Chem Phys. 2014;141:074313.

[130] Winner JD, West NA, McIlvoy MH, Buen ZD, Bowersox RDW, North SW. The role of near resonance electronic energy transfer on the collisional quenching of NO $(A^2\Sigma^+)$ by C_6H_6 and C_6F_6 at low temperature. Chem Phys. 2018;501:86–92.

[131] Settersten TB, Patterson BD, Carter CD. Collisional quenching of NO between 125 and 294 K. J Chem Phys. 2009;130:204302.

[132] Settersten TB, Patterson BD, Gray JA. Temperature — and species-dependent quenching of NO probed by two-photon laser-induced fluorescence using a picosecond laser. J Chem Phys. 2006;124:234308.

[133] Rellergert WG, Sullivan ST, Schowalter SJ, Kotochigova S, Chen K, Hudson ER. Evidence for sympathetic vibrational cooling of translationally cold molecules. Nature. 2013;495:490–94.

[134] Stoecklin T, Halvick P, Gannouni MA, Hochlaf M, Kotochigova S, Hudson ER. Explanation of efficient quenching of molecular ion vibrational motion by ultracold atoms. Nat Commun. 2016;7:11234.

[135] Arthurs A, Dalgarno A. The theory of scattering by a rigid rotator. Proc R Soc Lond Ser A. 1960;256:540–51.

[136] Faure A, Jankowski P, Stoecklin T, Szalewicz K. On the importance of full-dimensionality in low-energy molecular scattering calculations. Sci Rep. 2016;6:28449.

[137] Stoecklin T, Denis-Alpizar O, Halvick P, Dubernet ML. Rovibrational relaxation of HCN in collisions with He: rigid bender treatment of the bending-rotation interaction. J Chem Phys. 2013;139:124317.

[138] Stoecklin T, Denis-Alpizar O, Halvick P. Rovibrational energy transfer in the He-C_3 collision: rigid bender treatment of the bending-rotation interaction. Mon Not R Astron Soc. 2015;449:3420–25.

[139] Stoecklin T, Denis-Alpizar O, Clergerie A, Halvick P, Faure A, Scribano Y. Rigid-bender close-coupling treatment of the inelastic collisions of H_2O with para-H_2. J Phys Chem A. 2019;123:5704–12.

[140] James PL, Sims IR, Smith IWM. Rate coefficients for the vibrational self-relaxation of NO($X^2\Pi$, $\nu = 3$) at temperatures down to 7 K. Chem Phys Lett. 1997;276:423–9.

[141] Nikitin EE, Umanski SY. Effect of vibronic interaction upon the vibrational relaxation of diatomic molecules. Faraday Discuss Chem Soc. 1972;53:7–17.

[142] Forst W, Barker JR. Collisional energy transfer and macroscopic disequilibrium. Application to azulene. J Chem Phys. 1985;83:124–32.

[143] Lenzer T, Luther AK. Intermolecular potential effects in trajectory calculations of collisions between large highly excited molecules and noble gases. J Chem Phys. 1996;105:10944–53.

[144] Wright SMA, Sims IR, Smith IWM. Vibrational relaxation of highly excited toluene in collisions with He, Ar, and N_2 at temperatures down to 38 K. J Phys Chem A. 2000;104:10347–55.

[145] Chefdeville S, Kalugina Y, van de Meerakker SYT, Naulin C, Lique F, Costes M. Observation of partial wave resonances in low-energy O_2–H_2 inelastic collisions. Science. 2013;341:1094–6.

[146] Chefdeville S, Stoecklin T, Naulin C, Jankowski P, Szalewicz K, Faure A, Costes M, Bergeat A. Experimental and theoretical analysis of low energy CO + H_2 inelastic collisions. Astrophys J Lett. 2015;799:L9.

[147] Stoecklin T, Faure A, Jankowski P, Chefdeville S, Bergeat A, Naulin C, Morales SB, Costes M. Comparative experimental and theoretical study of the rotational excitation of CO by collision with *ortho* — and *para*-D_2 molecules. Phys Chem Chem Phys. 2017;19:189–95.

[148] Mertens LA, Labiad H, Denis-Alpizar O, Fournier M, Carty D, Le Picard SD, Stoecklin T, Sims IR. Rotational energy transfer in collisions between CO and Ar at temperatures from 293 to 30 K. Chem Phys Lett. 2017;683:521–8.

[149] Launay JM. Molecular collision processes. I. Body-fixed theory of collisions between two systems with arbitrary angular momenta. J Phys B At Mol Phys. 1977;10:3665–72.

[150] Alexander MH. Rotationally inelastic collisions between a diatomic molecule in a $^2\Sigma^+$ electronic state and a structureless target. J Chem Phys. 1982;76:3637–45.

[151] Alexander MH. Propensity rules for rotationally inelastic collisions of symmetric top molecules and linear polyatomic molecules with structureless atoms. J Chem Phys. 1982;77:1855–65.

[152] Alexander MH. Rotationally inelastic collisions between a diatomic molecule in a $^2\Pi$ electronic state and a structureless target. J Chem Phys. 1982;76:5974–88.

[153] Alexander MH, Dagdigian PJ. Propensity rules in rotationally inelastic collisions of diatomic molecules in $^3\Sigma$ electronic states. J Chem Phys. 1983;79:302–10.

[154] Alexander MH, Dagdigian PJ. Collision-induced transitions between molecular hyperfine levels: quantum formalism, propensity rules, and experimental study of $CaBr(X^2\Sigma^+)$+Ar. J Chem Phys. 1985;83: 2191–200.

[155] Le Picard SD, Honvault P, Bussery-Honvault B, Canosa A, Laubé S, Launay JM, Rowe B, Chastaing D, Sims IR. Experimental and theoretical study of intramultiplet transitions in collisions of $C(^3P)$ and $Si(^3P)$ with He. J Chem Phys. 2002;117:10109–120.

[156] Lique F, Kłos J, Le Picard SD. Fine-structure transitions of interstellar atomic sulfur and silicon induced by collisions with helium. Phys Chem Chem Phys. 2018;20:5427–34.

[157] Lique F, Werfelli G, Halvick P, Stoecklin T, Faure A, Wiesenfeld L, Dagdigian PJ. Spin-orbit quenching of the $C^+(^2P)$ ion by collisions with *para* — and *ortho*-H_2. J Chem Phys. 2013;138:204314.

[158] Alexander MH, Manolopoulos DE, Werner H-J, Follmeg B, Dagdigian PJ. 2011. Available from: www2.chem.umd.edu/groups/alexander/hibridon/hib43.

[159] Prezhdo OV. Quantized hamilton dynamics. Theor Chem Acc. 2006; 116:206–18.

[160] Semenov A, Babikov D. Mixed quantum/classical theory for molecule-molecule inelastic scattering: derivation of equations and application to N_2+H_2 system. J Phys Chem A. 2015;119: 12329–38.

[161] Feynman RP. Simulating physics with computers. Int J Theor Phys. 1982;21:467–88.

[162] Cao Y, Romero J, Olson JP, Degroote M, Johnson PD, Kieferova M, Kivlichan ID, Menke T, Peropadre B, Sawaya NPD, Sim S, Veis L, Aspuru-Guzik A. Quantum chemistry in the age of quantum computing. Chem Rev. 2019;119:10856–915.

Chapter 12

Conclusions: Future Challenges and Perspectives

Bertrand R. Rowe[*,§], André Canosa[†,¶] and Dwayne E. Heard[‡,‖]

*Rowe Consulting, 22 chemin des moines,
22750 Saint Jacut de la Mer, France
†CNRS, IPR (Institut de Physique de Rennes)-UMR 6251,
Université de Rennes, F-35000 Rennes, France
‡School of Chemistry, University of Leeds, LS2 9JT Leeds, UK
§bertrand.rowe@gmail.com
¶andre.canosa@univ-rennes1.fr
‖d.e.heard@leeds.ac.uk

Abstract

The present chapter highlights the main discoveries and breakthroughs described in the previous chapters of this book that uniform supersonic flows have allowed during the last forty years in the field of physical chemistry. The accumulation of experimental data in the domain of very-low-temperature reaction kinetics has stimulated new developments in theoretical quantum chemistry and astrochemical modeling, some of which will be further emphasized. Although a considerable body of knowledge has been explored during these four decades, we will comment about the many remaining challenges and open questions that could be tackled further using CRESU environments. Exploring these fields should be facilitated in the near future thanks to the significant number of new instrumental diagnostic tools that are in development in several research groups worldwide.

Keywords: CRESU; Challenges; Open questions; Absolute zero; Product detection; Spectroscopy; Nucleation; Reaction kinetics; Energy transfer; Astrochemistry, Quantum theory.

12.1 Principal Impacts of Using Uniform Supersonic Flows in Chemical Physics

12.1.1 *General thoughts*

Throughout the previous eleven chapters of this book, we and our co-authors and colleagues (whom we wish to thank warmly for their contributions) have attempted to gather the state of the art of knowledge from the last four decades for the use of uniform supersonic flows in the fields of chemical physics/physical chemistry, which have applications, for example, in chemical kinetics, quantum chemistry, and astrophysics. In this section, we will briefly review and discuss the major achievements and breakthroughs covered in the preceding chapters which were made possible thanks to the adaptation of a number of instrumental techniques to the supersonic environment. It is difficult to be comprehensive within a short summary, and so the examples chosen are necessarily selective but cover the major topics and results presented in this book.

The main impact of CRESU has been to allow the study of chemical kinetics at very low temperatures, which is crucial to allow the measurement of rate coefficients under or close to the conditions that will be encountered in important environments, such as in the interstellar medium. The requirements are quite stringent; it is necessary to have a uniform supersonic flow in terms of both temperature and density which enable kinetic parameters to be extracted, yet which is sufficiently remote from any apparatus wall in order to prevent condensation of the reagents. Extrapolation of kinetic data obtained at higher temperatures is fraught with difficulties, as the dependence of the rate coefficient upon temperature is very often not represented by the well-known Arrhenius equation, which may apply to data at higher temperatures. Even if some low-temperature data are already available from which a parameterization of the rate coefficients has been developed, extrapolation to temperatures around 10–30 K representative of some interstellar environments goes well beyond the intended range of the parameterization, and has been shown to lead to highly erroneous predictions. The problem of extrapolation is particularly acute for reactions which display a U shape in the rate coefficient with temperature, where the rate of increase of $k(\text{T})$ per unit of temperature increases at the lowest temperatures, or indeed flattens out after a significant increase. Although theory is able to provide some predictive capability and has been shown to reproduce the general form of $k(\text{T})$ at low temperatures, theory as yet is not able to calculate quantitatively the value of k at very low temperatures. This inability is largely due to the gas phase reactions which are important, for example, for astrochemical applications, possessing barrier heights and/or well depths of pre-reactive complexes and intermediates on their potential energy surfaces which are not so different from the energy uncertainty of the theoretical calculations. Experimental determinations using the CRESU method are therefore critical to advance our understanding of low-temperature kinetics.

A short summary is given below in the next section of the major achievements and highlights that are covered in each chapter. A wide range of topics is explored in the book, for example, the properties, motivation, and challenges for working in a very-low-temperature environment close to absolute zero, a historical perspective of the development of the CRESU method, determination of the properties of the CRESU uniform supersonic flows, chemical kinetics involving both neutral and ionic reagents (rate coefficients and product branching ratios), studies of nucleation, energy transfer, and molecular spectroscopy, the implications of the new data for astrochemistry in general and specifically for the development of chemical networks, and finally (in terms of this list but not in terms of importance) how theory has addressed some of these topics and significantly enhanced our fundamental understanding of many of the processes CRESU was designed to study.

12.1.2 *Summary and highlights from individual chapters*

Chapter 1 by Bertrand Rowe and André Canosa introduces the ultracold environment close to absolute zero which the CRESU supersonic uniform flow tries to emulate. A key question, which is critical to understanding chemical kinetics in such a flow, and to whether data obtained from such a flow are relevant to cold astrophysical environments, is the extent to which Local Thermodynamic Equilibrium (LTE) is applicable. LTE is discussed within a highly rigorous framework which links microscopic and macroscopic quantities, with the latter primarily being the type of experimental observable derived from a CRESU apparatus. The concept of temperature and what it means under relevant conditions in the laboratory and in space are discussed, together with a rigorous mathematical description of compressible flows and supersonic expansions which are central to the CRESU method. An overview is given of the astrophysical media in which astrochemistry emerges as a key focus, together with a historical perspective of the development of apparatus of different types with the common desire to achieve ultralow temperatures suitable for laboratory studies in chemistry and physics.

Chapter 2 by André Canosa and Elena Jiménez is a very comprehensive account of the history and development of the de Laval nozzle, contrasting the designs and performance of Laval nozzles that are in use worldwide for laboratory applications, both for continuous and pulsed operations. The underlying theory of the supersonic flow which develops within a Laval nozzle is explored in detail, and how the nozzle expansion is characterized experimentally to determine the temperature and density along the reactive flow, knowledge of which is absolutely critical for the determination and interpretation of kinetic parameters. The details given about Laval nozzles are probably unsurpassed regarding their physical characteristics, the properties of the supersonic flows they generate, their usage, and versatility, and are unique in comparing all of the instruments that have been used and are currently being used worldwide. Photographs of the Laval nozzles and surrounding apparatus in situ in their respective laboratories are given. The uniform and pulsed variants of the CRESU method each have their advantages and disadvantages, and these are discussed as part of a detailed comparison.

Chapter 3 by Ludovic Biennier, Sophie Carles, François Lique, and James Brian Mitchell describes the chemistry of ions in CRESU uniform supersonic flows. Ion processes are important for the ionosphere of Earth, atmospheres of bodies in our solar system, and in interstellar molecular clouds. Ion molecule reactions are usually fast and exothermic, possessing no overall barrier on the potential energy surface (PES) for reaction, and were one of the first types of reactions studied within a CRESU flow. The PESs are often

Chapter 6 by Barbara Wyslouzil and Ruth Signorell is concerned with the study of nucleation of aerosols, which has very important applications in both atmospheric chemistry and industrial settings. Here, nucleation is followed in real time on the microsecond timescale within either the CRESU supersonic uniform flow itself, or within the divergent section of the Laval nozzle preceding the uniform flow where the total temperature and density are changing. Mass spectrometric methods are used to monitor the temporal evolution of clusters from the onset of condensation to the growth or large clusters. A number of nucleation steps are considered, namely, homogeneous nucleation from the supersaturated vapor to the supercooled liquid, nucleation of a solid phase from the supercooled liquid, and heterogeneous nucleation of other species onto preexisting particles. A thorough description of nucleation theory is given for each of these mechanisms, including the critical role of the critical cluster size, as well as a comprehensive coverage of the CRESU supersonic nozzle method used to determine the uniform flow, and the different optical, mass spectrometric, and other techniques (e.g. X-ray) used to characterize the size, thermodynamic phase, and chemical composition of the clusters formed. A key observable is the rate of nucleation of a given cluster, and approaches to determining this are given together with comparison with theory. Several examples of systems studied in a supersonic flow are given, displaying different types of nucleation mechanisms. A notable example discussed is the determination of the freezing rate of supercooled liquid water to ice which has been studied by a range of groups over the temperature range $\sim$180–240 K, and the subsequent heterogeneous nucleation of CO_2 onto water ice particles over the temperature range 60–140 K, which occurs well downstream of the formation of the ice particles, and which was studied downstream of the throat of the Laval nozzle.

Chapter 7 by Ilsa Cooke and Ian Sims discusses the study of collisional energy transfer using supersonic uniform flows in a CRESU apparatus. As well as data for reactive collisions, the rates and products of inelastic collisions are critical inputs for astrochemical models, in particular to accurately describe radiative transfer schemes for models of astrophysical environments. Molecules in space are often formed under non-LTE conditions and collisions provide the mechanism to establish LTE (although radiative decay may also compete at the low densities of space), and hence a Boltzmann distribution of molecular-level populations, the assumption of which underpins many astronomical observations of molecular abundances using radio and optical telescopes. Inelastic collisions also probe unique parts of the potential energy surface, with comparison of experimental results with theory allowing how well we understand these parts to be gauged. The range of inelastic processes covered is quite wide ranging, and includes electronic, vibrational, and rotational energy transfer, and the chapter is comprehensive in collating the previous studies performed in each of these areas. There is a theoretical underpinning for each type of energy transfer, and the adaptation of the CRESU apparatus for the measurement of both rate coefficients and cross sections for energy transfer processes is given. For each type of collisional energy transfer, the chapter provides a very useful collation of results from different groups, over a wide range of temperatures down to as low as 5.5 K. Examples are then discussed for rotational energy transfer (RET), with a focus on detection with a CRESU apparatus using LIF, which enables measurement of state-to-state cross sections giving information on the final rotational state that is populated following the RET collision, as well as rate coefficients for removal of an initially populated rotational level. Vibrational energy transfer (VET) occurs on a much slower timescale than RET, and is difficult to study within the uniform supersonic flow of a CRESU apparatus, as the timescale for monitoring species is limited. However, VET involving V-V transfer or V-T,R transfer via the formation of a strongly bound complex, or VET of larger, highly excited species is faster and examples for VET of CH, NO, and toluene are given. Studies of vibrational energy transfer using CRESU have taken advantage of newly developed monitoring techniques such as chirped-pulse microwave absorption spectroscopy. Furthermore, the study of electronic energy transfer at very low temperatures within supersonic uniform flows is discussed, with examples including relaxation of excited atoms and small radical species, in some cases which can take place in competition with reactive removal (e.g. $^1CH_2$). Finally, the understudied spin-orbit energy transfer is discussed involving fine structure states in atoms. The comparison of these rates of energy transfer with theory is covered more in Chapter 11.

    Chapter 8 by Sébastien Le Picard, Abdessamad Benidar, and Fabien Goulay is concerned with the measurement of the product branching ratio (BR) for a chemical reaction at low temperatures, a critical parameter, but which is very difficult to measure and often not quantified for many reactions. Rate coefficients over a wide of temperatures are determined more routinely using supersonic uniform flows, but determination of the BR requires knowledge of absolute concentrations, or use of a reference reaction where the BR is known in advance for comparison. Theoretical methods may not be able to capture well any change in mechanism and hence the products of a reaction as the temperature is lowered, and direct determination of the BR is needed but sadly usually lacking. For astrochemical models, knowledge of the BR is as important as the rate coefficient, but often educated guesses are all that can be made, perhaps even just using thermochemical data. The database of BRs determined using CRESU is expanding, and this chapter discusses a range of examples using different techniques, for example, reactions where H atom BRs have been quantified using VUV-LIF at 121 nm. However, it is mass spectrometry that has become the dominant method for the determination of the BR, particularly newly emerging multiplex mass spectrometry methods that are more universal in nature. It is a considerable engineering challenge though to interface a supersonic uniform flow CRESU apparatus with a mass spectrometer, and the chapter discusses in detail several approaches that have been made, and how it is checked, for example, that the design of the sampling skimmer or aerofoil which directs gas to the ionizing region of the mass spectrometer does not perturb the supersonic flow in any way. A particular focus of the chapter is the use of tunable VUV radiation from the Paris SOLEIL and Berkeley ALS synchrotron sources to identify products from chemical reactions, with the CRESUSOL apparatus developed at Rennes enabling product detection down to 20 K. The wavelength tunable nature of the ionizing radiation enables the production of an ion of a given mass to be monitored as a function of ionizing wavelength, which when combined with theoretical simulations allows the identity of different isomers to be performed. A number of examples are chosen to illustrate the power of mass spectrometry for product identification, in particular for reactions of carbon-based radicals (e.g. C_2H). Chirped-pulse microwave absorption spectroscopy is able to detect and identify polyatomic products of chemical reactions with a CRESU environment, including different isomers, and offers significant promise as the very cold temperatures significantly simplify the absorption spectra that are measured. Optical frequency comb spectroscopy is an exciting development which may allow the identification of products of reactions initiated at very low temperatures in a Laval nozzle expansion. It is fair to say, that although detection and identification of products have become more common owing to the techniques mentioned above (previously only LIF was commonly used to identify some products), the measurement of BRs, which is sorely needed as input to astrochemical models, is still a rarity.

Chapter 9 by Robert Georges, Eszter Dudás, Nicolas Suas-David, and Lucile Rutkowski is dedicated to the discussion of the use of high-resolution infrared absorption spectroscopy in supersonic uniform flows generated by Laval nozzles, and also in under-expanded sonic jets containing shock features. The very significant cooling and therefore reduction in the molecular partition function make these flows ideal for infrared spectroscopy of polyatomic molecules whose spectra would otherwise be very congested and in some cases intractable for analysis. It makes possible the study of the molecules from their lowest rotational levels, spectra are further simplified from the narrowing of the Doppler spectral linewidth, and cooling allows the formation and stabilization of weakly bound complexes. Infrared spectroscopy in this very cold environment is illustrated for water vapor and methane as well as larger molecules, for example, butadiene and the aromatic species naphthalene. The adaptation of the CRESU apparatus for use with various IR absorption techniques, which use either single or multi-pass arrangements, is discussed, for example, setups which use tunable diode laser absorption spectroscopy (TDLAS), Fourier transform infrared (FTIR) spectroscopy, and the cavity methods of cavity ring down spectroscopy (CRDS) and cavity enhanced absorption spectroscopy (CEAS). An emerging technique for studying IR spectroscopy is that of optical frequency combs, and its potential is briefly discussed. Non-local thermodynamic equilibrium (non-LTE) effects (e.g. from the appearance of rotationally cold hot vibrational bands) occurring in hypersonic jets created from high-temperature Laval nozzles are discussed. Interestingly, the system can be driven back to LTE by inserting a blunt body within the flow in order to generate a normal shock wave. Behind it, spectral observations at high LTE temperature can then be carried out. Non-LTE effects are relevant for the determination of molecular abundances from high spectral resolution telescope observations in astrophysical environments. It also simplifies considerably the spectrum of hot bands. The homogeneous nucleation of clusters as studied using IR spectroscopy is also discussed, with examples shown including the formation of the water dimer and larger clusters of water vapor, and clusters of CO_2 studied at very high spectral resolution ($0.005\,\mathrm{cm}^{-1}$).

Chapter 10 by Eric Herbst sets the astrochemical context of chemistry occurring at very low temperatures. It stresses the very large contributions made by both laboratory and theoretical research in the study of ion-neutral and neutral-neutral reactions, which has aided our understanding of interstellar chemistry. Of particular importance are the kinetics measurements made by the CRESU technique at temperatures reaching as low as $\sim$10 K, which is applicable to the coldest part of interstellar clouds, as well as higher temperatures in the regions where stars and planets are created, including hot cores exceeding 100 K. A wide variety of organic molecules have been observed in these regions, including prebiotic molecules. The chapter provides an overview of the astrochemistry in interstellar clouds, which controls the abundances of various types of interstellar molecules which have been observed in the gas phase, such as simple species (e.g. H_2, H_3^+ CO), carbon-chain molecules (e.g. $HC_{2n}CN$, where n can be up to 5), complex organic molecules (ranging in size from 6 to 13 atoms and containing a carbon atom) which contain a range of functional groups (alcohols, aldehydes, ethers, carbonyls, cyanides), aromatics (e.g. benzene and cyano-benzene), fullerenes (e.g. C_{60}, C_{60}^+, and C_{70}), and inorganic species (e.g. PN, HCP, and metal-containing species, such as NaCN and TiO_2). Some molecules have also been observed in and on the surface of icy mantles, for example, methanol. The chapter outlines the various classes of reactions responsible for the formation and destruction of these types of molecules, covering gas phase ion-molecule chemistry for both positively and negatively charged ions, including radiative association and electron attachment, and gas phase neutral-neutral chemistry. The results of CRESU experiments which determine the rates of these processes are continually interleaved with the description of the chemistry, with the completeness of the kinetic database that exists for each type of reaction being quite different, and in some cases very sparse, especially under relevant cold conditions. The two principal databases for astrochemical kinetic data, KIDA and udfa (UMIST), are discussed, which in some cases provide evaluations of the various data provided by CRESU experiments or calculated using theory. The formation of cold COMs is discussed, as these have now been observed in the gas phase in several cold core regions where the temperature at around 10 K is too cold for desorption from grains to be their source. A number of reactions of COMs have been studied in CRESU experiments, with some displaying highly unusual variations of the rate coefficient (k) with temperature, for example, rapid upturns in k at temperatures below 150 K for H-atom abstraction reactions of OH with COMs followed by leveling out of k at even lower temperatures. One conclusion from the chapter is that "It is likely that the observation of more COMs in cold cores will drive a renaissance of low-temperature gas phase chemistry," which includes more experiments using CRESU with a wider range of detection techniques to monitor a wider range of reactive species.

In Chapter 11 by Philippe Halvick and Thierry Stoecklin, an account is given of the theoretical methods used to calculate rate coefficients (k), with an emphasis on the comparison with experimental determinations of k using the CRESU method. Theoretical approaches are vital, as often it is simply just not possible to either initiate a reaction of interest, for lack of a suitable precursor, or make a detection of a certain reactant or product molecule within the CRESU supersonic flow by using available spectroscopic techniques. Alternatively, it may be possible to determine the rate coefficient experimentally down to a certain temperature, but the real conditions of interest lie at temperatures beyond those which can be reached in the laboratory, and extrapolation beyond the intended use of any parameterization of $k(\mathrm{T})$ can be risky and lead to highly erroneous results. Finally, the determination of product branching ratios is still only rarely achieved, and so theory sometimes provides the only estimate available for which channels may likely be active at different temperatures. The theory chapter covers methods for the calculation of the potential energy surface on which the reaction occurs, which furnishes parameters such as activation barrier heights, the geometries of intermediates and transition states, and their internal energy level structure (allowing calculation of partition functions and densities of states needed for statistical rate theory methods), as well as the binding energies of weakly bound complexes which may form ahead of any barriers (positive or submerged) and which may be key for a reaction to occur (e.g. if it is sufficiently long lived for quantum mechanical tunneling to occur to form products). The key parameters though for astrochemical models are the rate coefficient for a given reaction and its product branching ratio, both as a function of temperature, and the chapter also covers theoretical methods for the calculation of these kinetic parameters, for example, using long-range adiabatic capture theory (for reactions without a barrier such as ion–molecule reactions) or transition state theory (TST), for which various approaches (e.g. variational TST) are discussed. Where possible, comparisons are made between theoretical and CRESU experimental determinations of $k(T)$, with plenty of examples provided for a variety of chemical systems, to illustrate the varying degree of agreement. Both reactive and inelastic collisions are considered and a variety of reaction mechanisms discussed, including spin-orbit effects and quantum mechanical tunneling. An interesting example is the reaction of OH with methanol, which displays a complex U-shaped variation of $k(T)$ with lowering temperature followed by a leveling off for $k(T)$ at the very lowest temperatures. Transition state theory is able to reproduce the general shape of $k(T)$, but at the lowest temperatures, good agreement with experiment requires the system to be close to the high-pressure limit, which is not the case from experimental evidence. Novel reaction dynamics approaches making use of quantum methods such as ring polymer molecular dynamics have been used to try to improve agreement, with roaming-type trajectories suggested to play a role for the reactions of OH with formaldehyde and methanol. Theoretical methods to calculate the rates of collisional relaxation of excited electronic, vibrational, rotational, and spin-orbit levels are examined, and compared with experimental data, some of which are available at very low temperatures determined in a CRESU apparatus. For example, rate coefficients for state-to-state rotational energy transfer for CO in collision with Ar have been measured in a CRESU environment and the agreement with Close Coupling calculations is evaluated.

12.2 Future Challenges and Open Questions

Although a lot of work has already been done using uniform supersonic flows and the CRESU technique, many possibilities of new developments still exist, some of which have already been discussed [1]. We have gathered a few in the next sections. Some of these developments would require technological improvements and breakthroughs.

12.2.1   *Extending the temperature range toward absolute zero*

In Chapter 1, some techniques (such as laser cooling and Stark decelerator) that have been developed with the main purpose of cooling down atoms and molecules as close as possible to absolute zero (i.e. well below 1 K) have been described. There is now a tendency to apply these techniques to chemistry as discussed in this chapter. However, there are in fact very few chemical reactions that have been studied in this way and this situation has not evolved yet [2].

At present, the lowest temperature achieved with a CRESU reactor to study reaction kinetics is 5.8 K [3], although the absolute record, obtained in a hydrogen supersonic flow, is slightly lower: 5.2 K as shown in Fig. 2.31 in Chapter 2. The question then is first to decide if there is an interest to develop the CRESU technique down to temperatures closer to absolute zero and if yes how to do so. From the astrochemical point of view, most of the "cold" media in the Universe are around temperatures close to 10 K that are routinely obtained in CRESU experiments, especially since the successful implementation of an aerodynamic chopper to pulse the supersonic flows as detailed in Chapter 2. Therefore, the interest of experiments conducted below say 1 K is mainly theoretical, with the hope to see enhanced quantum effects in the reactivity and to improve theory. Note that the ability to predict reactivity by theory is very useful in the astrochemical fields as experiments cannot be conducted for the thousands of reactions involved in models.

In the quantum mechanics framework, particles can behave like a wave with an associated wavelength λ depending on their mass m and velocity v:

$$\lambda = \frac{h}{mv} \tag{12.1}$$

The mechanical behavior of a particle depends on the typical length scale associated with the situation. Quantum effects can be ignored only for typical length scales much larger than the de Broglie wavelength. For a collection of thermalized particles, the thermal de Broglie wavelength defined as

$$\lambda_{th} = \frac{h}{\sqrt{2\pi m k_B T}} \qquad (12.2)$$

(where h is the Planck constant, k_B the Boltzmann constant, m the particle mass, and T the temperature) is, at a temperature of $0.5\,K$, larger or of the same order of magnitude than typical lengths of molecular interactions (a fraction of a nanometer). Therefore, specific quantum effects can be expected for chemical reactivity below $1\,K$. Also, at such temperatures, the collisional angular momentum is highly restricted, eventually reaching the limit where $l = 0$, which restrains the wave scattering for reactions, and tunneling effects will become increasingly important. The excited vibrational and rotational states will be completely depopulated, and the kinetic energy will be comparable to the interaction potential energy at long range.

Some molecules present spin isomers known as ortho and para forms which do not relax from one form to the other following temperature changes. This can have an effect on rotational state population leading to modifications in reactivity. Such effects have been observed with hydrogen at temperature well above $10\,K$ [4] (see also Chapter 3) due to the large energy spacing of hydrogen rotational levels, but are expected to be seen also at much lower temperatures for other molecules.

All these points justify any attempt to extend the temperature range of the CRESU technique below $1\,K$, considering the high chemical versatility of the technique. However, such an extension will not be straightforward, and the needed technological improvements are discussed in Section 12.3.

12.2.2 *Extending toward the study of reactions between unstable or condensable species*

Most of the work performed so far with CRESU apparatuses concerns reactions of ions or radicals with stable gaseous species. To date, very few radical-radical reactions have been studied and work performed in this way is described in Chapter 5. The difficulty here is either to produce, or to introduce directly within the flow, a large amount of an unstable radical. Remember that any tubing introduced within the supersonic flow will generate shock waves which in turn will destroy the flow uniformity, making it unsuitable for chemical reactivity studies.

On the contrary, it has been shown that the CRESU technique can be well suited to the investigation of reactions with condensable species, as demonstrated by the studies of the reactions of a few radicals with Polycyclic Aromatic Hydrocarbons (hereafter PAHs). The first direct measurement of the reaction rate coefficient of a PAH in the gas phase at very low temperature was reported by Rowe's research group in 2005 and concerned the OH + anthracene reaction [5] studied down to 58 K. To avoid condensation due to a large flow rate of reacting PAHs, the reservoir and the Laval nozzle had to be heated up to 470 K. Using a variety of nozzles, including a subsonic one, the temperature range of this dedicated CRESU apparatus was then 58–470 K [5, 6].

However, the difficulty that remains in such experiments is the large amount of condensate that can be formed on the room temperature part of the experimental chamber, especially downstream of the uniform flow and within the exhaust pipes to the pumps, therefore leading to an obligation of frequent and difficult cleaning. This has prevented to date further work in this way.

The possible technical advances in these two fields will be also discussed in Section 12.3.

12.2.3 *The special case of electrons*

Even if interesting effects due to the vibrational state of attaching molecules have been highlighted by a few CRESU experiments as shown in Chapter 4, the problem of electron thermalization has precluded possible electron attachment studies at very low electron temperature using the CRESU technique. Although it can be shown that electron thermal relaxation with the translational/rotational temperature of a nitrogen buffer is fast, the key mandatory point here is to avoid the creation of any excited species within the flow that could warm up electrons by super-elastic collisions. This is the case, for example, if a small part of a nitrogen buffer remains vibrationally hot after excitation by an electron beam used as a plasma source. Then, the electrons downstream of the source remain in equilibrium between two reservoirs of energy corresponding to the nitrogen rotational and vibrational energy levels. Therefore, they achieve a Maxwell–Boltzmann equilibrium at an intermediate temperature much higher than the flow translational/rotational temperature.

This is true whatever the chosen electron source. A possible way to overcome the problem would be to create electrons by laser photoionization near the threshold of a suitable precursor. For example, photoionization of trimethylamine (vertical ionization potential of 7.85 eV) by an excimer laser at 157 nm (equivalent energy of 7.897 eV) will yield photoelectrons with energy far below that of the first vibrational energy levels of nitrogen. Another solution could be the drifting of cold electrons in the flow, as it has been done for positive ions (see Chapter 3). "Cold" does not mean that the electron swarm in both cases needs to be created directly at the CRESU flow temperature but at an energy below the v = 1 vibrational level of nitrogen. Note that in order to improve the accessible temperature range, various mixtures of helium and nitrogen could be used as buffer gas [7].

12.2.4 *Challenges in product detection and spectroscopy*

In the field of chemistry, most of the CRESU work has been devoted to the study of chemical kinetics with special attention to the measurement of the temperature dependence of total rate coefficients. It is however of great interest for applications and from the fundamental point of view to determine the nature of reaction products and channel branching ratios.

As discussed in Chapter 8, the main difficulty here is to find detection techniques which are sufficiently universal together with a high sensitivity. For example, Laser Induced Fluorescence, although very sensitive, is also specific and restricted to a few species. Mass spectrometry is an extremely sensitive technique, but it requires ionizing products in a very controlled way and this before their destruction by secondary reactions. The coupling of synchrotron radiation to a uniform supersonic flow was achieved for the first time in the late 2000s when Leone's research group in Berkeley connected the ALS (Advanced Light Source) to their CRESU machine (see Chapters 2 and 8) to photoionize neutral products at threshold and beyond. The products were ionized after sampling by a pinhole of 0.45 mm set at the upstream top of an airfoil-shaped housing of a mass spectrometer. Then, they were mass analyzed by a quadrupole filter. Mass spectra were recorded at different reaction times and at different VUV photon energies. This allowed product isomers to be identified. This technique was applied to the determination of branching ratios in the reaction of C_2H with ethene and propene [8] and a series of butenes C_4H_8 [9] at 79 K, the lowest temperature achieved by this apparatus. After these few studies, the device was no longer used and was dismantled eventually. A new apparatus of the same kind has been developed since by the Rennes group under the coordination of S.D. Le Picard in order to be connected to the SOLEIL synchrotron in Saclay (France) and was named CRESUSOL. As detailed in Chapter 8, a few results, serving as proof of concept, have been obtained yet, for example, the identification of C_4H_2, a product issued from the reaction of C_2H with C_2H_2 [10]. The main difficulty in such an experiment is again to obtain a sufficient sensitivity and an appropriate density of products in order to assign an observed photo ion to a reaction channel. Following Chapter 8, possible improvements in this field are presented in Section 12.3.

A very promising spectroscopic technique is the Chirped Pulse Fourier Transform Micro-Wave (CP-FTMW) spectroscopy developed by B.H. Pate [11]. As described in Chapters 2 and 8, it was adapted by the Suits' research group at Wayne State University in the USA (WSU hereafter) to probe a pulsed CRESU flow perpendicularly [12, 13]. Such an apparatus could be extremely useful for the determination of branching ratios and has indeed been applied at WSU for the determination of branching ratios for the reaction CN + CH_3CCH at 20 K [14]. It can deliver isomer- and conformer-specific, quantitative detection. Two new apparatuses of this type are being built presently, one at the Department of Molecular Physics of the "Institut de Physique de Rennes" (France, IPR hereafter) through the direction of I.R. Sims (CRESUCHIRP project) and the other one at the "Institut des Sciences Moléculaires" at the University of Bordeaux (France) under the leadership of K. Hickson.

Very recently, infrared radiation was also coupled to a CRESU reactor to study reaction kinetics taking advantage of the Cavity Ring Down technique (CRD hereafter). This spectroscopic method which is discussed in detail in Chapter 9 has proven its high sensitivity since it is able to detect molecular concentrations as low as

10^9 cm^{-3}. It was combined with a pulsed CRESU machine named UF-CRDS (standing for Uniform Flow–Cavity Ring Down Spectroscopy) developed in A. Suits' group at the University of Missouri (USA) [15]. This new tool has been designed to probe radicals and reaction intermediates but also to follow the chemistry of hydrocarbon chains and Polycyclic Aromatic Hydrocarbons. As a first kinetic study, it was applied to the reaction CN(v = 1) + C$_3$H$_6$ (propene) with the CRD operating in the near infrared (1280–1420 nm) to follow the depletion of the CN radical rovibrational line R1e(5.5) of the A$^2\Pi$ – X$^2\Sigma^+$ electronic transition at 7070.2422 cm^{-1} (~1404 nm). Although products were not identified in this study, the method provides a way to quantify other species than the CN radical especially those having a CH group. More details can be found in Chapter 9 and in Suas-David *et al.* [15].

Considering now ion–molecule reactions, their ionic products are readily obtained by mass spectrometry. But, in order to obtain quantitative values for the reaction branching ratios, it is important to be able to select a single reactive ion as discussed in Chapter 3. A large amount of work remains to be done in this field using the new version of the CRESUS apparatus developed in Rennes (see Chapter 3).

12.2.5 *Extending the technique toward chemical complexity, particle nucleation, destruction, and growth*

Although a considerable amount of work has been performed in the field of homogeneous nucleation of a vapor, either for a pure substance or for a mixture (see Chapter 6), the question of aerosols' formation, growth, destruction, and chemistry remains a very open question in chemical physics and largely a "terra incognita". Indeed, aerosols could be considered as a specific state of matter and the Universe a "dusty" Universe.

A considerable number of open interrogations remain in this field. Considering nucleation, it is most often studied in the laboratory as a transition from the gas phase to liquid or solid state for pure substances. It is therefore mainly described by physical phenomena, including CNT (Classical Nucleation Theory) described in Chapter 6. However, it appears that in the real world the first nucleation step is often chemically driven or ion induced, two ways to overcome the barrier to nucleation. Once formed, aerosols also participate in the complicated chemistry of a variety of media such as planetary atmospheres (including Earth), interstellar clouds, and circumstellar shells, as well as in combustion processes with the crucial problem of soot formation.

For the Earth's atmosphere, the very nature of the cloud formation process is not yet fully understood: It is now largely admitted that species such as sulfuric acid, terpenes, which are a class of chemical compounds emitted by vegetation, and other minor species such as ions are seeds for nucleation. These various species participate in complicated chemical networks and therefore atmospheric nucleation is, at least partly, a chemical process. The question is still a matter of intense debate due to its clear link with climate problems. Especially the ion-induced nucleation mentioned in Chapter 3, and therefore the possible influence of cosmic rays and solar activity on cloud formation and climate, is a highly controversial subject.

The same problematic is found for planetary atmospheres as, for example, for the hazes observed in Titan's atmosphere, largely constituted hydrocarbon compounds and involving a rich plasma chemistry. In other astrophysical media such as, within others, interstellar clouds, cometary atmospheres, or circumstellar shells, a large variety of particles are observed, composed of carbonaceous materials, silicates, and condensable species such as water.

Closer to our usual world, the problem of soot formation in combustion cannot be considered as solved despite a considerable amount of work performed in the field. Here again, a lot of controversy can be observed in the literature: is there a role for ion nucleation? Is the fractal structure of the largest soot grains due to the agglomeration of the smallest spherical grains or to the carbonization of liquid PAHs droplets formed in the combustion process? These open questions are relevant for the problem of carbonaceous particle formation in the envelopes of carbon-rich stars, which uses quite often models derived from the combustion ones. Note that work performed at Rennes using the dedicated CRESU apparatus [16] on the dimerization of simple PAHs has demonstrated that the "standard" and widely accepted model of soot formation relies on the wrong hypothesis that dimerization of pyrene is possible in combustion processes.

For the evolution of all the media mentioned above, heterogeneous chemistry plays a key role as it is well known, for example, in the depletion of stratospheric ozone over Antarctica in the spring months, due to the presence of polar stratospheric clouds [17]. Of course, the grain chemistry is also thought to be largely involved in molecules synthesis in interstellar clouds as discussed in Chapter 10 of this book.

By carrying out experiments with Rayleigh scattering of a UV laser beam for detection, the Rennes research group showed the formation of a few nanometer-sized CO_2-based particles in a CRESU flow [18]. Calculations show that small nanoparticles (i.e. a few tens of nanometers or better a few nanometers) accommodate temperature and velocity within the flow. These particles can be either directly formed in the supersonic flow or seeded upstream from nanoparticle powders. They can be formed in a reasonable density, i.e. allowing easy use of spectroscopy and optical detection. Therefore, it can be thought that chemistry involving particle-laden flows could be studied in a dedicated CRESU apparatus.

12.3 Prospective of Needed Technological Developments

Many of the developments described above will imply technological breakthroughs. Below, we will focus on some specific points, most of them directed toward extension of chemical versatility and of temperature and density range. Beyond this, new detection techniques will be also evoked.

12.3.1 *Toward lower temperature: cooling nozzle walls and reservoir*

The case of temperature extension below 1 K has been discussed at length in the paper by Potapov *et al.* [1]. It was shown that in order to obtain a temperature in the range 0.1–1 K, extremely high Mach numbers are mandatory, implying a static pressure P which, with the required mass flow rate (see Chapter 1), is not compatible with continuous pumping, even using the differential pumping supersonic diffuser technique developed and tested in Rennes [19] but never used since. It was also shown that cryogenic cooling of the reservoir to liquid nitrogen temperature was a way to reduce the needed flow rate and therefore pumping capacity, although it will reduce the chemical versatility of the CRESU. Cooling the nozzle walls, possibly at liquid helium temperature, could be a better (or complementary) improvement. Consequently, it will be possible to keep a denser flow in the boundary layer, therefore avoiding a transition from a low to high Knudsen regime (i.e. avoiding transition and molecular free regimes) which will make the nozzle calculation extremely difficult.

Tables 12.1 and 12.2 give an overview of the possible parameters corresponding to various experimental arrangements for two reservoir temperatures, 300 K and 77 K, respectively. Parameters n, T, and P correspond to the physical conditions within the uniform flow, Q is the mass flow rate, P_{res} the reservoir pressure, T_{wall} the wall temperature of the divergent section of the Laval nozzle, and $\phi_{isent,exit}$ the isentropic diameter at the nozzle exit. Note that "volume buffer" corresponds to the volume of the buffer tank needed to make pressure oscillations negligible in operating conditions. The pulsed regime has been considered to allow a gain of one hundred in terms of gas consumption with respect to the continuous flow conditions. Pumping speeds have been indicated in l/s since for the proposed mass flow rates and pressures, roots blowers will not be adequate and booster pumps will have to be employed instead. From these tables, it can be seen that reaching a target of a 0.1–1-K flow is challenging but possible in a pulsed supersonic flow.

The problem when cooling the wall of the divergent part of the nozzle is that possible heterogeneous condensation of any supersaturated species could occur. This problem concerns most molecules when liquid helium cooling is used for the "below 1 K" target. It appears also for the cooling at room temperature of the divergent section of a nozzle dedicated to the study of highly condensable molecules such as PAHs since in this latter case the reservoir and convergent section of the nozzle are heated up at a temperature which avoids condensation. For both cases, different possibilities exist to keep the boundary layer free of condensable molecules. The first one would be to have porous walls for the divergent section and to blow some amount of a gas identical to the buffer one through them. However, the calculation of such a nozzle would be probably very tedious and not straightforward. A better way is probably to use a technique of "jet inside jet" such as described in Chapter 9 and in reference [20] (see also insert (I) in Fig. 12.1). In this case, an annular disposition of convenient concentric walls in the subsonic part of the apparatus

Table 12.1. Estimation of practical values to be achieved to run a CRESU apparatus in the temperature range of 0.2–1 K for $T_{res} = 300\,\mathrm{K}$.

n cm^{-3}	T K	Mach Number	$P\mu$ bar	P_{res} bar	T_{wall} K	Q g/s	$\varphi_{isent,\,exit}$ cm	Pumping l/s cont.	Pumping l/spulsed	Buffer Vol. m^3
3×10^{15}	1	29.9	0.41	0.65	300	0.15	7.4	2.0×10^6	2.0×10^4	2.0
3×10^{15}	1	29.9	0.41	0.65	77	0.1	6.0	1.4×10^6	1.4×10^4	1.4
3×10^{15}	1	29.9	0.41	0.65	4	0.05	4.3	6.8×10^5	6.8×10^3	0.7
3×10^{15}	0.2	67.1	0.083	7.2	300	0.15	7.4	1.0×10^7	1.0×10^5	10.1
3×10^{15}	0.2	67.1	0.083	7.2	77	0.1	6.0	6.8×10^6	6.8×10^4	6.8
3×10^{15}	0.2	67.1	0.083	7.2	4	0.05	4.3	3.4×10^6	3.4×10^4	3.4
1×10^{16}	0.2	67.1	0.28	24	300	0.15	4.0	3.0×10^6	3.0×10^4	3.0
1×10^{16}	0.2	67.1	0.28	24	77	0.1	3.3	2.0×10^6	2.0×10^4	2.0
1×10^{16}	0.2	67.1	0.28	24	4	0.05	2.3	1.0×10^6	1.0×10^4	1.0

Source: Adapted from Ref. [1].

Table 12.2.   Same as for Table 12.1 with $T_{res} = 77$ K.

n cm^{-3}	T K	Mach Number	$P\mu$ bar	P_{res} bar	T_{wall} K	Q g/s	$\varphi_{isent, exit}$ cm	Pumping l/s cont.	Pumping l/s pulsed	Buffer Vol. m^3
3×10^{15}	1	15.1	0.41	0.022	300	0.15	10.4	2.0×10^6	2.0×10^4	2.0
3×10^{15}	1	15.1	0.41	0.022	77	0.1	8.5	1.4×10^6	1.4×10^4	1.4
3×10^{15}	1	15.1	0.41	0.022	4	0.05	6.0	6.8×10^5	6.8×10^3	0.7
3×10^{15}	0.2	33.9	0.083	0.24	300	0.15	10.4	1.0×10^7	1.0×10^5	10.1
3×10^{15}	0.2	33.9	0.083	0.24	77	0.1	8.5	6.8×10^6	6.8×10^4	6.8
3×10^{15}	0.2	33.9	0.083	0.24	4	0.05	6.0	3.4×10^6	3.4×10^4	3.4
1×10^{16}	0.2	33.9	0.28	0.80	300	0.15	5.7	3.0×10^6	3.0×10^4	3.0
1×10^{16}	0.2	33.9	0.28	0.80	77	0.1	4.6	2.0×10^6	2.0×10^4	2.0
1×10^{16}	0.2	33.9	0.28	0.80	4	0.05	3.3	1.0×10^6	1.0×10^4	1.0

Source: Adapted from Ref. [1].

(converging section) could allow creating a uniform isentropic core inside an outer flow where transport phenomena occur (boundary layer). The outer flow would be created from a pure buffer and, as diffusion through the boundary layer is mainly inhibited in its supersonic part, it is anticipated that this will tackle the problem.

12.3.2 *Toward lower temperature: aerodynamic chopper*

Most possible advances discussed in this conclusion require to pulse the flow, either to reduce the needed pumping speed (see Tables 12.1 and 12.2) and the gas consumption or to avoid some condensation problems. It has been shown in Chapters 1 and 2 that pulsing downstream of the reservoir is the better option to inhibit the turbulence problem. The device known as "aerodynamic chopper" [21], which is in fact a pulsed knife valve, has been used in UCLM (University of Castilla-La Mancha, Spain) with great success and their apparatus is able to provide flows with useful lengths much higher than other pulsed CRESUs. In this apparatus, the chopper described in Chapter 2 is inserted in the divergent part of the nozzle. We suggest here that such a system could be inserted in a tubing just upstream of nozzle throat. The tubing should have to be sufficiently short to avoid the development of a boundary layer, i.e. with a length much shorter than the entrance length discussed in Chapter 1. Such an arrangement would have the following advantages:

- With respect to chopping within the divergent, which is the solution adopted at UCLM, this design, if successful, will allow easier use of a large variety of nozzles.
- Downstream of the chopper, it will be possible to mix within the tube (indicated by A in Fig. 12.1) unstable species such as radicals, ions, and electrons. Any of these species would possibly require another pulsing system to modulate their injection as well as a device allowing an efficient mixing in the flow.
- Nanoparticles as well as highly condensable species could be injected directly in the reservoir upstream of the chopper. In both cases, technological advances would be required to have a chopper working under these kinds of conditions.

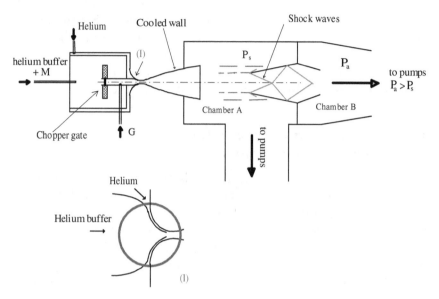

Fig. 12.1.   Some possible technological improvements.

Figure 12.1 summarizes the various possibilities presented above. A supersonic diffuser with differential pumping analog to the one developed in Rennes is shown. It uses two separate group of pumps operating at different pressures for chamber A and B. Note that the design of the tubing between these two chambers needs to be improved and could use a convergent part in order to recompress the flow in a better way. In insert (I), a possible design of a throat developed for the creation of a "jet inside jet" is shown. Its purpose is protection of the nozzle walls from condensation as discussed above. Also shown is the insertion of a tubing with an aerodynamic chopper just upstream of the throat.

12.3.3   *Toward advances in detection techniques and spectroscopy*

Concerning the identification and quantification of reaction products, as already mentioned, several new developments have risen in recent years based on mass spectrometry or spectroscopic methods in the microwave or infrared wavelength ranges. There is still room however for improving these various techniques and a few directions will be commented below.

The CRESUSOL apparatus has delivered its first results recently (see Chapter 8 for details). The detection method is based on the PEPICO technique which is a PhotoElectron-PhotoIon COincidence spectrometer. At present, the S/N ratio is rather limited due to the presence of a significant number of false coincidences. In order to minimize these, the ion detection in the time of flight mass spectrometer will be improved in the short term by the implementation of a position sensitive detector (PSD) which should increase significantly the ion contrast. In a second step, this PSD will be coupled to a temporal ion deflection device which eventually will remove the false coincidences entirely, thus increasing the sensitivity of the detection by many orders of magnitude. These developments will be of particular interest for the investigation of radical-neutral reactions since it will facilitate considerably the detection of products whose concentration is limited by the radical production methodology.

As described in detail in Chapter 8, the CP-FTMW technique is based on the polarization of molecules having a permanent dipole moment by a short chirped pulse of microwave frequencies. After excitation, the polarized sample of molecules can relax either spontaneously (the so-called Free Induction Decay, hereafter FID) or by collisions with the neighboring species. This latter situation is clearly not desirable since it provokes a pressure broadening of the emitted lines which can significantly degrade the S/N ratio and reduce the duration of the detected signal. Hence, every effort must be undertaken to minimize this effect. The CRESUCHIRP project mentioned earlier and described in Chapter 8 aims at tackling this difficulty from various directions. In this project, the reaction of interest will be induced by a 100-Hz photolysis laser in a continuous CRESU flow in a first step and later in a high-frequency pulsed CRESU reactor in development. Employing such repetition rates aims at increasing the accumulated signal and then improving the S/N ratio. Furthermore, improvements in signal processing are being implemented to compensate for the short duration of FIDs in high-pressure environments. From another side, the influence of the surrounding molecules can be obviously minimized if the pressure in the detection area is reduced. This can be achieved using lower-density uniform flows, which however will require much higher pumping capacities. A pulsed

CRESU reactor with the large pumping facilities available in Rennes is likely to match this exigency. Another solution, also presently being explored in Rennes, is to skim the uniform flow into a lower-pressure chamber and probe the species downstream of the skimmer. Besides the pressure reduction, this configuration could allow to cool down molecules even lower than in the uniform environment and then concentrate the targets of interest into a smaller number of rotational states, thus increasing their density and by extension the detected signal. However, cooling down requires converting thermal energy into kinetic energy and therefore collisional processes. In the continuous flow regime, shock waves could be generated in the inner part of the skimmer that could be counterproductive. A skimmer in the transition regime (Knudsen $\sim O(1)$) could be used but is extremely difficult to predict. On the contrary, we think that the possible technological advances described in the previous sections will make it possible to tune both the density and temperature at its best value for the required optimization.

From that point of view, achieving ultralow temperatures at convenient density for the development of CP-FTMW could be also extremely useful for spectroscopy research, especially in the microwave range, since, as mentioned, it considerably reduces the number of implied rotational levels.

In the infrared domain, a promising technique, mentioned in Chapters 8 and 9, is worth a brief discussion here. This is the so-called Time-Resolved Frequency Combs (TRFCs hereafter) method which has revolutionized the field of molecular spectroscopy by providing a new broadband light source offering a spectral and spatial coherence comparable to continuous lasers while benefiting from the developments carried out on Fourier Transform InfraRed (FTIR) spectroscopy [22]. TRFCs are generated by mode-locked femtosecond lasers and are well suited in the microsecond timescale [23]. They can scan the wavelength range between 300 and 9000 nm by relatively broad wavelength intervals which depend on the class of lasers employed. More particularly, the spectral range around 3000 nm is of great interest since it embeds fundamental bands of the C-H, N-H, and O-H stretch modes. By nature, this optical diagnostics is multiplexed so that the time-dependent concentration of several species can be monitored simultaneously. Implementation of such a technique can be achieved perpendicularly to the flow axis of a CRESU reactor to study reaction kinetics. Such a development is in progress at IPR under the direction of L. Rutkowski and at the School of Chemistry at the University of Leeds (United Kingdom) under the supervision of J.H. Lehman.

The technique is not restricted to reaction kinetics since it has already been used to characterize the acetylene rovibrational spectra in a supersonic jet issued from a Laval nozzle expansion [24]. However, the field is essentially virgin and many investigations on pure cold infrared spectroscopy should be carried out in the next years. Another spectroscopic technique, namely, the Cavity-Enhanced Absorption Spectroscopy (CEAS), is also in development at IPR under the coordination of A. Benidar in order to be combined with a planar uniform supersonic jet to investigate the spectroscopy of cold molecular ions and van der Waals ionic complexes. More details can be found in Chapter 9.

12.4 General Conclusion

In the present book, an extensive compilation of instrumental developments and experimental results has been described. However, as illustrated in this ending chapter, a variety of improvements and breakthroughs either in the CRESU technique itself or in production and detection techniques are still possible. We hope that this will motivate a new generation of researchers to jump on the CRESU bandwagon to push the present boundaries even further. Note that, as in the original CRESU, it will require one to develop a multidisciplinary work between a variety of fields: aerodynamics, physics, chemistry, and spectroscopy.

References

[1] Potapov A, Canosa A, Jiménez E, Rowe BR. Uniform supersonic chemical reactors: 30 years of astrochemical history and future challenges. Ang Chem Int Ed. 2017;56(30):8618–40.

[2] Heazlewood BR, Softley TP. Towards chemistry at absolute zero. Nat Rev Chem. 2021;5:125–40.

[3] Berteloite C, Lara M, Bergeat A, Le Picard SD, Dayou F, Hickson KM, Canosa A, Naulin C, Launay JM, Sims IR, Costes M. Kinetics

and dynamics of the $S(^1D_2) + H_2$ reaction at very low temperatures and collision energies. Phys Rev Lett. 2010;105(20):203201.

[4] Marquette JB, Rebrion C, Rowe BR. Reactions of $N^+(^3P)$ ions with normal, para and deuterated hydrogens at low temperatures. J Chem Phys. 1988;89(4):2041–7.

[5] Goulay F, Rebrion-Rowe C, Le Garrec JL, Le Picard SD, Canosa A, Rowe BR. The reaction of anthracene with OH radicals: an experimental study of the kinetics between 58 K and 470 K. J Chem Phys. 2005;122(10):104308.

[6] Goulay F, Rebrion-Rowe C, Biennier L, Le Picard SD, Canosa A, Rowe BR. The reaction of anthracene with CH radicals: an experimental study of the kinetics between 58 K and 470 K. J Phys Chem A. 2006;110(9):3132–7.

[7] Canosa A, Ocaña AJ, Antiñolo M, Ballesteros B, Jiménez E, Albaladejo J. Design and testing of temperature tunable de laval nozzles for applications in gas-phase reaction kinetics. Exp Fluids. 2016;57(9):152.

[8] Bouwman J, Goulay F, Leone SR, Wilson KR. Bimolecular rate constant and product branching ratio measurements for the reaction of C_2H with ethene and propene at 79 K. J Phys Chem A. 2012;116(15):3907–17.

[9] Bouwman J, Fournier M, Sims IR, Leone SR, Wilson KR. Reaction rate and isomer-specific product branching ratios of $C_2H + C_4H_8$: 1-butene, cis-2-butene, trans-2-butene, and isobutene at 79 K. J Phys Chem A. 2013;117(24):5093–105.

[10] Durif O, Capron M, Messinger J, Benidar A, Biennier L, Bourgalais J, Canosa A, Courbe J, Garcia GA, Gil JF, Nahon L, Okumura M, Rutkowski L, Sims IR, Thiévin J, Le Picard SD. A new instrument for kinetics and branching ratio studies of gas phase collision processes at very low temperatures. Rev Sci Inst. 2021;92(1):014102.

[11] Brown GG, Dian BC, Douglass KO, Geyer SM, Shipman ST, Pate BH. A broadband Fourier transform microwave spectrometer based on chirped pulse excitation. Rev Sci Inst. 2008;79(5):053103.

[12] Abeysekera C, Zack LN, Park GB, Joalland B, Oldham JM, Prozument K, Ariyasingha NM, Sims IR, Field RW, Suits AG. A chirped-pulse Fourier-transform microwave/pulsed uniform flow spectrometer. II. Performance and applications for reaction dynamics. J Chem Phys. 2014;141(21):14203.

[13] Oldham JM, Abeysekera C, Joalland B, Zack LN, Prozument K, Sims IR, Park GB, Field RW, Suits AG. A chirped-pulse Fourier-transform microwave/pulsed uniform flow spectrometer. I. The low-temperature flow system. J Chem Phys. 2014;141(15):54202.

[14] Abeysekera C, Joalland B, Ariyasingha N, Zack LN, Sims IR, Field RW, Suits AG. Product branching in the low temperature reaction of CN with propyne by chirped-pulse microwave spectroscopy in a uniform supersonic flow. J Phys Chem Lett. 2015;6(9):1599–604.

[15] Suas-David N, Thawoos S, Suits AG. A uniform flow-cavity ring-down spectrometer (UF-CRDS): a new setup for spectroscopy and kinetics at low temperature. J Chem Phys. 2019;151(24):244202.

[16] Sabbah H, Biennier L, Sims IR, Rowe BR, Klippenstein SJ. Exploring the role of PAHs in the formation of soot: pyrene dimerization. J Phys Chem Lett. 2010;1(17):2962–7.

[17] Molina MJ. Heterogeneous chemistry on polar stratospheric clouds. Atmos Environ Part A-Gen Top. 1991;25(11):2535–7.

[18] Bonnamy A, Georges R, Benidar A, Boissoles J, Canosa A, Rowe BR. Infrared spectroscopy of $(CO_2)_N$ nanoparticles $(30 < N < 14500)$ flowing in a uniform supersonic expansion. J Chem Phys. 2003; 118(8):3612–21.

[19] Le Page V. Conception et Mise au Point d'un Moyen d'Essai CRESU (Cinétique de Réaction en Ecoulement Supersonique Uniforme) pour l'Etude des Réactions Ion-Molécule et Application aux Températures Ultra-Basses (10 K) [PhD Thesis]. France: Université de Rennes 1; 1995.

[20] Canosa A, Georges R, Morales S, Rowe BR. Procédé et Dispositif pour la Séparation d'Isotopes à partir d'un Ecoulement Gazeux. Patent: FR 2975920 A1 — WO 2012164230 A1; 2012 Dec 7.

[21] Jiménez E, Ballesteros B, Canosa A, Townsend TM, Maigler FJ, Napal V, Rowe BR, Albaladejo J. Development of a pulsed uniform supersonic gas expansion system based on an aerodynamic chopper for gas phase reaction kinetics studies at ultra-low temperatures. Rev Sci Inst. 2015;86(4):045108.

[22] Roberts FC, Lewandowski HJ, Hobson BF, Lehman JH. A rapid, spatially dispersive frequency comb spectrograph aimed at gas phase chemical reaction kinetics. Molec Phys. 2020;118(16):e1733116.

[23] Fleisher AJ, Bjork BJ, Bui TQ, Cossel KC, Okumura M, Ye J. Mid-infrared time-resolved frequency comb spectroscopy of transient free radicals. J Phys Chem Lett. 2014;5(13):2241–6.

[24] Thorpe MJ, Adler F, Cossel KC, de Miranda MHG, Ye J. Tomography of a supersonically cooled molecular jet using cavity-enhanced direct frequency comb spectroscopy. Chem Phys Lett. 2009;468(1–3):1–8.

Index